Handbook of Semiconductors

This book provides readers with state-of-the-art knowledge of established and emerging semiconducting materials, their processing, and the fabrication of chips and microprocessors. In addition to covering the fundamentals of these materials, it details the basics and workings of many semiconducting devices and their role in modern electronics and explores emerging semiconductors and their importance in future devices.

- Provides readers with latest advances in semiconductors.
- Covers diodes, transistors, and other devices using semiconducting materials.
- Covers advances and challenges in semiconductors and their technological applications.
- Discusses fundamentals and characteristics of emerging semiconductors for chip manufacturing.

This book provides directions to scientists, engineers, and researchers in materials engineering and related disciplines to help them better understand the physics, characteristics, and applications of modern semiconductors.

Ram K. Gupta is an Associate Vice President for Research and Support and a Professor of Chemistry at Pittsburg State University. Dr. Gupta has been recently named by Stanford University as being among the top 2% of research scientists worldwide. Before joining Pittsburg State University, he worked as an Assistant Research Professor at Missouri State University, Springfield, MO, then as a Senior Research Scientist at North Carolina A&T State University, Greensboro, NC. Dr. Gupta's research spans a range of subjects critical to current and future societal needs, including semiconducting materials and devices, biopolymers, flame-retardant polymers, green energy production and storage using nanostructured materials and conducting polymers, electrocatalysts, optoelectronics and photovoltaics devices, organic-inorganic heterojunctions for sensors, nanomagnetism, biocompatible nanofibers for tissue regeneration, scaffolds and antibacterial applications, and bio-degradable metallic implants. Dr. Gupta has published over 290 peer-reviewed journal articles, made over 420 national/international/regional presentations, chaired/organized many sessions at national/international meetings, wrote several book chapters (100+), worked as Editor for many books (50+), and received several million dollars for research and educational activities from external agencies. He is also serving as Editor, Associate Editor, Guest Editor, and editorial board member for various journals.

Handbook of Semiconductors
Fundamentals to Emerging Applications

Edited by
Ram K. Gupta

CRC Press
Taylor & Francis Group
Boca Raton London New York

CRC Press is an imprint of the
Taylor & Francis Group, an **Informa** business

Designed cover image: Shutterstock
First edition published 2025
by CRC Press
2385 NW Executive Center Drive, Suite 320, Boca Raton FL 33431

and by CRC Press
4 Park Square, Milton Park, Abingdon, Oxon, OX14 4RN

CRC Press is an imprint of Taylor & Francis Group, LLC

ISBN: 978-1-032-58455-3 (hbk)
ISBN: 978-1-032-78906-4 (pbk)
ISBN: 978-1-003-45014-6 (ebk)

DOI: 10.1201/9781003450146

Typeset in Times LT Std
by KnowledgeWorks Global Ltd.

Dedicated to Dr. Shawn Naccarato, whose unwavering dedication to the exploration of semiconducting materials has not only inspired but also ignited innovation and motivation within the realm of advanced applications. Your guidance and vision have illuminated paths of discovery, shaping the future of technology with brilliance and purpose.

Contents

Preface

Semiconducting materials dominate many industries and are used in almost all electronic devices. The current chip shortage and its effect on the worldwide economy are concerns for many countries. Semiconducting materials provide unique and tunable electrical, optical, and electronic properties. Along with these unique characteristics, they also possess high thermal, mechanical, and environmental stability. Apart from traditional silicon-based semiconductors, many emerging semiconductors are being considered for chips and microelectronic manufacturing. Their properties can be further improved by doping, growing them in thin-film structures, and creating good heterojunctions. A fundamental understanding of the charge-transport mechanism in semiconducting is crucial.

This book provides state-of-the-art knowledge in semiconducting materials, their processing, and the fabrication of chips and microprocessors. Applications of semiconductors in electronic devices, chips, and microprocessors are covered in detail, along with the fundamentals of semiconductors and their processing. This book provides the basics and workings of many semiconducting devices and their role in modern electronics. This book also explores emerging semiconductors and their importance in future devices. The future and challenges of many emerging semiconductors are also explored. Experts in these areas have covered all the chapters, making this a suitable textbook for students and researchers.

Ram K. Gupta, Professor
Department of Chemistry
National Institute for Materials Advancement
Pittsburg State University
Pittsburg, Kansas, USA

List of Contributors

Arpana Agrawal
Department of Physics
Shri Neelkantheshwar Government Post-
    Graduate College
Khandwa, India

Farida A. Ali
ITER, Sikhsha 'O' Anusandhan (deemed to be
    University)
Bhubaneswar, India

Farhad Alizadegan
Department of Colour, Resin and Surface
    Coatings
Iran Polymer and Petrochemical Institute
Tehran, Iran

K. S. Anantharaju
Department of Chemistry
Dayananda Sagar College of Engineering
Shavige Malleshwara Hills
Bengaluru, India

Arnet Maria Antony
Centre for Nano and Material Sciences
Jain (Deemed-to-be University), Jain Global
    Campus
Bangalore, India

Erdi C. Aytar
Department of Horticulture
Uşak University
Uşak, Türkiye

Nasrin Babazadeh
Department of Polymer Engineering and Color
    Technology
Amirkabir University of Technology
Tehran, Iran

R. Geetha Balakrishna
Centre for Nano and Material Sciences
Jain (Deemed-to-be University), Jain Global
    Campus
Bangalore, India

Dipak Barman
Department of Physics
Nanophysics Laboratory
Gauhati University
Guwahati, India

Kshitij Bhargava
Department of Electrical Engineering
Annasaheb Dange College of Engineering and
    Technology (ADCET)
Ashta, India

Arunava Bhattacharyya
Department of Chemical Engineering
University of Calcutta
Kolkata, India

Chinmoy Bhattacharya
Department of Chemistry
Indian Institute of Engineering Science &
    Technology, (IIEST)
Howrah, India

Sanjib Bhattacharya
UGC-HRDC
University of North Bengal
Darjeeling, India

Jnanraj Borah
Department of Physics
Nanophysics Laboratory
Gauhati University
Guwahati, India

Mourad Boughrara
Department of Sciences
Physics of Materials and Systems Modeling
    Laboratory (PMSML)
Moulay Ismail University
Zitoune, Morocco

Arpita Paul Chowdhury
Department of Chemistry
Dayananda Sagar College of Engineering
Shavige Malleshwara Hills
Bengaluru, India

Subhajit Das
Division of Functional Materials and Devices
CSIR-Central Glass & Ceramic Research
    Institution
Jadavpur, Kolkata, India

Allen Davis
Department of Chemistry
National Institute of Materials Advancement
Pittsburg State University
Pittsburg, Kansas, USA

Alper Durmaz
Ali Nihat Gökyiğit Botanical Garden
    Application and Research Center
Artvin Çoruh University
Artvin, Türkiye

Amir Ershad-Langroudi
Department of Colour, Resin and Surface
    Coatings
Iran Polymer and Petrochemical Institute
Tehran, Iran

Navid Nasajpour Esfahani
Department of Materials Science and
    Engineering
Georgia Institute of Technology
Atlanta, Georgia, USA

Hamid Garmestani
Department of Materials Science and Engineering
Georgia Institute of Technology
Atlanta, Georgia, USA

Sangeeta Ghosh
Department of Chemistry
Indian Institute of Engineering Science &
    Technology, (IIEST)
Howrah, India

Ram K. Gupta
Department of Chemistry
National Institute for Materials Advancement
Pittsburg State University
Pittsburg, Kansas, USA

Jeffery Horinek
Department of Chemistry
National Institute for Materials Advancement
Pittsburg State University
Pittsburg, Kansas, USA

Priya A. Hoskeri
Department of Physics
Dayananda Sagar College of Engineering
Bangalore, India

İbrahim M. Kahyaoğlu
Department of Chemistry
Ondokuz Mayıs University
Samsun, Türkiye

Thangavel Kanagasekaran
Department of Physics
Organic Optoelectronics Laboratory
Indian Institute of Science Education and
    Research (IISER)
Tirupati, India

Selcan Karakuş
Department of Chemistry
Istanbul University-Cerrahpaşa
Istanbul, Türkiye

Mohamed Kerouad
Department of Sciences
Physics of Materials and Systems Modeling
    Laboratory (PMSML)
Moulay Ismail University
Zitoune, Morocco

K. Keshavamurthy
Department of Physics
Dayananda Sagar College of Engineering
Shavige Malleshwara Hills
Bengaluru, India

Adem Kocyigit
Department of Electronics and Automation
Vocational High School, Bilecik Şeyh Edebali
    University
Bilecik, Turkey

Amir Koohbor
School of Engineering
Georgia State University
Atlanta, Georgia, USA

Karuna Kumari
Department of Physics
Indian Institute of Technology
Patna, India

Mohsen Lashgari
Department of Chemistry
Institute for Advanced Studies in Basic
    Sciences (IASBS)
Zanjan, Iran

Ajay Lathe
Department of Chemistry
Mahatma Phule ASC College
Mumbai, India

Steven Y. Liang
George W. Woodruff School of Mechanical
    Engineering
Georgia Institute of Technology
Atlanta, Georgia, USA

Xi Lin
Frontiers Science Center for Flexible
    Electronics (FSCFE)
Shaanxi Institute of Flexible Electronics
    (SIFE)
Northwestern Polytechnical University
Xi'an, China

Chia-Jyi Liu
Department of Physics
National Changhua University of
    Education
Changhua County, Taiwan

Xinghui Liu
Science and Technology on Aerospace
    Chemical Power Laboratory
Hubei Institute of Aerospace
    Chemotechnology
Xiangyang, China
and
Department of Materials Science and
    Engineering
City University of Hong Kong
Hong Kong, China

Xiaoguang Luo
Frontiers Science Center for Flexible
    Electronics (FSCFE)
Shaanxi Institute of Flexible Electronics
    (SIFE)
Northwestern Polytechnical University
Xi'an, China

Joseph C. M.
Department of Physics
Dayananda Sagar College of Engineering
Bangalore, India

Abdelhamid Ait M'hid
Department of Sciences
Physics of Materials and Systems Modeling
    Laboratory (PMSML)
Moulay Ismail University
Zitoune, Morocco

Sabita Mali
ITER, Sikhsha 'O' Anusandhan (deemed to be
    University)
Bhubaneswar, India

Neeraj Mehta
Department of Physics
Institute of Science, Banaras Hindu
    University
Varanasi, India

R. Mohan
Department of Physics
Sree Sevugan Annamalai College, (Affiliated
    to Alagappa University, Karaikudi)
Devakottai, India

Seyed Mehdi Mousaei
Department of Processing
Iran Polymer and Petrochemical Institute
Iran Institute
Tehran, Iran

Akhila Muhammed
Sree Narayana College (Affiliated to University
    of Kerala)
Kollam, India
and
Government College Attingal (Affiliated to
    University of Kerala)
Thiruvananthapuram, India

Sujit Mukherjee
Department of Chemical Engineering
University of Calcutta
Kolkata, India

Vinuth Raj T. N.
Department of Physics
Faculty of Engineering and Technology
JAIN (Deemed-to-be) University
Bangalore, India

Sumi V. Sasidharan Nair
Sree Narayana College (Affiliated to University
of Kerala)
Kollam, India
and
Government College Attingal (Affiliated to
University of Kerala)
Thiruvananthapuram, India

André E. Nogueira
Department of Chemistry
Division of Fundamental Sciences (IEF)
Aeronautics Institute of Technology - ITA,
São José dos Campos
São Paulo, Brazil

Francisco G. E. Nogueira
Department Chemical Engineering
University of São Carlos-UFSCar
São Paulo, Brazil

Debmalya Pal
Department of Chemical Engineering
University of Calcutta
Kolkata, India

Anil M. Palve
Department of Chemistry
Mahatma Phule ASC College
Mumbai, India

Siddappa A. Patil
Centre for Nano and Material Sciences
Jain (Deemed-to-be University), Jain Global
Campus
Bangalore, India

Santosh V. Patil
Department of Electrical and Computer
Science Engineering
Institute of Infrastructure, Technology,
Research and Management (IITRAM)
Ahmedabad, India

Asmita Poddar
Department of Electrical Engineering
Dream Institute of Technology
Kolkata, India

K. Pramoda
Centre for Nano and Material Sciences
Jain (Deemed-to-be University), Jain Global
Campus
Bangalore, India

Periyasamy Angamuthu Praveen
Department of Physics
Organic Optoelectronics Laboratory
Indian Institute of Science Education and
Research (IISER)
Tirupati, India

K. Ravichandran
Department of Physics, PG & Research
AVVM Sri Pushpam College,
(Affiliated Bharathidasan University,
Tiruchirappalli)
Thanjavur, India

P. Ravikumar
Department of Physics
Tagore Government Arts and Science College,
(Affiliated to Pondicherry University)
Puducherry, India

Debasish Ray
Department of Chemistry
Indian Institute of Engineering Science &
Technology, (IIEST)
Howrah, India

Soumya J. Ray
Department of Physics
Indian Institute of Technology
Patna, India

Lucas S. Ribeiro
Department of Chemistry
University of São Carlos-UFSCar
São Paulo, Brazil

Arpita Roy
Department of Physics
Indian Institute of Technology
Patna, India

Madhab Roy
Department of Electrical Engineering
Jadavpur University
Kolkata, India

Subhasis Roy
Department of Chemical Engineering
University of Calcutta
Kolkata, India

Himanshu S. Sahoo
Department of Chemistry
Indian Institute of Engineering Science &
    Technology, (IIEST)
Howrah, India

Bimal K. Sarma
Department of Physics
Nanophysics Laboratory
Gauhati University
Guwahati, India

Devi Bala Saraswathi Sethuraman
Department of Physics
National Changhua University of Education
Changhua County, Taiwan

Rijith Sreenivasan
Sree Narayana College (Affiliated to University
    of Kerala)
Kollam, India
and
Government College Attingal (Affiliated to
    University of Kerala)
Thiruvananthapuram, India

S. Suvathi
Department of Physics, PG & Research
AVVM Sri Pushpam College, (Affiliated
    Bharathidasan University, Tiruchirappalli)
Thanjavur, India

Juliana A. Torres
Department of Chemistry
University of São Carlos
UFSCar-São Carlos
São Paulo, Brazil

Haizhong Weng
School of Physics, CRANN and AMBER
Trinity College Dublin
Dublin, Ireland

Shaoteng Wu
State Key Laboratory of Superlattices and
    Microstructures
Institute of Semiconductors, Chinese Academy
    of Sciences
Beijing, China

Shiheng Xin
School of Physics and Electronic Information
Yan'an University
Yan'an, China

Fuchun Zhang
School of Physics and Electronic Information
Yan'an University
Yan'an, China

Chunyi Zhi
Department of Materials Science and
    Engineering
City University of Hong Kong
Hong Kong, China

1 Semiconductors
An Introduction

André E. Nogueira, Lucas S. Ribeiro,
Francisco G. E. Nogueira, and Juliana A. Torres

1.1 INTRODUCTION

The first semiconductor devices and materials emerged in the late 19th and early 20th centuries. German physicist Ferdinand Braun developed the first semiconductor diode, known as the "cat's whisker" detector, in 1874, which was used as a signal detector in a crystal radio. Semiconductors ushered in a transformative era in technology during the nineteenth century, and their significance in contemporary society is undeniable, since they have a wide range of applications such as in electronic circuits, solar cells, light-emitting diodes (LEDs), laser technologies, catalysis, and photocatalysis [1, 2].

Nowadays, a deeper understanding of the properties of the crystal structure of semiconductors is essential in most branches of science. This involves a detailed examination of the atomic arrangement of these materials and its influence on their electronic, optical, and physical properties (Table 1.1).

Inorganic semiconductors are composed of inorganic elements, such as silicon (Si), germanium (Ge), gallium arsenide (GaAs), and titanium dioxide (TiO_2), among others. The arrangement of a substantial number of atoms into a long-range ordered atomic structure results in the overlapping

TABLE 1.1
Conductor and Semiconductor Properties

Characteristics	Semiconductor	Conductor
Conductivity	Moderate	High
Resistivity	Moderate	Low
Forbidden gap	Small forbidden gap	No forbidden gap
Temperature coefficient	Negative	Positive
Conduction	Very small number of electrons for conduction	Large number for electrons conduction
Conductivity value	The conductivity of materials can range from 10^{-7} to 10^{-13} mho/m, distinguishing conductors from insulators	Very high 10^{-7} mho/m
Resistivity value	Materials can have electrical conductivity ranging from 10^{-5} to 10^5 Ω·m	Negligible; less than 10^{-5} Ω·m
Current flow	Due to holes and unbound electrons	Due to free electrons
Number of carries at normal temperature	Low	Very high
Zero Kelvin behavior	Acts like an insulator	Acts like a superconductor
Formation	Formation by covalent bonding	Formation by metallic bond
Valence electrons	The outermost shell contains a total of four valence electrons	One valence electron in outermost shell

Source: Adapted with permission from [3], copyright (2023), CRC Press.

DOI: 10.1201/9781003450146-1

of adjacent orbitals to form bonds. This three-dimensional lattice extends throughout the material. Thus, the linear combination of two atomic orbitals yields two molecular orbitals: a bonding molecular orbital and an antibonding molecular orbital. As more atoms are aligned, additional molecular orbitals are created through the interference of the wave functions of their atomic orbitals. This results in the establishment of a set of non-degenerate orbitals characterized by a minimal energy difference between consecutive levels [4].

The band that is occupied by electrons is referred to as the valence band (VB). It is composed of lower-energy orbitals and thus has a higher probability of electron occupation. However, the conduction band (CB) comprises higher-energy orbitals and consequently exhibits lower levels of electron occupancy probability. The energy gap between these two bands is known as band gap energy (E_g) (Figure 1.1). The band structure of a semiconductor is a critically important parameter for its application as a photocatalyst, since it plays a crucial role in defining the material's light absorption properties and redox capacity.

The probability of occupation of these bands by a free electron is given by the Fermi function, which is related to the absolute temperature of the system, the Boltzmann constant, and the Fermi level (E_f). The latter is defined as the reference energy level of the material, corresponding to the energy level, at which there is a probability of 1/2 for electrons to be occupied, and it is an important parameter in predicting its electrical behavior [5].

In semiconductors, the Fermi level is typically positioned near the center of the band gap ($E_f = 1/2\ E_g$). This placement allows electrons located at the top of the VB to readily transition to the lower levels of the CB at temperatures below 0 K. When the semiconductor absorbs photons with energy equal to or greater than the band gap energy, electrons become excited from the VB to the CB, generating electron-hole (e^-/h^+) pairs as charge carriers on the semiconductor's surface.

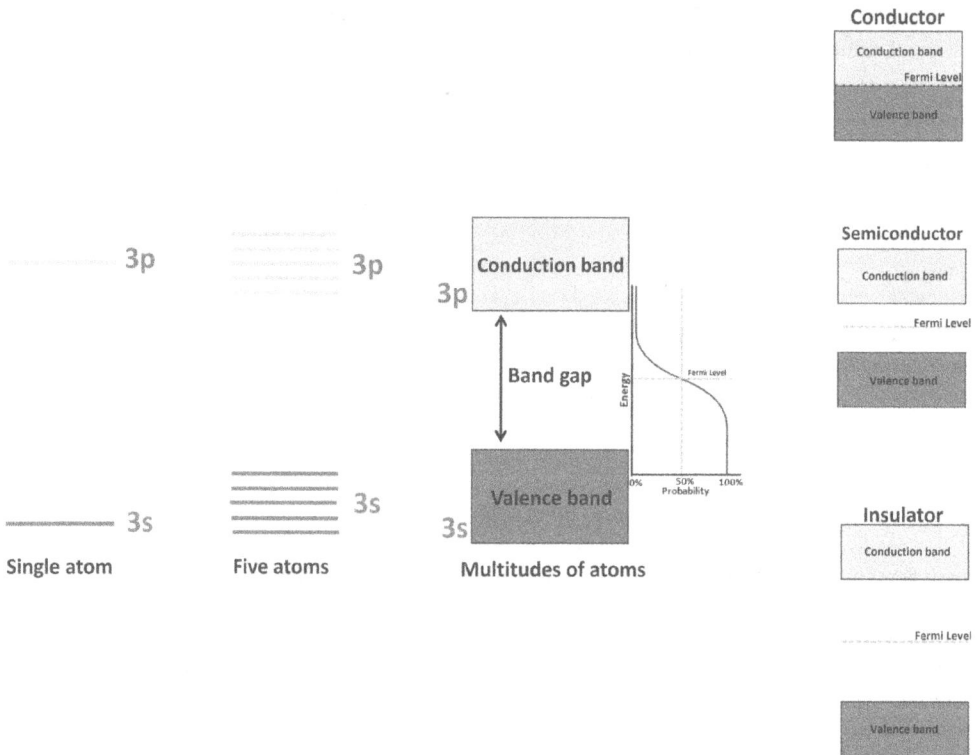

FIGURE 1.1 Schematic representation of the overlap of atomic orbitals to form an energy band and the band structure of different materials (conductor, semiconductor, and insulator).

These e^-/h^+ pairs can catalyze redox reactions, and their behavior is directly influenced by the positions of the VBs and CBs [6–8]. For redox reactions to take place, it is essential that the charge carriers possess a sufficient lifetime. Conversely, electrons can undergo recombination with holes, which diminishes the catalytic efficiency of semiconductors [5].

In the case of chemically pure semiconductors, electrons are promoted from the VB to the CB, leaving behind unoccupied states in the VB, which are referred to as holes (h^+). In such scenarios, conduction occurs intrinsically. Si and Ge are well-known examples of intrinsic semiconductors [4, 9]. However, this type of semiconductor has limited application due to its low conductivity. In this context, several strategies have been employed to enhance the electrical properties of these materials. Notably, one such approach is doping, which involves the introduction of atoms, referred to as impurities, into the crystal lattice of the semiconductor. These semiconductors, modified through doping, are commonly referred to as extrinsic semiconductors [7].

1.2 SEMICONDUCTOR DOPING

Semiconductors can be divided into two categories: intrinsic and extrinsic. Intrinsic semiconductors are those in which their electronic properties are based only on the pure structure of the material, that is, a completely full VB separated from a completely empty CB [10]. Furthermore, in this type of semiconductor, for every excited electron in the CB, there is a hole in the VB. However, practically all semiconductors are extrinsic since impurities in the concentration of 1×10^{-12} are already able to affect the electrical properties through the insertion of excess electrons and holes [11].

Doping is the process of introducing impurities into a semiconductor crystal to deliberately alter its conductivity due to electron deficiency or excess [12]. This procedure can happen during semiconductor manufacturing, in which the semiconductor is completely doped, or doped partially after wafer fabrication. In the case of Si and Ge elemental semiconductors, the wafer can be partially doped by diffusion and ion implantation processes (Figure 1.2) [13].

In diffusion doping, impurities are transported from a region of high concentration to a region of low concentration by random molecular motion. The dopant material can move through the semiconductor structure filling the empty spaces in the crystal lattice, which always exist, even in perfect crystals. They can also move in between the atoms in the crystal lattice, or the impurity atoms are located in the crystal lattice and are exchanged with the semiconductor atoms. The diffusion movement therefore depends on the temperature, the concentration difference between the phases, and the nature of the dopant [14]. In ion implantation, ions are accelerated in an electric field and beamed directly into the semiconductor. The ions swiftly enter the semiconductor by occupying interstitial sites, and the depth of penetration can be accurately controlled

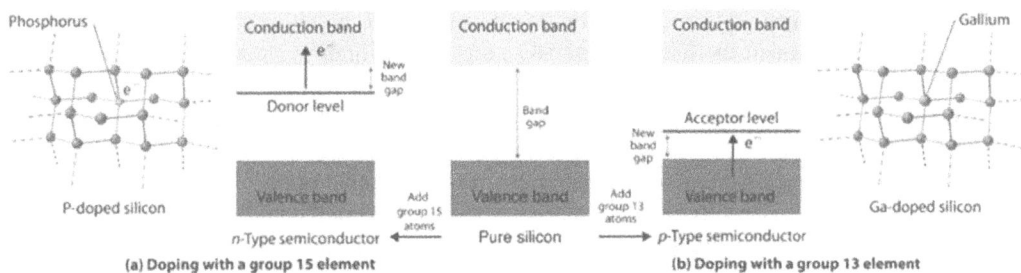

FIGURE 1.2 Structures and band diagrams of n-type and p-type semiconductors. (**a**) Doping Si with phosphorus (P) element results in a new filled level between the VBs and CBs of the host. (**b**) Doping Si with gallium element results in a new empty level between the VBs and CBs of the host. (Figure adapted with permission from Reference [18], copyright by the authors, some rights reserved; exclusive licensee [LibreTexts]. Distributed under a Creative Commons Attribution License 4.0 (CC BY) https://creativecommons.org/licenses/by/4.0/.)

by adjusting the voltage needed to drive the ions. Unlike the diffusion process, this form of doping takes place at room temperature. Regions that should not be doped can be covered with a photoresist layer [15].

In elemental semiconductors, which have four electrons, we can add impurities that have more or fewer electrons, thus creating an excess of electrons or holes in the structure. In the case of doping using elements with valence 5, such as P, As, and Sb, only four of the five electrons of the dopant will be able to form covalent bonds with the atoms of the semiconductor [16]. As a consequence, the unbound electron is only weakly bound to the region around the impurity by electrostatic attraction. Its binding energy is therefore very small, allowing it to become a free electron in the structure and create an energy level between the VBs and CBs. For each free electron in the semiconductor, a new energy level is created in the band gap region close to the CB [17]. Furthermore, when these electrons are excited, holes are not created in the VB, causing these semiconductors to have a greater number of electrons in the conduction layer than holes in the valence layer. This type of doping results in n-type semiconductors.

The opposite effect can be obtained by doping Si and Ge with trivalent atoms, such as Al, B and Ga. In this case, the impurities are electron-deficient and are unable to form the four bonds as the semiconductor atoms. As a consequence, a hole is formed in the structure, which can be rapidly filled by an electron from an adjacent bond. Thus, this hole begins to move through the structure, changing places with electrons participating in the conduction process. This phenomenon can also be explained by the creation of an energy level in the band gap region just above the VB that allows the easy excitation of an electron and the creation of holes in the VB. In this case the semiconductor is denominated p-type, and no free electrons are created in the CB of the semiconductor [19].

The doping process can also be performed for metal oxide semiconductors. Many of these semiconductors are used as photocatalysts for processes involving, for example, photodegradation of organic pollutants and photoreduction of carbon dioxide into value-added molecules [20]. The problem is that these oxides often have a high band gap value and high recombination rates of photogenerated charges, which end up limiting their use. One way to reverse these problems, including allowing the material to be activated by visible light, is through doping, which can occur in different ways.

Self-doping involves the introduction of cations of the same metal as the oxide, albeit with a distinct charge, into the material. A case in point is observed in titanium oxide semiconductors containing Ti^{3+} cations within their structure. This doping technique is predominantly employed to narrow the band gap of the semiconductor, while inducing minimal structural perturbation. The manipulation of heating duration and temperature stands as a pivotal facet of this process [21]. Alternatively, doping can be achieved using cationic metals, anionic non-metals, or non-metallic molecules. For instance, TiO_2 can be doped with a range of cations like Ag^+, Fe^{2+}, and Fe^{3+}, as well as elements like nitrogen, sulfur, and boron. Analogous to the previous scenario, these dopants are introduced to diminish the band gap of TiO_2, thereby facilitating its activation under visible light, and enhancing its efficacy in breaking down pollutants [22]. Another illustration involves the incorporation of copper oxide into photocatalysts employed in CO_2 photoreduction. This addition of copper empowers certain semiconductors to facilitate the conversion of CO_2 into molecules such as carbon monoxide, a process previously hindered by their band gap position [23]. Nonetheless, these introduced impurities frequently function as sites for charge recombination, thereby diminishing the efficiency of these semiconductors. An alternative approach involves co-doping with two or more types of dopants, which serves to mitigate the quantity of recombination centers. Instances of this include the simultaneous doping of TiO_2 with sulfur, nitrogen, and carbon, along with the co-doping of other oxides utilizing a combination of N and C [24].

As highlighted earlier, the process of doping facilitates not only the enhanced stimulation of electrons and holes through the reduction of the band gap and the establishment of energy levels within it, but also leads to a diminished rate of recombination for these excited charges. Consequently, these doped semiconductors exhibit heightened electronic conductivity, thereby enabling more

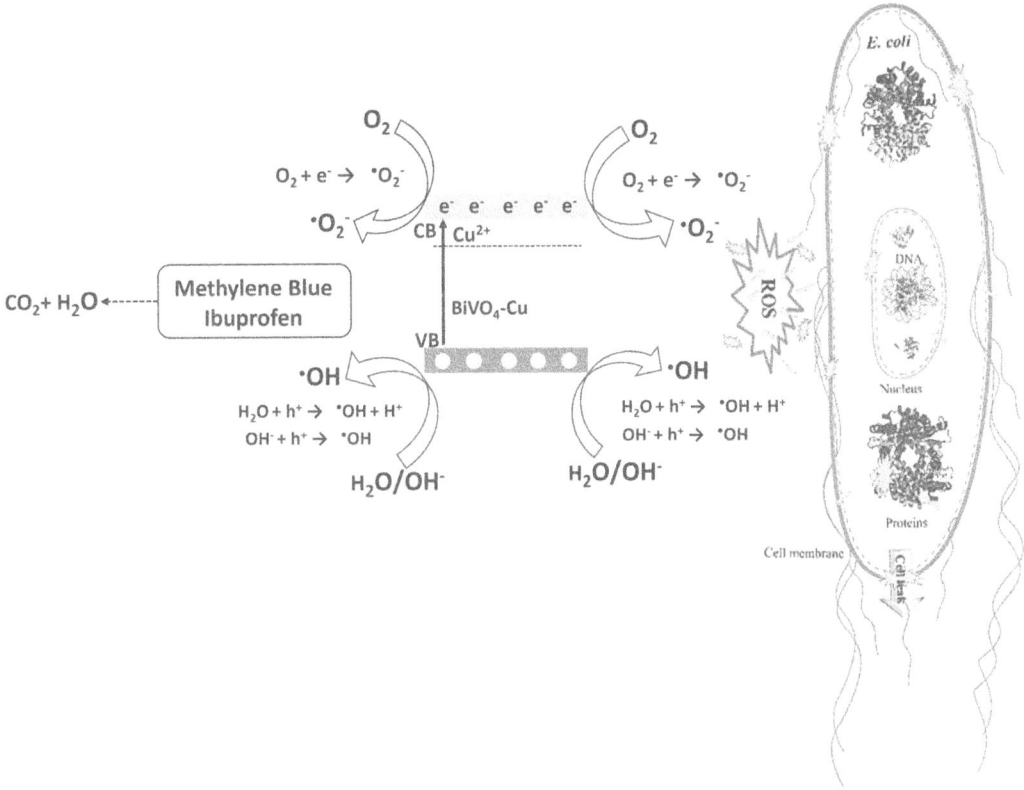

FIGURE 1.3 Schematic diagram showing the photocatalytic degradation mechanism of methylene blue (MB), ibuprofen (IBP), and inactivation of *Escherichia coli* by a Cu-doped $BiVO_4$-based semiconductor under visible light irradiation. (Adapted with permission from Reference [30], copyright 2023, Elsevier.)

substantial performance tailored to specific applications. Additionally, doping yields improved structural characteristics, including increased surface area, reduced crystallite size, and heightened crystallinity [25].

Studies show that semiconductor doping with noble metals significantly enhances the activity of semiconductors for environmental applications. Furthermore, noble metals can effectively capture the photo-generated charges between VB and CB, resulting in a high efficiency of charge separation and a decrease of recombination rate. Consequently, noble metals contribute to enhancing photocatalytic performance by generating surface plasmon resonance (SPR) and serving as electron traps [26–29].

According to Regmi et al. [30] the Cu-doped $BiVO_4$ semiconductor has better efficiency than undoped $BiVO_4$, and the 1 wt. % Cu-doped $BiVO_4$ sample showed the best efficiency to degradation of methylene blue (dye) and ibuprofen, as well as the inactivation of *Escherichia coli* (bacteria) (Figure 1.3). According to the authors, the incorporation of Cu ions in the $BiVO_4$ lattice creates an in-gap state, which facilitates the mobility of the charge carrier and inhibits the recombination of electron and hole pairs, leading to improved photocatalytic activities.

1.3 P-N JUNCTION AND SEMICONDUCTOR DEVICES

The p-n junction is one of the fundamental principles that underlie the technological revolution in electronics and semiconductor devices. It is an essential component in the manufacture of devices

(b) Forward bias of the p-n junction

(a) p-n junction

(c) Reverse bias of the p-n junction

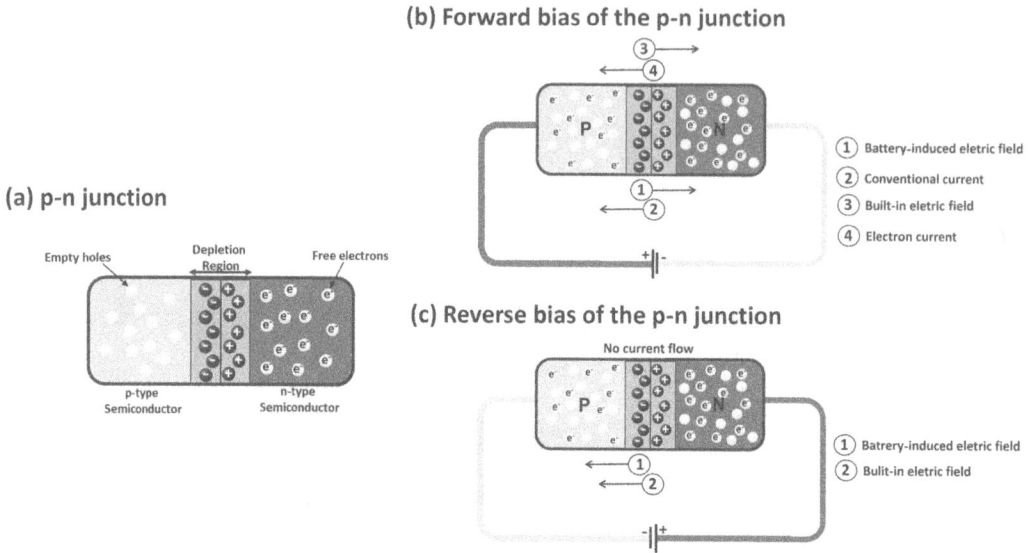

FIGURE 1.4 (a) Illustrative depiction of a p-n junction formed by semiconductors. (b) Illustrative depiction of the forward bias condition of a p-n junction (Application of a positive potential to the p-type semiconductor induces electron migration to the left, resulting in an abundance of free electrons in the conduction band (CB) of the n-type semiconductor, thus facilitating current flow). (c) Illustrative depiction of the reverse bias condition of a p-n junction (Application of a positive potential to the n-type semiconductor triggers electron movement toward the right, while the absence of free electrons in the p-type semiconductor hinders current flow)

such as diodes and transistors, which, in turn, form the basis for much of modern electronics. The p-n junction is formed by the combination of p-type and n-type semiconductor materials. When these two materials are brought together, electron and hole diffusion take place at the interface, creating a depletion zone that acts as a natural barrier to an electrical current (Figure 1.4) [3, 31]. This depletion zone forms the basis for the functionality of diodes, which are electronic components that permit the flow of electrical current in only one direction. When a positive voltage is applied to the p-type material and a negative voltage to the n-type material, the barrier is overcome, allowing a current to pass through the diode. Conversely, if the polarity is reversed, the depletion zone widens, blocking current flow and preventing undesired electrical flow [3, 31].

Transistors, in turn, are essential semiconductor devices used for amplifying and controlling electrical currents. They are constructed using multiple p-n junctions, enabling precise manipulation of current between these junctions. Bipolar junction transistors (BJTs) and field-effect transistors (FETs) are examples of transistors that directly benefit from the properties of the p-n junction. Both rely on the ability to regulate current between p-n junctions. In BJTs, the current between the collector and emitter is controlled by the base current. In FETs, the current between the drain and the source is governed by the electric field generated at the p-n junction between the drain and the substrate [3, 31].

The significance of the p-n junction in the fabrication of semiconductor devices is unquestionable. It serves as the foundation for creating electronic components that have made possible the development of communication systems, computers, smartphones, efficient power sources, and numerous other devices that have shaped the modern world. The capacity to regulate the flow of electrical current through the p-n junction has catalyzed a technological revolution that continues to progress, propelling innovation in various fields of science and industry [32].

Furthermore, the continuous improvement of semiconductor manufacturing technology, enabling the reduction of component sizes and increased efficiency, has been one of the primary catalysts of technological advancement in recent decades. This is exemplified by Moore's law, which observes the exponential growth in the number of transistors on a chip and the concurrent reduction in costs over time [33].

1.4 EMERGING SEMICONDUCTOR APPLICATIONS

As previously mentioned, semiconductors have generated significant interest in various areas such as nanophotonics, nanoelectronics, miniaturized sensors, energy conversion, nonlinear optics, detectors, imaging devices, quantum applications, solar cells, catalysis, water treatment, and biomedicine [34].

Semiconductor materials have introduced a new paradigm in scientific development. Consequently, the development of highly efficient semiconductor materials plays a crucial role in the sustainable and economic use of solar energy, mainly where sunlight is readily available. Thus, nanomaterials characterized by a high specific surface area, strong visible light absorption, and efficient charge separation have attracted growing attention.

A review article by Fang et al. [35] in 2022 showed the potential application of hollow semiconductor photocatalysts for solar energy conversion. This potential arises from the capacity to modify their structures, thereby enhancing energy conversion through increased specific surface area utilization, improved solar light absorption, and greater exposure of active centers. Furthermore, multiple articles have demonstrated that when light radiation penetrates a hollow semiconductor, it undergoes multiple refractions on the thin inner surface of the shell. This phenomenon enhances the light utilization efficiency of the semiconductor, leading to greater energy absorption and the separation efficiency of charges [36].

There has been a growing interest in the hydrogen evolution reaction (HER) from water using semiconductor materials such as ZnO, TiO_2, Nb_2O_5, NiS, SnO_2, and CdS. This interest arises from the possibility of utilizing inexhaustible sources such as water and sunlight for this purpose [37, 38]. Nevertheless, significant challenges persist in improving conversion efficiency. This is primarily due to the limitations associated with most semiconductors, including their inability to provide active sites, high rates of recombination of charge carriers, and efficient harnessing of visible light for the HER. Thus, the incorporation of a co-catalyst or the development of the heterostructure can be necessary to improve the conversion efficiency [39].

For instance, numerous studies demonstrate that the electronic coupling of various semiconductors with graphitic carbon nitride ($g-C_3N_4$) materials enhances the transport of photogenerated charges and improves the HER process. Moreover, the formation of different types of heterojunctions also influences the photocatalytic pathway for HER, thereby impacting the overall efficiency of the photocatalyst [40].

The progress made in charge manipulation within nanoscale semiconductor-based devices has opened a pioneering avenue for the development of photonic quantum-computation approaches since flying qubits are typically associated with photons. Flying qubits constitute an integral component of global initiatives aimed at establishing secure data transmission networks, commonly referred to as the quantum internet [41]. In semiconductor devices, a single electron is manipulable in a surface-gate-defined nanoscale structure such as a quantum dot or a waveguide [42].

In the biomedical field, two-dimensional nanomaterials, including graphene (2D NBG), have emerged as a captivating class of materials possessing noteworthy antibacterial properties. These materials, exemplified by two-dimensional transition metal dichalcogenides (TMDCs), have found applications in anticancer, antimicrobial, and antiviral therapies [43]. The efficacy of these treatments' hinges on numerous factors, encompassing size, morphology, crystalline orientation, concentration, stability, and surface functionalization. Furthermore, the non-toxic nature of semiconductor materials like TiO_2 and ZnO has rendered them of significant interest as sensing materials in medical and pharmaceutical systems since they exhibit biocompatibility and can readily bind with biological entities to serve as biosensors for detecting the severity of certain chronic diseases, the presence of the Zika virus in urine, or even the coronavirus in fluid samples [44].

1.5 CHALLENGES AND FUTURE PERSPECTIVES

Semiconductors face a series of complex and interconnected challenges in the present day, with two of the most pressing being miniaturization and energy efficiency. Miniaturization, as an unrelenting pursuit of smaller and denser components, has reached a critical point where physical barriers are beginning to emerge, making it imperative to explore new materials and architectures to maintain the pace of advancement. Concurrently, energy efficiency is a central concern, as the power consumption of electronic devices continues to rise.

It is important to note that the semiconductor industry is in a constant state of growth and evolution. The global semiconductor market was valued at $591.8 billion in 2022 and is expected to reach approximately $1,883.7 billion by 2032, poised to grow at a compound annual growth rate (CAGR) of 12.28% during the forecast period from 2023 to 2032 [45]. In 2030, the two largest sectors in this market will be servers, data centers, and storage, accounting for 22%, and smartphones, accounting for 19.4% (Figure 1.5). This expansion of the semiconductor market reflects the increasing demand

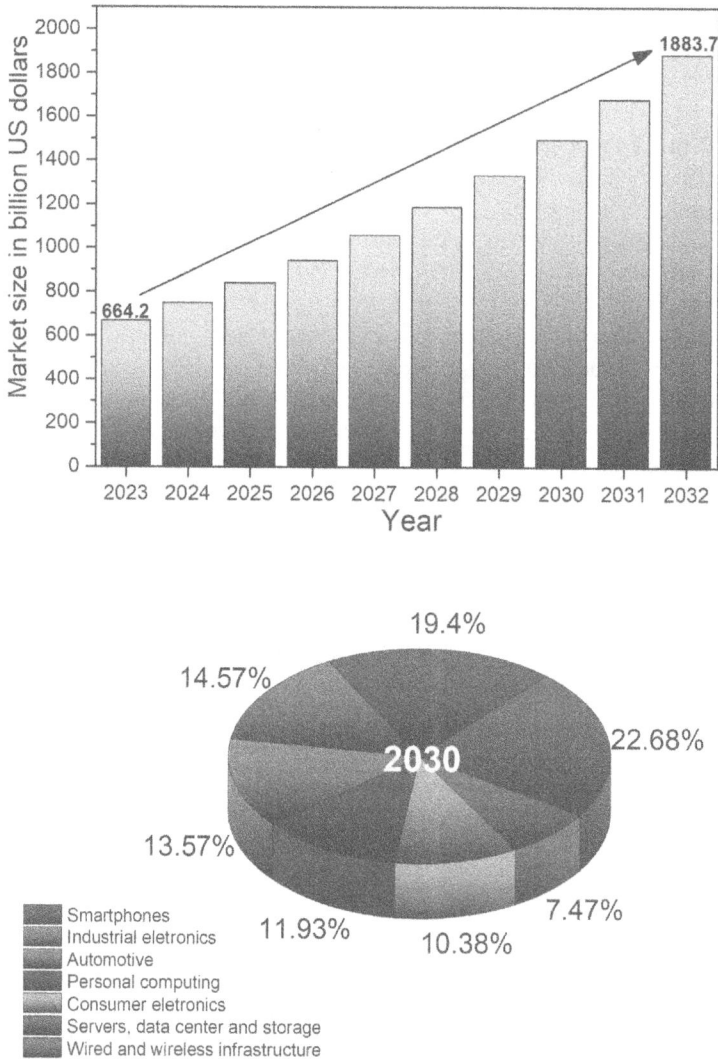

FIGURE 1.5 Semiconductor market size, 2023 to 2032 (billion US dollars) and the percentage of different sectors in 2030 [31, 46].

for technology and electronics in an increasingly connected world, underscoring the continued importance of semiconductors in future innovations and the global economy [46].

In this context, future prospects for semiconductors point toward the adoption of innovative semiconductor materials, such as those based on two-dimensional materials, as well as the exploration of new computational paradigms like quantum and neuromorphic computing. Additionally, advancements in component integration, such as the creation of heterogeneous systems on a single chip, also promise to play a fundamental role in overcoming these challenges and ushering in a new era of more powerful and efficient electronic devices. In conclusion, the journey of semiconductors, from the submicron era to the nano era, has revolutionized technology and communication, creating a vast market for the internet and wireless communication. The unique properties of semiconductor nanomaterials have opened up new frontiers in energy conversion, storage, quantum applications, water treatment, biomedical applications, sensors, and electronics. The semiconductor industry remains at the forefront of technological innovation, shaping the future of various industries and enhancing the quality of life for people around the world.

ACKNOWLEDGMENTS

The authors are grateful to the National Council for Scientific and Technological Development (CNPq) (grants #308823/2022 and #406860/2022-0).

REFERENCES

1. M. Ordu, The role of semiconductors in the future of optical fibers, Front Phys. 11 (2023) 1141795.
2. K. Mishra, N. Devi, S.S. Siwal, V.K. Gupta, V.K. Thakur, Hybrid semiconductor photocatalyst nanomaterials for energy and environmental applications: Fundamentals, designing, and prospects, Adv. Sustain. Syst. 7 (2023) 2300095.
3. M.K. Majumder, V.R. Kumbhare, A. Japa, B.K. Kaushik, Introduction to Microelectronics to Nanoelectronics Design and Technology, CRC Press, 2020.
4. R. Bueno, O.R. Lopes, K. Carvalho, C. Ribeiro, H. Mourão, Semicondutores heteroestruturados: uma abordagem sobre os principais desafios para a obtenção e aplicação em processos fotoquímicos ambientais e energéticos, Quim. Nova. 42 (2019) 661–675.
5. S. Bai, J. Jiang, Q. Zhang, Y. Xiong, Steering charge kinetics in photocatalysis: Intersection of materials syntheses, characterization techniques and theoretical simulations, Chem Soc Rev. 44 (2015) 2893–2939.
6. J. Zeng, Z. Li, H. Jiang, X. Wang, Progress on photocatalytic semiconductor hybrids for bacterial inactivation, Mater. Horiz. 8 (2021) 2964–3008.
7. M. Tahir, S. Tasleem, B. Tahir, Recent development in band engineering of binary semiconductor materials for solar driven photocatalytic hydrogen production, Int. J. Hydrogen Energy. 45 (2020) 15985–16038.
8. J. Lu, S. Gu, H. Li, Y. Wang, M. Guo, G. Zhou, Review on multi-dimensional assembled S-scheme heterojunction photocatalysts, J. Mater. Sci. Technol. 160 (2023) 214–239.
9. A. Baccaro, I. Gutz, Fotoeletrocatálise em semicondutores: dos princípios básicos até sua conformação à nanoescala, Quim. Nova. 41 (2017) 326–339.
10. Q. Li, Q. Chen, B. Song, Giant radiative thermal rectification using an intrinsic semiconductor film, Mater. Today Phys. 23 (2022) 100632.
11. W.D. Callister Jr, D.G. Rethwisch, Callister's Materials Science and Engineering, John Wiley & Sons, 2020.
12. N. Umezawa, J. Ye, Role of complex defects in photocatalytic activities of nitrogen-doped anatase TiO_2, Phys. Chem. Chem. Phys. 14 (2012) 5924.
13. C. Jacoboni, P. Lugli, Review of Semiconductor Devices. In: The Monte Carlo Method for Semiconductor Device Simulation. Comp Microelectronics. (1989) 162–217. https://doi.org/10.1007/978-3-7091-6963-6_4
14. X. Li, J. Yang, H. Sun, L. Huang, H. Li, J. Shi, Controlled synthesis and accurate doping of wafer-scale two-dimensional semiconducting transition metal dichalcogenides, Adv. Mater. 2305115 (2023) 1–14.
15. T. Narita, T. Kachi, K. Kataoka, T. Uesugi, P-type doping of GaN by magnesium ion implantation, Appl. Phys. Express. 10 (2017) 016501.

16. B. Singha, C.S. Solanki, N-type solar cells: Advantages, issues, and current scenarios, Mater. Res. Express. 4 (2017) 072001.

17. L.J. Geerligs, I.G. Romijn, A.R. Burgers, N. Guillevin, A.W. Weeber, J.H. Bultman, H. Wang, F. Lang, W. Zhao, G. Li, Z. Hu, J. Xiong, A. Vlooswijk, Progress in low-cost n-type silicon solar cell technology, in: 2012 38th IEEE Photovoltaic Specialists Conference, IEEE, (2012) 001701–001704.

18. LibreTexts Chemistry, Metals and Semiconductors, (2019). https://chem.libretexts.org/@go/page/42093 (accessed September 25, 2023).

19. G.A. Armin, B.B. Matthew, H. Bram, M. Thomas, Industrial silicon wafer solar cells – Status and trends, Green. 2 (2012) 135–148.

20. L.S. Ribeiro, I.M. Pinatti, J.A. Torres, A.S. Giroto, F. Lesse, E. Longo, C. Ribeiro, A.E. Nogueira, Rapid microwave-assisted hydrothermal synthesis of $CuBi_2O_4$ and its application for the artificial photosynthesis, Mater. Lett. 275 (2020) 128165.

21. W. Fang, M. Xing, J. Zhang, A new approach to prepare Ti^{3+} self-doped TiO_2 via $NaBH_4$ reduction and hydrochloric acid treatment, Appl. Catal. B. 160–161 (2014) 240–246.

22. C. Sahoo, A.K. Gupta, Characterization and photocatalytic performance evaluation of various metal ion-doped microstructured TiO_2 under UV and visible light, Journal of Environmental Science and Health, Part A. 50 (2015) 659–668.

23. A.L.A. Faria, H.A. Centurion, J.A. Torres, R.V. Gonçalves, L.S. Ribeiro, C. Riberio, C. da Cruz, F.G.E. Nogueira, Enhancing Nb_2O_5 activity for CO_2 photoreduction through Cu nanoparticles cocatalyst deposited by DC-magnetron sputtering, J. CO_2 Util. 53 (2021) 101739.

24. S.M. El-Sheikh, T.M. Khedr, A. Hakki, A.A. Ismail, W.A. Badawy, D.W. Bahnemann, Visible light activated carbon and nitrogen co-doped mesoporous TiO_2 as efficient photocatalyst for degradation of ibuprofen, Sep. Purif. Technol. 173 (2017) 258–268.

25. S.S. Mohtar, F. Aziz, A.F. Ismail, N.S. Sambudi, H. Abdullah, A.N. Rosli, B. Ohtani, Impact of doping and additive applications on photocatalyst textural properties in removing organic pollutants: A review, Catalysts. 11 (2021) 1160.

26. M. Arumugam, R. Koutavarapu, K.K. Seralathan, S. Praserthdam, P. Praserthdam, Noble metals (Pd, Ag, Pt, and Au) doped bismuth oxybromide photocatalysts for improved visible light-driven catalytic activity for the degradation of phenol, Chemosphere. 324 (2023) 138368.

27. M. Qin, K. Jin, X. Li, R. Wang, Y. Li, H. Wang, Novel highly-active Ag/Bi dual nanoparticles-decorated BiOBr photocatalyst for efficient degradation of ibuprofen, Environ. Res. 206 (2022) 112628.

28. H.T. Ren, S.Y. Jia, Y. Wu, S.H. Wu, T.H. Zhang, X. Han, Improved photochemical reactivities of Ag_2O/ g-C_3N_4 in phenol degradation under UV and visible light, Ind Eng Chem. Res. 53 (2014) 17645–17653.

29. S.J. Lee, H.J. Jung, R. Koutavarapu, S.H. Lee, M. Arumugam, J.H. Kim, M.Y. Choi, ZnO supported Au/ Pd bimetallic nanocomposites for plasmon improved photocatalytic activity for methylene blue degradation under visible light irradiation, Appl. Surf. Sci. 496 (2019) 143665.

30. C. Regmi, Y.K. Kshetri, R.P. Pandey, T.H. Kim, G. Gyawali, S.W. Lee, Understanding the multifunctionality in Cu-doped $BiVO_4$ semiconductor photocatalyst, J. Environ. Sci. (China). 75 (2019) 84–97.

31. G. Domingo, Semiconductor Basics: A Qualitative, Non-Mathematical Explanation of How Semiconductors Work and How They Are Used, Wiley, 1st edition, 2020.

32. N. Hossain, M.H. Mobarak, M.A. Mimona, M.A. Islam, A. Hossain, F.T. Zohura, M.A. Chowdhury, Advances and significances of nanoparticles in semiconductor applications – A review, Results in Eng. 19 (2023) 101347.

33. G.E. Moore, Cramming more components onto integrated circuits, Reprinted from Electronics. 38 (1965) 114.

34. R.V. Nair, V.S. Gummaluri, M.V. Matham, C. Vijayan, A review on optical bandgap engineering in TiO_2 nanostructures via doping and intrinsic vacancy modulation towards visible light applications, J. Phys. D. Appl. Phys. 55 (2022) 313003.

35. B. Fang, Z. Xing, D. Sun, Z. Li, W. Zhou, Hollow semiconductor photocatalysts for solar energy conversion, Adv. Powder Mater. 1 (2022) 100021.

36. X. Li, D. Chen, N. Li, Q. Xu, H. Li, J. He, J. Lu, AgBr-loaded hollow porous carbon nitride with ultrahigh activity as visible light photocatalysts for water remediation, Appl. Catal. B. 229 (2018) 155–162.

37. J. Bai, R. Shen, W. Chen, J. Xie, P. Zhang, Z. Jiang, X. Li, Enhanced photocatalytic H2 evolution based on a Ti_3C_2/$Zn_{0.7}Cd_{0.3}S$/Fe_2O_3 Ohmic/S-scheme hybrid heterojunction with cascade 2D coupling interfaces, Chem. Eng. J. 429 (2022) 132587.

38. K. Chang, X. Hai, H. Pang, H. Zhang, L. Shi, G. Liu, H. Liu, G. Zhao, M. Li, J. Ye, Targeted synthesis of 2H- and 1T-phase MoS_2 monolayers for catalytic hydrogen evolution, Adv. Mater. 28 (2016) 10033–10041.

39. H. Lin, K. Zhang, G. Yang, Y. Li, X. Liu, K. Chang, Y. Xuan, J. Ye, Ultrafine nano 1T-MoS_2 monolayers with NiO_x as dual co-catalysts over TiO_2 photoharvester for efficient photocatalytic hydrogen evolution, Appl. Catal. B. 279 (2020) 119387.

40. Y. Rajput, P. Kumar, T.C. Zhang, D. Kumar, M. Nemiwal, Recent advances in g-C_3N_4-based photocatalysts for hydrogen evolution reactions, Int. J. Hydrogen Energy. 47 (2022) 38533–38555.

41. H. Edlbauer, J. Wang, T. Crozes, P. Perrier, S. Ouacel, C. Geffroy, G. Georgiou, E. Chatzikyriakou, A. Lacerda-Santos, X. Waintal, D.C. Glattli, P. Roulleau, J. Nath, M. Kataoka, J. Splettstoesser, M. Acciai, M.C. da Silva Figueira, K. Öztas, A. Trellakis, T. Grange, O.M. Yevtushenko, S. Birner, C. Bäuerle, Semiconductor-based electron flying qubits: Review on recent progress accelerated by numerical modelling, EPJ Quantum Technol. 9 (2022) 21.

42. A. Peruzzo, J. McClean, P. Shadbolt, M.-H. Yung, X.-Q. Zhou, P.J. Love, A. Aspuru-Guzik, J.L. O'Brien, A variational eigenvalue solver on a photonic quantum processor, Nat. Commun. 5 (2014) 4213.

43. S.R. Laraba, W. Luo, A. Rezzoug, Q. ul ain Zahra, S. Zhang, B. Wu, W. Chen, L. Xiao, Y. Yang, J. Wei, Y. Li, Graphene-based composites for biomedical applications, Green Chem. Lett. Rev. 15 (2022) 724–748.

44. A.D. Terna, E.E. Elemike, J.I. Mbonu, O.E. Osafile, R.O. Ezeani, The future of semiconductors nanoparticles: Synthesis, properties and applications, Mater. Sci. Eng. B. 272 (2021) 115363.

45. Precedence Research, Advanced Optics Market (By Technology: Ray, Wave, Quantum; By Application: LiDAR, Lighting Solution, Optical Communication, Intelligence, Surveillance & Reconnaissance, Medical Equipment, Camera, Metrology Devices, 3D Scanner, Others; By End-Use Industry: Commercial, Industrial, Defense, Medical, Aviation (Drone), Automotive, Space, Others) – Global Industry Analysis, Size, Share, Growth, Trends, Regional Outlook, and Forecast 2023–2032, https://www.precedenceresearch.com/Advanced-Optics-Market. (2023).

46. Thomas Alsop, Semiconductor market revenue worldwide from 2020 to 2030, by application, https://www.statista.com/statistics/498265/cagr-main-semiconductor-target-markets/. (2023).

2 Diluted Magnetic Semiconductors

*Mourad Boughrara, Abdelhamid Ait M'hid,
and Mohamed Kerouad*

2.1 INTRODUCTION

Following the appearance of the first transistor in 1954, microelectronics became a major technological revolution of the late twentieth century. Improvements in the performance of computer chips
have been made possible by a further reduction in component sizes, which are now of the order of
a few atomic distances. At this scale, fundamental physical obstacles arise, and the quantum nature
of electrons needs to be considered. At this level, the researchers in this field are trying to find new
ways of research that can supplant conventional electronics at nanometric dimensions. One interesting alternative is spintronics.

Spintronics has already had many successes. The discovery in 1988 by Albert Fert and Peter
Grunberg (Nobel Prize in 2007) of the changing magnetic-resistive effect (GMR) in the alternating
multilayers of ferromagnetic metal and insulating oxide enabled major technological advancements
in the microcomputing industry. The development of new hard disk reading head architectures was
based on the principle of the giant magnetoresistive effect. This new technology made it possible to
increase the density of hard disks in computers. As a result, hard disk capacities grew rapidly from
just a few gigabytes in the early 1990s to over one terabyte in current times, and they continue to
increase. Hard disk drives offer attractive characteristics that combine speed, rewrite capability, and
non-volatile information storage even when the supply voltage is absent. Other applications have
also been developed, such as magnetoresistive random-access memories (MRAMs), which make
it possible to replace dynamic random access memory (DRAM) in the random access memory
(RAM) of current computers with much lower access times. Unlike DRAM memories, the information in MRAMs is no longer stored in the form of electric loads but in the form of magnetic
moments, thanks to the technology of magnetic junctions and tunnels.

To produce innovative spintronic components, researchers are currently looking for ferromagnetic semiconductors at room temperature. Diluted magnetic semiconductors (DMSs) of type II-VI
form an important class of DMSs in which electric and magnetic doping can be controlled independently; the first to be identified were alloys like $Zn_{1-x}Mn_xTe$ and $Cd_{1-x}Mn_xTe$ [1]. More recently, the
compounds ZnO [2–4], ZnS [5–7], and ZnSe [8–10] have been investigated and need more study to
understand the nature of the magnetic interactions that control room temperature ferromagnetism.
The p-type doping of type II-VI DMSs is a challenge, and the control of magnetic interactions
through electrical doping could lead to a ferromagnetic DMS at room temperature. In addition to
that, defects can induce long-range ferromagnetic interactions.

This chapter provides a comprehensive overview of DMSs, encompassing both experimental
synthesis and characterization as well as theoretical investigations. The chapter is structured as follows: Initially, common experimental techniques utilized to fabricate DMS materials are surveyed,
including hydrothermal, co-precipitation, sol-gel, sputtering, and electrodeposition methods. The
subsequent section reviews key findings from recent experimental studies examining the impacts
of parameters such as semiconductor type, dopant concentration, and growth conditions on the
electronic and magnetic characteristics of select DMSs. This is followed by an in-depth analysis of
experimental results on the electronic structure and magnetism of specific DMS systems based on

DOI: 10.1201/9781003450146-2

ZnO, ZnS, and GaN. The second part of the chapter focuses on theoretical approaches of modeling DMS materials using density functional theory and Monte Carlo simulations (MCSs), offering insights into the physical underpinnings behind observed electronic and magnetic phenomena.

2.2 EXPERIMENTAL INVESTIGATION OF DMS

2.2.1 Synthesis Techniques

Researchers utilize specialized synthesis techniques to explore novel semiconductor materials and applications. In this subsection, we will provide an overview of common DMS fabrication methods like hydrothermal method, co-precipitation method, etc. Understanding these fundamental processes lays the groundwork for examining new experimental findings about the electronic and magnetic behaviors of DMSs.

2.2.1.1 Hydrothermal Method

Hydrothermal synthesis has emerged as a widely used technique for fabricating semiconductor nanostructures [11]. This approach involves heating aqueous solutions containing precursor materials in a sealed autoclave at elevated temperatures (130–250°C) and autogenous pressures. Under these conditions, the precursors undergo dissolution and recrystallization to form nanocrystals with well-defined size and morphology. Key advantages of hydrothermal synthesis include its simplicity, low cost, and environmental friendliness compared to other methods. By tuning parameters like temperature, pressure, reaction time, and solution pH, researchers can exercise precise control over nucleation and growth. A wide variety of semiconductor nanomaterials like metal oxides, sulfides, and selenides have been synthesized hydrothermally. The technique has also enabled fabrication of complex heterostructures by separating nucleation and growth stages.

2.2.1.2 Co-Precipitation Method

Co-precipitation is a straightforward synthesis technique for obtaining semiconductor nanoparticles [12]. In this approach, solutions containing precursor salts are mixed together, causing the precursors to precipitate out simultaneously as an intimate mixture. Adding a precipitation agent like ammonia adjusts the pH and induces co-precipitation. The precipitate is then filtered, washed, dried, and calcined to form the desired nanoparticles. Compared to other methods, co-precipitation is simple, economical, and easily scalable. However, it usually provides less control over particle size distribution and morphology. Still, co-precipitation allows rapid synthesis of nanoparticles from a wide range of semiconductor materials, including metal oxides, chalcogenides, and even doped semiconductors. By tuning factors like pH, temperature, and precipitation agent, some control over morphology and size can be achieved.

2.2.1.3 Sol-Gel Method

The sol-gel technique is an important synthesis route for fabricating semiconductor nanomaterials [13]. This method involves the transition of a colloidal solution "sol" into a solid "gel" phase. The sol-gel process typically starts with metal alkoxide precursors that undergo hydrolysis and condensation reactions to form a colloid. This colloid can be cast into shapes or films and dried to produce oxide gels. Further heating results in pore-free ceramic materials. The sol-gel approach provides molecular-level mixing of precursors, enabling fabrication of unique structures like nanocomposites, nanorods, and quantum dots. Key advantages include mild synthesis conditions, simple equipment, and fine control over composition and morphology. However, limitations exist like long processing times and cracking during drying.

2.2.1.4 RF-Magnetron Sputtering

Sputtering has proven to be an effective technique for fabricating high-quality DMS thin films [14]. This method relies on ejecting atoms from a target source material using energetic ion bombardment and depositing the sputtered atoms onto a substrate. For DMS synthesis, sputtering enables precise control over the dopant and host semiconductor composition in the growing film, which is critical for achieving the desired magnetic and electronic properties. Specifically, co-sputtering from multiple target sources allows accurate mixing of the dopant and semiconductor atoms during deposition. Another advantage of sputtering is the ability to deposit at lower temperatures, enabling formation of single crystal films on amorphous substrates. The sputtered films exhibit smooth surfaces, uniform thickness, good adhesion, and fewer defects compared to other fabrication approaches. However, achieving p-type doping during sputtering remains challenging due to compensation effects.

2.2.1.5   Electrodeposition Method

Electrodeposition has emerged as a versatile technique for synthesizing semiconductor nano-structures [15]. This method relies on applying a voltage between two electrodes immersed in an aqueous precursor solution. When a semiconductor material is used as the cathode, deposition occurs as precursor ions migrate and undergo reduction at the cathode surface. Key parameters like applied potential, temperature, pH, and precursor concentration can be tuned to control the morphology and properties of the deposited material. Electrodeposition enables formation of adherent films, free-standing structures, and nanocomposites on conductive substrates. It is a simple and inexpensive approach amenable to scaling up. However, the need for conducting substrates and mass transport limitations are drawbacks. Generally, electrodeposition allows facile synthesis of semiconductors like metal chalcogenides, oxides, and doped materials. It has been utilized to produce nanowires, nanorods, and thin films for diverse applications in photovoltaics, sensing, and electronics. With further development, electrodeposition holds promise as a sustainable nanofabrication technique.

2.2.2   EXPERIMENTAL RESULTS

To further understand the electronic and magnetic properties of semiconductors, this subsection will explore recent experimental findings on how these characteristics are impacted by parameters including semiconductor type, dopant concentration, and synthesis conditions. Specifically, we will highlight research on the magnetic attributes of select group II-VI and III-V semiconductors. Examining studies investigating factors like composition, doping, and growth techniques will provide critical insights into the physics governing the electrical and magnetic phenomena in these materials. This focused analysis of key experimental works will elucidate the complex interplay between band structure, crystal structure, and magnetism that determines semiconductor behaviors.

In this part, we provide a compact overview of the key electronic and magnetic characteristics of transition metal-doped ZnO DMSs. The experimental results of the electronic structure and carrier-mediated magnetism are summarized. The outlook on potential spintronic applications is also discussed. The aims are to review ZnO DMS for spintronics and highlight the opportunities and open questions surrounding this emerging class of multifunctional semiconductor materials.

Muhammad Tariq et al. [16] have studied the impact of growth conditions on (Fe, Cu) co-doped ZnO nanoparticles by applying a high pulsed magnetic field (4T) during hydrothermal synthesis. The researchers dissolved zinc acetate along with ferric nitrate and copper chloride in deionized water. After adding sodium hydroxide solution under stirring, the mixture was sealed in a Teflon-lined autoclave and heated at 140°C for four hours. Following purification and drying, the resulting nanoparticles were analyzed with and without exposure to the magnetic field during synthesis. The introduction of a high pulsed magnetic field during hydrothermal synthesis significantly altered the electronic and magnetic properties of the (Fe, Cu) co-doped ZnO nanoparticles. XRD and Raman analysis indicated that applying a 4T field led to increased substitutional doping of Fe^{3+} and Cu^{2+} ions along with enhanced lattice distortions and defects, especially oxygen vacancies, as evidenced by XPS. The higher oxygen vacancy concentration for the 4T sample is attributed to the reduction in compressive strain during growth under the magnetic field. The increase in defects and dopant incorporation contributed to a three-fold enhancement in the room temperature ferromagnetic saturation magnetization from 0.033 emu/g for the 0T sample to 0.041 emu/g for the 4T sample. This significant improvement in ferromagnetism is explained by the bound magnetic polaron model [17], where the magnetic exchange interactions between oxygen vacancies and transition metal ions create magnetic polarons that can overlap to stabilize long-range ferromagnetic order. The 4T sample exhibited more magnetic polarons with improved overlap, leading to superior room-temperature ferromagnetism.

Sourav Nayak et al. [18] have synthesized cobalt-doped ZnO nanoparticles using various methods including sol-gel, co-precipitation, hydrothermal, and solvothermal with different doping concentrations varied from 0 to 15%. In their investigation, the authors observed that the electronic structure and magnetism of the cobalt-doped ZnO nanoparticles were significantly tuned by the doping level. Increasing substitutional incorporation of Co^{2+} ions in place of Zn^{2+} led to a systematic redshift and narrowing of the band gap from 3.37 eV down to 2.9 eV. This reduction of the band gap is attributed to sp-d exchange interactions between the transition metal Co ions and the host ZnO band. The magnetic properties exhibited a strong dependence on Co concentration as well. At low doping around 5%, robust room temperature ferromagnetism was observed, with magnetic hysteresis measurements indicating strong ferromagnetic ordering mediated by bound magnetic polarons. These polarons involve coupling between oxygen vacancies induced by Co doping and the Co ions' magnetic moments. However, with further increase in Co doping beyond 5%, antiferromagnetic interactions between adjacent Co ions became dominant, leading to a transition from ferromagnetic to paramagnetic behavior. The electronic structure was thus effectively tuned via the Co concentration to achieve a narrowed band gap, while the magnetic polarity could be controlled between ferromagnetic and paramagnetic by carefully managing the complex magnetic interactions related to defects in the material.

To study the effect of dopant concentration and growth conditions on the electronic and magnetic characteristics of the ZnO and CZO films, B.L. Zhu et al. [19] have prepared ZnO and Cu-doped ZnO (CZO) thin films using RF-magnetron sputtering. ZnO and $Zn_{0.98}Cu_{0.02}O$ ceramic targets were used to deposit films on glass substrates at 300°C. The sputtering was done in either pure Ar or Ar+H2 ambient at 1.0 Pa pressure. The authors demonstrated that both Cu doping and the introduction of H2 during sputter deposition exerted substantial influence on the electronic and magnetic characteristics of the ZnO and CZO films. Resistivity values showed that CZO films had over 50 times higher resistivity than ZnO films in both Ar (ZnO: 3.44 Ωcm, CZO: 183.47 Ωcm) and Ar+H2 (ZnO: 0.10 Ωcm, CZO: 67.54 Ωcm) ambients, stemming from Cu acting as an acceptor and reducing carrier concentration. However, Cu incorporation induced strong room temperature ferromagnetism, with CZO films exhibiting saturation magnetization (Ms) values around 0.5 emu/g compared to negligible Ms for ZnO films. The addition of H2 during growth increased the concentration of donors like oxygen vacancies and interstitial Zn, lowering resistivity of ZnO and CZO to 0.10 Ωcm and 67.54 Ωcm, respectively. Moreover, the higher oxygen vacancy density introduced by H2 resulted in an enhancement of Ms to 0.033 emu/g for ZnO and 0.641 emu/g for CZO films deposited in Ar+H2 compared to Ar ambient. The ferromagnetic interactions are mediated by bound magnetic polarons formed by coupling between the increased oxygen vacancies and the Cu ions in CZO.

In an analogous study, Hassan Ahmoum et al. [20] undertook the synthesis of nickel-doped zinc oxide (ZnO) nanoparticles utilizing the sol-gel method. The Ni doping concentration varied from 0 to 12%. Their primary objective was to investigate the influence of dopant concentration on the electronic and magnetic properties of the ZnO material. The authors found that the electronic structure and magnetism of the Ni-doped ZnO nanoparticles were significantly modulated by the Ni concentration. XRD patterns showed the hexagonal wurtzite structure was retained after doping, with systematic peak shifting to higher angles indicating successful Ni^{2+} substitution on Zn^{2+} sites.

The corresponding monotonic decrease in lattice parameters from a = 3.259 Å and c = 5.218 Å for undoped ZnO to a = 3.247 Å and c = 5.192 Å for 12% Ni-doped ZnO confirms the incorporation of smaller Ni^{2+} ions (r = 0.69 Å) versus larger Zn^{2+} ions (r = 0.74 Å). Meanwhile, broadening of the (101) peak signified reduced crystallite size from 25.1 nm down to 21.2 nm and 17.1 nm for 6% and 12% Ni doping, respectively, implying constrained nanostructure growth. Detailed magnetic characterization revealed emergence of robust room temperature ferromagnetism with Ni incorporation, in contrast to the diamagnetic undoped ZnO, which showed negligible saturation magnetization (Ms) of 0.033 emu/g. For 6% Ni doping, Ms dramatically increased to 0.94 emu/g while the squareness of the hysteresis loop was evidenced by coercivity of 155 Oe and remanence of 0.148 emu/g. At higher 12% Ni content, Ms was lowered but still substantial at 0.76 emu/g, with coercivity of 142 Oe and remanence of 0.124 emu/g as shown in Figures 2.1 and 2.2. Temperature-dependent remanence confirmed the critical ferromagnetic transition temperature was around or exceeding 300K. The variations in ferromagnetism with Ni clearly demonstrate its efficacy in controlling the magnetism.

Cr-doped ZnO nanoparticles with Cr doping concentrations that varied from 0 to 5% were also prepared using the solution combustion method [21]. It is found that the electronic structure and magnetism of the Cr-doped ZnO nanoparticles were markedly affected by the Cr concentration. XRD revealed the systematic reduction in crystallite size from 31 ± 2 nm to 19 ± 1 nm with increasing Cr doping up to 5%, indicative of constrained nanostructure growth. Magnetic characterization uncovered the emergence of robust room temperature ferromagnetism in contrast to the diamagnetic undoped ZnO. The saturation magnetization (Ms) was substantial at 0.867 emu/g for undoped ZnO and initially increased to 0.674 emu/g for 3% Cr doping. However, higher 5% Cr incorporation

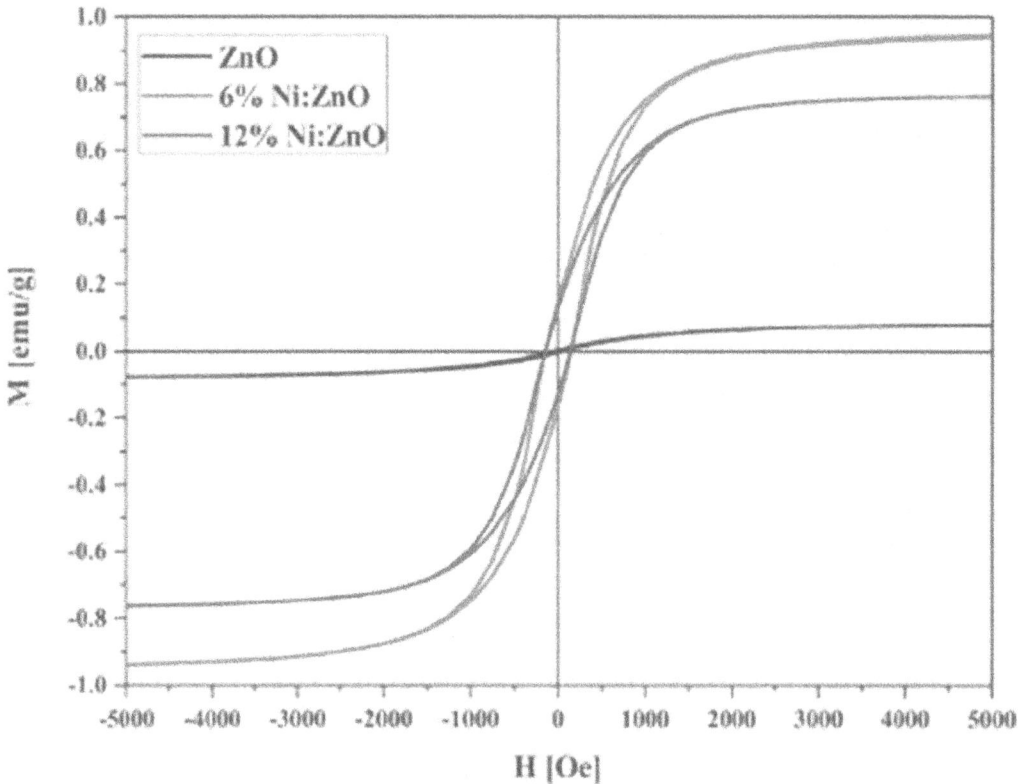

FIGURE 2.1   Experimental magnetization versus applied magnetic field for different samples at 300K [20].

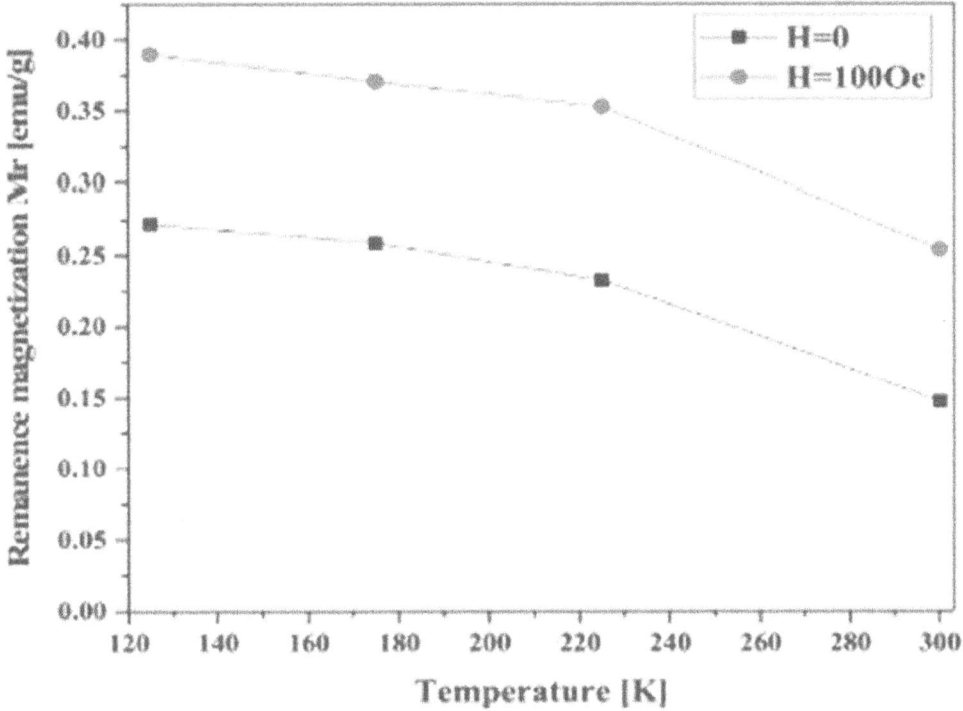

FIGURE 2.2 Remanence magnetization versus temperature for 6% of Ni doped ZnO [20].

reduced Ms to 0.627 emu/g. The hysteresis loop squareness also decreased with increasing Cr doping, with the coercivity declining from 228 Oe to 190 Oe for 3% Cr and then rising slightly to 224 Oe for 5% Cr. Remanence followed the same trend, decreasing from 0.13 emu/g for undoped ZnO to 0.07 and 0.05 emu/g for 3% and 5% Cr-doped ZnO, respectively. At lower chromium (Cr) doping concentrations, ferromagnetism is initially enhanced due to the formation of bound magnetic polarons. These polarons involve a coupling between Cr^{3+} ions and zinc vacancies, which promotes the ferromagnetic alignment of spins. However, as the Cr doping concentration increases further, the ferromagnetism begins to reduce. This subsequent reduction in ferromagnetic behavior at higher Cr doping levels is also explained by the same mechanism involving bound magnetic polarons. However, excessive Cr doping leads to increased antiferromagnetic couplings between adjacent Cr ions that lower the overall magnetization.

In the same context, S. Narasimman et al. [22] have synthesized pristine ZnO and Mn-doped ZnO nanorods using the hydrothermal method. The authors have studied the effect of Mn doping concentration on the electronic and magnetic properties of the ZnO material. Mn doping concentrations of 5%, 10%, 15%, and 20% were prepared. Through their rigorous analysis, the authors provided a comprehensive understanding of how the electronic structure and magnetism of hydrothermally grown ZnO nanorods were methodically adjusted through variations in Mn doping concentration. XRD patterns revealed that the hexagonal wurtzite structure was retained after Mn incorporation up to 20%, with the emerging impurity peak indicating the formation of $ZnMnO_3$ clusters. The (101) peak intensity increased while crystallite size concurrently decreased from 65 nm down to 44 nm, with higher Mn levels as calculated using Scherrer's equation, indicative of constrained nanostructure growth. As Mn content increased, SEM micrographs confirmed the rod-shaped morphologies, with average diameter and length reduced from ~13 μm × 1 μm to ~8 μm × 0.8 μm for pure ZnO to 20% Mn-doped ZnO, respectively. EDS spectra verified the elemental composition through the changing Zn:Mn:O ratios. XPS further analyzed the chemical states, with binding energies suggesting Mn^{4+}, Zn^{2+}, and O^{2-} as the dominant species. Magnetic characterization was performed by coating the Mn-doped ZnO nanorods in an optical fiber to construct a sensor. The transmission spectrum exhibited predictable increases with an applied magnetic field from 17.2 mT to 190.6 mT, owing to positive magneto-refractive effects. A maximum sensitivity of 27.2% was attained for 15% Mn doping, with fast response and recovery times of 7s and 12s, respectively. The changing morphology, magneto-optical activity, and rapid-response kinetics demonstrate the potential of hydrothermal Mn-doped ZnO nanorods for tunable and sensitive fiberoptic magnetic field sensing.

To study the impact of the crystallite size and dopant type on dopant concentration, Amarjyoti Kalita [23] synthesized Co-doped and Cu-doped ZnO nanocrystals using a chemical co-precipitation method. Hexagonal wurtzite crystal structure stability was affirmed through XRD analysis after the doping process. TEM imaging showed the formation of agglomerated, roughly spherical nanocrystals with sizes of 15–20 nm, agreeing with the crystallite sizes determined from XRD. EDS analysis indicated the actual Co and Cu doping levels were lower than the targeted amounts, likely due to thermodynamic equilibrium effects during growth. The optical band gap increased from 3.39 eV for undoped ZnO to 3.60 eV for 2% Co-doped and 3.43 eV for 2% Cu-doped samples, showing a clear band gap expansion. The ESR spectra for Co-doped samples exhibited broad resonance peaks with g-factors around 2.3, while Cu-doped samples showed narrowed peaks with g-factors of 2.1 along with hyperfine splitting from Cu ions. This confirms a reduction in ferromagnetic interactions for Cu doping. The room temperature saturation magnetization values measured by VSM were approximately 5×10^{-3} emu/g for undoped ZnO, 3×10^{-3} emu/g for 2% Co-doped, and 7×10^{-4} emu/g for 2% Cu-doped samples. The substantial decrease in magnetization for Cu-doped ZnO agrees with the ESR results showing weakened ferromagnetic coupling. Meanwhile, the Co-doped samples retained decent ferromagnetic behavior, though the magnetization also declined somewhat compared to undoped ZnO due to antiferromagnetic Co-Co interactions. Generally, the electronic structure was modified by the dopant ions through band gap changes, enhanced defect formation, and mediation of magnetic interactions.

The next part of this synthesis will be devoted to the electronic and magnetic properties of zinc sulfide (ZnS) as an important wide band gap II-VI semiconductor, which has received much interest as a DMS material. Doping ZnS with transition metals (TM) results in a material that exhibits ferromagnetism at certain temperature, making it attractive for spintronics, which requires control over both the charge and spin of electrons. In this part, we will provide an overview of the electronic and magnetic properties of TM-doped ZnS. Key results from experimental studies on the magnetic and electronic structure of ZnS:TM are summarized. Vaishali Shukla et al. [24] have synthesized ZnS nanoparticles and TM (Mn, Co, Ni)-doped ZnS nanoparticles by a chemical precipitation method in aqueous cetyltrimethylammonium bromide (CTAB) micellar solutions. The magnetic measurements revealed a transition from weak diamagnetic behavior to ferromagnetism after doping. The saturation magnetization (Ms) at room temperature increased from 19.21×10^{-3} emu/g for undoped ZnS to 13.09×10^{-3} emu/g, 9.91×10^{-3} emu/g, and 13.47×10^{-3} emu/g for Mn-, Co-, and Ni-doped ZnS, respectively. The coercivity also decreased to 177.34 G, 176.59 G, and 137.58 G compared to 183.42 G in undoped ZnS. The remnant magnetization followed a similar trend. The observed ferromagnetism arises from p-d exchange interactions between the localized d electrons of the Mn^{2+}, Co^{2+}, and Ni^{2+} ions and the hybridized p electrons of the S^{2-} ions. In summary, the magnetic measurements quantitatively demonstrated that transition metal doping introduces robust room temperature ferromagnetism in ZnS by modifying the electronic structure.

To study the effect of co-doping on the magnetic properties of ZnS, B. Poornaprakash et al. [25] have synthesized Cr-doped ZnS and (Cr,V)-co-doped ZnS nanoparticles by a hydrothermal method using polyvinylpyrrolidone (PVP) as a surfactant. It is shown that the number of spins increased from 2.14×10^9 in the 1% Cr doped sample to 5.03×10^9 and 5.25×10^9 spins for the 1% Cr+1% V and 1% Cr+2% V co-doped samples, confirming additional unpaired electrons were introduced by the V co-doping. The magnetic measurements showed a clear transition from dia-magnetism in undoped ZnS to robust room-temperature ferromagnetism after doping. The satura-tion magnetization (Ms) increased from 0.275 emu/g for 1% Cr doping to 0.431 emu/g and 0.657 emu/g for the 1% Cr+1% V and 1% Cr+2% V co-doped samples. The remnant magnetization fol-lowed a similar increasing trend. The enhanced ferromagnetism with higher V doping indicates the critical role of V in modulating the exchange interactions. In general, the quantitative magnetic characterization conclusively demonstrated the tuned electronic structure and successful introduc-tion of strong room temperature ferromagnetism in the transition metal doped ZnS nanoparticles, indicating enhanced ferromagnetism with higher V doping. The results confirm successful tuning of the electronic structure and introduction of robust room temperature ferromagnetism via transi-tion metal doping.

In a similar investigation, N. Manivannan et al. [26] have synthesized Cr-doped ZnS nanopar-ticles using a chemical precipitation method. The authors studied the effect of Cr-doping concen-tration on the electronic and magnetic properties of the ZnS material. Powders with 0.025, 0.05, and 0.075 mol% Cr doping were prepared. The vibrating sample magnetometer measurements revealed a clear transition from diamagnetism in undoped ZnS to room-temperature ferromag-netism after Cr doping. The saturation magnetization (Ms) increased from 0 for undoped ZnS to 1.57×10^{-4} emu/g for 0.025 mol% Cr doping. Further increasing the Cr doping to 0.05 and 0.075 mol% resulted in Ms values of 2.97×10^{-4} emu/g and 3.68×10^{-4} emu/g, respectively. The remnant magnetization followed a similar increasing trend, from 0 for undoped ZnS to 7.54×10^{-5}, 1.45×10^{-4}, and 1.83×10^{-4} emu/g for 0.025, 0.05, and 0.075 mol% Cr-doping. The coercivity increased as well, from 0 Oe in undoped ZnS to 1593, 1298, and 396 Oe for 0.025, 0.05, and 0.075 mol% Cr, respectively. The quantitative magnetic characterization conclusively demonstrates that the electronic structure is successfully tuned by Cr doping to introduce robust room temperature fer-romagnetism, with the strength of ferromagnetism systematically enhanced as the Cr concentra-tion increases.

The last part of this section will be devoted to the magnetic properties of the GaN DMS as another promising candidate for spintronic applications. GaN, a III-V group semiconductor with a wide band gap, is renowned for its exceptional electronic properties and has already made signifi-cant inroads in optoelectronics and high-power electronic devices. However, what truly sets GaN apart is its intrinsic ability to integrate magnetism into its electronic structure, thus paving the way for the development of next-generation spintronic devices. Yana Grishchenko et al. [27] syn-thesized nickel-gallium nitride (Ni-GaN) composites through a multi-step process. Initially, they grew n-type gallium nitride (GaN) layers using the metalorganic vapor phase epitaxy (MOVPE) technique. Subsequently, these n-type GaN layers underwent electrochemical etching to make them porous. After creating the porous GaN layers, the researchers incorporated nickel into the porous structure to form the desired nickel-GaN composite materials. The synthesis approach involved first producing n-type GaN layers by MOVPE, then making them porous via electrochemical etching, and finally introducing nickel to obtain the nickel-GaN composites. The nickel-GaN composites showed typical ferromagnetic behavior, but with properties tuned by the synthesis approach. For the thin nickel film, thickness was measured by SEM as 155 ± 38 nm. The saturation magnetization at 2K was 56.5 ± 15 emu/g, agreeing well with the expected bulk nickel value of ~58 emu/g. This demonstrated the high quality of the electrodeposited nickel. The thin film showed strong in-plane anisotropy with 2K coercivity of <50 Oe in-plane versus >300 Oe out-of-plane. Remanence fol-lowed an unusual trend, increasing to 0.6 at 300K in-plane. The XRD patterns revealed the film was (111) textured with crystallite sizes around 40 nm and low strain (−0.08%). For infiltrated nickel, isotropic-like behavior was seen with coercivity <10 Oe and remanence ~0.2 for both orientations at 300K. The saturation magnetizations were lower, around 0.5 emu/g at 300K for longer depositions.

XRD showed smaller grains (~30 nm) but slightly higher strain (0.06%) for infiltration. Blocking temperatures were estimated as >700K. In summary, tailoring of the anisotropy and coercivity was achieved between thin film and infiltrated composites by controlling the deposition in the porous GaN templates.

2.3 THEORETICAL INVESTIGATIONS OF DILUTED MAGNETIC SEMICONDUCTORS

In physics, the use of numerical simulation makes it possible to explore the behavior of a model but also to obtain results which will be compared to experimental data. With the development of the computational tools and power, an important advance is shown in electronic structure theory and algorithmic development. Theoretical investigations became an essential tool to predict and understand experimental phenomena. In this section, we will present a review of theoretical studies on DMSs.

The theoretical study focused on examining the magnetic properties of semiconductors and understanding how the introduction of dopant elements influences the physical characteristics of these materials that are intrinsically insulating when no dopants are present, and the researchers use MCSs and density functional theory, which are known as powerful techniques that give accurate results.

Boron nitride (BN) and aluminum nitride are extensively studied because of their possible uses as diluted magnetic semiconductors in spintronic applications. In this optic, Ahmoum et al. [28] have studied structural, electronic, and magnetic properties of the Al-doped BN with the presence of N vacancy by using the density functional theory. It is shown that the band gap decreases with the increase of the concentration of Al impurities, and these impurities and the N vacancy induce magnetic properties. The total magnetization is found to be 0.09 μ_B for undoped BN with N vacancy and 0.12 μ_B for Al-doped BN with N vacancy. From this work, it is concluded that the magnetic properties are due to Al impurities and the four B atoms that surrounded the N vacancy.

The magnetic properties of (Mn, C) co-doped AlN with N vacancy are also investigated [29]. In this work, we have also used the density functional theory to discover the effect of co-doping on the magnetic properties of this material (AlN). In a previous work [30], the authors theoretically studied the carbon-doped AlN, and they found that the substitution of nitrogen by carbon induced the magnetic moment by about 1 μ_B. Our work demonstrates that pure aluminum nitride (AlN) does not exhibit magnetic properties. However, when AlN is co-doped with manganese and carbon in the presence of nitrogen vacancies, it acquires the characteristics of a magnetic semiconductor. Specifically, the AlN co-doped with manganese, carbon, and nitrogen vacancies displays a band gap of approximately 2.24 eV for the spin-down electrons. It is also shown that the magnetic properties are due to the Mn^{3+} incorporated in the AlN lattice and to the nitrogen vacancy.

ZnO doped with TM is currently studied as part of research done on the magnetic properties of DMSs. Due to its characteristics, such as wide band gap (3.4 eV) [31], very large exciton binding energy (60 meV) [32], and high thermal and mechanical stability at room temperature [33], ZnO is considered an ideal semiconductor for several applications, such as light-emitting diodes [34], solar cells [35], and photodetectors [36].

For magnetic applications, the aim is to obtain a ferromagnetic doped ZnO with a Curie temperature above temperature ambient. In recent works [20, 37], we have investigated the effect of nickel and vanadium concentration on the magnetic properties of doped ZnO.

In the first work, we have studied the Ni-doped ZnO for different nickel concentrations (5.55%, 8.33%, and 12.5%). We have found the pure ZnO exhibits a weak saturation magnetization of 0.033 emu/g, which is similar to the results obtained by Chithira et al. [38]. This weak magnetization can only be explained by defects that may be present in the sample. A recent work of Fedorov et al. [39]

TABLE 2.1

Bond Length [Å], the Different Energy between AFM and FM [meV], the Total Moment Magnetic [μ_B], and the Localized Moment Magnetic at Ni [μ_B] [20]

Doping Concentration	Configuration	d[Å]	ΔE[meV]	m_{tot} [μ_B]	m_{Ni}[μ_B]
5.55% Ni:ZnO	Ni-O-Ni	3.16	19.92	3.90	3.22
8.33% Ni:ZnO	Ni-O-Ni	3.26	16.79	4.00	3.19
12.5% Ni:ZnO	Ni-O-Ni	2.93	−4.612	0.01	0.09

shows that O interstitials and Zn vacancies induce magnetic properties and generate a magnetic moment of 1.98 μ_B and 1.26 μ_B, respectively. Similar phenomena regarding the weak magnetization are also reported in BN with the presence of N defects [28].

By using density functional theory and for low doping concentration of nickel, it is found that the ferromagnetic state is the most stable, with a total magnetic moment per cell of around 4.00 μ_B (Table 2.1). It is also shown that the main contribution to the magnetic moment comes from Ni atoms (3.2 μ_B), and a minor contribution of 0.26 μ_B comes from the O atoms surrounding the Ni atoms. For the high concentration of Li (12.5%), the material becomes antiferromagnetic, which is due to the increase of Ni content that leads to a reduction of the exchange coupling interactions between Ni atoms.

To understand the origin of the magnetic properties of Ni doped ZnO, we have calculated the partial density of state for the different concentrations (Figure 2.3). For the low concentration (Figures 2.3a and 2.3b), the impurity band is located at the middle of the band gap for the spin-down channel, which is formed essentially by 3d-Ni states. These findings indicate that Ni-doped

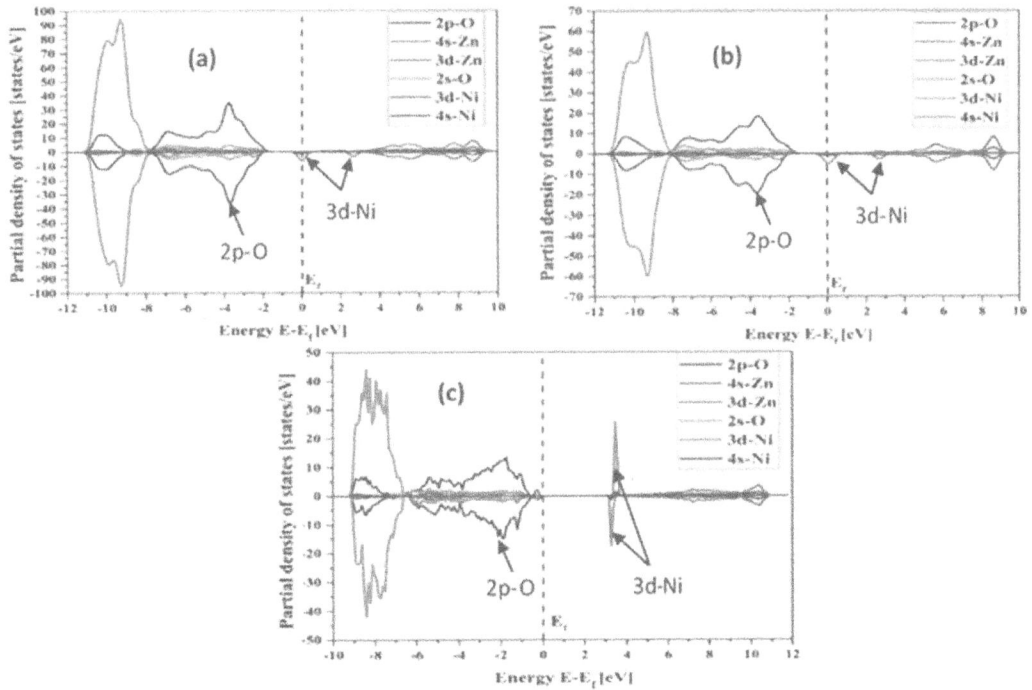

FIGURE 2.3 Partial densities of states of (**a**) 5.55% Ni-doped ZnO, (**b**) 8.33% Ni-doped ZnO, and (**c**) 12.5% Ni-doped ZnO [20].

ZnO with 5.55% and 8.33% concentration are half-metallic and ferromagnetic. For the 12.5% concentration (Figure 2.3c), the band impurity composed of $3d$-Ni states is found to be shifted deeply inside the conduction band, and the Fermi level is found to cross an impurity level located above the valence band maximum, which is formed essentially by $2p$-O states. The origin of this impurity band can only be explained by the distortion effect caused by the incorporation of Ni impurities. This indicates that the material obtained is an antiferromagnetic, non-degenerated semiconductor.

After getting the exchange interaction from the calculus made by the density functional theory, MCSs based on the heat bath algorithm are used to calculate the magnetization and the hysteresis loop. The phase transition from ferromagnetic to paramagnetic phases is second order, and the Curie temperatures are found to be 360K and 303K for 5.55% and 8.33% of Ni-doped ZnO respectively. Concerning the hysteresis behavior, the area of the loop decreases with the increasing of the Ni concentration.

In other work, the effect of vanadium (V) concentration on the magnetic properties of ZnO by using the density functional theory and MCSs are investigated [37]. It is found that V-doped ZnO presents a Curie temperature above the room temperature for high doping concentration (Figure 2.4). To understand the origin of the ferromagnetism in these samples, the authors have calculated the total and partial density of the state of vanadium-doped ZnO (Figure 2.5). The results show the emergence of magnetic bands near the Fermi level due to hybridization between the V-3d and O-2p orbitals, indicating that the V impurities induce ferromagnetism in ZnO.

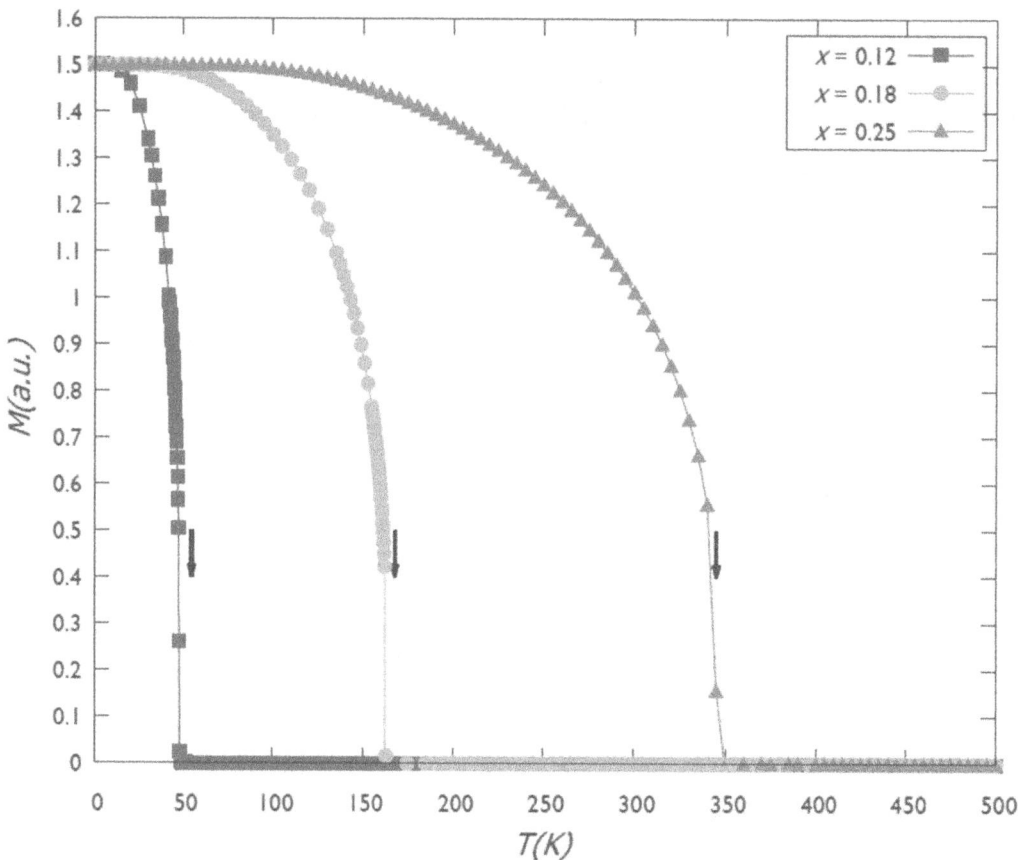

FIGURE 2.4 Total magnetization versus temperature of the $Zn_{1-x}V_xO$ system for x = 0.12, 0.18, and 0.25 [37].

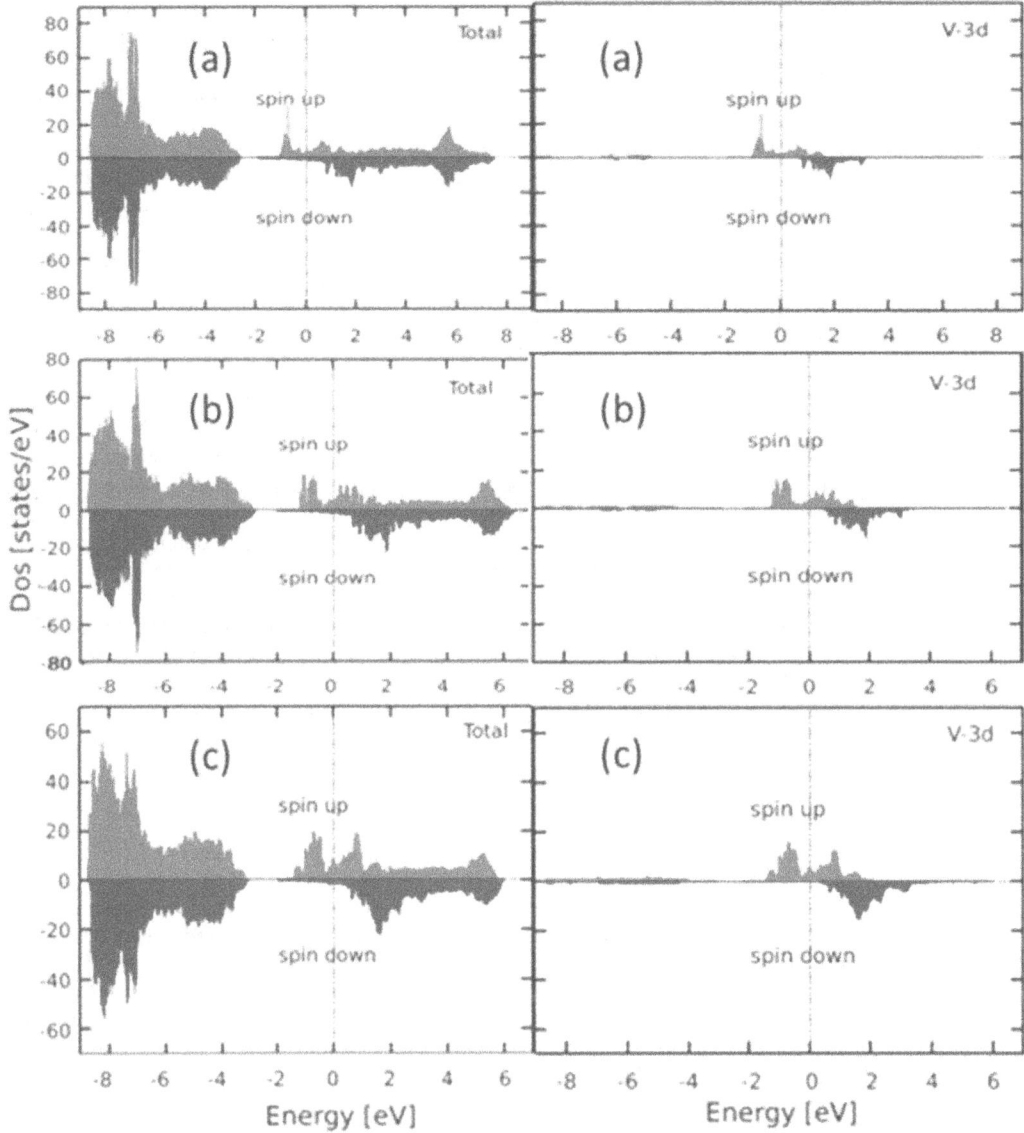

FIGURE 2.5 Total (right) and partial (left) density of states (DOS) of $Zn_{1-x}V_xO$; (a) for x = 0.12, (b) for x = 0.18, and (c) for x = 0.25 [37].

REFERENCES

1. J.K. Furdyna, Diluted magnetic semiconductors, J. Appl. Phys. 64 (1988) R29–R64.
2. K. Sato, H. Katayama-Yoshida, Stabilization of ferromagnetic states by electron doping in Fe-, Co- or Ni-doped ZnO, Jpn. J. Appl. Phys. 40 (2001) L334.
3. Y. Jiang, W. Wang, C. Jing, C. Cao, J. Chu, Sol-gel synthesis, structure and magnetic properties of Mn-doped ZnO diluted magnetic semiconductors, Mater. Sci. Eng. B. 176 (2011) 1301–1306.
4. K.P. Sheng qiang Zhou, Q. Xu, G. Talut, M. Lorenz, W. Skorupa, M. Helm, J. Fassbender, M. Grundmann, H. Schmidt, Ferromagnetic transition metal implanted ZnO: A diluted magnetic semiconductor?, Vacuum. 83 (2009) S13–S19.
5. D.A. Reddy, C. Liu, R.P. Vijayalakshmi, B.K. Reddy, Effect of Al doping on the structural, optical and photoluminescence properties of ZnS nanoparticles, J. Alloys Compd. 582 (2014) 257–264.

6. A. Roychowdhury, S.P. Pati, S. Kumar, D. Das, Effects of magnetite nanoparticles on optical properties of zinc sulfide in fluorescent-magnetic Fe3O4/ZnS nanocomposites, Powder Technol. 254 (2014) 583–590.

7. J. Cao, D. Han, B. Wang et al., Low temperature synthesis, photoluminescence, magnetic properties of the transition metal doped wurtzite ZnS nanowires, J. Solid State Chem. 200 (2013) 317–322.

8. N.S. Myoung, V.V. Fedorov, S.B. Mirov, L.E. Wenger, Temperature and concentration quenching of mid-IR photoluminescence in iron doped ZnSe and ZnS laser crystals, J. Lumin. 132 (2012) 600–606.

9. S.P. Nehra, M. Singh, Effect of vacuum annealing and hydrogenation on ZnSe/Mn multilayer diluted magnetic semiconductor thin films, Vacuum. 85 (2011) 719–724.

10. T. Li, W. Wang, Q. Shi, J. Zhang, L. Zhao, Transition from ferromagnetism to superparamagnetism in diluted magnetic Fe (II)-doped ZnSe microspheres, J. Magn. Magn. Mater. 543 (2022) 168625.

11. S.S. Ghosh, A. Sil, Effect of microwave processing on Mn doped ZnO diluted magnetic semiconductor characteristics, Mater. Today Commun. 32 (2022) 103941.

12. M.A. Mahmood, K. Althubeiti, S.S. Abdullaev, N. Rahman, M. Sohail, S. Iqbal, K. Safeen, A. Safeen, A. Khan, R. Khan, Diluted magnetic semiconductor behavior in Co- and Gd-Co-doped ZnO nanotubes for spintronic applications, J. Mater. Sci: Mater. Electron. 34 (2023) 1784.

13. T.A. Taha, E.M. Ahmed, A.I. El-Tantawy, A.A. Azab, Investigation of the iron doping on the structural, optical, and magnetic properties of Fe-doped ZnO nanoparticles synthesized by sol–gel method, J. Mater. Sci: Mater. Electron. 33 (2022) 6368–6379.

14. X. Duan, X. Chen, D. Wu, C. Lu, X. He, S. Ye, F. Lin, R. Wang, C. Wang, Te doping effects on the ferromagnetic performance of the MnGe/Si quantum dots grown by ion beam sputtering deposition, J. Alloys Compd. 34 (2023) 172047.

15. M.C. Haciismailoglu, M. Ahmetoglu, M. Haciismailoglu, M. Alper, T. Batmaz, Electrical and optical properties of Schottky diodes fabricated by electrodeposition of Ni films on n-GaAs, Sens. Actuator A Phy. 347 (2022) 113931.

16. M. Tariq, Y. Zaman, M. Shahzad, K. Ahmad, A.B. Siddique, H. Zaman, O-vacancies induced in co-doped ZnO:(Cu, Fe) synthesized via hydrothermal method by pulsed magnetic field, Mater. Sci. Engin. B. 294 (2023) 116549.

17. A.C. Durst, R.N. Bhatt, P.A. Wolff, Bound magnetic polaron interactions in insulating doped diluted magnetic semiconductors, Phy. Rev. B. 65 (2002) 235205.

18. S. Nayak, P. Kumar, Review on structure, optical and magnetic properties of cobalt doped ZnO nanoparticles, Mater. Today: Proceed. 01 (2023) 318.

19. B.L. Zhu, X.M. Cao, M. Xie, J. Wu, X.W. Shi, Effects of Cu doping and deposition atmosphere on structural, electrical, optical and magnetic properties of ZnO films, Physica B: Condensed Matter. 658 (2023) 414844.

20. H. Ahmoum, G. Li, Y. Piao, S. Liu, R. Gebauer, M. Boughrara, M.S. Su'ait, M. Kerouad, Q. Wang, Ab-initio, Monte Carlo and experimental investigation on structural, electronic and magnetic properties of $Zn_{1-x}Ni_xO$ nanoparticles prepared via sol-gel method, J. Alloys Compd. 854 (2021) 157142.

21. H.S. Lokesha, A.R.E. Prinsloo, P. Mohanty, C.J. Sheppard, Impact of Cr doping on the structure, optical and magnetic properties of nanocrystalline ZnO particles, J. Alloys Compd. (2023) 170815.

22. S. Narasimman, L. Balakrishnan, Z.C. Alex, Highly sensitive magnetic field sensor based on uniform core fiber using Mn doped ZnO nanorods as cladding, Mater. Sci. Semicond. Process. 166 (2023) 107732.

23. A. Kalita, Crystallite size controlled doping of transition metals (Co, Cu) in ZnO nanocrystals to investigate microstructural, optical, magnetic and photocatalytic properties. Curr. Appl. Phys. 52 (2023) 65–79.

24. V. Shukla, M. Singh, Room temperature luminescence and ferromagnetism from transition metal ions: Mn-, Co- and Ni-doped ZnS nanoparticles, Mater. Sci. Engin. B. 280 (2022) 115685.

25. B. Poornaprakash, P. Puneetha, K. Subramanyam, M.W. Kwon, D.Y. Lee, M.S.P. Reddy, S. Sangaraju, B.A. Al-Asbahi, Y.L. Kim, Synthesis of diluted magnetic semiconductor ZnS: Cr and ZnS:(Cr+V) nanoparticles for spintronic applications, Mater. Sci. Semicond. Process. 161 (2023) 107479.

26. N. Manivannan, B.C. Shekar, P. Matheswaran, M.M. Ibrahim, C.S. Kumaran, Induced ferromagnetic behavior of Cr doped ZnS nano particles, Mater. Today: Proceed. 48 (2022) 258–262.

27. Y. Grishchenko, J. Dawson, S. Ghosh, A. Gundimeda, B.F. Spiridon, N.L. Raveendran, R.A. Oliver, S. Kar-Narayan, Y. Calahorra, Magnetic properties of nickel electrodeposited on porous GaN substrates with infiltrated and laminated connectivity, J. Magnet. Magnet. Mater. 580 (2023) 170877.

28. H. Ahmoum, M. Boughrara, M. Kerouad, Electronic and magnetic properties of Al doped (w-BN) with intrinsic vacancy, Superlattices Microstruct. 127 (2019) 186–190.

29. H. Ahmoum, M. Boughrara, M.S. Su'ait, M. Kerouad, Electronic and magnetic properties of Mn-doped and (Mn,C)-codoped w-AlN with the presence of N vacancy, J. Supercond. Nov. Magnet. 32 (2019) 3691.

30. C.O. López, DFT predictions of ferromagnetism in the $AlC_{0.0625}N_{0.9375}$ and $AlC_{0.125}N_{0.875}$ compounds, Results Phys. 5 (2015) 281–285.

31. K. Irikura, F. Marken, P.J. Fletcher, G. Kociok-Kohn, M.V. Boldrin Zanoni, Direct and indirect light energy harvesting with films of ambiently deposited ZnO nanoparticles, Appl. Surf. Sci. 527 (2020) 146927.

32. J. Pachiyappan, N. Gnanasundaram, G.L. Rao, Preparation and characterization of ZnO, MgO and ZnO-MgO hybrid nanomaterials using green chemistry approach, Results Mater. 7 (2020) 100104.

33. G. Li, Y. Gao, S. Liu, Z. Wang, S. Liu, Q. Wang, Microstructural evolution of the oxidized ZnO: Cu films tuned by high magnetic field, J. Alloys Compd. 753 (2018) 673–678.

34. Y. Peng, M. Que, H.E. Lee, R. Bao, X. Wang, J. Lu, Z. Yuan, X. Li, J. Tao, J. Sun, J. Zhai, K.J. Lee, C. Pan, Achieving high-resolution pressure mapping via flexible GaN/ZnO nanowire LEDs array by piezo-phototronic effect, nanomater, Energy. 58 (2019) 633–640.

35. Y. Zhang, G. Zhai, L. Gao, Q. Chen, J. Ren, J. Yu, Y. Yang, Y. Hao, X. Liu, B. Xu, Y. Wu, Improving performance of perovskite solar cells based on ZnO nanorods via rod-length control and sulfidation treatment, Mater. Sci. Semicond. Process. 117 (2020) 105205.

36. M. Zheng, P. Gui, X. Wang, G. Zhang, J. Wan, H. Zhang, G. Fang, H. Wu, Q. Lin, C. Liu, ZnO ultraviolet photodetectors with an extremely high detectivity and short response time, Appl. Surf. Sci. 481 (2019) 437–442.

37. A. Ait M'hid, M. Boughrara, M. Kerouad, Ab-initio and Monte Carlo studies of the structural, electronic and magnetic properties of V-doped ZnO compound, Physica B: Condensed Matter. 654 (2023) 414717.

38. P.R. Chithira, T.T. John, Defect and dopant induced room temperature ferromagnetism in Ni doped ZnO nanoparticles, J. Alloys Compd. 766 (2018) 572–583.

39. A.S. Fedorov, M.A. Visotin, A.S. Kholtobina, A.A. Kuzubov, N.S. Mikhaleva, H.S. Hsu, Investigation of intrinsic defect magnetic properties in wurtzite ZnO materials, J. Magn. Magn Mater. 440 (2017) 5–9.

3 Types and Properties of Semiconductors

Ajay Lathe and Anil M. Palve

3.1 INTRODUCTION

Semiconductors have a rich history that traces back to the invention of the rectifier in 1874. However, it wasn't until the late 1940s that groundbreaking advancements in semiconductor technology occurred [1]. In 1947, Bell Laboratories in the United States saw the point-contact transistor introduction by John Bardeen and Walter Brattain; then in in 1948, William Shockley's development of the junction transistor [2]. These pivotal inventions marked the onset of the transistor era. Before the transistor's emergence, computing technology relied heavily on vacuum tubes. In 1946, the University of Pennsylvania constructed a computer using vacuum tubes, an enormous machine that consumed required electricity and emitted substantial heat. But this landscape began to change dramatically with the introduction of the innovative transistor calculator, signifying a turning point in computing technology. In 1956, the collective efforts of Shockley, Bardeen, and Brattain were rewarded with the Nobel Prize in Physics, recognizing their pioneering contributions to semiconductor research and the creation of the transistor [3].

Following the transistor's inception, the semiconductor industry embarked on a rapid expansion. By 1957, the industry had already exceeded $100 million. A pivotal moment arrived in 1959 when Jack Kilby of Texas Instruments and Robert Noyce of Fairchild Semiconductor in the United States independently invented the bipolar integrated circuit (IC), marking the dawn of the IC era. These ICs, known for their compact size and lightweight nature, found widespread applications across various electronic devices [4]. In 1967, Texas Instruments achieved a significant milestone by developing the electronic desktop calculator, powered by IC technology. Meanwhile, in Japan, electronic equipment manufacturers engaged in a fierce competition known as the "calculator wars," continuously releasing calculators through the late 1970s. As IC integration continued to advance, the era of large-scale integrated circuits (LSIs) emerged, further expanding the horizons of semiconductor technology. Progress in the following decades led to the development of very large-scale integrated circuits (VLSIs) in the 1980s, housing anywhere from 100,000 to 10 million electronic components per chip [5, 6]. The subsequent decade witnessed the creation of ultra-large-scale integrated circuits (ULSIs), featuring more than 10 million electronic components per chip. In the 2000s, the production of system LSIs, multifunctional chips integrating diverse capabilities into a single unit, became widespread. As ICs evolved to offer heightened performance and multifaceted functions, their applications proliferated, making semiconductors ubiquitous in our society, playing a vital role in supporting our daily lives across various domains [7].

Semiconductors, often overlooked but integral to modern technology, quietly power the digital age, permeating every aspect of our lives. These unassuming materials, situated between conductors like metals and insulators like rubber, hold a distinctive place in material sciences and electronics. In this introduction, we embark on a journey to uncover the realm of semiconductors, beginning with their definition and tracing their remarkable historical evolution. We emphasize the importance of understanding the diverse types and properties of semiconductors, laying the foundation for comprehensive exploration. Semiconductors, at their core, are materials that occupy a captivating middle ground in terms of electrical conductivity. Unlike conductors such as copper or aluminum, which facilitate the unimpeded flow of electrical current, or insulators like rubber or

DOI: 10.1201/9781003450146-3

wood, which staunchly resist its passage, semiconductors possess the extraordinary ability to finely modulate their electrical conductivity. This intrinsic trait renders them indispensable in our modern electronic landscape [6]. Semiconductors, including silicon nanowire, gallium nitride, carbon nanotubes, quantum dot (QD), indium arsenide, and organic semiconductor material, exhibit distinctive electrical behavior due to their atomic and electronic structures. Within these materials, electrons can precisely traverse a critical energy gap, known as the band gap, enabling controlled electrical conduction. This property is fundamental to their versatility, allowing them to function as switches, amplifiers, and detectors in various electronic devices [1]. The summary of a variety of semiconductor materials used to date is tabulated in Table 3.1.

Semiconductor technology, driven by research in materials like carbon nanotubes, QDs, and organic semiconductor materials, has continually evolved. This evolution, exemplified by the integration of nanoscale transistors and advanced fabrication techniques, has led to smaller, faster, and more efficient devices. These innovations have revolutionized communication, computation, and interaction with the world. Semiconductors remain at the forefront of scientific and technological progress, opening new frontiers in an era of limitless possibilities [8]. In today's world, semiconductors are integral to nearly every aspect of our daily lives, from smartphones and computers to renewable energy systems and medical devices. Understanding their types and properties is not merely an academic exercise but a fundamental knowledge that underpins the design, development,

TABLE 3.1
Semiconductor Materials

General Classification	Semiconductor Materials	
	Symbol	Name
1. Elemental	Si	Silicon
	Ge	Germanium
2. Compounds		
	SiC	Silicon carbide
	AlP	Aluminum phosphide
	AlAs	Aluminum arsenide
	AlSb	Aluminum antimonide
	GaN	Gallium nitride
	GaP	Gallium phosphide
	GaAs	Gallium arsenide
	GaSb	Gallium antimonide
	InP	Indium phosphide
	InAs	Indium arsenide
	InSb	Indium antimonide
	ZnO	Zinc oxide
	ZnS	Zinc sulfide
	ZnSe	Zinc selenide
	ZnTe	Zinc telluride
	CdS	Cadmium sulfide
	CdSe	Cadmium selenide
	CdTe	Cadmium telluride
	HgS	Mercury sulfide
	PbS	Lead sulfide
	PbSe	Lead selenide
	PbTe	Lead telluride

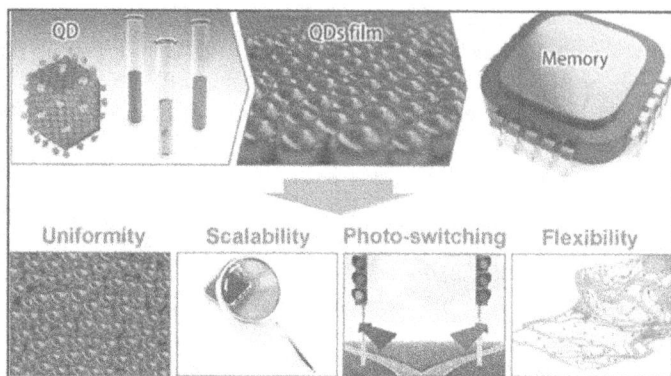

FIGURE 3.1 Schematic illustrations describing the assembly and important aspects of quantum dot film as the building blocks. (Reproduced with permission from [11]. Copyright 2020 American Chemical Society.)

and operation of the technology that defines our world. Semiconductor materials and their diverse properties serve as the bedrock of innovation and progress. They are the foundation upon which the digital revolution stands, and their continued evolution promises to unlock new frontiers in electronics, energy, and beyond. Profoundly grasping the types and properties of semiconductors equips engineers, scientists, and innovators with the tools to shape the future. It is a journey not only of historical significance, but one teeming with boundless potential [9, 10]. The assembly and important aspects of QD film as building blocks are shown in Figure 3.1.

3.2 SEMICONDUCTOR FUNDAMENTALS: UNVEILING ATOMIC STRUCTURE AND BAND THEORY

Semiconductors, those unassuming materials that bridge the gap between conductors and insulators, are the backbone of modern electronics and technology. To truly grasp their significance and versatility, we must embark on a journey into the fundamental principles that govern their behavior. A hydrothermal method of synthesis was used for WO_3 nanoparticles and magnesium doped WO_3 nanoparticles wherein the presence of magnesium dopant slightly increased the d-spacing (0.0069 nm) in WO_3 nanoparticles [12]. The conduction band electrons are quite easily excited to valence band by absorption of natural visible light energy by CdS nanowires [13]. Similarly, the CdS nanoparticles and their binary and ternary composites were utilized for ultrafast photoreduction of hexavalent chromium, Cr(VI) ions in solution [14]. In this section, we delve into the semiconductor fundamentals, peeling back the layers of atomic structure, energy bands, valence and conduction bands, and the all-important band gap. The n-type memory device (a-b) tunneling of an electron and (c-d) band-to-band tunneling for an electron are shown in Figure 3.2.

3.2.1 ATOMIC STRUCTURE AND SEMICONDUCTOR CRYSTALS

At the heart of semiconductor behavior lies the arrangement of atoms within their crystalline structure. Semiconductors are typically crystalline in nature, meaning their atoms are arranged in a highly ordered, repetitive pattern. The most prevalent semiconductor materials, such as silicon (Si) and germanium (Ge), form crystalline lattices composed of covalently bonded atoms. In these lattices, each atom shares electrons with its neighboring atoms, creating a stable structure. However, this orderly arrangement also sets the stage for the unique electrical properties of semiconductors. When external influences, such as heat or light, are introduced, they can disrupt the balance of electrons within the crystal lattice, leading to changes in electrical conductivity [15]. The hexagonal crystallized (002) ZnO layers deposited on crystalline, mesoporous, and nanoporous silicon by spin

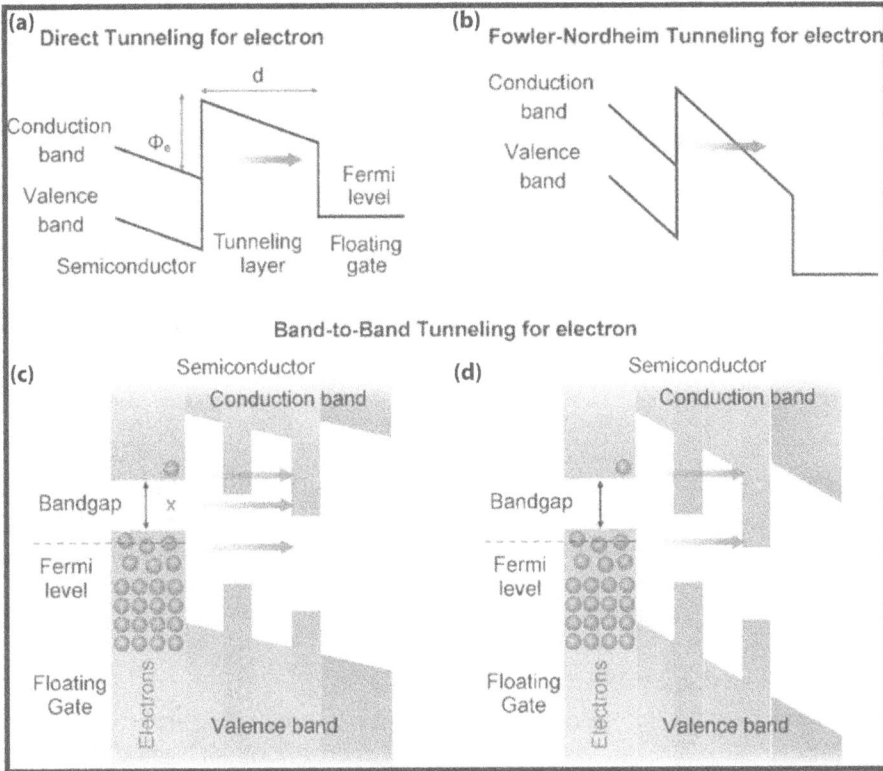

FIGURE 3.2   n-type memory device (a–b) tunneling of an electron and (c–d) band-to-band tunneling for an electron. (Reproduced with permission from [11]. Copyright 2020 American Chemical Society.)

coating technique can used to study atomic structure of the semiconductor. The low damping factor value defines the crystallinity nature. The spectroscopic ellipsometry (SE) was found to be 29 meV for ZnO layer composites [16]. The presence of oxygen in solid silicon forms Si-O-Si interstitial bonding. The bivalent nature and the small size have higher solubility than nitrogen and carbon due to an anomalous distribution coefficient [17].

3.2.2   ENERGY BANDS AND THE ELECTRON CONFIGURATION

To comprehend the electrical behavior of semiconductors, imagine a range of energy levels in which electrons are permitted to exist within a semiconductor. This range, known as the energy band, can be subdivided into two primary regions: the valence band and the conduction band. The relation between electron energies and optical transitions is tabulated in Table 3.2. The reflectance data is in the range of photon energies between 1.5 eV to 25 eV for Si, Ge, GaP, GaAs, InAs, and InSb [18].

TABLE 3.2
Relation between Electron Energies and Optical Transitions

Energy of the Electrons	Optical Transitions
High → Low	Emission
Low → High	Absorption

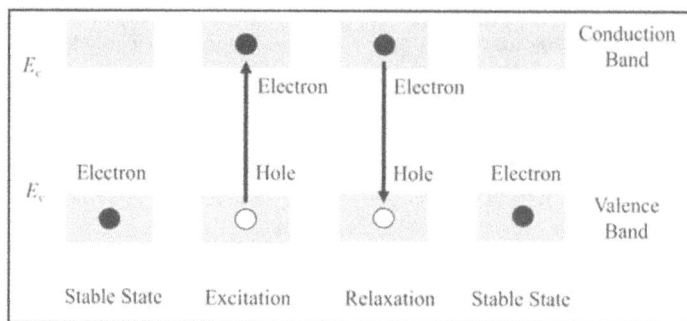

FIGURE 3.3 Schematic illustration of excitation and valence to conduction band electron relaxation.

3.2.2.1 Valence Band

The valence band is the lower-energy region of the energy band. Electrons in this band are present with their parent atoms tightly bound, and they contribute to the overall stability of the crystal structure. In this state, electrons are restricted to move and contribute to electrical conduction. A schematic illustration of excitation and valence to conduction band electron relaxation is shown in Figure 3.3.

3.2.2.2 Conduction Band

Above the valence band lies the conduction band, an energy level where electrons have mobility throughout the crystal lattice. Electrons in the conduction band carry an electric current, making CB essential for electrical conductivity.

3.3 TYPES AND PROPERTIES OF SEMICONDUCTOR

The key to understanding semiconductor behavior lies in the energy gap that splits the valence and conduction band—the band gap. The nanocomposite device of a hetero p-n CuO-ZnO acts as a photo-generated electron reservoir. The use of a wide band gap ZnO nanorod semiconductor along with a narrow band gap CuO reduces photogenerated electron-hole pairs to combine. The rightly engineered band gap of nanocomposites can enhance the semiconductor activity for a variety of purposes, for instance, improved visible light photo-reduction of Cr(VI) capacity [19]. The energy gap represents an impassable barrier for electrons in their natural state. Electrons within the valence band are forbidden from entering the conduction band unless they acquire sufficient energy to overcome this gap. This band gap property serves as the linchpin of semiconductor functionality. It is the foundation upon which semiconductors modulate their electrical conductivity in response to external stimuli [20, 21]. When electrons gain energy, they can jump across the band gap into the conduction band, creating charge carriers that contribute to electrical current. The use of multi-lanthanoid equiatomic oxide with $Gd_{0.2}La_{0.2}Ce_{0.2}Hf_{0.2}Zr_{0.2}O_2$ and $Gd_{0.2}La_{0.2}Y_{0.2}Hf_{0.2}Zr_{0.2}O_2$ compositions were utilized efficiently for complete photoreduction of hexavalent chromium due to their distinct band gap values [22]. Conversely, without the necessary energy, electrons remain confined to the valence band, rendering the material an insulator.

3.3.1 INTRINSIC SEMICONDUCTORS

3.3.1.1 Introduction of Intrinsic Semiconductors

Intrinsic semiconductors are those materials that inherently possess semiconductor properties without the need for intentional doping or modification. Silicon (Si) and germanium (Ge) are prime

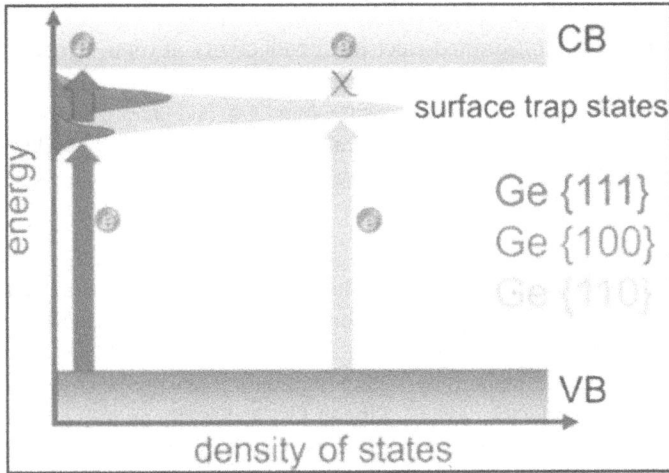

FIGURE 3.4 A diagram showing the amount of trap states and the trap state energies. (Reproduced with permission from [24]. Copyright 2020 American Chemical Society.)

examples of intrinsic semiconductors. These materials have a well-defined band gap, allowing them to conduct electricity under specific conditions [17, 23].

3.3.1.2 The Properties and Characteristics

In the fascinating world of semiconductors, intrinsic materials stand as fundamental pillars. These semiconductors, such as silicon (Si) and germanium (Ge), possess intrinsic properties that allow them to conduct electricity under specific conditions (Figure 3.5). In this exploration, we delve deep into the properties and characteristics of intrinsic semiconductors, shedding light on their unique behavior, intrinsic carrier concentration, and their intriguing temperature-dependent conductivity (Figure 3.4).

3.3.1.3 Properties and Characteristics of Intrinsic Semiconductors

Intrinsic semiconductors are materials that inherently possess semiconductor properties without any deliberate doping or modification. Silicon (Si) and germanium (Ge) are exemplary intrinsic semiconductors, and they have played pivotal roles in the advancement of electronics. The properties and characteristics that define intrinsic semiconductors include the following.

3.3.1.3.1 Band gap

One of the defining features of intrinsic semiconductors is their well-defined band gap. For silicon, this band gap is approximately 1.1 electron volts (eV), while for germanium, it is approximately 0.67 eV. This energy gap dictates the conditions under which these materials can conduct electricity. A technique for transforming infrared light into more intense visible light, photochemical upconversion has potential uses in photovoltaics, photocatalysis, biological imaging, and drug delivery. By employing sensitizers made of nanocrystal PbS semiconductors, a composition for upconversion absorbs photons below silicon's band gap, populating triplet energy states below oxygen singlet states. The energy required for visible spectrum luminescence of the triplet-state chromophoric violanthrone [25] is provided by two singlet oxygen molecules. Conversely, the utilization of bulk germanium with its indirect band gap (0.67 eV) poses challenges due to its suboptimal optical properties including photoluminescence. Overcoming this longstanding obstacle to germanium's application in optoelectronics remains elusive. Potential solutions include synthesizing ultrathin 2-D layers of germanium on silicon-based substrates or inducing significant structural modifications in the crystal lattice [26].

3.3.1.3.2 Limited Conductivity

Intrinsic semiconductors exhibit limited electrical conductivity at room temperature. Most of the electrons reside in the valence band, tightly constrained to their parent atoms. Only a small fraction of electrons possesses sufficient energy to cross the band gap and enter the conduction band, creating charge carriers that contribute to an electrical current. This limited conductivity is a crucial characteristic, allowing precise control of electrical behavior [27, 28].

3.3.1.3.3 Temperature Sensitivity

The electrical conductivity of intrinsic semiconductors is highly temperature-dependent. As temperature increases, more electrons acquire the energy needed to cross the band gap, leading to increased electrical conductivity. Conversely, at lower temperatures, fewer electrons can bridge the gap, resulting in reduced conductivity. This temperature sensitivity is a vital property that is exploited in various semiconductor devices. Silicon (Si) and germanium (Ge) are prime examples and are essential intrinsic semiconductors, each with its own set of properties and applications.

1. *Silicon (Si)*: This is the most widely used semiconductor material, forming the backbone of modern electronics. Silicon's abundance and compatibility with existing manufacturing processes have propelled it to the forefront of the semiconductor industry. It serves as the foundation for microprocessors, memory devices, and a host of integrated circuits that power computers, smartphones, and countless electronic gadgets.
2. *Germanium (Ge)*: This was one of the first materials to exhibit semiconductor behavior, which led to development of transistors. It has a narrower band gap of approximately 0.67 eV, which makes it less efficient for certain high-speed applications compared to silicon. However, germanium remains important in niche applications such as infrared detectors and some high-frequency devices [29].

3.3.1.3.4 Intrinsic Carrier Concentration

Intrinsic semiconductors have an inherent concentration of charge carriers—electrons and holes—that is determined by their temperature and the band gap. At absolute zero temperature (0K), the number of charge carriers is minimal, as very few electrons possess the energy to overcome the band gap. As temperature increases, the intrinsic carrier concentration also increases exponentially, following a mathematical relationship described by the intrinsic carrier concentration equation [30].

3.3.2 EXTRINSIC SEMICONDUCTORS

Beyond intrinsic materials, researchers and engineers have devised a method to enhance the electrical characteristics of semiconductors through a process known as doping. Extrinsic semiconductors involve the intentional introduction of controlled amounts of specific impurities into the crystalline lattice structure of a semiconductor material. This strategic introduction of impurities alters the magnitude of charge carriers in the matter, influencing its electrical conductivity. Doping gives rise to two primary categories of extrinsic semiconductors: n-type and p-type. n-type semiconductors involve the deliberate inclusion of electron-rich impurities, such as phosphorus or arsenic [31]. This process results in an abundance of liberated electrons, thereby augmenting the material's electron conductivity. On the other hand, p-type semiconductors entail the incorporation of electron-deficient impurities, like boron or aluminum. This yields an excess of "holes," or positively charged carriers, facilitating efficient hole conduction within the material [32]. The semiconductor fundamentals encompass the intricate interplay between atomic structure, energy bands, and the all-important band gap. These principles lay the groundwork for understanding the behavior of semiconductors, which serves as the cornerstone of modern technology (Figure 3.5) [33].

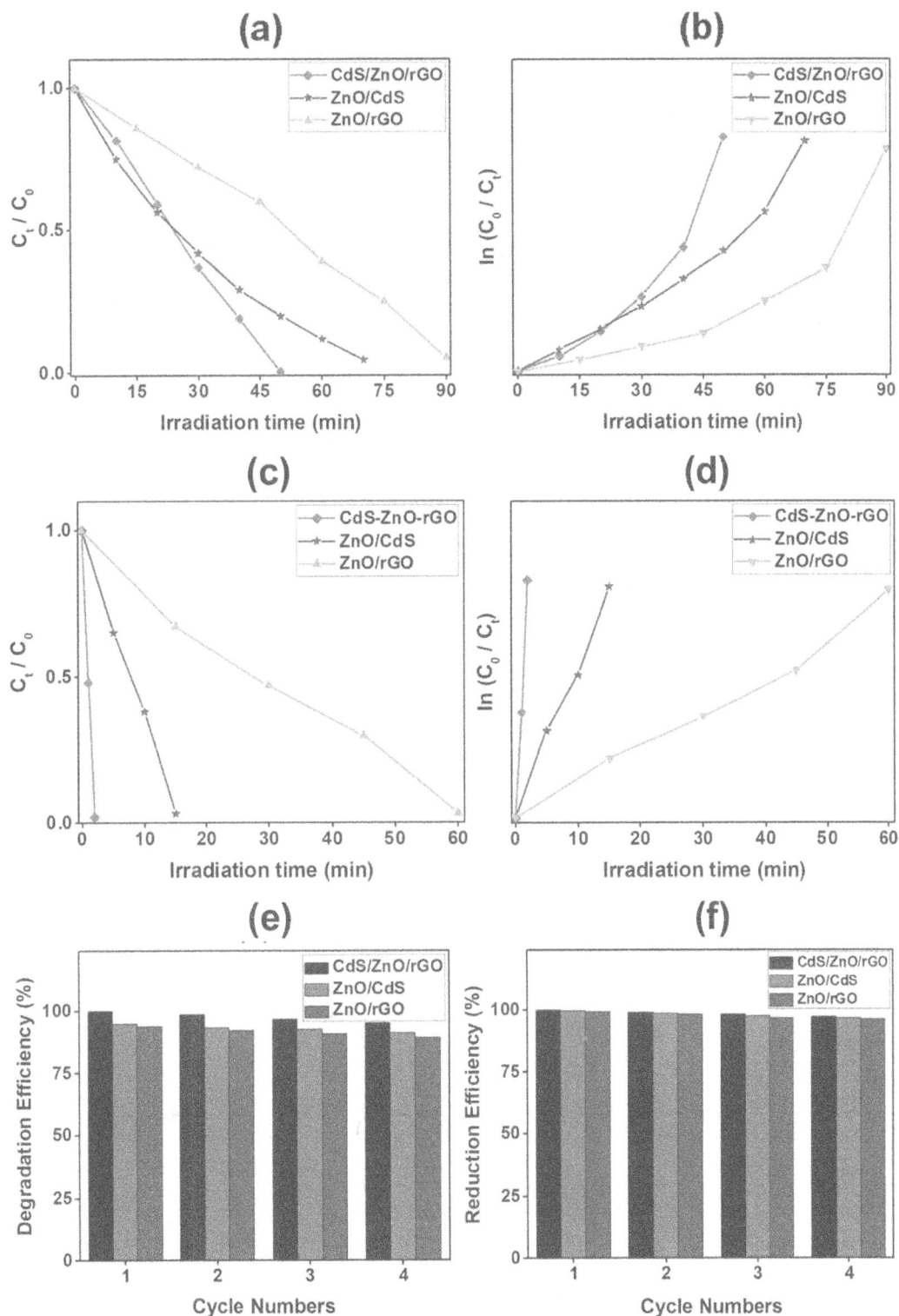

FIGURE 3.5 Plots of Ct/C0 vs. irradiation time ln (C0/Ct) vs. irradiation time. The recyclability test are shown for the nanocomposites (a–f). (Reproduced with permission from [14]. Copyright 2021 Royal Society of Chemistry.)

3.3.2.1   The Impact of Dopants on Electrical Conductivity

In the realm of semiconductor technology, extrinsic semiconductors, also known as doped semiconductors, play a pivotal job in tailoring the electrical behavior of materials to meet specific application requirements. In this exploration, we delve into the world of extrinsic semiconductors, introducing the concept of doping, elucidating the distinctions between n-type and p-type semiconductors, and shedding light on critical role that dopants play in influencing electrical conductivity. Extrinsic semiconductor materials were intentionally modified with precise impurities to enhance their electrical behavior. These impurities, known as dopants, are carefully introduced into the crystalline lattice structure of the semiconductor material. The strategic incorporation of dopants allows for precise control over the concentration and mobility of charge carriers within the material, leading to tailored electrical behavior [33].

3.3.2.2   N-type and P-type Semiconductors

3.3.2.2.1   N-type Semiconductors

The deliberate inclusion of electron-rich dopants such as phosphorus (P) or arsenic (As) in n-type semiconductors introduces additional free electrons into the crystal lattice. These extra electrons are the common charge carriers, contributing to electron conductivity within the material. As a result, n-type semiconductors exhibit enhanced electron mobility and conductivity compared to their intrinsic counterparts [31, 34].

3.3.2.2.2   P-type Semiconductors

Conversely, p-type semiconductors involve the incorporation of electron-deficient dopants, such as boron (B) or aluminum (Al). This introduces "holes" or positively charged carriers into the crystal lattice, which facilitate hole conduction. In p-type semiconductors, holes are the common charge carriers, leading to improved hole mobility and conductivity. p-type materials exhibit characteristics that are complementary to those of n-type semiconductors [35–37].

3.3.2.3   Role of Dopants and Their Impact on Electrical Conductivity

The choice of dopants and their concentration is a critical factor in shaping the electrical properties of extrinsic semiconductors. Dopants are strategically selected based on their electron donor or acceptor characteristics, depending on whether n-type or p-type behavior is desired.

3.3.2.3.1   Electron Donor Dopants

Dopants with extra electrons in their outer atomic shells, such as phosphorus or arsenic, act as electron donors when introduced into the crystal lattice. These extra electrons are mobile in the conduction band, thus increasing the electron volume and thereby enhancing conductivity in n-type semiconductors [38].

3.3.2.3.2   Electron Acceptor Dopants

Dopants with fewer electrons in their outer atomic shells, such as boron or aluminum, act as electron acceptors when added to the crystal lattice. They create "holes" in the valence band, where electrons are absent due to electron deficiency of the dopant. These holes facilitate hole conduction, increasing hole mobility and conductivity in p-type semiconductors. The net result of doping is the generation of charge carriers—either mobile electrons in n-type or holes in p-type — that significantly contribute to electrical current flow. By controlling the type and concentration of dopants, engineers and scientists can finely tune the electrical behavior of extrinsic semiconductors to suit a wide range of applications. Extrinsic semiconductors or doped semiconductors are integral to the world of semiconductor technology, enabling precise control over electrical conductivity and charge carrier behavior. n-type and p-type semiconductors, each with their distinct dopants and charge carriers, offer versatile solutions for a multitude of electronic and semiconductor device applications (Figure 3.6).

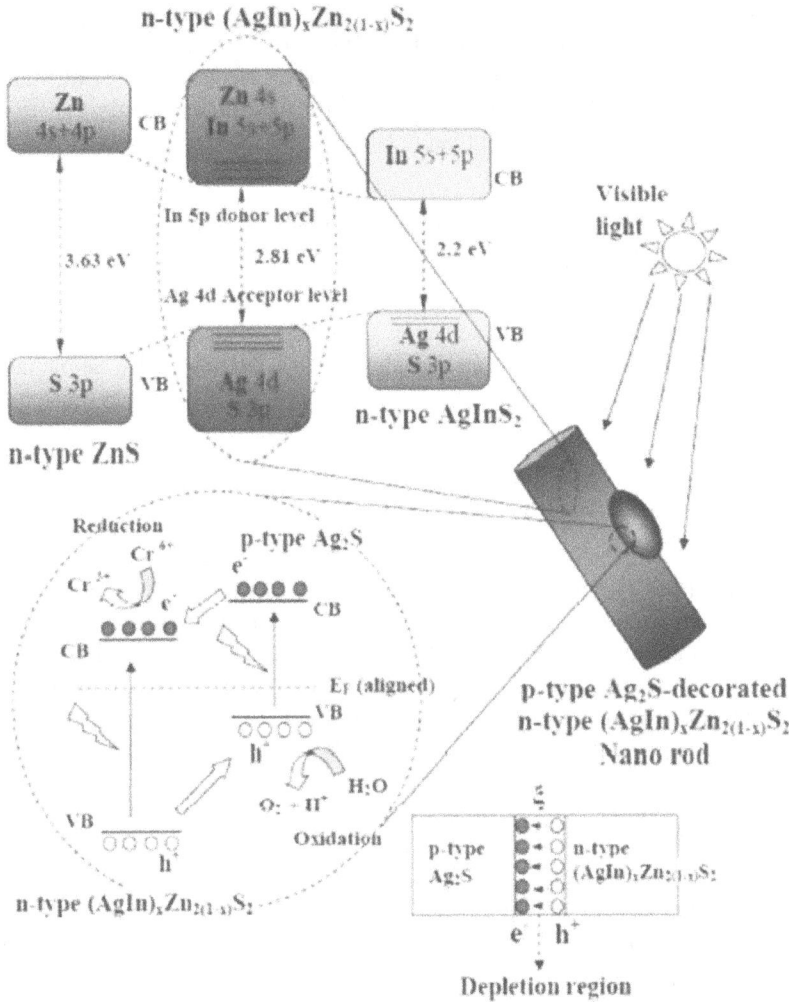

FIGURE 3.6 A schematic photoreduction mechanism of n-type $(AgIn)_xZn_{2(1-x)}S_2$ and p-type Ag_2S nanocomposite semiconductor material. (Reproduced with permission from [35]. Copyright 2015 American Chemical Society.)

3.4 BAND GAP: MODULATION OF CONDUCTIVITY

The band gap property is the requirement of semiconductor functionality, influencing how these materials respond to external factors and stimuli. The band gap dictates the conditions under which a semiconductor can conduct electricity. Electrons within the valence band are effectively trapped unless they acquire sufficient energy to bridge the gap and enter the conduction band. This inherent property allows for precise control of electrical conductivity, making semiconductors exceptionally versatile. The ability of semiconductors to modulate their electrical conductivity in response to external factors is a direct consequence of the band gap. When subjected to alterations in temperature, exposure to light, or variation in voltage, electrons can gain or lose energy, thereby changing their status from bound to mobile and vice versa.

3.4.1 Effects of Temperature Variations on the Band Gap

Temperature has a significant part in influencing the band gap and, consequently, the electrical behavior of semiconductors. As temperature increases, the thermal energy provided to electrons allows an increasing number of them to acquire the energy needed to cross the band gap into the conduction band. This phenomenon leads to enhanced electrical conductivity in semiconductors at higher temperatures. Conversely, at lower temperatures, fewer electrons can bridge the gap, resulting in reduced conductivity. Temperature sensitivity is a vital characteristic that is exploited in semiconductor device design, especially in temperature sensors and thermistors [39].

3.4.2   EFFECTS OF EXPOSURE TO LIGHT ON THE BAND GAP

Another intriguing aspect of the band gap is its interaction with light. When photons of energy equal to or larger than the energy band gap strike a semiconductor, they can excite electrons from the valence to the conduction band. This phenomenon, called the photovoltaic effect, is the foundation of light-sensitive semiconductor devices such as photodiodes and solar cells. Semiconductors with wider band gaps are more suitable for capturing higher-energy photons, such as those from sunlight, making them ideal for efficient energy conversion [40].

3.4.3   EFFECTS OF VOLTAGE MODULATIONS ON THE BAND GAP

Applying an external voltage across a semiconductor material can also impact the band gap. In certain devices, such as laser diodes and light-emitting diodes (LEDs), the application of voltage leads to electron-hole recombination in the conduction and valence bands. This recombination releases energy in the form of photons, resulting in light emission. Thus, voltage manipulation can be used to control the light output in these devices, making them crucial for optical communication and displays. The band gap phenomenon serves as the cornerstone of semiconductor behavior, allowing these materials to finely tune their electrical conductivity in response to external influences [41].

3.4.4   EFFECT OF IONIZING RADIATION ON SEMICONDUCTORS

When subjected to ionizing radiation, electrons in a semiconductor may be elevated to the conduction band, provided they acquire energy greater than the band gap (E_{gap}). Typically, heavy-charged particles primarily transfer energy through two mechanisms: (1) excitation, where the energy is imparted to the atom by the charged particle, which then causes the electrons to make a transition to higher energy levels; and (2) ionization, where the charged particle possesses sufficient energy to eject an electron from the atom, resulting in formation of ion pairs in the material [42]. Stopping power is a parameter that conveniently characterizes the ionization properties of medium, with an expression for specific energy loss described by the Bethe formula. Heavy particles like alpha (α) have high stopping powers in most materials, leading to short travel distances [43]. An additional interaction where beta particles may release energy via a radiative course is known as bremsstrahlung, where the charged particle emits energy as it accelerates or decelerates. This radiation is referred to as braking radiation. The photons, including X-rays and gamma rays, ionize an atom directly, although they are electrically neutral, primarily through the Compton scattering and photoelectric effect. Three crucial mechanisms are: Compton scattering, photoelectric effect, and pair production. Ionizing radiation leaves some of its energy behind in each of these situations. As an example, a silicon detector with a thickness of around 300 μm will generate approximately 3.2×10^4 pairs of electron holes when a minimum ionizing particle (MIP) goes via it perpendicularly. This count is relatively small in comparison with the entirety of mobile transporters in an intrinsic semiconductor with the same surface area and thickness [44–46]. For instance, at 20°C, pure germanium contains 1.26×10^{21} atoms and consistently generates free electrons (7.5×10^{11}) and holes (7.5×10^{11}) due to thermal energy. Consequently, the signal-to-noise ratio (S/N) is insignificant. Introducing a minute amount (0.001%) of arsenic adds extra free electrons (10^{15}) within the same capacity, increasing conductivity by 10,000 folds. In doped materials, the S/N ratio remains low. One approach to addressing this is by cooling the semiconductor. An enhancement can be realized by employing reverse-bias to p-n junction, a fundamental principle in maximum silicon radiation sensors. In this setup, to the p-side, a negative voltage, and to the n-side, a positive voltage, are applied. This arrangement causes holes in the p-region to migrate away from the junction toward p-contact, while electrons move alike toward the n-contact, effectively depleting the detector of free carriers.

3.5 CONCLUSION

In our comprehensive exploration of semiconductor types and properties, we have embarked on a fascinating journey through the heart of modern technology. From the inception of semiconductors to their pivotal role in today's world, we have uncovered a multitude of insights that underscore the enduring significance of understanding these remarkable materials. As we conclude our expedition, let us summarize the key takeaways, recognize the profound and lasting importance of semiconductor knowledge, and glimpse the boundless possibilities that lie ahead in semiconductor technology. (1) Semiconductor fundamentals: We have unraveled the fundamental concepts underpinning semiconductors, delving into atomic structures, energy bands, conduction and valence bands, and the critical role of the band gap in regulating electrical behavior. These foundational principles are the building blocks of semiconductor science. (2) Intrinsic and extrinsic semiconductors: Our exploration led us to intrinsic semiconductors like silicon (Si) and germanium (Ge), where we discovered their unique properties and the temperature-dependent behavior of intrinsic carrier concentration. We also explored extrinsic semiconductors, shaped by the art of doping to enhance electrical characteristics. (3) The band gap phenomenon: We delved into the band gap, a defining feature that sets semiconductors apart. Its role in governing electrical conductivity and its responsiveness to external factors like temperature, light, and voltage were thoroughly examined. (4) Excitation by ionizing radiation: We explored how semiconductors respond to ionizing radiation, shedding light on processes such as excitation and ionization. By optimizing signal detection while minimizing noise, we unveiled the practical implications of these interactions. (5) Significance: Understanding semiconductor types and properties is not an academic pursuit; it is a linchpin of contemporary technological advancement. Semiconductors underpin the digital age, powering our computers, smartphones, and an ever-expanding array of electronic devices. Their significance extends beyond electronics, permeating renewable energy systems, medical devices, and communication networks. Semiconductors are the invisible architects of modern society. Semiconductors enable revolutions in computing and information technology. They drive miniaturization, boosting performance and energy efficiency. As we step into the era of quantum computing, artificial intelligence, and the Internet of Things (IoT), semiconductors will continue to steer innovation, reshaping our world in ways we can scarcely imagine.

REFERENCES

1. Al-Ahmadi N A 2020 Metal oxide semiconductor-based Schottky diodes: A review of recent advances *Mater. Res. Express* **7** 032001
2. Andrews E G 1963 Telephone switching and the early Bell Laboratories computers *Bell Syst. Tech. J* **42** 341–53
3. Huff H R 2001 John Bardeen and transistor physics *AIP Conference Proceedings* vol 550 (AIP) pp 3–32
4. Khan H N, Hounshell D A and Fuchs E R H 2018 Science and research policy at the end of Moore's law *Nat. Electron* **1** 14–21
5. Brazier J and McMorrow K 2021 Instrumental: Collections from Science. Calculating and Computing, Chau Chak Wing Museum, The University of Sydney, 2021.
6. O'Regan G 2018 A Concise History of Computing. In *World of Computing* (Cham: Springer International Publishing) pp 29–74
7. Iwai H, Kakushima K and Wong H 2006 Challenges for future semiconductor manufacturing *Int. J. High Speed Electron. Syst* **16** 43–81

8. Graham A P, Duesberg G S, Hoenlein W, Kreupl F, Liebau M, Martin R, Rajasekharan B, Pamler W, Seidel R, Steinhoegl W and Unger E 2005 How do carbon nanotubes fit into the semiconductor roadmap? *Appl. Phys. A Mater. Sci. Process* **80** 1141–51

9. Lathe A and Palve A M 2023 A review: Engineered nanomaterials for photoreduction of Cr(VI) to Cr(III) *J. Hazard. Mater. Adv* **12** 100333

10. Gupta R, Peveler W J, Lix K and Algar W R 2019 Comparison of semiconducting polymer dots and semiconductor quantum dots for smartphone-based fluorescence assays *Anal. Chem.* **91** 10955–60

11. Lv Z, Wang Y, Chen J, Wang J, Zhou Y and Han S-T 2020 Semiconductor quantum dots for memories and neuromorphic computing systems *Chem. Rev.* **120** 3941–4006

12. Thwala M M and Dlamini L N 2020 Photocatalytic reduction of Cr(VI) using Mg-doped WO3 nanoparticles *Environ. Technol* **41** 2277–92

13. Ganesh R S, Durgadevi E, Navaneethan M, Sharma S K, Binitha H S, Ponnusamy S, Muthamizhchelvan C and Hayakawa Y 2017 Visible light induced photocatalytic degradation of methylene blue and rhodamine B from the catalyst of CdS nanowire *Chem. Phys. Lett.* **684** 126–34

14. Lathe A and Palve A M 2021 Reduced graphene oxide-decorated CdS/ZnO nanocomposites for photoreduction of hexavalent chromium and photodegradation of methylene blue *Dalt. Trans* **50** 14163–75

15. von Behren J, van Buuren T, Zacharias M, Chimowitz E H and Fauchet P M 1998 Quantum confinement in nanoscale silicon: The correlation of size with bandgap and luminescence *Solid State Commun* **105** 317–22

16. Bouzourâa M-B, Naciri A E, Moadhen A, Rinnert H, Guendouz M, Battie Y, Chaillou A, Zaïbi M-A and Oueslati M 2016 Effects of silicon porosity on physical properties of ZnO films *Mater. Chem. Phys* **175** 233–40

17. Ye H and Yu J 2014 Germanium epitaxy on silicon *Sci. Technol. Adv. Mater* **15** 024601

18. Wilson N P, Lee K, Cenker J, Xie K, Dismukes A H, Telford E J, Fonseca J, Sivakumar S, Dean C, Cao T, Roy X, Xu X and Zhu X 2021 Interlayer electronic coupling on demand in a 2D magnetic semiconductor *Nat. Mater.* **20** 1657–62

19. Yu J, Zhuang S, Xu X, Zhu W, Feng B and Hu J 2015 Photogenerated electron reservoir in hetero-p–n CuO–ZnO nanocomposite device for visible-light-driven photocatalytic reduction of aqueous Cr(VI) *J. Mater. Chem. A* **3** 1199–207

20. Lathe A, Ansari A, Badhe R, Palve A M and Garje S S 2021 Single-step production of a TiO 2 @MoS 2 heterostructure and its applications as a supercapacitor electrode and photocatalyst for reduction of Cr(VI) to Cr(III) *ACS Omega* **6** 13008–14

21. Gupta A, Sardana S, Dalal J, Lather S, Maan A S, Tripathi R, Punia R, Singh K and Ohlan A 2020 Nanostructured polyaniline/graphene/Fe 2 O 3 composites hydrogel as a high-performance flexible supercapacitor electrode material *ACS Appl. Energy Mater* **3** 6434–46

22. Anandkumar M, Lathe A, Palve A M and Deshpande A S 2021 Single-phase $Gd_{0.2}La_{0.2}Ce_{0.2}Hf_{0.2}Zr_{0.2}O_2$ and $Gd_{0.2}La_{0.2}Y_{0.2}Hf_{0.2}Zr_{0.2}O_2$ nanoparticles as efficient photocatalysts for the reduction of Cr(VI) and degradation of methylene blue dye *J. Alloys Compd* **850** 156716

23. Zhou W and Coleman J J 2016 Semiconductor quantum dots *Curr. Opin. Solid State Mater. Sci* **20** 352–60

24. Tan C-S, Lu M-Y, Peng W-H, Chen L-J and Huang M H 2020 Germanium possessing facet-specific trap states and carrier lifetimes *J. Phys. Chem. C* **124** 13304–9

25. Gholizadeh E M, Prasad S K K, Teh Z L, Ishwara T, Norman S, Petty A J, Cole J H, Cheong S, Tilley R D, Anthony J E, Huang S and Schmidt T W 2020 Photochemical upconversion of near-infrared light from below the silicon bandgap *Nat. Photonics* **14** 585–90

26. Liu Y, Yang D, Xu T, Shi Y, Song L and Yu Z-Z 2020 Continuous photocatalytic removal of chromium (VI) with structurally stable and porous Ag/Ag_3PO_4/reduced graphene oxide microspheres *Chem. Eng. J* **379** 122200

27. Mitchell E W J and Sillars R W 1949 Observations of the electrical behaviour of silicon carbide contacts *Proc. Phys. Soc. Sect. B* **62** 509–22

28. Anon 1948 Electronic applications of germanium *Nature* **162** 982–3

29. Becker J A, Green C B and Pearson G L 1947 Properties and uses of thermistors-thermally sensitive resistors Bell *Syst. Tech. J* **26** 170–212

30. Shockley W, Pearson G L and Haynes J R 1949 Hole injection in germanium-quantitative studies and filamentary transistors Bell *Syst. Tech. J* **28** 344–66

31. Pon A, Bhattacharyya A and Rathinam R 2021 Recent developments in black phosphorous transistors: A review *J. Electron. Mater* **50** 6020–36

32. Cai W, Mu X, Pan Y, Guo W, Wang J, Yuan B, Feng X, Tai Q and Hu Y 2018 Facile fabrication of organically modified boron nitride nanosheets and its effect on the thermal stability, flame retardant, and mechanical properties of thermoplastic polyurethane *Polym. Adv. Technol* **29** 2545–52

33. Liu A, Phang S P and Macdonald D 2022 Gettering in silicon photovoltaics: A review *Sol. Energy Mater. Sol. Cells* **234** 111447

34. Mauersberger T, Ibrahim I, Grube M, Heinzig A, Mikolajick T and Weber W M 2020 Size effect of electronic properties in highly arsenic-doped silicon nanowires *Solid. State. Electron* **168** 107724

35. Abdullah H and Kuo D-H 2015 Facile synthesis of n-type (AgIn)xZn2(1– x)S2/p-type Ag2S nanocomposite for visible light photocatalytic reduction to detoxify hexavalent chromium *ACS Appl. Mater. Interfaces* **7** 26941–51

36. Kozhakhmetov A, Stolz S, Tan A M Z, Pendurthi R, Bachu S, Turker F, Alem N, Kachian J, Das S, Hennig R G, Gröning O, Schuler B and Robinson J A 2021 Controllable p-type doping of 2D WSe2 via vanadium substitution *Adv. Funct. Mater* **31** 2105252

37. Tsao J Y, Chowdhury S, Hollis M A, Jena D, Johnson N M, Jones K A, Kaplar R J, Rajan S, Van de Walle C G, Bellotti E, Chua C L, Collazo R, Coltrin M E, Cooper J A, Evans K R, Graham S, Grotjohn T A, Heller E R, Higashiwaki M, Islam M S, Juodawlkis P W, Khan M A, Koehler A D, Leach J H, Mishra U K, Nemanich R J, Pilawa-Podgurski R C N, Shealy J B, Sitar Z, Tadjer M J, Witulski A F, Wraback M and Simmons J A 2018 Ultrawide-bandgap semiconductors: Research opportunities and challenges *Adv. Electron. Mater* **4** 1600501

38. Chaves A, Azadani J G, Alsalman H, da Costa D R, Frisenda R, Chaves A J, Song S H, Kim Y D, He D, Zhou J, Castellanos-Gomez A, Peeters F M, Liu Z, Hinkle C L, Oh S-H, Ye P D, Koester S J, Lee Y H, Avouris P, Wang X and Low T 2020 Bandgap engineering of two-dimensional semiconductor materials *npj 2D Mater. Appl* **4** 29

39. Gunawan S D, Djunaidi M C and Haris A 2022 Variation of annealing temperature with excess of NaOH concentration on Ag2S synthesis from argentometry titration waste as NTC thermistor *Mater. Today Proc* **63** S385–90

40. Wagner E P 2016 Investigating bandgap energies, materials, and design of light-emitting diodes *J. Chem. Educ* **93** 1289–98

41. Lashkov I, Krechan K, Ortstein K, Talnack F, Wang S-J, Mannsfeld S C B, Kleemann H and Leo K 2021 Modulation doping for threshold voltage control in organic field-effect transistors *ACS Appl. Mater. Interfaces* **13** 8664–71

42. Almora O, García-Batlle M and Garcia-Belmonte G 2019 Utilization of temperature-sweeping capacitive techniques to evaluate band gap defect densities in photovoltaic perovskites *J. Phys. Chem. Lett.* **10** 3661–9

43. Blase X, Duchemin I, Jacquemin D and Loos P-F 2020 The Bethe-Salpeter equation formalism: From physics to chemistry *J. Phys. Chem. Lett.* **11** 7371–82

44. Rogers F, Xiao M, Perez K M, Boggs S, Erjavec T, Fabris L, Fuke H, Hailey C J, Kozai M, Lowell A, Madden N, Manghisoni M, McBride S, Re V, Riceputi E, Saffold N and Shimizu Y 2019 Large-area Si(Li) detectors for X-ray spectrometry and particle tracking in the GAPS experiment *J. Instrum* **14** P10009–P10009

45. Abubakr E, Ohmagari S, Zkria A, Ikenoue H and Yoshitake T 2022 Laser-induced novel ohmic contact formation for effective charge collection in diamond detectors *Mater. Sci. Semicond. Process* **139** 106370

46. Tsai H, Ghosh D, Panaccione W, Su L-Y, Hou C-H, Wang L, Cao L R, Tretiak S and Nie W 2022 Addressing the voltage induced instability problem of perovskite semiconductor detectors *ACS Energy Lett* **7** 3871–9

4 Wide Band Gap Semiconductors
Preparation, Tunable Properties, and Applications

K. Ravichandran, S. Suvathi, P. Ravikumar, and R. Mohan

4.1 INTRODUCTION

Research in various areas of materials science throws light on the properties of materials under different conditions and leads to the development of new technologies for producing materials with tuned properties suitable for a wide range of applications. Materials scientists have established that the characteristics of several kinds of materials can be modified as required by appropriately changing the preparation methods, process parameters, and/or by adding other suitable materials (such as dopants/composite or alloy partners) to make them fit for the application of interest.

In recent decades, wide band gap semiconductors (WBSs), materials having band energy gaps greater than those of conventional semiconductors, have been increasingly used in various technological applications as their properties can be tuned desirably. For example, tunable wide band gap semiconductors like zinc oxide (ZnO) and tin oxide (SnO_2) are considered desirable candidates in many fields/devices that include cost-effective solar cells, sensors, LEDs, super capacitors, photocatalytic and antibacterial systems, etc. The optical, electrical, structural, magnetic, thermal, mechanical, and photocatalysis characteristics of these materials can be tuned by adding compatible dopants with optimum concentrations [1]. These wide band gap transparent conducting oxides have specific characteristics: a wide band gap ranging from 2.0 to 3.60 eV, abundance in nature, long shelf life, chemical stability, and non-toxicity, to name a few. The addition of transition metals in optimum proportion and the preparation of nanocomposites with compatible composite partners like carbonaceous materials can tune the band gap of the TCOs, making them photocatalysts under visible light.

4.2 TECHNOLOGICAL SIGNIFICANCE OF WBS MATERIALS

Wide band gap semiconducting materials are a class of materials that includes zinc oxide, tin oxide, cadmium sulfide, cadmium selenide, gallium selenide, copper oxide, zinc telluride, zinc sulfide, zinc selenide, gallium phosphide, aluminum arsenide, aluminum phosphide, and diamond. These wide band gap materials are widely used in various technological and environmental applications like solar cells [2, 3], photocatalysis [4], supercapacitors [5], opto-electronics [6], gas sensors [7], antibacterial systems [8], LEDs [9], fuel cells, spintronics, photonics, biomedical applications, bioimaging, biotags, giant magnetic resistance sensors, and rechargeable batteries. Certain representative WBS materials, methods that are commonly used for the preparation of these WBS materials, their tunable properties, and applications in which they are used owing to their tunable properties are depicted in Figure 4.1.

The band gaps of common WBS materials and their important applications as reported in the literature are presented in Table 4.1. As given in the table, the band gap of the WBS ranges from

DOI: 10.1201/9781003450146-4

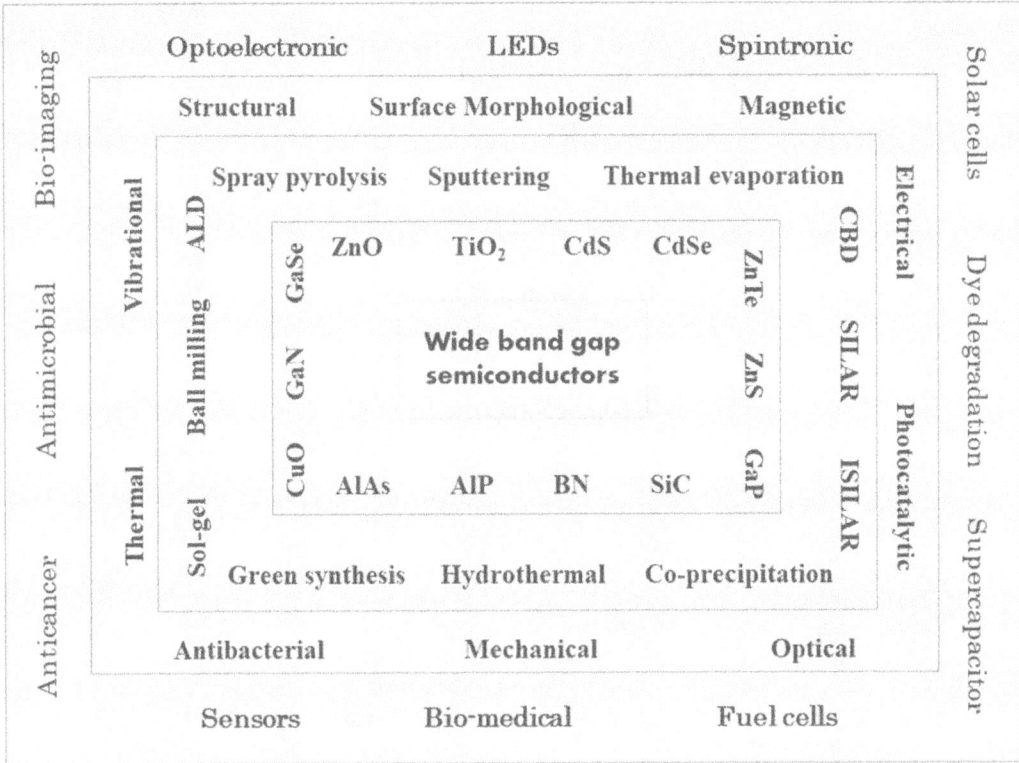

FIGURE 4.1   Certain important wide band gap semiconductors, their preparation methods and applications.

1.24 eV for copper oxide to 5.95 eV for boron nitride. This wide variation in band gap makes this class of materials fit for different applications. Moreover, this band gap can be varied suitably by changing the process parameters like dopant concentration and preparation methods, as presented in Table 4.2 for zinc oxide and Table 4.3 for cadmium sulfide.

4.3   ZINC OXIDE, TIN OXIDE, AND CADMIUM SULFIDE– CERTAIN REPRESENTATIVE WBS MATERIALS

Being one of the technologically important wide band gap semiconductors, ZnO offers many advantages, which include low cost [10], abundancy, non-toxicity, superior durability, thermal and chemical stability, tunable surface morphology, and environmentally friendly features [11]. It can be used as a cost-effective sensor in industry for monitoring toxic gases. Due to the wider band gap of 3.37 eV, ZnO can act as an effective photocatalytic material only under UV irradiation. However, its tunable band gap makes it a cost-effective alternative to titanium oxide (TiO$_2$), which is a commercially common photocatalytic material. ZnO is used in sunscreen lotions, as it has UV blocking ability. When prepared in thin film form, ZnO exhibits good optical and electrical properties desirable for solar cell window layers [12]. When ZnO is prepared in nano-structured forms, its surface morphology and surface-to-volume ratio can be tuned for a wide range of applications. Being a chemically inert antibacterial agent, it is used in food packaging [13]. In general, properties like resistivity, band gap, crystallite size, and surface morphology of ZnO prepared in thin films, nanomaterial, or other forms depend on the synthesis method and the process parameters employed.

TABLE 4.1
Important Applications of Wide Band Gap Semiconductors

S. No.	Wide Band Gap Semiconductors	Band Gap (eV)	Applications	Reference
1.	Zinc oxide	3.37	Solar cells Gas sensors UV light emitters Photocatalysis Luminescent materials Heat mirrors and coatings	[14]
2.	Tin oxide	3.60	Solar cell electrodes Opto-electronics, Light emitting diodes, Flat panel displays and transparent electronics	[15]
3.	Cadmium sulfide	2.42	Solar cells	[16]
4.	Cadmium selenide	1.74	Supercapacitors	[17]
5.	Gallium selenide	2.10	Photovoltaics	[18]
6.	Copper oxide	1.24	Solar selective absorbers	[19]
7.	Zinc telluride	2.26	Opto-electronics	[20]
8.	Zinc sulfide	3.60	Gas sensors Bio-sensors Flat panel displays	[21]
9.	Zinc selenide	2.82	Optical temperature sensors	[22]
10.	Gallium phosphide	2.24	Light emitting diodes Gas captors and solar cells	[23]
11.	Gallium nitride	3.40	Photoelectric detectors	[24]
12.	Aluminum arsenide	2.12	Opto-electronics	[25]
13.	Aluminum phosphide	2.50	Gas sensors	[26]
14.	Boron nitride	5.95	Opto-electronic devices	[27]
15.	Silicon carbide	3.26	Photoelectronic devices	[28]
16.	Diamond	5.30	Bio-sensors Solar cells	[29]

TABLE 4.2
Preparation Methods and Tunable Properties of Zinc Oxide Nanomaterial

S. No.	Material	Methods	Crystallite Size	Morphology	Band Gap	Applications	Ref.
1.	ZnO	Green synthesis	28.12 nm	Hexagonal	3.01 eV	Photocatalysis	[30]
2.	ZnO	Hydrothermal precipitation	24.70 nm 23.09 nm	Pseudospherical Spherical	3.25 eV	Photocatalysis	[31]
3.	ZnO: Tb	Co-precipitation	18.06 nm 20.00 nm	Spherical	2.61 eV	Gas sensors	[32]
4.	ZnO: Co	Spray pyrolysis	19.20 nm	Packed and faded-like structure	3.22 eV	Gas sensors	[33]
5.	ZnO: Al	Sol-gel		Porous	3.40 eV	Perovskite solar cell	[34]
6.	ZnO: Ni	Chemical vapor deposition	32.90 nm	Nanorods	3.22 eV	Solar cells	[35]
7.	ZnO: Sn	Simplified spray pyrolysis	26.68 nm	Hexagonal with micropore	3.23 eV	Antibacterial study	[36]

TABLE 4.3

Preparation Methods and Tunable Properties of Cadmium Sulfide Nanomaterial

S. No.	Material	Methods	Crystallite Size	Morphology	Band Gap	Applications	Ref.
1.	CdS: Zn	Co-precipitation	44.00 nm	Nano-flacks		Supercapacitor	[37]
2.	CdS	Atomic layer deposition	8.70 nm	Nanorods	1.93 eV	Dye-sensitized solar cells	[38]
3.	CdS	Biosynthesis	4.00 nm	Spherical	3.31 eV	Photocatalysis	[39]
4.	CdS	SILAR	11.90 nm	Spherical	2.35 eV	Photovoltaic	[40]
		CBD	13.29 nm		2.46 eV		

4.4 IMPORTANT PROPERTIES OF ZnO, SnO$_2$, AND CdS

Generally, ZnO and CdS appear in one of the two crystal structures, *viz.* cubic (zinc blende) and hexagonal (wurtzite), whereas SnO$_2$ has a tetragonal (rutile) structure (Figure 4.2). Some important physical, structural, electrical, and optical characteristics of ZnO, SnO$_2$, and CdS are presented in Table 4.4.

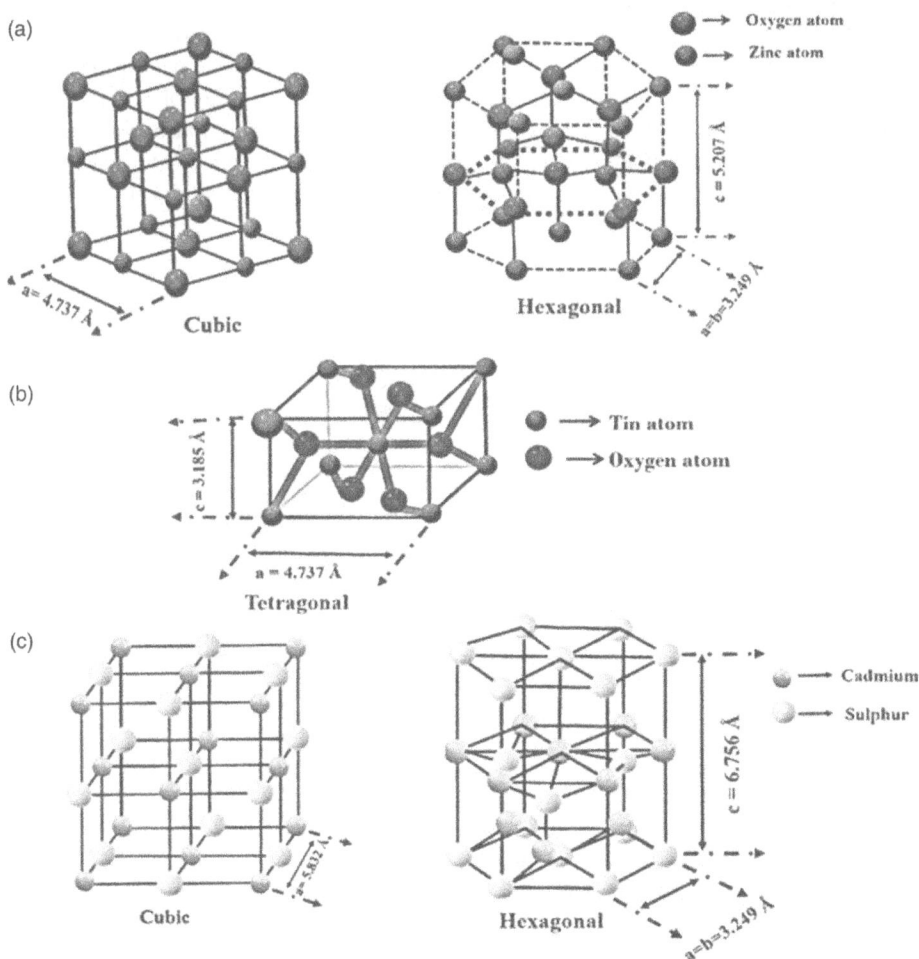

FIGURE 4.2 Crystal structure of (**a**) ZnO, (**b**) SnO$_2$, and (**c**) CdS.

TABLE 4.4

Properties of Three Important WBS Materials, *viz.* ZnO, SnO$_2$, and CdS

Property	Zinc Oxide	Tin Oxide	Cadmium Sulfide
Molecular formula	ZnO	SnO$_2$	CdS
Band gap	3.37 eV	3.80 eV	2.42 eV
Space group	P6$_3$mc	P4$_2$/mnm	P6$_3$/mc
Volume of unit cell	47.62 (Å)^3	75.73 (Å)^3	104.86 (Å)^3
Conductivity	n-type	n-type	n-type
Melting point	1974°C	1630°C	1750°C
Ionic radius	1.53 Å	1.18 Å	0.97 Å
Molecular weight	81.38 g/mol	150.71 g/mol	144.47 g/mol
Density	5.606 g/cm^3	6.95 g/cm^3	4.826 g/cm^3

4.5 PREPARATION OF WIDE BAND GAP SEMICONDUCTORS

Wide band gap semiconductors can be prepared using various physical, chemical, and green synthesis methods as depicted in Figure 4.3. Sputtering, thermal evaporation, lithography, and ball-milling are some of the common physical methods used for the preparation of wide band gap semiconductors. Similarly, chemical methods like chemical bath deposition, spray pyrolysis, and successive ionic layer adsorption and reaction (SILAR) are adopted for the preparation of WBS. Nowadays, the green synthesis route greatly attracts the attention of materials scientists, as it does not require chemicals that are hazardous to the environment and living beings. In green synthesis, extracts of

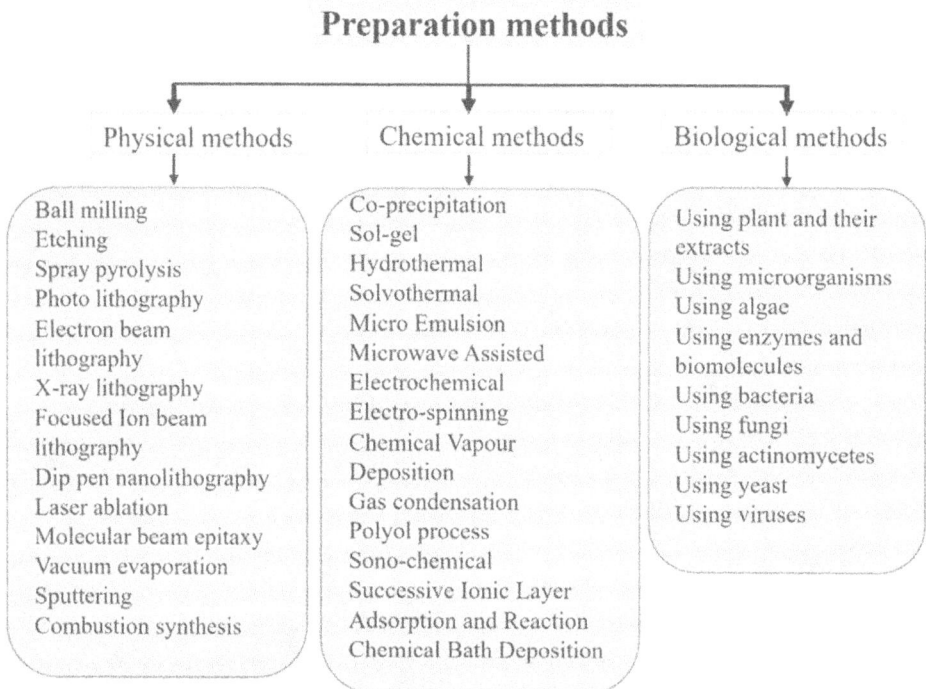

FIGURE 4.3 Different methods for the synthesis of diverse wide band gap semiconductor materials.

different parts of plants like leaves, flowers, and petals are used for reducing and capping agents during the synthesis process. Interestingly, biowastes such as fruit peels, vegetable wastes, and food wastes are also used for biosynthesis. Certain microorganisms are also used for biosynthesizing WBS materials. The chart given in Figure 4.3 lists the different methods categorized under these three broad classifications of preparation methods generally adopted to prepare WBS materials. Schematic diagrams of certain equipment used for the preparation of WBS materials via certain physical methods (like ball milling, vacuum evaporation, DC sputtering, and laser ablation), chemical methods (such as chemical vapor deposition and electrochemical deposition), and biological methods (like green synthesis and biosynthesis) are shown in Figure 4.4 (a)–(c).

By choosing suitable preparation methods and by optimizing the control/process parameters, wide band gap semiconductors with required properties can be obtained. For instance, cadmium sulfide (CdS) prepared by employing different methods shows different crystallite size, different surface morphology with grain shapes, and different band gaps.

4.6 TUNABLE PROPERTIES OF WBS MATERIALS

By appropriately varying the process parameters (like the concentration of the precursor solution, type of solvent, dopants added, concentration of the dopants, process temperature, post-synthesis/deposition heat treatments, pH of the starting solution, and atmospheric conditions), the properties of WBS can be varied accordingly. For instance, the electrical and optical properties of zinc oxide and cadmium sulfide can be tuned depending on the applications for which the material is prepared.

4.6.1 Zinc Oxide

4.6.1.1 Structural Properties

The structural properties of WBS materials are generally investigated by employing the most common analytical technique, X-ray diffraction. In 1908, Swiss physicist Paul Scherrer developed a formula known as Scherrer's formula to estimate the size of the crystallites. From this formula, we can realize that the wider the FWHM, the smaller the size of the crystallite.

It is established that by using different source materials and changing process parameters, the crystallite size can be modified. For example, Lee et al. [8], synthesized one of the most common WBS, zinc oxide, through a precipitation method using $Zn(CH_3CO_2)_2 \cdot 2H_2O$ as the source material and found that the crystallite size of the resultant product was nearly 20 nm, whereas Sharma et al. prepared ZnO using anhydrous $ZnCl_2$ with a crystallite size of 8 nm [22]. Similarly, Anandhi et al. using $Zn(CH_3CO_2)_2$ tuned the crystallite size of ZnO from 25 to 46 nm by changing the concentration of the precipitating agent [41]. The size of the crystallite, surface area, growth rate, and surface morphology of the synthesized ZnO can be tuned by fixing process parameters like pH, process temperature, and proportion of the dopants added. The lattice constants a, b, and c, the dislocation density, and the number of crystallites per unit area can also be tuned by suitably varying the crucial process parameters.

Some of the reported properties of ZnO tuned by doping with different dopants for enhancing photocatalytic, antibacterial, opto-electronics, and gas sensing applications are presented in Tables 4.2 and 4.3. Even the lattice constants of the hexagonal structure of ZnO can be modified by adding certain dopants like Al and F to the ZnO lattice. As seen in Table 4.5, when Al is doped with ZnO, the lattice constant a changes remarkably from 3.239 to 3.408. Similarly, when fluorine is doped with ZnO the lattice constant c decreases from 5.204 to 5.193. Changes are also observed in other WBS like ZnO and CdS for different dopants (Table 4.5).

Vacuum annealing improves crystal quality of the transparent conducting oxides of doped zinc oxide films. Such annealing-induced modification makes this material more suitable for optoelectronic devices.

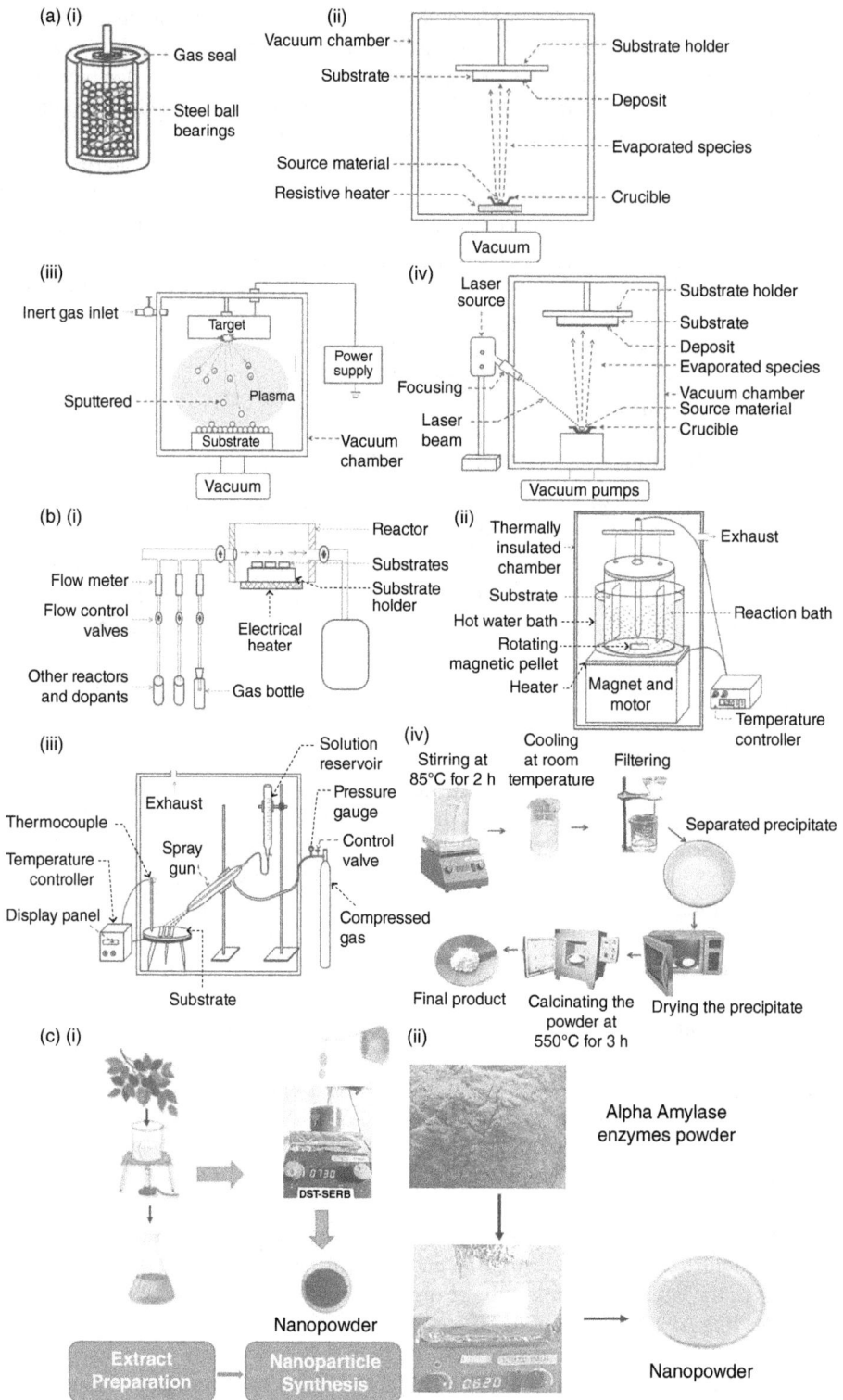

FIGURE 4.4 (a) Schematic diagram of certain physical methods: (i) ball milling, (ii) vacuum evaporation, (iii) DC sputtering, and (iv) laser ablation method. (b) Schematic diagram of certain chemical methods: (i) chemical vapor deposition, (ii) chemical bath deposition, (iii) spray pyrolysis, (iv) SILAR, and (v) coprecipitation/soft-chemical method. (c) Schematic diagram of certain biological methods: (i) green synthesis and (ii) biosynthesis.

TABLE 4.5
Tunable Properties of Zinc Oxide Nanomaterial

S. No.	Material	Crystallite Size (nm)	Lattice Constants (a & c)	Ref.
1.	ZnO	39.970	3.239 & 5.191	[42]
	ZnO:Al	35.217	3.408 & 5.189	
2.	S:Cd (1:1)	11.900	5.787	[40]
	S:Cd (7:1)	33.000	5.821	
3.	ZnO	30.000	3.247 & 5.205	[43]
	ZnO:F	35.000	3.244 & 5.193	

The incorporation of dopants and addition of composite partners generally cause changes in the intensity of the diffraction peaks and/or shift in the peak positions. The former suggests changes in the crystalline quality, and the latter indicates changes in the respective inter-planar distances of ZnO lattice. For instance, the addition of graphene oxide (GO) and Ag leads to a reduction in the peak intensities, indicating a decrease in the crystalline quality.

In general, ZnO thin films have hexagonal grains on the surface. Sometimes, spherical, linear, and polygon like grains can also be observed when the process parameters and material compositions are changed. It is well-known that smaller grains are desirable for better photocatalytic and sensor applications as they lead to an enhanced reactive surface area.

4.6.1.2 Surface Morphological Properties

Surface morphology with porous structure is also a desirable factor for better gas sensing and photocatalytic abilities of zinc oxide thin films, as they can improve the adsorption of dye molecules/gas molecules during dye degradation/gas sensing processes. When recording the surface morphology of prepared nanomaterials, many different grain shapes have been observed, of which some images are shown in Figure 4.5.

FIGURE 4.5 Tunable morphologies of the synthesized WBS nanomaterials by different methods.

Doping/co-doping of certain materials, vacuum annealing, variation in solvent volume, and aging treatment strongly influence the surface morphology (grain size, shape, and surface roughness) of WBS materials.

Surface morphology of the fluorine doped ZnO (ZnO:F) thin films deposited using a simplified spray pyrolysis is strongly influenced by aging time. Fresh starting solution results in non-uniform spherical grains with widely varied diameters, whereas four-days-aged solution leads to uniformly sized spherical grains. Decreased grain size was observed for four-days-aged solution due to F incorporation caused by aging [41]. When the aging period is increased further, cylindrical grains replace the spherical ones, and at 16 days of aging, cylindrical grains are formed. Similarly, the variation in solvent volume also affects the morphology of ZnO:F films. The grain size increases progressively as the solvent volume is increased, suggesting a progressive increase in the crystallinity of the film. The gradual decrease in the F incorporation into the ZnO lattice with the increase in solvent volume may be the reason for this increase in grain size. Generally, increase in F incorporation in the ZnO:F lattice decreases the crystallinity of the material as observed by Tsin et al. [12]. Large grain size is desirable for optoelectronic applications, as larger grains support the release of stresses. The post-deposition annealing process plays an important role in the surface morphology of ZnO-based films. The increase in annealing temperature causes an increase in the grain size, which can be correlated with the Ostwald ripening effect [44].

The porous nature of the surface of WBS materials in thin film form is suitable for gas-sensing applications. For instance, a possible mechanism for the enhanced ammonia gas-sensing ability of chromium-doped zinc oxide thin films prepared using spray pyrolysis method can be explained as follows: The oxygen vacancies present in the sample serve as the adsorption sites, which leads to a stimulated adsorption of oxygen molecules on the surface. Hence, when the sample is exposed to ammonia gas, more ammonia molecules will react with the largely available oxygen molecules on the surface, resulting in an enhanced sensitivity. The porous nature of the sample provides more active sites for the interaction of the gas molecules and thereby the inner surface also contributes by sensing the ammonia gas along with the top surface. The results reveal that even a small amount of leakage of ammonia gas (10 ppm) can effectively be sensed by this ZnO:Cr sensor under light and dark conditions [45].

4.6.1.3 Transparent Conducting Properties

Wide band gap semiconductors like SnO_2 and ZnO are transparent conducting oxide materials. For transparent conducting oxides in thin film form, high visible transmittance and low sheet resistance (sheet resistance is the resistance offered by a square of any size on the film's surface) are the parameters required for getting a good quality factor. The quality factor is one of the most crucial factors needed for solar cell window layer applications. The Hackee's quality factor depends on both optical transmittance (T) and sheet resistance R_{sh}. The optical transmittance of nebulizer-sprayed ZnO film is in the range of 70–96%. The transmittance of a thin film mainly depends on its texture. Low scattering owing to structural homogeneity and high crystalline quality of the films are attributed to high transmittance, whereas the metal richness leads to an adverse effect on transmittance. The addition of graphene oxide (GO) and silver (Ag) causes a decrease in transmittance of ZnO film [46]. The study on Mo+F-doped ZnO film demonstrated that fluorine doping increases the transmittance of ZnO films slightly with increasing doping level because of the proper substitution of F^- ions into the O^{2-} sites that suppresses the metal richness of the system. However, at the high doping concentrations (say 20 at.%), the visible transmittance decreases due to scattering of photons that present at large numbers. A notable decrease in the IR transmittance upon fluorine doping is strong evidence for the scattering by charge carriers. The transmittance increases with the increase in the substrate temperature reported by Subha et al. [14]. The absorption edge of undoped ZnO is generally sharp, and this sharpness is affected by the influence of certain dopants and composite partners, as the addition of these materials introduce defects in the system.

The bandgap of nebulizer-sprayed ZnO found via Tauc's plot technique is in the range of 3.12–3.39 eV is reported by Christuraj et al. [37]. When appropriate dopants are added, the band gap reduces substantially. As per our study, the band gap is reduced from 3.2 to 2.6 eV when Ag and g-C_3N_4 are added with ZnO.

As per the observations, the electrical resistivity of ZnO film decreases upon the addition of suitable anionic/cationic dopants like F, Mo, and Ta [46]. The addition of higher ionic state cations and suitable anions can increase the carrier concentration of the films and thereby results in reduced sheet resistance: The Mo + F doping increases the quality factor of ZnO film by nearly two orders of magnitude. The synergistic effects of the dopants result in a low resistivity of 5.12×10^{-3} Ω cm with good mobility (2.00 cm^2/Vs) and carrier concentration (24.5×10^{20} cm^{-3}) at the Mo and F doping concentrations of 2 and 15 at.%, respectively. The results on the influence of air/vacuum annealing on the properties of Mo+F co-doped ZnO thin films showed that the resistivity decreases upon vacuum annealing, whereas it increases upon air annealing. The vacuum annealing treatment favorably influences the optical transmittance, too, i.e., the vacuum annealing yields high visible transmittance and thereby has high quality factor.

4.6.2　Tin Oxide (SnO_2)

Tin oxide is another wide band gap (3.8 eV) n-type semiconducting material that attracts the attention of materials scientists owing to its unique combined properties, *viz.* excellent optical access, good electrical conductivity, high infrared reflectance, good thermal and chemical stability, non-toxicity, and abundance in nature. These characteristic features along with its tunable properties make SnO_2 a fitting material for display devices, defrosting/defogging windows, solar cell window layers, and certain other optoelectronic devices. The photocatalytic ability of SnO_2 makes it a good choice for photocatalytic water purification processes. It can also act as an antibacterial and antifungal agent. Recently, p-type tin oxide films have been prepared by adding suitable acceptor dopants, which paves the way for the development of an interesting new field called transparent electronics. Tin oxide can be used in photoconductors, photo switches, ultraviolet detectors, electromagnetic shields, functional glasses, organic light emitting devices, varistors, flat panel displays, touch panels, heat mirrors, non-volatile memory devices, and electrochemical lithium-ion storage, to name a few.

4.6.3　Cadmium Sulfide (CdS)

Cadmium sulfide is a promising candidate of wide band gap semiconductor for solar cells with high efficiency heterojunction thin films. It is a semiconductor (n-type) having band gap of 2.42 eV with excellent optical properties. It is one of the promising candidates for various applications. Cadmium sulfide is a good sensitizer. It is considered for reliable, low-cost, cost-effective photovoltaics. Although it is a toxic material, its modification in properties attracts the attention of researchers.

Stoichiometry is one of the crucial parameters of CdS thin films required for many applications like CdS/CdTe thin film solar cells. Sulfur deficiency is a common defect that leads to poor stoichiometry in CdS films deposited by chemical methods. The simple modification made in the SILAR deposition procedure can rectify this sulfur deficiency defect (stoichiometric deviation) [40].

Studies on wide band gap semiconductors revealed that by adopting different preparation methods, adding suitable dopants, and appropriately changing process parameters, the properties of WBS materials can be tuned for environmental applications like photocatalytic degradation of toxic organic dye molecules, prevention of bacterial growth, gas sensing, super capacitors, bioimaging, fuel cells, biotags, cell labeling, tissue engineering, low-cost rechargeable batteries, GMRs, and cost-effective solar cells [14–17]. Other materials of this class, which include gallium phosphide (p-GaP), gallium nitride (GaN), boron nitride (BN), zinc selenide (ZnSe), silicon carbide (SiC), zinc telluride (ZnTe), and zinc sulfide (ZnS), also have tunable properties suitable for various applications [14–20]. All of these materials are promising for various environmental, energy, and opto-electronic applications and can be prepared by cost-effective methods. For instance, gallium phosphide can be used as a photocathode material for photocatalytic hydrogen evolution and photo voltage generation. Boron nitride nanotube is a potential material for wastewater treatment and antibacterial applications. Gallium nitride is a good gas sensor for sensing gases like hydrogen (H_2), methane (CH_4), benzene, nitric oxide (NO), sulfur-dioxide (SO_2), ammonia (NH_3), and carbon dioxide (CO_2). It is a promising material for cost-effective lighting, too. Zinc sulfide is a potential photocatalyst for water remediation. Zinc selenide and zinc telluride can be used for saving energy, solar cells, sensors, and photocatalytic applications.

4.7   APPLICATION OF WBS MATERIALS

4.7.1   Photocatalytic Dye Degradation Ability of WBS Materials

Photocatalytic efficiency of WBS materials in thin film form is independent of film thickness, whereas the intensity of irradiated light and pH of the test dye solution play considerable roles in enhancing the dye degradation ability of the photocatalytic material. For wide band semiconductor material, knowing the point of zero charge (PZC) and the pH values corresponding to the maximum surface negative/positive charge will be useful for achieving the maximum dye (cationic/anionic) degradation efficiency for a photocatalytic material. Addition of g-C_3N_4, Ag, Co, La, Mo, Cu, Mn, Ta, W, F, and N enhances the photocatalytic dye degradation ability of ZnO thin film deposited on glass/stainless steel substrates.

4.7.2   Gas Sensing Ability

Addition of Cr, La, Tb, Y, and Sn improves the ammonia gas sensitivity of ZnO thin films.

4.7.3   Antibacterial Ability of Wide Band Gap Semiconductors

Addition of certain carbonaceous materials and metals, *viz.* Cu+Graphene, Ag+ Graphene, Sr+Graphene, Co+rGO, Mg+rGO, and g-C_3N_4, improves the photocatalytic dye degradation and antibacterial abilities of ZnO nanopowder/nancomposites. The efficient separation of photoinduced charge carriers and the consequent delay in their recombination caused by the graphene layers may be the main reasons for the improved photocatalytic and antibacterial efficacy of the photocatalytic material. The addition of Mn, Mn+Ni, Mn+Co, Mn+F, and Mg+Co enhances the magnetic and antibacterial properties of ZnO nanopowder. When fluorine is added as a dopant or co-dopant along with certain metals, *viz.* Co, Fe, Ag and Mg, the antibacterial efficiency of ZnO increases. The ZnO nanomaterial activated with enzymes like amylase, phosphatase, urease, and protease through the addition of vermiwash exhibits enhanced photocatalytic activity.

REFERENCES

1. Fatemi, A., Tohidi, T., Jamshidi-Galeh, K., Rasouli, M., & Ostrikov, K. (2022). Optical and structural properties of Sn and Ag-doped PbS/PVA nanocomposites synthesized by chemical bath deposition. *Scientific Reports*, *12*(1), 12893.
2. Jang, S., Jang, J.S., Karade, V., Jo, E., Kim, J., Suryawanshi, M.P., He, M., Park, J., & Kim, J.H. (2021). Evolution of structural and optoelectronic properties in fluorine-aluminum co-doped zinc oxide (FAZO) thin films and their application in CZTSSe thin-film solar cells. *Solar Energy Materials and Solar Cells*, *232*, 111342.
3. Palanchoke, U., Kurz, H., Noriega, R., Arabi, S., Jovanov, V., Magnus, P., Aftab, H., Salleo, A., Stiebig, H., & Knipp, D. (2014). Tuning the plasmonic absorption of metal reflectors by zinc oxide nanoparticles: Application in thin film solar cells. *Nano Energy*, *6*, 167–172.

4. Qumar, U., Hassan, J.Z., Bhatt, R.A., Raza, A., Nazir, G., Nabgari, W., & Ikram, M. (2022). Photocatalysis vs adsorption by metal oxide nanoparticles. *Journal of Materials Science & Technology*, *131*, 122–166.

5. Berrabah, S.E., Benchettara, A., Smaili, F., Benchettara, A., & Mahieddine, A. (2023). High performance hybrid supercapacitor based on electrochemical deposed of nickel hydroxide on zinc oxide supported by graphite electrode. *Journal of Alloys and Compounds*, *942*, 169112.

6. Hu, H., Xiong, Z., Khang, C., Cui, Y., & Chen, L. (2023). Hydroxyl-functionalized ZnO monolayers for optoelectronic devices: Atomic structures and electronic properties. *Vacuum*, *208*, 111721.

7. Tseng, S.F., Chen, P.S., Hsu, S.H., Hsiao, W.T., & Peng, W.J. (2023). Investigation of fiber laser-induced porous graphene electrodes in controlled atmospheres for ZnO nanorod-based NO_2 gas sensors. *Applied Surface Science*, *620*, 156847.

8. Lee, P.-M., Wu, Y.-J., Hsieh, C.-Y., Liao, C.-H., Liu, Y.-S., & Liu, C.-Y. (2015). Observation of Ni3+ acceptor in P-type Ni(P):SnO2. *Applied Surface Science*, *337*, 33–37. https://doi.org/10.1016/j.apsusc.2015.02.060

9. Gunshor, R.L., & Nurmikko, A.V. (1996). Wide band gap semiconductors and their application to light emitting devices. *Current Opinion in Solid State and Materials Science*, *1*(1), 4–10.

10. Puri, N., & Gupta, A. (2023). Water remediation using titanium and zinc oxide nanomaterials through disinfection and photocatalysis process: A review. *Environmental Research*, *227*, 115786.

11. Ishwarya, R., Tamilmani, G., Jayakumar, R., Albeshr, M.F., Mahboob, S., Shshid, D., Riaz, M.N., & Govindarajan, M., Vaseeharan. (2023). Synthesis of zinc oxide nanoparticles using *Vigna munga* seed husk extract: An enhanced antibacterial, anticancer activity and eco-friendly bio-toxicity assessment on algae and zooplankton. *Journal of Drug Delivery Science and Technology*, *79*, 104002.

12. Tsin, F., Venerosy, A., Vidal, J., Collin, S., Clatot, J., Lombez, L., Paire, M., Borensztajn, S., Broussillou, C., Grand, P.P., Jaime, S., Lincot, D., & Rousset, J. (2015). Electrodeposition of ZnO window layer for chalcogenide solar cell. *Scientific Reports*, *5*(1), 8961. https://doi.org/10.1038/srep08961

13. Meindrawan, B., Suyatma, N.E., Wardana, A.A., & Pamela, V.Y. (2018). Nanocomposite coating based on carrageenan and ZnO nanoparticles to maintain the storage quality of mango. *Food Packaging and Shelf Life*, *18*, 140–146. https://doi.org/10.1016/j.fpsl.2018.10.006

14. Debanath, M.K., & Karmakar, S. (2013). Study of blueshift of optical band gap in zinc oxide (ZnO) nanoparticles prepared by low-temperature wet chemical method. *Materials Letters*, *111*, 116–119.

15. Rana, M.P.S., Singh, F., Negi, S., Gautam, S.K., Singh, R.G., & Ramola, R.C. (2016). Band gap engineering and low temperature transport phenomenon in highly conducting antimony doped tin oxide thin films. *Ceramics International*, *42*(5), 5932–5941.

16. Liu, J., Cao, M., Feng, Z., Ni, X., Zhang, J., Qu, J., Zhang, S., Guo, H., Yuan, N., & Ding, J. (2022). Thermal evaporation-deposited hexagonal CdS buffer layer with improved quality, enlarged band gap, and reduced band gap offset to boost performance of $Sb_2(S,Se)_3$ solar cells. *Journal of Alloys and Compounds*, *920*, 165885.

17. Shah, M.S.U., Zue, X., Shah, M.Z.U., Hou, H., Shah, M.K., Ahmad, I., Sajad, M., & Shah, A. (2023). Nickel selenide nano-cubes anchored on cadmium selenide nanoparticles: First-ever designed as electrode material for advanced hybrid energy storage applications. *Journal of Energy Storage*, *63*(2023), 107065.

18. Mousavi, S.H., Muller, T.S., & Oliveria, P.W.D. (2012). Synthesis of colloidal nanoscaled copper-indium-gallium-selenide (CIGS) particles for photovoltaic applications. *Journal of Colloid and Interface Science*, *182*(1), 48–52.

19. Aakib, H.E., Rochdi, N., Tchenka, A., & Pierson, J.F.., Outzourhit. (2023). Copper oxide coatings deposited by reactive radio-frequency sputtering for solar absorber applications. *Materials Chemistry and Physics*, *296*, 121796.

20. Hu, C., Chen, J., Wang, Y., H, Y., & Wang, S. (2022). A telluride-doped porous carbon as highly efficient bifunctional catalyst for rechargeable Zn-air batteries. *Electrochimica Acta*, *404*, 139606.

21. Onochie, U.P., Ikpeseni, S.C., Owamah, H.J., Igweoko, A.E., Ukala, D.C., Nwigwe, H.J., & Augustine, C. (2021). Analysis of optical band gap and Urbach tail of zinc sulphide coated with aqueous and organic dye extracts prepared by chemical Bath deposition technique. *Optical Materials*, *114*, 110970.

22. Sharma, V., & Mehata, M.S. (2022). A parallel investigation of un-doped and manganese ion-doped zinc selenide quantum dots at cryogenic temperature and application as an optical temperature sensor. *Material Chemistry and Physics*, *276*, 125349.

23. Benalia, S., Merabet, M., Rached, D., Douri, Y.A., Abidri, B., Khenata, R., & Labair, M. (2015). Band gap behavior of scandium aluminum phosphide and scandium gallium phosphide ternary alloys and superlattices. *Materials Science in Semiconductor Processing*, *31*, 493–500.

24. Amir, H.A.A.A., Fakhri, M.A., Alwahib, A.A., Salim, E.T., Alsultany, F.H., & Hashim, U. (2022). Synthesis of gallium nitride nanostructure using pulsed laser ablation in liquid for photoelectric detector. *Materials Science in Semiconductor Processing, 150*, 106911.

25. Zhang, Z., Fan, X., Zhu, J., Yuan, K., Zhou, J., & Tang, D. (2023). Pressure-driven anomalous thermal transport behaviors in gallium arsenide. *Journal of Materials Science & Technology, 142*, 89–97.

26. Kosar, N., Asgar, M., Ayub, K., & Mahmood, T. (2019). Halides encapsulation in aluminum/boron phosphide nanoclusters: An effective strategy for high cells voltages in Na-ion battery. *Material Science in Semiconductor Processing, 97*, 71–79.

27. Legesse, M., Rashkeev, S.N., Saidoui, H., Mellouhi, F.E., Ahzi, S., & Alharbi, F.H. (2020). Band gap tuning in aluminum doped two-dimensional hexagonal boron nitride. *Material Chemistry and Physics, 250*, 123176.

28. Liu, Z.Y., Yang, D.C., Eglitis, R.I., Jia, R., & Zhang, H.X. (2022). Penta-silicon carbide: A theoretical investigation. *Materials Science And Engineering: B, 281*, 115740.

29. Deng, Z., Zhu, R., Ma, L., Zhou, K., Yu, Z., & Wei, Q. (2022). Diamond for antifouling applications: A review. *Carbon, 196*, 923–939.

30. K K, S., P M, P.N., & Vasundhara, M. (2023). Enhanced photocatalytic activity in ZnO nanoparticles developed using novel *Lepidagathis ananthapuramensis* leaf extract. *RSC Advances, 13*(3), 1497–1515. https://doi.org/10.1039/D2RA06967A

31. Yusoff, N., Ho, L.-N., Ong, S.-A., Wong, Y.-S., & Khalik, W. (2016). Photocatalytic activity of zinc oxide (ZnO) synthesized through different methods. *Desalination and Water Treatment, 57*(27), 12496–12507. https://doi.org/10.1080/19443994.2015.1054312

32. Hastir, A., Kohli, N., & Singh, R.C. (2016). Temperature dependent selective and sensitive terbium doped ZnO nanostructures. *Sensors and Actuators B: Chemical, 231*, 110–119. https://doi.org/10.1016/j.snb.2016.03.001

33. Mani, G.K., & Rayappan, J.B.B. (2015). A highly selective and wide range ammonia sensor—Nanostructured ZnO:Co thin film. *Materials Science and Engineering: B, 191*, 41–50. https://doi.org/10.1016/j.mseb.2014.10.007

34. Zhao, X., Shen, H., Zhou, C., Lin, S., Li, X., Zhao, X., Deng, X., Li, J., & Lin, H. (2016). Preparation of aluminum doped zinc oxide films with low resistivity and outstanding transparency by a sol–gel method for potential applications in perovskite solar cell. *Thin Solid Films, 605*, 208–214. https://doi.org/10.1016/j.tsf.2015.11.001

35. Mudusu, D., Nandanapalli, K.R., Dugasani, S.R., Park, S.H., & Tu, C.W. (2016). Zinc oxide nanorods shielded with an ultrathin nickel layer: Tailoring of physical properties. *Scientific Reports, 6*(1), 28561. https://doi.org/10.1038/srep28561

36. Vasanthi, M., Ravichandran, K., Jabena Begum, N., Muruganantham, G., Snega, S., Panneerselvam, A., & Kavitha, P. (2013). Influence of Sn doping level on antibacterial activity and certain physical properties of ZnO films deposited using a simplified spray pyrolysis technique. *Superlattices and Microstructures, 55*, 180–190. https://doi.org/10.1016/j.spmi.2012.12.011

37. Christuraj, P., Dinesh Raja, M., Pari, S., Madankumar, D., & Shankar, V.U. (2022). Synthesis and characterization of Zn doped CdS nanoparticles as electrode material for supercapacitor application. *Materials Today: Proceedings, 50*, 2691–2694. https://doi.org/10.1016/j.matpr.2020.08.220

38. Alkuam, E., Mohammed, M., & Chen, T.-P. (2017). Fabrication of CdS nanorods and nanoparticles with PANI for (DSSCs) dye-sensitized solar cells. *Solar Energy, 150*, 317–324. https://doi.org/10.1016/j.solener.2017.04.056

39. Bhadwal, A.S., Tripathi, R.M., Gupta, R.K., Kumar, N., Singh, R.P., & Shrivastav, A. (2014). Biogenic synthesis and photocatalytic activity of CdS nanoparticles. *RSC Advances, 4*(19), 9484. https://doi.org/10.1039/c3ra46221h

40. Senthamilselvi, V., Saravanakumar, K., Jabena Begum, N., Anandhi, R., Ravichandran, A.T., Sakthivel, B., & Ravichandran, K. (2012). Photovoltaic properties of nanocrystalline CdS films deposited by SILAR and CBD techniques—a comparative study. *Journal of Materials Science: Materials in Electronics, 23*(1), 302–308. https://doi.org/10.1007/s10854-011-0409-7

41. Anandhi, R., Mohan, R., Swaminathan, K., & Ravichandran, K. (2012). Influence of aging time of the starting solution on the physical properties of fluorine doped zinc oxide films deposited by a simplified spray pyrolysis technique. *Superlattices and Microstructures, 51*(5), 680–689. https://doi.org/10.1016/j.spmi.2012.02.006

42. Saravanakumar, K., & Ravichandran, K. (2012). Synthesis of heavily doped nanocrystalline ZnO:Al powders using a simple soft chemical method. *Journal of Materials Science: Materials in Electronics, 23*(8), 1462–1469. https://doi.org/10.1007/s10854-011-0612-6

43. Ravichandran, K., Snega, S., Jabena Begum, N., Swaminathan, K., Sakthivel, B., Rene Christena, L., Chandramohan, G., & Ochiai, S. (2014). Enhancement in the antibacterial efficiency of ZnO nanopowders by tuning the shape of the nanograins through fluorine doping. *Superlattices and Microstructures*, *69*, 17–28. https://doi.org/10.1016/j.spmi.2014.01.020

44. Ravichandran, K., Vasanthi, M., Thirumurugan, K., Sakthivel, B., & Karthika, K. (2014). Annealing induced reorientation of crystallites in Sn doped ZnO films. *Optical Materials*, *37*, 59–64. https://doi.org/10.1016/j.optmat.2014.04.045

45. Manivasaham, A., Ravichandran, K., & Subha, K. (2017). Light intensity effects on the sensitivity of ZnO:Cr gas sensor. *Surface Engineering*, *33*(11), 866–876. https://doi.org/10.1080/02670844.2017.1331724

46. Ravichandran, K., Uma, R., Sriram, S., & Balamurgan, D. (2017). Fabrication of ZnO:Ag/GO composite thin films for enhanced photocatalytic activity. *Ceramics International*, *43*(13), 10041–10051. https://doi.org/10.1016/j.ceramint.2017.05.020

5 Doping Methods and Their Effects

Devi Bala Saraswathi Sethuraman and Chia-Jyi Liu

5.1 INTRODUCTION

One of the primary drivers behind the success of the microelectronics industry lies in its capacity to modify the physical and electrical properties of semiconductor materials through a process known as doping, involving the introduction of substitutional impurities into the crystal lattice. This capability has enabled the continuous pursuit of Moore's law [1], which has, in turn, led to the relentless miniaturization of semiconductor-based microelectronic components. This trend has resulted in devices with improved performance, reduced power consumption, and more cost-effective production.

In 2012, the traditional planar design of metal oxide semiconductor field effect transistor (MOSFET) transistors encountered an obstacle due to short channel effects, posing a fundamental limitation to further downsizing. Consequently, the semiconductor industry made a pivotal shift from planar technology to three-dimensional (3D) solutions, such as FinFET transistors. These FinFET transistors offered superior control over current flow in the conduction channel, increased speed, and reduced parasitic capacitance. By 2020, the semiconductor industry had commenced mass production of fin field effect transistor (FinFET) devices at the 5 nm node [2].

Today, research is focused on investigating alternative stacked architectures such as gate-all-around field-effect transistors (GAAFETs). In this innovative design, the gate completely surrounds a silicon nanosheet used as the channel, resulting in an even more impressive performance. The adoption of this groundbreaking device structure is expected to begin at the 2 nm technology node in the coming period.

The doping of these 3D devices has given rise to new technological and fundamental challenges. In the mass production of highly miniaturized transistors, the conventional doping method involves ion implantation, which entails bombarding the material with high-energy dopant ions. However, this approach is rendered inefficient for nanoscale electronic devices due to the damage inflicted on the crystal lattice as ions traverse through it. Even employing high-temperature thermal treatments to restore the silicon lattice's structure does not fully address this issue. Furthermore, ion implantation is ill-suited for complex 3D geometries because the ion beam has a highly directional nature.

In contrast, diffusion from solid and gaseous sources offers a gentler doping solution but faces challenges in precisely controlling doping concentration and uniformity, which are crucial considerations for these extremely miniaturized devices. Additionally, alternative top-down doping techniques that have shown success in research contexts [3] are not practical for industrial applications.

To address this, Ho et al. introduced self-assembled monolayers (SAMs) in 2007, allowing non-destructive, uniform, and conformal doping via molecular diffusion method — a promising bottom-up approach ideal for complex 3D structures.

With the dominance of FinFETs and the shift toward gate all around (GAA) nanowire architecture, exploring new materials and processes is crucial, given the high surface-to-volume ratios in thin-body structures. Understanding doping methods is pivotal for assessing these structures' suitability for future applications. This chapter seeks to shed light on the intricate interplay between emerging materials, advanced processes, and doping strategies within the dynamic landscape of semiconductor technology, aiming to guide researchers and engineers toward optimized solutions for the technology challenges of tomorrow.

DOI: 10.1201/9781003450146-5

5.2 SEMICONDUCTOR MATERIALS

Solid-state materials can be categorized into three primary classes: insulators, semiconductors, and conductors. Figure 5.1 illustrates the range of electrical conductivity (σ), associated with key materials within each of these three categories [4]. Insulators, exemplified by materials like fused quartz and glass, exhibit extremely low conductivities, typically in the range of 10^{-18} to 10^{-8} S/cm. In contrast, conductors such as aluminum and silver showcase high conductivities, typically ranging from 10^4 to 10^6 S/cm [4].

Semiconductors, occupying an intermediate position between insulators and conductors, display conductivities that fall within this range. The conductivity of a semiconductor is notably responsive to various factors, including temperature, illumination, magnetic fields, and the presence of trace quantities of impurity atoms—typically, on the order of 1 microgram to 1 gram of impurity atoms per kilogram of semiconductor material. This sensitivity in conductivity renders semiconductors among the most crucial materials in the realm of electronic applications [4].

5.2.1 FUNDAMENTALS OF DOPING

Pure silicon or germanium are rarely used as semiconductors. In order to be practically useful, semiconductors need to undergo a deliberate introduction of impurities. This addition of impurities changes their conductivity characteristics, essentially converting them into semiconductors. This process of introducing impurities into an initially pure substance is called doping [5], and the added substance is a dopant. After the doping process, an initially pure material transforms into an impure one. It's only after doping that these materials become suitable for practical applications.

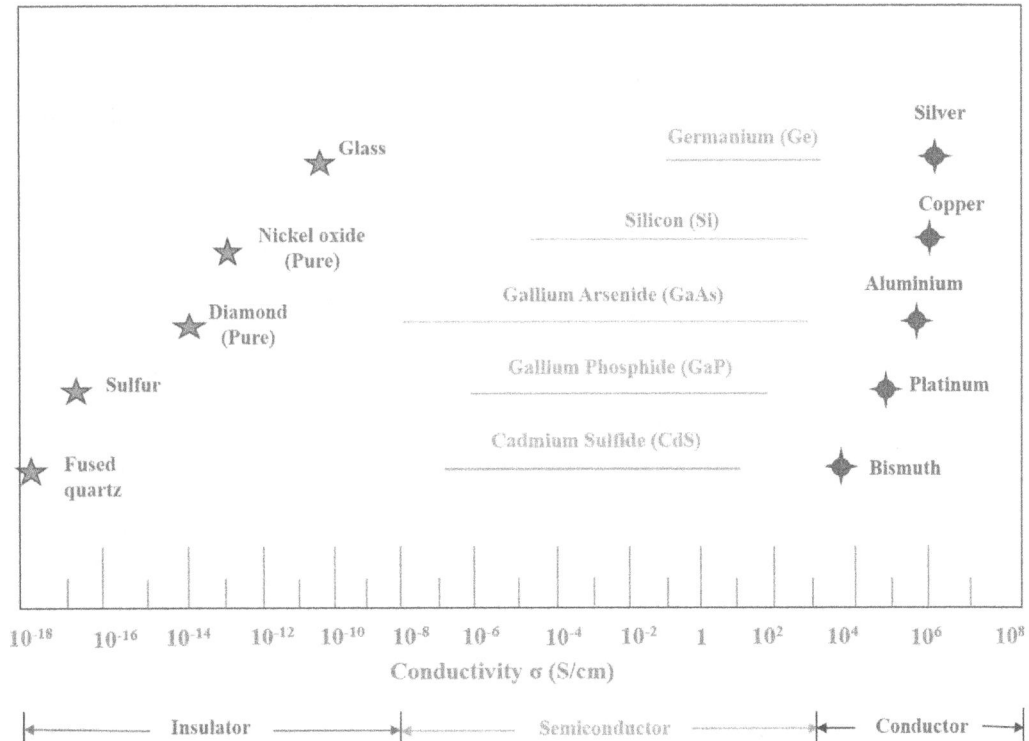

FIGURE 5.1 Typical conductivity ranges for insulators, semiconductors, and conductors [4].

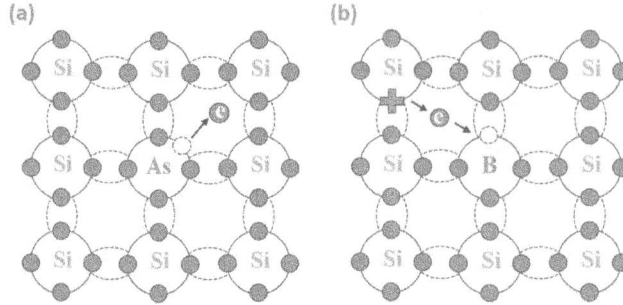

FIGURE 5.2 (a) N-type doping: The arsenic atom contributes its fifth valence electron, functioning as a mobile charge carrier; (b) P-type doping: An electron fills the vacant position on the boron atom, leading to the formation of a new hole.

5.2.2 N-Type Doping

Figure 5.2a depicts n-type doping in silicon using arsenic, where donor impurities possess higher valence than the host material's atoms. Upon ionization, a donor impurity contributes an electron to the conduction band, increasing mobile electron count and classifying the material as n-type. This process involves introducing Group 5 impurity atoms (e.g., phosphorus or arsenic) into the semiconductor with four valence electrons. Substituting a silicon atom with a five-valence electron impurity generates a surplus electron, elevating electrical conductivity by adding free electrons to the material's structure.

5.2.3 P-Type Doping

Figure 5.2b illustrates the p-type doping of silicon with boron. Acceptors are impurities that have a lower valence compared to the host material, resulting in incomplete atomic bonding within the lattice. Consequently, they capture electrons, essentially creating holes in the valence band. This process causes the acceptor centers to acquire a negative charge, and the semiconductor is designated as p-type. In other words, materials whose valence band minimum (VBM) is close to the vacuum level promote "p-type doping" [5]. This is achieved by introducing impurity atoms derived from Group 3 of the periodic table, such as boron or aluminum, into the semiconductor material, which inherently has four valence electrons. These impurity atoms have only three valence electrons, and when they replace a silicon atom within the crystal lattice, they create a vacancy where an electron is missing. These vacancies, known as holes, also have the capacity to influence the electrical conductivity of the material.

In a semiconductor's energy band diagram (Figure 5.3a), donor and acceptor levels lie within the forbidden energy gap [6]. Impurity energy levels within this gap are categorized as shallow or deep. Donors are referred to as shallow when their energy levels are in proximity to the bottom of the conduction band, while acceptors are considered shallow when their energy levels are close to the top of the valence band. Shallow impurities are those that usually require energies roughly equivalent to thermal energy for ionization to take place. Conversely, deep impurities demand higher energies for ionization, making them generally less likely to produce free carriers. These deep impurities, however, can serve as effective recombination centers where electrons and holes recombine and neutralize each other.

5.3 THE EFFECTS OF DOPING ON THE PROPERTIES OF SEMICONDUCTORS

5.3.1 Background

Doping has a significant impact on the Fermi energy (also known as the Fermi level) of semiconductors by introducing additional electronic states into the material's energy band structure. The Fermi energy represents the highest energy level within a material that electrons can occupy at absolute

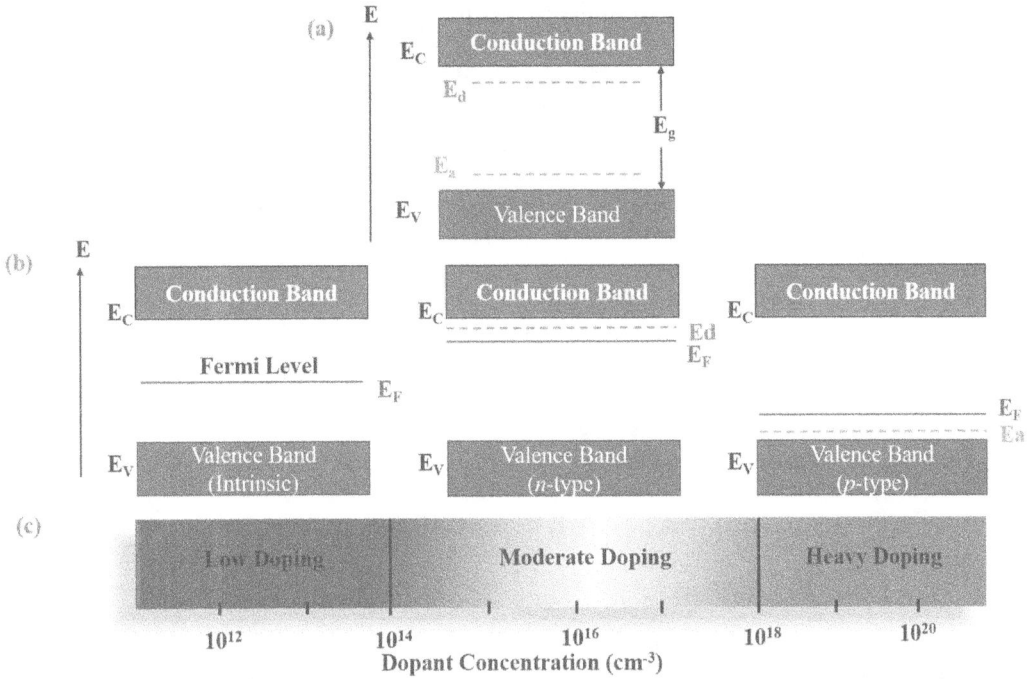

FIGURE 5.3 (a) Energy band diagram of semiconductors. (b) Alteration in the Fermi energy position and the incorporation of allowed energy levels within the bandgap resulting from the process of doping [6]. (c) The range of doping levels used in c-Si [6].

zero temperature and serves as a reference point for understanding the distribution of electrons and holes within the semiconductor. Here's how doping affects the Fermi energy of semiconductors.

5.3.1.1 Intrinsic Semiconductor (Pure Semiconductor)

In an intrinsic semiconductor, the Fermi energy lies within the energy band gap between the valence band and the conduction band [5]. At absolute zero temperature, all electronic states below the Fermi energy are occupied, while those above it remain empty. In this state, there is an equal number of electrons and holes, resulting in no net charge density.

5.3.1.2 Influence of N-Type Doping on the Fermi Energy Position

N-type doping involves adding impurity atoms, typically from Group 5 of the periodic table (e.g., phosphorus or arsenic), which have more valence electrons than the host semiconductor material (e.g., silicon or germanium). These impurity atoms introduce additional electrons into the crystal lattice, which become part of the conduction band due to their higher energy compared to the valence band electrons in the host material [5]. Consequently, the Fermi energy shifts closer to the conduction band because of the increased electron concentration in the conduction band resulting from the presence of dopant atoms.

5.3.1.3 Influence of P-Type Doping on the Fermi Energy Position

P-type doping involves adding impurity atoms, typically from Group 3 of the periodic table (e.g., boron or aluminum), which have fewer valence electrons than the host semiconductor material. These impurity atoms create holes in the crystal lattice, which act as positively charged mobile carriers in the valence band. The Fermi energy shifts closer to the valence band because of the increased hole concentration in the valence band due to the presence of dopant atoms [5].

Figure 5.3b depicts Fermi energy shifts and introduced energy levels in the band gap due to doping [6]. Donor and acceptor atoms inserted into silicon create permissible energy levels within the forbidden band gap. For instance, the weakly bound fifth valence electron of a phosphorus (P) atom easily transitions to the conduction band (E_C). These energy levels, denoted as E_d, relate to loosely bound valence electrons of donors, placed near E_C as dashed lines, indicating localized electrons around donor atoms. Similarly, acceptor atoms introduce allowable energy levels, E_a, near the valence band (E_V) [6].

Semiconductors modify electrical conductivity via doping, relying on charge carrier concentration and mobility. Figure 5.3c shows the typical doping range for crystalline silicon (c-Si) [6]. Semiconductor conductivity, whether p-type or n-type, relies on hole or electron predominance. The abundant carriers are majority carriers (holes in p-type, electrons in n-type), while less prevalent ones are minority carriers (electrons in p-type, holes in n-type) [6]. This carrier asymmetry defines a semiconductor's conductivity behavior.

5.4   ADVANCED DOPING TECHNIQUES

5.4.1   Ion Implantation Method

5.4.1.1   Fundamentals of Ion Implantation

Ion implantation was initially introduced in semiconductor technology more than three decades ago to enable precise incorporation of n- and p-type dopants at specific depths below the material's surface. Today, it has become a vital and irreplaceable step in integrated circuit production, serving as a key doping method where high-energy dopant-containing ions are directed at the semiconductor, replacing atoms in the lattice through high-temperature processing. Its advantage lies in precise control over dopant amount and depth distribution within the host semiconductor, ensuring consistent doping across vast semiconductor areas, making it a cornerstone in microelectronics.

However, ion implantation technology has several drawbacks. Firstly, it can cause damage to the crystal lattice due to ion bombardment. Additionally, defects may arise during subsequent thermal treatments, leading to transient enhanced diffusion. Moreover, many of the source gases commonly used in ion implantation are considered hazardous from both health and environmental perspectives. Finally, this technique is not suitable for 3D nanostructured materials since it cannot provide uniform dopant incorporation for non-planar nanostructures.

5.4.1.2   Plasma Immersion Ion Implantation – Advanced Technique

5.4.1.2.1   Introduction

Plasma immersion ion implantation (PIII), often referred to as plasma doping, is an innovative method involving the extraction of accelerated ions from plasma using high voltage. These ions are then implanted into a semiconductor substrate. This approach offers significant advantages over traditional ion implantation, typically resulting in more conformal doping profiles than conventional ion implantation techniques. Conrad et al. [7] and Tendys et al. [8] introduced the PIII technique in the late 1980s. Originally known as plasma source ion implantation (PSII), this method is now commonly referred to as PIII to distinguish it from traditional beamline implantation involving plasma ion sources.

5.4.1.2.2   Plasma-Surface Modification Mechanisms

Plasma discharges, often referred to as the fourth state of matter, exhibit exceptional physical characteristics. These properties are profoundly influenced by electromagnetic fields due to the quasi-neutral equilibrium maintained between electrons and ions and the dominance of electrostatic

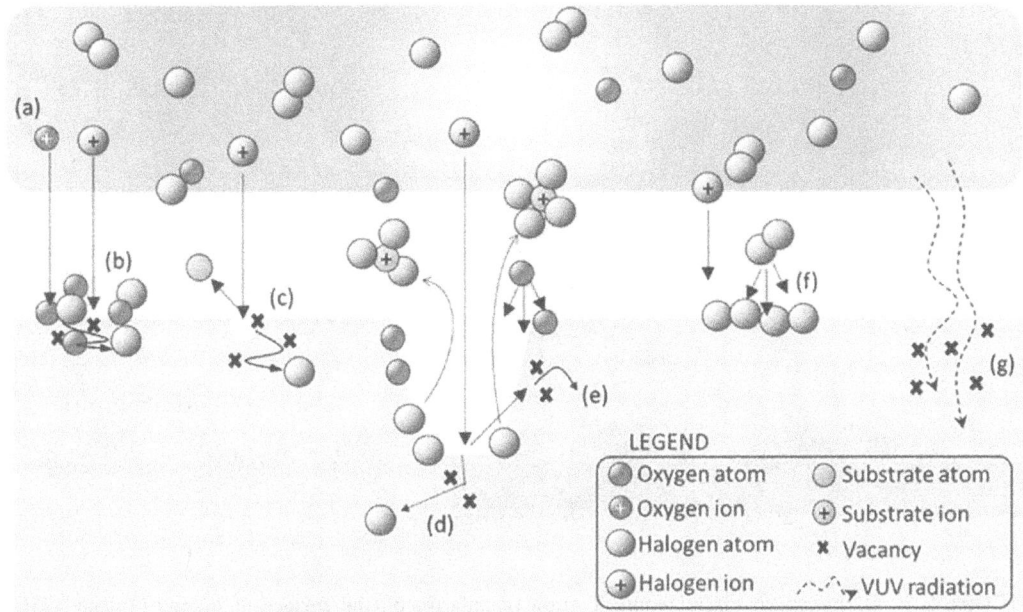

FIGURE 5.4 Diagram illustrating the mechanisms of damage and reactions induced by plasma: (a) Introduction of oxygen from reactor components; (b) surface oxidation through interactions with ion bombardment; (c) sputtering and creation of vacancies caused by ion bombardment; (d) lateral straggling at the etch front of features; (e) ion reflection along the sidewalls of features; (f) alteration of the surface layer due to low-energy ions; and (g) generation of vacancies caused by vacuum ultraviolet (VUV) emissions. (Adapted with permission from [10]. Copyright [2021] AIP Publishing.)

forces over typical gas dynamics. Low electron temperature (T_e) plasmas are extensively used in material processing, harnessing these phenomena to facilitate chemical reactions at gas-solid interfaces and induce directional etching processes [9]. However, the complex interplay of reactions within plasma discharges leads to various pathways for material structural deterioration, including the formation of defects, voids, grain enlargement, and other effects. Consequently, these transformations adversely affect the electrical properties of devices, resulting in increased leakage current and shifts in threshold voltage, among other issues. A schematic representation of such effects is presented in Figure 5.4 [10].

5.4.1.2.3 *Ion Energy Control*

One of the most foundational yet highly impactful observations in the realm of plasma discharges revolves around the manipulation of positively charged ions by applying a negative potential difference, achieved through either direct current (DC) or radio frequency (RF) generators. Groundbreaking research conducted by Coburn and Winters showcased the immense potential of these ions in driving chemical reactions, exemplified by the synergistic etching behavior of silicon (Si) in a system involving argon (Ar) and xenon difluoride (XeF_2) [9]. Consequently, the role of ion energy in various etching mechanisms, such as sputter yield and ion-enhanced chemical etching, has been extensively studied for key material systems. These investigations have significantly contributed to advancements in logic and memory technologies.

As the rapid downscaling of complementary metal-oxide-semiconductor (CMOS) technology progressed and intricate architectures like the FinFET emerged, addressing plasma-induced damage (PID) became increasingly crucial. In response, alternative solutions were introduced, including inductive coupling (ICP) and electron cyclotron resonance (ECR), along with the incorporation of multiple RF sources to power the plasma. These innovations led to the achievement of lower

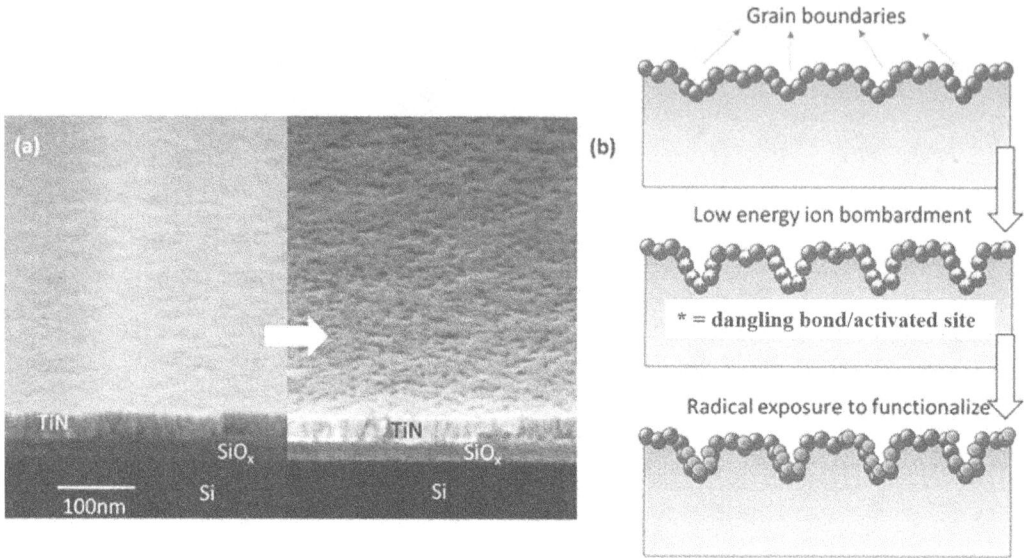

FIGURE 5.5 (a) Enhanced visualization of grain boundaries on the surface of titanium nitride (TiN) produced through physical vapor deposition (PVD) after undergoing modification using O_2/Ar plasma. (b) Schematic illustration depicting the creation of selectively functionalized growth templates for self-assembled monolayers at sub-lithographic length scales. (Adapted with permission from [10]. Copyright [2021] AIP Publishing.)

electron temperatures at high plasma densities, enabling a partial decoupling between ionization fraction and ion energy. This level of control was previously unattainable using reactive ion etching (RIE).

Precise control of ion energy distribution functions has gained even greater importance with the emergence of plasma-enhanced atomic layer etching (PEALE) processes. In PEALE, achieving self-limited behavior regarding the formation and removal of a modified surface is only possible within a narrow ion energy window.

Energetic species within plasma discharges have played pivotal roles extending beyond the realm of etching processes. They have been instrumental in enabling advancements in plasma-enhanced chemical vapor deposition (PECVD) and plasma-enhanced atomic layer deposition (PEALD) techniques. More recently, researchers have explored the application of bias power as a means to induce geometry-dependent phase variations in materials deposited through PEALD. For instance, Figure 5.5a presents results from Marchack et al.'s research group [10], where oxygen plasma was used to induce surface modifications in titanium nitride (TiN) films produced via physical vapor deposition (PVD). Scanning electron microscopy (SEM) images reveal that grain boundaries at the film's surface are accentuated without removing bulk material. This morphological alteration, in conjunction with films featuring precisely engineered grain sizes, holds the potential to facilitate the ordered growth of nanoparticles or the self-assembly of polymers with a level of control that was previously unimaginable, as illustrated in Figure 5.5b [10].

5.4.1.2.4 Applications

5.4.1.2.4.1 Material Functionalization through Plasma Doping   In addition to its role in material etching and deposition, plasmas have found applications in surface treatment and modification to enhance functionality. One notable process in this category is plasma doping (PD), where ionized species are implanted into the surface of a wafer through bias application. By carefully selecting process conditions and gas precursors, these impacting ions become part of the target substrate as either interstitial or substitutional dopants, thus altering the material's physical properties. The use of PD in the semiconductor industry dates back to the early 1990s, primarily driven by the need for ultra-shallow junctions in integrated circuits (ICs). Traditional ion implantation methods typically required high powers exceeding 10 kV, making them unsuitable for achieving ultra-shallow junctions. However, the flexibility provided by source and bias power generators, along with precise control over substrate temperature, allowed for the attainment of the required dopant depth profiles with minimal channel damage. Through extensive development and refinement, PD has been successfully applied to tight-pitch 3D FinFET structures, which come with stringent requirements for conformality and uniformity. This highlights the adaptability and effectiveness of PD as a versatile technique for tailoring semiconductor material properties to meet the evolving demands of the semiconductor industry.

Significant research efforts have been dedicated to III-V devices, primarily due to their remarkable carrier mobility. However, a key challenge has been reducing source/drain contact resistance in these materials. To address this challenge, researchers have explored Si, Sn, and Ge as n-type dopants. Conventional ion implantation methods have proven less effective for III-V materials like InGaAs due to the need for high-temperature dopant activation. Group V species tend to evaporate under these conditions, leading to stoichiometry loss and surface degradation. Additionally, the native oxide of InGaAs is resistant to typical etchants like HCl or HF solutions, which can hinder dopant incorporation.

PD performed within a manufacturing-compatible plasma reactor offers unique advantages in overcoming these challenges. These advantages include independent control of dose rate regardless of implant energy and the ability to switch between implantation, deposition, and etching regimes. The research group led by Marchack et al. [10] conducted plasma ion doping of Si and Sn into InGaAs using high-density plasma systems to investigate the effects of key PD process parameters, such as plasma source power, gas pressure, applied bias power, and processing time. Their study aimed to characterize the influence on dopant depth profiles, concentration, surface damage, and structural modifications. They confirmed that ion density and energy play a crucial role in determining the dopant depth profile. Of particular significance, applied bias power emerged as the most reliable and repeatable input parameter for defining the dopant depth profile and concentration. Moreover, it had the capability to modulate the transition between doping, deposition, and etching processes. Figure 5.6a illustrates the exposed In, Ga, and As bonds on the surface of InGaAs following the elimination of the native oxide layer through Ar/He ion bombardment [10], and Figure 5.6b shows the process flow demonstrating an integrated in situ approach involving plasma surface cleaning, plasma ion doping, and plasma deposition [10].

5.4.1.2.4.2 ALD/ALE: The Role of Plasma Processing in Atomic Layer Deposition Atomic layer deposition (ALD) has become an indispensable technique in the CMOS industry for depositing thin films precisely with exceptional control over their thickness [11]. ALD stands out due to the self-saturating surface reactivity of the precursor molecules it uses and its sequential, cyclical dosing of these molecules. These inherent qualities make ALD particularly valuable for achieving uniform and conformal coatings on 3D structures with high aspect ratios.

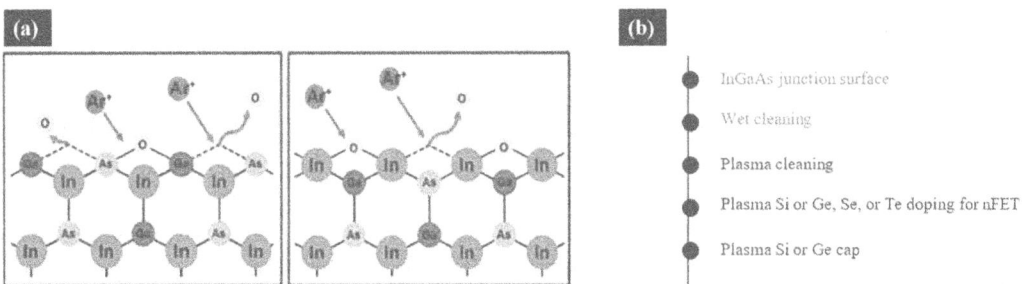

FIGURE 5.6 (a) Illustration showing exposed In, Ga, and As bonds on the InGaAs surface subsequent to the removal of the native oxide layer through Ar/He ion bombardment. (b) Process flow diagram outlining the integrated in situ approach involving plasma-based surface cleaning, ion doping, and capping. (Adapted with permission from [10]. Copyright [2021] AIP Publishing.)

Plasma plays a significant role in enhancing the microstructural characteristics of thin films within the generic ALD process. This improvement in microstructure subsequently enhances other physical properties, and this innovative technique is known as atomic layer annealing (ALA) [12]. In ALA, a brief exposure to a chemically inert plasma, typically argon, is performed at the end of each ALD cycle. During this exposure, kinetic energy from the plasma is transferred to the substrate surface, resulting in surface annealing. This annealing effect leads to the rearrangement of surface atoms and an improvement in the crystallographic ordering of the material. One practical example is the deposition of aluminum nitride (AlN) using ALA. Traditional PEALD of AlN involves cyclical exposures to trimethylaluminum (TMA) and nitrogen plasma at temperatures around 250°C. However, incorporating ALA at the end of each cycle significantly enhances the crystallinity of the AlN film. Initially developed by Shih et al., this technique has since been successfully applied by various research groups to other material systems, resulting in films with improved optical, electrical, and templating properties. For instance, ALA has been integrated into TiN PEALD processes, reducing the resistivity of the film by up to 50%.

While ALD is known for its conformal coating properties, there are situations where the opposite is desired—where ALD nucleation is intentionally inhibited on certain surfaces while promoted on others. This technique, known as area-selective deposition (ASD), allows for unique process flows and enables the fabrication of micro- and nanostructures that would otherwise be unattainable. The core concept of ASD is to achieve ALD coating exclusively on specific, predefined regions of the sample surface. This naturally requires the utilization of a material system with at least two chemically distinct surfaces. For instance, surfaces terminated with hydroxyl groups can readily facilitate chemical reactions with many ALD precursor molecules, enabling selective deposition on these hydroxyl-terminated regions.

On the opposite end of the atomic-scale patterning spectrum is atomic layer etching (ALE). Similar to ALD, ALE processes involve nominally self-limited half-reaction steps, and these reactions can be driven either thermally or through plasmas, known as PEALE. The capability of PEALE to create modified surface layers significantly thinner than those achieved with conventional continuous wave (CW) plasma processes is a crucial factor in enabling next-generation technologies. In recent demonstrations, Marchack et al. [13] showcased precise control over the oxygen concentration within an ultra-thin ($\sim$ 2 nm) surface layer in tantalum nitride (TaN) through exposure to hydrogen (H_2) plasma in a cyclic PEALE process using a standard inductively coupled plasma (ICP) reactor. This achievement has significant implications for achieving etch uniformity and reducing sputter redeposition, especially in densely packed structures. Further exploration of such techniques, particularly with low electron temperature (T_e) sources, could potentially refine this level of control down to true monolayer length scales.

Cyclic ALE processes also offer the advantage of controlling the etch selectivity of materials, thereby expanding the possibilities of integration schemes without the need for additional lithographic masks or novel gas chemistries. Despite its advantages, ALE is not a cure-all for the challenges that lie ahead. As future technologies target reduced dimensions and intricate architectures, the impact of surface layer composition becomes more pronounced, necessitating a higher degree of tunability in plasma processes. Furthermore, the complexity introduced by device constituents, mask materials, and chamber components can inadvertently lead to unwanted reactions and affect final etch profiles. While undesired reactions, such as oxidation, pose challenges to achieving ideal ALE behavior, they should not discourage us from pursuing this line of research. Even unintended consequences can be leveraged to develop novel fabrication schemes, such as self-aligned masking for lithographic subdivision.

5.4.2 DIFFUSION METHODS AND ADVANCEMENTS

Alternatively, impurity introduction can be achieved through a diffusion process, which involves the movement of atoms from regions with a high concentration to regions with a low concentration.

5.4.2.1   Gas Phase or Solid Phase Diffusion – Advanced Techniques

Other advanced methods based on gas-phase or solid-phase diffusion have also been explored. However, these approaches encounter fundamental challenges in managing dopant concentration near the surface and achieving uniformity. In gas-source doping, a gas is utilized to deliver the dopant to the substrate, providing equal access to all surfaces of the silicon device. On the other hand, solid-source doping involves applying a solution to silicon surfaces, followed by a thermal treatment to facilitate diffusion. The solution containing the dopant typically consists of a mixture of SiO_2 and dopant atoms or Si-based polymers with dopant atoms incorporated into the polymer chains, such as phosphosilicates or borosilicates [14]. Unfortunately, while this method is straightforward and non-destructive, it does not ensure precise dose control over large substrate areas. Moreover, residues from the dopant precursor solution are challenging to remove from the surface, leading to the formation of a chemically modified layer that is difficult to strip off.

To address these limitations, self-assembly techniques followed by molecular grafting have demonstrated their effectiveness in producing a consistent monolayer of dopant-containing molecules attached to pristine crystalline silicon substrates over large areas. Assuming there are enough reactive sites on the substrates, the actual density of grafted molecules is primarily determined by the molecular footprint. This self-limiting reaction has been utilized in conjunction with conventional spike annealing procedures to facilitate the creation of ultra-shallow junctions in silicon doped with boron, phosphorus, and arsenic, achieving depths below 5 nm [15].

5.4.2.2   Molecular Monolayer Doping (MLD) – Advanced Technique

Molecular monolayer doping (MLD) offers exceptional versatility by enabling precise adjustments of various parameters, including surface preparation, molecular footprint, capping layer, and thermal treatment, to optimize both the surface coverage of molecules and the diffusion of dopants into the semiconductor. However, the wet chemistry commonly employed in the MLD process presents significant challenges in terms of its suitability for manufacturing [16]. Furthermore, the dopant concentrations typically achieved through MLD are in the range of $\sim10^{20}$ cm^{-3}. To make this approach viable for industrial applications, it is essential to expand its capabilities to achieve doping concentrations ranging from 10^{15} to 10^{21} cm^{-3}. One modified approach proposed to tackle this challenge involves blending dopant-containing molecules with dopant-free molecules and subsequently grafting them onto the silicon surface. The composition of the final monolayer is assumed to be proportional to the fraction of dopant-containing molecules in the solution. While this method allows for some variation in doping dose over an order of magnitude, concerns have been raised about its reproducibility and accuracy.

5.5   OVERVIEW OF THE DIFFERENT DOPING PROCESSES – PROS AND CONS

Table 5.1 provides an overview of the advantages and disadvantages of different doping methods, including (1) ion implantation, (2) plasma-assisted doping, (3) gas-source doping, (4) solid-source doping, and (5) MLD [17]. In the following paragraphs, we will examine each row of Table 5.1 separately, evaluating each assessment criterion independently.

 i. *High dose incorporation*: Incorporating a substantial amount of dopants is crucial for various device applications, particularly for reducing access resistance. Ensuring the effective activation of these dopants is a distinct research area that deserves comprehensive examination. When it comes to achieving high-dose dopant incorporation, ion implantation emerges as the preferred choice. Modern implantation tools can handle extremely high implant doses, reaching levels of approximately 10^{15} to 10^{16} atoms/cm^2 [18]. This results in chemical concentrations exceeding 10^{20} atoms/cm^3 or even higher, meeting the requirements of most semiconductor device applications. Plasma-assisted doping processes are also well-suited for achieving high doses, as specific conditions within the process can

TABLE 5.1

An Overview of the Strengths and Weaknesses of Various Doping Processes

	Assessment criteria	Ion implant	Plasma-assisted	Gas-source	Solid-source	MLD
(i)	High dose incorporation	Very good	Very good	Good	Good	Needs development
(ii)	Bulk crystal damage (physical)	Needs thermal treatment to repair	Needs thermal treatment to repair	Unlikely	Unlikely	Unlikely
(iii)	Surface crystal damage (chemical)	Unlikely	Unlikely	Needs care and attention	Unlikely	Needs care and attention
(iv)	Cleaning recipe development required after deposition	No	Yes, if operating in deposition mode	No	Yes	No
(v)	Barrier at surface that inhibits incorporation	Can inject beyond surface with energy	Can inject beyond surface with energy	Limiting	Limiting	Limiting
(vi)	Industry standard	Most used	Not used as much as ion implant	Not used as much as ion implant	Not used as much as ion implant	Needs development
(vii)	Suitability for 3D structures (conformality)	Shadowing can be an issue	Better than Ion Implant	Can be conformal	Can be conformal	Can be conformal
(viii)	Suitability for TMD materials	Unlikely	Unlikely	Possible	Possible	Possible

Source: Adapted with permission from [17]. Copyright (2023), Margarita Georgieva, Nikolay Petkov, Ray Duffy, some rights reserved; exclusive license (Elsevier Ltd.).

accelerate charged ions toward the target. In the case of gas or solid sources, they can be considered suitable for achieving chemical concentrations in the range of 10^{19} to 10^{20} atoms/cm^3 [19]. However, techniques relying on in-diffusion often face challenges, as the surface can act as a trapping site for common impurities.

Currently, MLD is not considered a high-dose process because its dopant incorporation levels are relatively lower compared to competing processes. Nevertheless, it's important to note that MLD is still in a less mature stage compared to other techniques, and ongoing research holds the potential for breakthroughs. For instance, strategies such as achieving denser packing of molecules after surface functionalization and overcoming surface barriers could advance this chemistry-based process technology. Additionally, exploring molecules containing polymers may offer a path to improved MLD concentration levels, making it an exciting and rapidly evolving research area [20].

ii. *Bulk crystal damage (physical)*: When considering the issue of bulk crystal damage, it is generally recommended to minimize it whenever feasible. Such damage can lead to sub-optimal device performance, giving rise to unwanted problems such as leakage current or reduced device-to-device reliability due to extended defects and stacking faults. Given this concern, processes with a significant physical component, namely ion implantation and plasma-assisted methods, may require more extensive optimization or innovative process integration approaches to overcome this limitation. In contrast, gas-source, solid-source, and MLD techniques do not encounter this specific challenge.

iii. *Surface crystal damage (chemical)*: Surface damage to the target substrate or structure can also occur, primarily due to unintended or unexpected chemical reactions. Ion implantation and plasma-assisted processes propel ions past the surface, while solid sources often consist of non-corrosive silicon oxides. Gas sources can potentially disassociate into other species, such as atomic hydrogen, which might etch the surface under certain conditions, demanding careful attention. MLD, being a chemistry-based process, involves wet treatments for cleaning and functionalization that can create rough or textured surfaces if not properly optimized.

iv. *Cleaning recipe development required after deposition*: The need for a cleaning step arises from any material deposition that must be removed before further device fabrication can proceed. For example, solid sources in the form of oxides must be eliminated after dopant drive-in, necessitating an additional round of testing and optimization. This, in turn, adds both cost and complexity to the overall process flow. In the case of plasma-assisted doping, a thin film of material can often be deposited, depending on the tool setup, and it also requires a subsequent cleaning step. MLD typically involves the use of an oxide-capping layer to encapsulate the molecules on the surface during thermal treatment, which promotes dopant incorporation. However, this cap oxide must be removed after the dopant drive-in phase.

v. *Barrier at the surface that inhibits incorporation*: The surface plays a critical role in semiconductor doping. It serves as a collector of point defects, which can be advantageous and a trapping site for dopant atoms, effectively preventing them from modifying the underlying semiconductor. This surface trapping can be viewed as an energy barrier that must be overcome to introduce impurities into the target material. Doping techniques that rely primarily on in-diffusion and have minimal physical components in their dopant delivery are particularly susceptible to this surface barrier. In the future, achieving a state of dopant supersaturation at the surface interface may help address this issue.

vi. *Industry standard*: The designation of a process or technology as an industry standard depends not only on its performance but also on its level of maturity, reliability, and reproducibility. Ion implantation stands out as a leader in this category due to decades of research, tool development, and extensive exploration of the associated physics. Its advantages include the ability to achieve high levels of dopant incorporation and a wide range of available elemental species. Ion implantation is undeniably a crucial part of the semiconductor device manufacturing value chain.

vii. *Suitability for 3D structures (conformality)*: While not as mature as ion implantation, PD offers solutions for applications where ion implantation may face challenges. This includes scenarios involving 3D structures that demand a high degree of dopant uniformity and conformality. It's worth noting that a lack of conformality can result in local resistance variations within a device, and the highly directional ion beam of ion implantation has limitations in this regard. For example, densely packed structures may cast shadows over neighboring regions during ion implantation. In contrast, plasma doping can deposit material uniformly on a surface, which can then serve as a dopant source during subsequent thermal treatment. In-diffusion-based processes have a distinct advantage in addressing shadowing concerns, and this should be considered when dealing with solid-source depositions to ensure they are conformal. ALD can offer a solution to enhance conformality in such cases. Gas sources can be adjusted using chamber pressures and gas flow rates, while MLD relies on surface reactions and is expected to provide excellent conformality to surfaces.

viii. *Suitability for TMD materials*: Lastly, when assessing doping solutions for transition metal dichalcogenide (TMD) thin-film devices, it's important to note that this research area is still in its infancy. Therefore, the conclusions drawn here may evolve as more research is conducted in the future. In general, subjecting TMDs to physical bombardment results in the creation of defects that are not easily remedied through thermal treatment, such as vacancies in metal or chalcogen atoms. It is the efficient repair mechanism for damage in group IV semiconductors that allows ion implantation to dominate the market. However, TMDs do not exhibit the same self-repair capability, and the solution to this issue is not yet clear. Further research is required to address this challenge. In any case, it is advisable to avoid the physical injection of ions into TMDs, and instead rely on gentler deposition or chemistry-functionalization processes in this specific application domain.

5.6   FUTURE OF SEMICONDUCTOR TECHNOLOGY

When considering the high-level evaluation of the advantages and disadvantages, epitaxial growth presents a notable benefit by allowing for the incorporation of high levels of dopants into the target substrate without causing physical harm to the semiconductor crystal or disturbing the semiconductor surface. It circumvents surface energy barriers associated with dopant inclusion, and the process has reached a sufficient level of maturity for utilization in manufacturing settings. Nevertheless, it necessitates the growth of a new layer or material, which may require careful consideration when dealing with limited spaces and distances between features. Additionally, its suitability for novel materials like TMDs warrants further exploration, as the idea of elevated source drains in TMD MOS (Transition Metal Dichalcogenide Metal Oxide Semiconductor) devices has not been firmly established. The potential advantages of proximity doping arising from surface layers should not be underestimated, particularly since 2D materials are significantly influenced by their surfaces and encapsulation.

5.7   SUMMARY

This chapter explores state-of-the-art doping techniques in semiconductor technology. It begins by examining plasma doping's role and delves into its synergy with ALD and ALE. Gas-phase and solid-phase diffusion methods are scrutinized for their impact. MLD precision is discussed, followed by tailored doping for TMDs. The significance of epitaxial growth and proximity doping in crafting advanced semiconductor structures is highlighted. This chapter serves as a comprehensive guide to the cutting-edge doping techniques that underpin the ever-evolving landscape of semiconductor technology within the context of this book.

REFERENCES

1. G. E. Moore, Cramming more components onto integrated circuits, Reprinted from Electronics, volume 38, number 8, April 19, 1965, pp.114 ff., IEEE Solid-State Circuits Soc. Newsletter. 11 (3) (2006) 33–35.
2. E. Sicard, L. Trojman, Introducing 5-nm FinFET technology in microwind. (2021).
3. E. S. Snow, P. M. Campbell, P. J. McMarr, Fabrication of silicon nanostructures with a scanning tunneling microscope, Appl. Phys. Lett. 63 (1993) 749–751.
4. S. M. Sze, M. K. Lee, K. Singleton, S. Mendel, Semiconductor Devices – Physics and Technology, Chapter 1, 2008, Hoboken, NJ, John Wiley & Sons, INC.
5. B. G. Yacobi, I. Brodie, A. Sher, Semiconductor Materials – An Introduction to Basic Principles, Chapter 4, 2003, New York, NY, Kluwer Academic Publishers.
6. A. Smets, K. Jäger, O. Isabella, R. Van Swaaij, M. Zeman, Solar Energy: The Physics and Engineering of Photovoltaic Conversion, Technologies and Systems, Chapter 6, 2016, Cambridge, UIT Cambridge Ltd.
7. J. R. Conrad, J. L. Radtke, R. A. Dodd, F. J. Worzata, N. C. Tran, Plasma source ion-implantation technique for surface modification of materials, J. Appl. Phys. 62 (1987) 4591–4596.
8. J. Tendys, L. J. Donnelly, M. J. Kenny, J. T. A. Pollock, Plasma immersion ion implantation using plasmas generated by radio frequency techniques, Appl. Phys. Lett. 53 (1988) 2143–2145.

9. J. W. Coburn, H. F. Winters, Plasma etching—A discussion of mechanisms, J. Vac. Sci. Technol. 16 (2) (1979) 391–403.

10. N. Marchack, L. Buzi, D. B. Farmer, H. Miyazoe, J. M. Papalia, H. Yan, G. Totir, S. U. Engelmann, Plasma processing for advanced microelectronics beyond CMOS, J. Appl. Phys. 130 (2021) 080901-2–080901-16.

11. B. C. Mallick, C.-T. Hsieh, K.-M. Yin, Y. A. Gandomi, K.-T. Huang, Review—On atomic layer deposition: Current progress and future challenges, ECS J. Solid State Sci. Technol. 8 (2019) N55–N78.

12. W.-H. Lee, Y.-T. Yin, P.-H. Cheng, J.-J. Shyue, M. Shiojiri, H.-C. Lin, M.-J. Chen, Nanoscale GaN epilayer grown by atomic layer annealing and epitaxy at low temperature, ACS Sustainable Chem. Eng. 7 (2019) 487–495.

13. N. Marchack, J. Innocent-Dolor, M. Hopstaken, S. Engelmann, Control of surface oxide formation in plasma-enhanced quasiatomic layer etching of tantalum nitride, J. Vac. Sci. Technol. A. 38 (2020) 022609.

14. M. L. Hoarfrost, K. Takei, V. Ho, A. Heitsch, P. Trefonas, A. Javey, R. A. Segalman, Spin-on organic polymer dopants for silicon, J. Phys. Chem. Lett. 4 (2013) 3741–3746.

15. C. Qin, H. Yin, G. Wang, Y. Zhang, J. Liu, Q. Zhang, H. Zhu, C. Zhao, H. H. Radamson, A novel method for source/drain ion implantation for 20 nm FinFETs and beyond, J. Mater. Sci.: Mater. Electron. 31 (2020) 98–104.

16. J. C. Ho, R. Yerushalmi, Z. Jacobson, Z. Fan, R. L. Alley, A. Javey, Controlled nanoscale doping of semiconductors via molecular monolayers, Nat. Mater. 7 (2008) 62–67.

17. M. Georgieva, N. Petkov, R. Duffy, 3D to 2D perspectives - traditional and new doping and metrology challenges at the nanoscale, Mater. Sci. Semicond. 163 (2023) 107584.

18. Y. Sasaki, K. Okashita, B. Mizuo, M. Kubota, M. Ogura, O. Nishijima, Conformal doping mechanism for fin field effect transistors by self-regulatory plasma doping with AsH3 plasma diluted with He, J. Appl. Phys. 111 (2012) 013712.

19. Y. Kiyota, T. Inada, Sticking coefficient of boron and phosphorus on silicon during vapor-phase doping, J. Vac. Sci. Technol. A. 19 (2001) 2441.

20. M. Perego, F. Caruso, G. Seguini, E. Arduca, R. Mantovan, K. Sparnacci, M. Laus, Doping of silicon by phosphorus end-terminated polymers: Drive-in and activation of dopants, J. Mater. Chem. C. 8 (2020) 10229–10237.

6 Emerging Materials in Semiconductor Devices

Arpana Agrawal

6.1 INTRODUCTION

The rapid advancement of technology and the ever-increasing demand for more efficient and powerful electronic devices have led to a constant search for new materials with enhanced properties. In the field of semiconductor devices, emerging materials have garnered significant attention due to their potential to revolutionize the performance and functionality of electronic components. These materials, which encompass a wide range of novel substances, offer unique characteristics that could surpass the limitations of traditional semiconductor materials.

The importance of emerging materials in semiconductor devices cannot be overstated as they hold the promise of enabling higher-speed and lower-power electronics, paving the way for the development of advanced technologies such as flexible electronics, wearable devices, and high-performance computing. By harnessing the distinctive properties of these materials, engineers and researchers can explore new design possibilities and overcome the bottlenecks faced by conventional semiconductor technologies. A few of the emerging materials for semiconductor technology are organic semiconductors, two-dimensional (2D) materials (i.e. graphene-based materials), transition metal dichalcogenides (TMDs) (MoS_2, WS_2, WSe_2, etc.), hybrid systems (combining inorganic/organic materials or 2D materials with traditional semiconductors), or perovskite materials, etc. [1–4].

Organic semiconductors represent an exciting avenue in semiconductor technology. Their versatile electronic properties, solution processability, and flexibility make them promising candidates for applications ranging from flexible displays to energy-efficient electronics. Ongoing research endeavors are aimed at overcoming their challenges and realizing their full potential, paving the way for a new era of electronic devices that blend advanced functionality with innovative design. As innovations unfold, 2D materials including graphene-based materials, TMDs, or hybrid systems are poised to reshape semiconductor technology and impact industries ranging from electronics to energy and beyond. Moreover, perovskite materials have also ushered in a new era of semiconductor device technology. Their remarkable optoelectronic properties, cost-effective synthesis methods, and versatility position them as contenders for next-generation solar cells, light-emitting devices (LEDs), and more. All these materials have revolutionized the field of semiconductor devices. However, the synthesis method and fabrication technology along with the integration of these emerging materials with the modern fabrication techniques also play a vital role in semiconductor device fabrication.

Accordingly, the present chapter aims to provide a comprehensive overview of the exciting field of emerging materials for semiconductor devices, including organic semiconductors, 2D materials, perovskite materials, their characteristics, integration with modern fabrication techniques, applications, and challenges. A brief discussion on traditional semiconductor materials and the limitations they present will also be discussed. Furthermore, the characterization techniques essential for evaluating their properties and performance will also breifly be described.

6.2 TRADITIONAL SEMICONDUCTOR MATERIALS

Traditional semiconductor materials have been the cornerstone of modern electronics for several decades. These materials, such as silicon (Si), germanium (Ge), InAs, gallium arsenide (GaAs), wide band gap semiconductors, etc. have proven to be highly reliable and versatile for a wide range

 DOI: 10.1201/9781003450146-6

of electronic applications [5–9]. They possess unique electrical properties that allow for the controlled flow of electrons, making them essential for the construction of transistors, integrated circuits, and other semiconductor devices.

Silicon, in particular, has been the dominant material in the semiconductor industry due to its abundance, excellent electrical properties, and well-established manufacturing processes. It has facilitated the exponential growth of computational power and the miniaturization of electronic components, leading to the development of modern computing devices. Germanium, though less widely used than silicon, has similar electrical properties and has found applications in niche areas such as infrared detectors and high-speed transistors. Gallium arsenide, with its superior electron mobility, has been employed in high-frequency applications, microwave devices, and optoelectronics.

Despite their long-standing success, traditional semiconductor materials are not without limitations. These limitations have motivated researchers to explore and develop emerging materials with improved properties. Some key challenges and limitations of traditional semiconductor materials include the performance limitations, physical limitations, manufacturing complexity and cost, environmetal impact, limited functional diversity, etc. As device dimensions shrink to nanoscale levels, traditional materials face challenges in maintaining performance and functionality. Issues like leakage currents, heat dissipation, and power consumption become more prominent, hampering further advancements in device performance.

Also, traditional materials have limited optical properties, restricting their application in optoelectronic devices such as LEDs and solar cells. The mechanical properties of these materials may also hinder the realization of flexible or stretchable electronic devices.

Apart from this, while silicon-based processes have been optimized over the years, the manufacturing of traditional semiconductor materials can still be complex and expensive. The fabrication of high-quality single-crystal substrates and precise doping processes can add significant costs to semiconductor device production. The extraction and purification of traditional semiconductor materials, such as silicon, can also have environmental consequences due to energy-intensive processes and the generation of hazardous waste materials. Addressing these limitations and challenges has become a driving force behind the exploration and development of emerging materials for semiconductor devices. By leveraging the unique properties of newly emerging materials, researchers aim to overcome the hurdles encountered by traditional materials and unlock new opportunities for next-generation electronic devices.

6.3 OVERVIEW OF EMERGING MATERIALS

Emerging materials refer to a diverse group of novel substances that exhibit unique properties and functionalities, distinct from traditional semiconductor materials. These materials are typically characterized by their exceptional electrical, optical, mechanical, or thermal properties, which make them promising candidates for applications in semiconductor devices. Emerging materials encompass various classes, including organic semiconductors, 2D materials, perovskite materials, etc.

The characteristics of emerging materials vary depending on their specific class. Organic semiconductors, for example, are carbon-based compounds that possess inherent flexibility, low cost, and the ability to be processed through solution-based techniques. 2D materials, such as graphene and TMDs, have a single or few atomic layers, offering exceptional mechanical strength, high carrier mobility, and unique electrical properties. Perovskite materials are a class of compounds with a distinctive crystal structure that exhibits excellent light-absorbing properties and high charge carrier mobility. The emergence of new materials for semiconductor devices holds immense importance due to their potential to overcome the limitations of traditional materials and drive advancements in various areas of technology. The importance of emerging materials lies in their ability to enable the development of high-performance electronic devices with enhanced capabilities. By harnessing their unique properties, researchers can explore new avenues for applications such as flexible electronics, wearable devices, bendable displays, high-speed transistors, energy harvesting devices, optoelectronics, and sensing technologies. For example, organic semiconductors offer the potential for flexible, lightweight, and low-cost electronics. They can be utilized in flexible displays, electronic skins, and wearable sensors. 2D materials exhibit exceptional electrical and thermal properties, making them suitable for high-speed transistors, sensors, and energy-efficient devices. Perovskite materials have shown tremendous potential in solar cells, LEDs, and photodetectors due to their excellent light-absorption and charge transport properties.

The integration of emerging materials in semiconductor devices has the potential to revolutionize various industries, including electronics, energy, healthcare, and communications. The exploration and utilization of these materials are crucial for pushing the boundaries of current technological capabilities and creating innovative solutions to address the demands of a rapidly evolving world.

6.4 CHARACTERIZATION TECHNIQUES FOR EMERGING MATERIALS

Characterization techniques play a crucial role in understanding the properties and behavior of emerging materials for semiconductor devices. These techniques allow researchers to analyze, measure, and evaluate the structural, electrical, optical, thermal, and mechanical properties of these materials. By gaining insights into these properties, researchers can optimize material synthesis, device fabrication processes, and performance. Characterization techniques typically involve a combination of experimental measurements, imaging, spectroscopy, and analysis and provide valuable information about the material's composition, crystal structure, surface morphology, charge transport, optical absorption, emission properties, and more. The data obtained from these techniques aids in the development and improvement of emerging materials for semiconductor applications. The techniques for analyzing and evaluating emerging materials for semiconductor devices include X-ray diffraction (XRD), scanning electron microscopy (SEM), transmission electron microscopy (TEM), atomic force microscopy (AFM), Raman spectroscopy, X-ray photoelectron spectroscopy (XPS), photoluminescence spectroscopy (PL), electrical characterization, thermal analysis, Z-scan technique, magnetic measurements, etc. [10–17].

XRD is used to determine the crystal structure, lattice parameters, and phase purity of materials. By analyzing the diffraction pattern resulting from X-ray scattering, researchers can identify the crystallographic phases present in the material, assess the quality of crystal growth, and verify the structural properties of emerging materials. SEM provides high-resolution imaging of material surfaces and enables researchers to visualize the morphology, surface features, and topography of emerging materials. This technique also allows for the characterization of nanostructures, such as nanowires and quantum dots, and aids in assessing the quality and uniformity of materials. TEM provides detailed information about the internal structure, crystal defects, and composition of materials at the nanoscale, offering high-resolution imaging and allowing researchers to visualize the atomic arrangement and interfaces within emerging materials. This technique is particularly useful for analyzing 2D materials, nanowires, and quantum dots. AFM is employed to investigate the surface topography, roughness, and mechanical properties of materials and uses a cantilever with a sharp tip to scan the material's surface, providing detailed information about its features at the nanoscale. This technique is valuable for assessing the quality and uniformity of emerging materials. Raman spectroscopy utilizes laser-induced scattering of light to analyze the vibrational modes of materials and provides information about the molecular structure, chemical composition, and crystallinity of emerging materials. This spectroscopy can be used to study the optical properties, strain, and doping effects in 2D materials and nanowires. XPS is a surface sensitive technique and allows the determination of electronic band alingments, core level and valence level studies, chemical composition, etc. PL spectroscopy measures the emission of light from materials upon excitation with a light source and is used to evaluate the optical properties, bandgaps, and defect states in emerging materials. PL spectroscopy is particularly useful for characterizing quantum dots, perovskite materials, and organic semiconductors. Vibrating-sample magnetometers, magnetoresistance, and superconducting quantum interference devices facilitate their magnetic and magnetotransport studies.

For semiconductor devices, electrical characterization techniques, such as Hall effect measurements, conductivity measurements, and transistor characterization are requiredto assess the electronic properties and charge transport behavior of emerging materials. These techniques provide insights into carrier concentration, mobility, resistivity, and device performance. Thermal analysis techniques, such as differential scanning calorimetry and thermogravimetric analysis, are used to investigate the thermal stability, phase transitions, and thermal properties of emerging materials. This information is critical for assessing the suitability of materials for high-temperature applications and understanding their thermal behavior. Z-scan techniques facilitate determining the nonlinear optical parameters, particularly nonlinear absorption and nonlinear refraction, which are prerequisites for adapting the material for optoelectronic device applications.

6.5   EMERGING MATERIALS FOR SEMICONDUCTING DEVICES

6.5.1   Organic Semiconductors

Organic semiconductors, a class of materials at the intersection of organic chemistry and semiconductor physics, have emerged as a versatile and promising alternative to traditional inorganic semiconductors. These carbon-based compounds exhibit unique electrical properties, making them ideal candidates for various electronic applications. The key advantage of organic semiconductors lies in their tunable electronic properties via chemical modification which allows fine control over their energy levels and charge carrier transport. This facilitates tailoring the properties of organic semiconductors for specific applications and also enhance device performance. One of their remarkable features is their compatibility with solution-based processing techniques. This enables large-area, low-cost, and flexible fabrication methods, which contrast with the energy-intensive processes used for traditional semiconductors. The lightweight and flexible nature of organic semiconductors makes them suitable for applications requiring conformable electronics, such as flexible displays, wearable sensors, and rollable screens.

Organic semiconductors have found applications in a diverse range of electronic devices. Organic LEDs stand out as one of the most successful applications, providing vivid displays with lower energy consumption. Organic photovoltaics (OPVs) offer a unique approach to solar energy harvesting, demonstrating potential for lightweight and semi-transparent solar panels. Organic thin-film transistors (OTFTs) are utilized in flexible electronics, enabling bendable and stretchable circuits. Wilbers et al. [18], have demonstrated the mechganism of charge transport in vertical pillar structures at nanoscale regime, which consists of a thin layer (~5 nm) of poly(3-hexylthiophene) (P3HT), a well known organic semiconductor. A high current density of 10^6 A/m^2 has been obtained. Figure 6.1 illustrates the sequential steps involved in producing vertical metal-P3HT-metal pillars. Initially, a layer of P3HT was applied through spin-coating onto clean Au bottom electrodes situated on Si/SiO$_2$ substrates as shown in Figure 6.1a and 6.1b). The top metal contacts of Au disks (diameters: 200 nm to 2 μm; thickness: 70 nm) were wedged onto the P3HT.

For wedging transfer, these top contacts were encapsulated within a hydrophobic polymer (cellulose acetate butyrate [CAB]) which was prepared by dissolving CAB in ethyl acetate. When the device is submerged in water, the water infiltrates between the hydrophilic SiO$_2$ substrate and the hydrophobic CAB, causing the polymer to detach. Due to the low adhesion between Au and SiO$_2$, the Au disks were lifted off along with the CAB polymer. Subsequently, the Au disks were wedged onto the new substrate featuring 5 μm wide bottom electrodes coated with P3HT. This was achieved

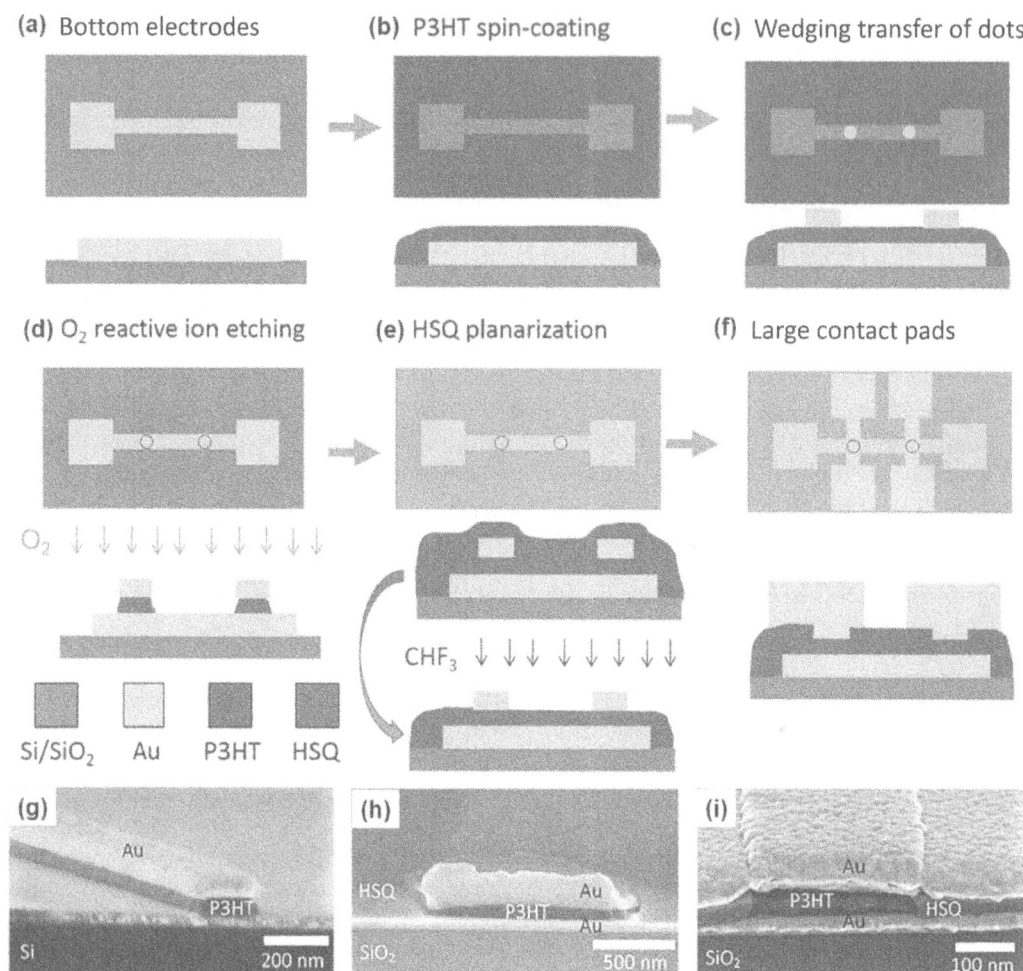

FIGURE 6.1 Diagram illustrating the fabrication process for vertical Au-P3HT-Au pillars, with both top and side views depicted: (a) Initiation process involving pattern creations for the bottom electrodes through photolithography on Si/SiO$_2$ substrates. (b) Following this, a layer of P3HT is uniformly applied through spin-coating. (c) Transferring of top contacts onto the thin P3HT film using a wedging technique. (d) Subsequently, the vertical pillars are formed through directional dry etching using oxygen plasma, with the top contacts functioning as an etch mask for the P3HT. (e) To complete the process, a layer of HSQ is spin-coated and then planarized through reactive ion etching, opening up the top Au contacts. (f) Finally, large top contacts are evaporated onto the structure, and their pattern is defined through photolithography. Cross-sectional SEM images: Pt-P3HT-Au test structure (g), Au-P3HT-Au test structure following the dry etching and spin coating of HSQ (h), and the final test structure (i). (Adapted with permission [18]. Copyright [2017] Copyright the Authors, some rights reserved; exclusive licensee [Nature Springer]. Distributed under a Creative Commons Attribution License 4.0 [CC BY] https://creativecommons.org/licenses/by/4.0/.)

by placing the substrate at an angle of approximately 45° on a grid holder within a beaker filled with Milli-Q water and slowly draining the water (Figure 6.1c). In this manner, the cellulose polymer initially made contact with the hydrophilic SiO$_2$ and then gently adhered to the P3HT. The next step involved performing directional reactive ion beam etching to create vertical pillars (Figure 6.1d). The Au disks served as an etch mask for the P3HT during this process. Subsequently, the pillars were enclosed within an electrically insulating layer of hydrogen silsesquioxane (HSQ), which was thinner on top of the pillars compared to the substrate. While maintaining the protection of the organic layer (as demonstrated in Figure 6.1e), the Au top contacts of the pillars were exposed through reactive ion beam etching. To establish contact with the pillar structures, a 100 nm thick metal layer was then deposited and patterned using photolithography to form substantial contact pads (Figure 6.1f). For a more comprehensive description of the fabrication procedure, please refer to the experimental section in the associated documentation. Figure 6.1g–6.1i gives the cross-sectional images obtained using SEM for the Pt-P3HT-Au test structure (g), the Au-P3HT-Au test structure following the dry etching, and spin coating of HSQ (h) and the final test structure (i).

Mensfoort et al. [19], have examined the current density voltage behavior of various orgaic semiconductorbased, sandwichtype structures. Wang et al. [20], have also studied the hopping transport behavior of electrochemically gated transistors based on P3HT. Polythiophene based thin-film transistorswere also fabricated by Kline et al. [21]. Apart from P3HT, PCBM is also a promising material for trasistor applications [22]. It is noteworthy to mention here that fabricating hybrid structures where inorganic materials were combined with organic materials also leads to various fascinating applications. Such a hybrid structure was also reported by Krieg et al. [23], combining GaN with oxidative chemical vapor deposited PEDOT for optoelectronic applications.

Apart from immense advantages of organic semiconductors, they also present challenges. One major concern is their relatively low charge carrier mobility compared to inorganic counterparts, limiting their application in high-speed electronic devices. Additionally, environmental stability and degradation over time pose obstacles for widespread commercial adoption. Developing efficient and stable manufacturing processes and addressing the issue of charge carrier mobility are focal points of ongoing research.

6.5.2 Two-dimensional Materials

2D materials, consisting of atomically thin layers, have sparked a revolution in materials science and semiconductor technology. With unique properties stemming from their nanoscale thickness, these materials offer remarkable opportunities for a wide array of applications as shown in Figure 6.2. At the forefront of 2D materials is graphene, a single layer of carbon atoms arranged in a hexagonal lattice. Beyond graphene, TMDs and other materials like h-BN, phosphorene, and black phosphorus exhibit diverse properties due to their distinct crystal structures and compositions. Huang et al. [24], presented a comprehensive review of various 2D semiconductors.

Two-dimensional materials exhibit extraordinary electrical, optical, mechanical, and thermal properties and high charge carrier mobility, making them potential candidates for high-speed transistors and logic devices. Their nanoscale thickness enables unique mechanical flexibility and strength, making them ideal for flexible electronics and wearable devices. Moreover, their direct bandgap in TMDs leads to strong light-matter interactions, promising breakthroughs in optoelectronics and photodetectors. In electronics, they are being explored for ultra-thin and high-performance transistors, memory devices, and logic circuits. Their exceptional thermal conductivity makes them appealing for thermal management in advanced electronics. In photonics, they enable efficient light emission and detection, paving the way for improved LEDs, photodetectors, and quantum devices. Mechanically flexible and transparent, 2D materials are indispensable in flexible electronics and transparent conductive films. Their gas-sensing capabilities are harnessed for environmental monitoring and medical diagnostics. Additionally, their large surface areas and tailored functionalization support advancements in energy storage and conversion devices, such as batteries, supercapacitors, and catalysts.

The advent of the first-ever monolayer MoS_2 transistor with an exceptionally high on/off ratio served as a catalyst for a surge in research on 2D semiconductor transistors [25]. While 2D semiconductors can easily achieve atomic layers thinner than 1 nm, it is their performance at extremely reduced channel lengths that dictates their potential applications in high-performance transistors. To ascertain their resistance to short-channel effects, experiments were conducted on MoS_2 field effect transistors (FETs) with channel lengths ranging from microns to tens of nanometers,

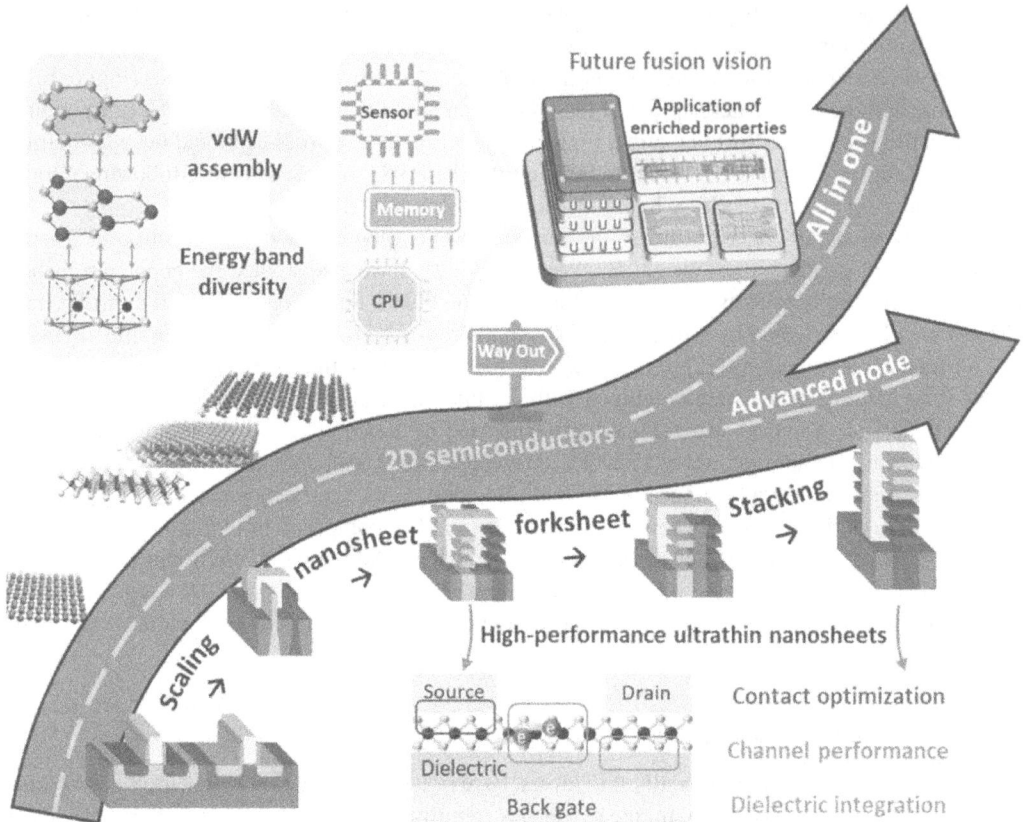

FIGURE 6.2 Schematic illustration showing the road map of several novel properties and applications of 2D semiconductors. (Adapted with permission [24]. Copyright [2022] Copyright the Authors, some rights reserved; exclusive licensee [Nature Springer]. Distributed under a Creative Commons Attribution License 4.0 [CC BY] https://creativecommons.org/licenses/by/4.0/.)

revealing MoS_2's immunity to short channels [26, 27]. Furthermore, a MoS_2 transistor employing carbon nanotubes as the gate electrode demonstrated remarkable switching performance with an on/off ratio exceeding 10^6, despite having a gate of only 1 nm [28]. In a conventional planar gate device setup, Xie et al. [29], managed to achieve physical channel lengths as short as 8 and 3.8 nm using graphene as the electrode.

The current performance is the fundamental metric for 2D semiconductors acting as channels, with the effective mass and bandgap being critical factors primarily determined by the material itself. The effective mass is an intrinsic parameter that gauges the maximum current achievable through the channel and is directly related to the material's band structure, representing the second derivative of the band dispersion [30]. The maximum ballistic current flowing through the channel is proportionate to the ballistic velocity of the carriers. A low effective mass facilitates achieving high carrier velocities, implying that materials exhibiting lower effective masses can offer higher current limits [31]. However, the effective mass should not be too low, as it may lead to carrier tunneling between the source and drain while failing to provide sufficient carrier density of states, ultimately compromising transistor performance [32, 33]. Different 2D semiconductors offer choices for selecting the appropriate effective mass to meet specific current performance requirements. For instance, MoS_2, with a substantial effective mass for electrons (about $0.5m_o$), effectively mitigates direct tunneling between source and drain electrodes in ultra-short channel devices, thus suppressing leakage currents. In contrast, WS2 has a relatively smaller effective mass for both electrons and

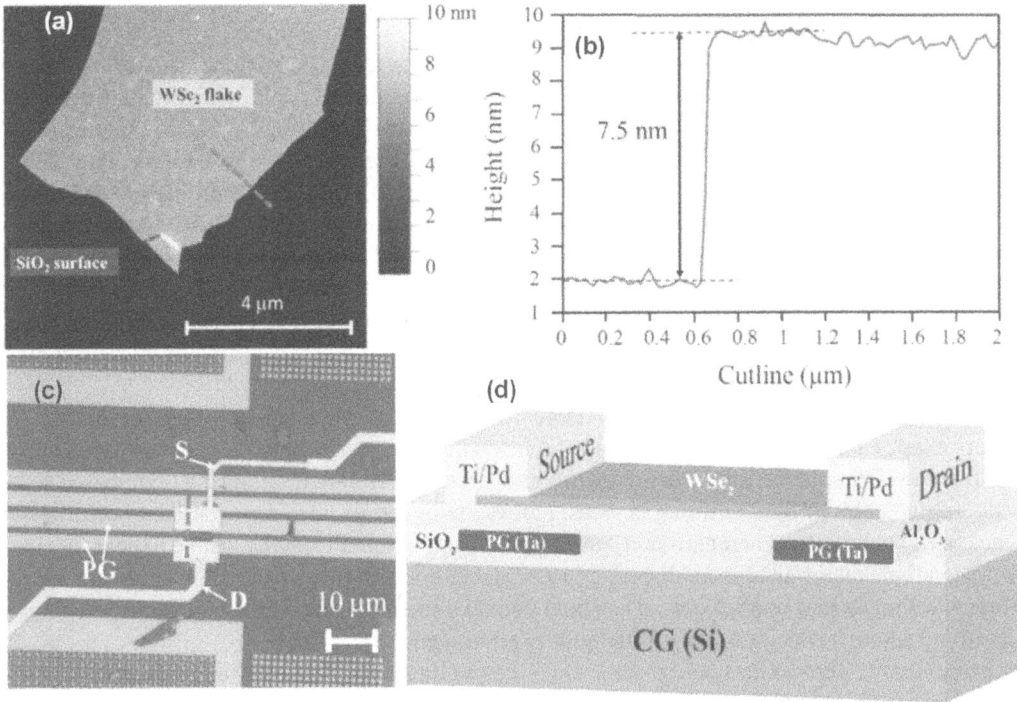

FIGURE 6.3 An overview of the WSe$_2$ flake characteristics and device fabrication: (a) An AFM topography image showing the cleaned WSe$_2$ flake with a 7.5 nm thickness, equivalent to approximately 10 monolayers. (b) The height profile of the flake confirming its thickness as mentioned in (a). (c) An optical image displaying the fabricated device with specific regions controlled by bulk-Si And program gates. (d) A 3D schematic cross-section of the device is presented across a red cutline in (c). (Adapted with permission [36]. Copyright [2016] Copyright the Authors, some rights reserved; exclusive licensee [Nature Springer]. Distributed under a Creative Commons Attribution License 4.0 [CC BY] https://creativecommons.org/licenses/by/4.0/.)

holes, making it a promising candidate for transistor channels [34]. The bandgap directly influences the on-state current to off-state leakage current ratio, with band structures evolving as thickness decreases due to interlayer coupling and confinement effects [35].

Resta et al. [36], have fabricated a polarity controllable device based on WSe$_2$ layers which were obtianed via a mechanical exfoliation method similar to graphene synthesis. TheAFM images of the mechanicxally exfoliated WSe$_2$ onto SiO$_2$/Si substrate after residue tape cleaning are depicted in Figure 6.3a. Figure 6.3b depicts the corresponding height profile to extract the thickness of WSe$_2$ flake across the cutline shown in Figure 6.3a. Figure 6.3c gives the optical image of the fabricated device consisting of a channel of length 1.5 μm. Herein, two gates are involved, namely the control gate (CG) and the program gate (PG). The former is used to perform the ON/OFF function by properly gating the channel region while the latter facilitates the device polarity or the carrier injection by tailoring the contact Schottky barrier. Figure 6.3d schematically illustrates the crosssectional image of the device across the cutline shown in Figure 6.3c.

It's important to note that gate-all-around devices are predominantly gifted due to their exceptional electrostatic control capabilities and space efficiency. Oh et al. [37], developed a vertical field-effect transistor utilizing position-controlled ZnO nanotubes grown on a graphene substrate. While 2D materials hold tremendous promise, challenges remain. Scalable production methods, reproducibility, and precise manipulation are central concerns. Moreover, preserving the pristine quality of 2D materials during device fabrication and integration requires careful attention.

6.5.3 PEROVSKITE MATERIALS

Perovskite materials have rapidly emerged as a captivating class of compounds with immense potential in the realm of semiconductor devices. Named after their crystal structure, these materials exhibit an astonishing range of properties that make them appealing for various applications, especially in photovoltaics and optoelectronics. Perovskite materials have a general formula ABX_3, where A and B are cations and X is an anion. They boast exceptional light-absorbing properties due to their high optical absorption coefficient and long carrier diffusion length. This property, coupled with their ability to be easily synthesized using solution-based methods, have spurred tremendous interest in perovskite solar cells. These materials also possess high charge carrier mobility and diffusion coefficients, contributing to their application in light-emitting devices and photodetectors. Their direct bandgap and tunable optical properties offer remarkable versatility for designing devices across the electromagnetic spectrum.

Perovskite solar cells, in particular, have revolutionized photovoltaics due to their cost-effectiveness, ease of fabrication, and rapid efficiency improvement. The power conversion efficiencies of perovskite solar cells have soared within a short span, rivaling traditional silicon-based solar cells. In the realm of light-emitting devices, perovskite materials have demonstrated potential for efficient light emission. Their emission wavelength can be tailored from the visible to near-infrared region, making them valuable for displays, LEDs, and lasers. Moreover, perovskite materials' versatility is leading to new applications. They hold potential in photodetectors, where their exceptional sensitivity across a wide range of wavelengths is advantageous.

Additionally, perovskite tandem solar cells, combining multiple layers of perovskite materials, offer a path toward achieving even higher efficiencies. Figure 6.4 schematically illustrates the perovskite framework possessing different dimensionalities along with their various optoelectronic applicatios [38]. Liu et al. [39], have reported the defects in perovskite based solar cells. Perovskite filled membranes are also important for the preparation of X-ray detector arrays [40]. LEDs based on perovskite materials have also been demonstrated by Cho et al. [41]. However, challenges related to long-term stability, material toxicity, and scalability need to be addressed for their commercial viability.

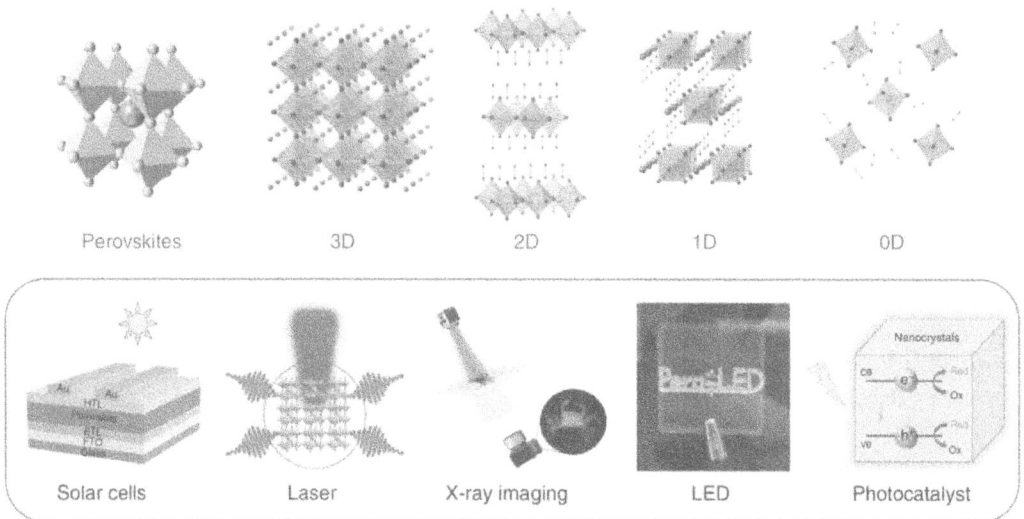

FIGURE 6.4 Schematic illustration of perovskite frameworks exhibiting various dimensionalities along with their optoelectronic applications. (Adapted with permission [38]. Copyright [2023] Copyright the Authors, some rights reserved; exclusive licensee [Nature Springer]. Distributed under a Creative Commons Attribution License 4.0 [CC BY] https://creativecommons.org/licenses/by/4.0/.)

6.6 SYNTHESIS, MODERN DEVICE FABRICATION TECHNIQUES, AND INTEGRATION CHALLENGES

Synthesis and fabrication techniques for emerging materials play a critical role in realizing their potential for semiconductor devices. These methods involve the synthesis of emerging materials and their deposition, patterning, and post-processing to create functional structures and devices. The choice of fabrication technique depends on the specific material, device requirements, and desired properties.

The emerging matterials for semiconductor devices can be synthesized via various methods including solution processing, chemical vapor deposition (CVD), physical vapor deposition (PVD), molecular beam epitaxy (MBE), etc. [42–46]. Solution processing techniques, such as spin coating, dip coating, and inkjet printing, are widely used for fabricating organic semiconductors and perovskite materials and involve the deposition of a solution containing the material onto a substrate, followed by drying or annealing to form the desired film. Solution processing enables large-area deposition, low-cost fabrication, and compatibility with flexible substrates. CVD is a commonly employed technique for growing thin films of emerging materials and involves the reaction of precursor gases on a heated substrate to deposit the material layer by layer. CVD can be used for the synthesis of 2D materials, nanowires, and quantum dots, allowing control over the film thickness, crystallinity, and doping. On the other hand, PVD techniques, including pulsed laser deposition, sputtering and evaporation, are employed for depositing thin films of emerging materials involving the physical vaporization of the material, followed by condensation onto a substrate. PVD is used for depositing metals, oxides, and other materials with precise control over film thickness and composition. MBE is a high-precision technique used for growing epitaxial thin films with atomic-level control and includes the deposition of materials under ultra-high vacuum conditions using molecular or atomic beams. MBE is commonly used for the growth of 2D materials and semiconductor heterostructures, providing excellent control over material quality and interface engineering.

Once the emerging materials are synthesized, they are employed for semiconductor device fabrications. Various modern device fabrication techniques have been succesfuly employed for this pupose, including, lithography, 3D printing, 4D printing, etc. [47]. Lithography techniques, such as photolithography and electron beam lithography, are employed for patterning emerging materials at the nanoscale and involve the use of masks or electron beams to selectively expose and etch the material, creating patterns and structures necessary for device fabrication. Patterning is crucial for defining device features and interconnections.

The integration of emerging materials into semiconductor devices presents several challenges due to differences in material properties, compatibility with existing technologies, and device performance requirements. Strategies such as interface engineering, strain engineering, formation of hybrid structures, device design optimization, scalability and manufacturing, etc. are employed to address these challenges and enable effective integration of emerging materials.

Emerging materials often require suitable interfaces with other materials to ensure efficient charge transport, minimize defects, and optimize device performance. Interface engineering techniques, such as interfacial layers, surface treatments, and passivation, are employed to enhance the compatibility and performance of emerging materials in semiconductor devices. Integrating emerging materials with existing semiconductor technologies may require addressing compatibility issues such as lattice mismatches, thermal expansion coefficients, and chemical interactions. Hence, inculsion of buffer layers, strain engineering, and hybrid structures to accommodate these differences can circumvent such issues and enable seamless integration.

Device design plays a crucial role in harnessing the unique properties of emerging materials. Design optimization involves tailoring the device structure, geometry, and dimensions to leverage the advantages of the material while mitigating any limitations. Simulation tools and modeling techniques aid in optimizing device performance and functionality. Ensuring the scalability and manufacturability of emerging materials is essential for their widespread adoption.

6.7   PERFORMANCE AND FUTURE PROSPECTS

Comparing the performance of emerging materials with traditional semiconductor materials provides valuable insights into their advantages, limitations, and potential for improvement. Several key parameters are commonly considered for performance evaluation, incuding electrical, optical, mechanical, and thermal performance. Emerging materials, such as organic semiconductors, 2D materials, and perovskites, often exhibit unique electrical properties compared to traditional materials like silicon and gallium arsenide. Parameters such as charge carrier mobility, on/off ratio, and sub-threshold swing are analyzed to assess the suitability of emerging materials for high-performance semiconductor devices. The optical properties of emerging materials, such as absorption and emission spectra, quantum yield, and exciton dynamics are crucial for applications such as LEDs, photodetectors, and solar cells, where the material's light-emitting or light-harvesting capabilities are critical.

Flexibility, stretchability, and mechanical robustness are also very important for emerging materials for their utility in flexible electronics and wearable devices. Comparing the mechanical properties, such as Young's modulus and fracture strength, of emerging materials with traditional materials helps understand their suitability for these applications. Emerging materials' thermal properties, including thermal conductivity and coefficient of thermal expansion, help in evaluating their ability to dissipate heat and withstand high-temperature operation and are particularly important for power electronics and high-performance devices.

The field of emerging materials for semiconductor devices continues to evolve rapidly, with ongoing research and development efforts driving advancements. Researchers are actively working on enhancing the performance of emerging materials by improving their structural quality, charge carrier mobility, thermal stability, and environmental robustness. This involves the discovery and synthesis of new material variants, exploring doping strategies, and developing novel synthesis and fabrication techniques. Efforts are underway to overcome integration challenges and enable seamless incorporation of emerging materials into existing semiconductor technologies. This includes developing new device architectures, optimizing interfaces, and exploring hybrid material combinations to improve compatibility and maximize device performance.

The advancements in the field of emerging materials and technologies may lead to breakthrough applications in areas like healthcare, Internet of Things (IoT), and energy harvesting. To enable widespread adoption, efforts are being made to improve the scalability and manufacturability of emerging materials by developing large-scale synthesis and deposition techniques, optimizing fabrication processes, and integrating emerging materials into existing semiconductor manufacturing infrastructure. Continued research aims to improve the energy conversion efficiency of solar cells, reduce power consumption in electronics, and explore emerging materials for energy storage devices. Additionally, efforts are being made to address environmental concerns associated with the use of certain materials, such as toxicity and recycling.

6.8   CONCLUSION

Various aspects of emerging materials including organic semiconductors, 2D materials, such as graphene and TMDs, and perovskite materials employed for semiconductor device applications have been explored in this chapter. The chapter began with an overview of emerging materials including types, their unique properties, applications, advantages over traditional materials, as well as the challenges associated with their use. Furthermore, various characterization techniques used to analyze and evaluate them, including XRD, SEM, TEM, Raman spectroscopy, AFM, XPS, PL spectroscopy, electrical characterization, thermal analysis, and Z-scan technique, have been briefly described. Various synthesis methods (e.g. solution processing, CVD, PVD, and MBE) adopted for the synthesis of emerging materials, along with their device fabrication techniques (lithography, 3D/4D printing), have also been highlighted. Integration challenges and strategies for incorporating emerging materials into semiconductor devices were also explored, including inter-face engineering, material compatibility, device design optimization, and scalability and manu-facturing considerations. Lastly, the performance of emerging materials compared to traditional materials has been examined and the potential future developments and advancements in emerg-ing materials for semiconductor devices have also been concluded.

REFERENCES

1. J.C. Blakesley, F.A. Castro, W. Kylberg, G.F. Dibb, C. Arantes, R. Valaski, M. Cremona, J.S. Kim, Towards reliable charge-mobility benchmark measurements for organic semiconductors. *Org. Electron.,* 15 (2014) 1263–1272.
2. S. Chu, Y. Zhang, P. Xiao, W. Chen, R. Tang, Y. Shao, T. Chen, X. Zhang, F. Liu, Z. Xiao, Large-area and efficient sky-blue perovskite light-emitting diodes via blade-coating. *Adv. Mater.,* 34 (2022) 2108939.
3. A. Agrawal, G.C. Yi, Database on the nonlinear optical properties of graphene based materials. *Data in Brief,* 28 (2020) 105049.
4. A. Agrawal, G.C. Yi (2023). Photonic and optoelectronic applications of graphene: nonlinear optical properties of graphene and its applications. In *Recent Advances in Graphene and Graphene-Based Technologies* (pp. 9-1 to 9-22). Bristol, UK: IOP Publishing.
5. R.A. Soref, Silicon-based optoelectronics. *Proceedings of the IEEE,* 81 (1993) 1687–1706.
6. A. Agrawal, T.A. Dar, P. Sen, Structural and optical studies of magnesium doped zinc oxide thin films. *J. Nano Electron. Phys.,* 5 (2013) 02025–28.
7. M. Fiederle, S. Procz, E. Hamann, A. Fauler, C. Fröjdh, Overview of GaAs and CdTe pixel detectors using medipix electronics. *Cryst. Res. Technol.,* 55 (2020) 2000021.
8. F.A. Tantray, A. Agrawal, M. Gupta, J.T. Andrews, P. Sen, Effect of oxygen partial pressure on the struc-tural and optical properties of ion beam sputtered TiO_2 thin films. *Thin Solid Films,* 619 (2016) 86–90.
9. A. Agrawal, Y. Tchoe, H. Kim, J.Y. Park, Qualitative analysis of growth mechanism of polycrystalline InAs thin films grown by molecular beam epitaxy. *Appl. Surf. Sci.,* 462 (2018) 81–85.
10. A. Agrawal, T.A. Dar, D.M. Phase, P. Sen, Anomalous band bowing in pulsed laser deposited Mg Zn1−O films. *J. Cryst. Growth,* 384 (2013) 9–12.
11. E. Ghanbari, S.J. Picken, J.H. van Esch, Analysis of differential scanning calorimetry (DSC): Determining the transition temperatures, and enthalpy and heat capacity changes in multicomponent systems by analytical model fitting. *J. Thermal Anal. Calorimetry,* 1–17 (2023).
12. T.A. Dar, A. Agrawal, R.J. Choudhary, P. Sen, Electrical and magnetic transport properties of undoped and Ni doped ZnO thin films. *Thin Solid Films,* 589 (2015) 817–821.
13. A. Agrawal, T.A. Dar, P. Sen, D.M. Phase, Transport and magneto transport study of Mg doped ZnO thin films. *J. Appl. Phys.,* 115 (2014).
14. A. Agrawal, T.A. Dar, D.M. Phase, P. Sen, Type I and type II band alignments in ZnO/MgZnO bilayer films. *Appl. Phys. Lett.,* 105 (2014).
15. R. Chouhan, P. Baraskar, A. Agrawal, M. Gupta, P.K. Sen, P. Sen, Effects of oxygen partial pressure and annealing on dispersive optical nonlinearity in NiO thin films. *J. Appl. Phys.,* 122 (2017).
16. A. Agrawal, T.A. Dar, J.T. Andrews, P.K. Sen, P. Sen, Negative thermo-optic coefficients and optical limiting response in pulsed laser deposited Mg-doped ZnO thin films. *JOSA B,* 33 (2016) 2015–2019.
17. A. Agrawal, T.A. Dar, R. Solanki, D.M. Phase, P. Sen, Study of nonlinear optical properties of pure and Mg-doped ZnO films. *Physica Status Solidi (b),* 252 (2015) 1848–1853.
18. J.G. Wilbers, B. Xu, P.A. Bobbert, M.P. de Jong, W.G. van der Wiel, Charge transport in nanoscale verti-cal organic semiconductor pillar devices. *Sci. Rep.,* 7 (2017) 41171.
19. S.L.M. Van Mensfoort, R. Coehoorn, Effect of Gaussian disorder on the voltage dependence of the cur-rent density in sandwich-type devices based on organic semiconductors. *Phys. Rev. B,* 78 (2008) 085207.
20. S. Wang, M. Ha, M. Manno, C. Daniel Frisbie, C. Leighton, Hopping transport and the Hall effect near the insulator–metal transition in electrochemically gated poly (3-hexylthiophene) transistors. *Nat. Commun.,* 3 (2012) 1210.
21. R. Joseph Kline, M.D. McGehee, M.F. Toney, Highly oriented crystals at the buried interface in poly-thiophene thin-film transistors. *Nat. Mater.,* 5 (2006) 222–228.
22. H. Aarnio, P. Sehati, S. Braun, M. Nyman, M.P. de Jong, M. Fahlman, R. Österbacka, Spontaneous charge transfer and dipole formation at the interface between P3HT and PCBM. *Adv. Energy Mater.,* 1 (2011) 792–797.

23. L. Krieg, F. Meierhofer, S. Gorny, S. Leis, D. Splith, Z. Zhang, H. von Wenckstern, M. Grundmann, X. Wang, J. Hartmann, C. Margenfeld, T. Voss, Toward three-dimensional hybrid inorganic/organic opto-electronics based on GaN/oCVD-PEDOT structures. *Nat. Commun.,* 11 (2020) 5092.

24. X. Huang, C. Liu, P. Zhou, 2D semiconductors for specific electronic applications: From device to system. *NPJ 2D Mater. App.,* 6 (2022) 51.

25. B. Radisavljevic, A. Radenovic, J. Brivio, V. Giacometti, A. Kis, Single-layer MoS_2 transistors. *Nat. Nanotechnol.,* 6 (2011) 147–150.

26. H. Liu, A.T. Neal, P.D. Ye, Channel length scaling of MoS_2 MOSFETs. *ACS Nano.,* 6 (2012) 8563–8569.

27. F. Zhang, J. Appenzeller, Tunability of short-channel effects in MoS_2 field-effect devices. *Nano Lett.,* 15 (2015) 301–306.

28. S.B. Desai, S.R. Madhvapathy, A.B. Sachid, J.P. Llinas, Q. Wang, G.H. Ahn, G. Pitner, M.J. Kim, J. Bokor, C. Hu, H.S.P. Wong, MoS_2 transistors with 1-nanometer gate lengths. *Science.,* 354 (2016) 99–102.

29. L. Xie, M. Liao, S. Wang, H. Yu, L. Du, J. Tang, J. Zhao, J. Zhang, P. Chen, X. Lu, G. Wang, Graphene-contacted ultrashort channel monolayer MoS_2 transistors. *Adv. Mater.,* 29 (2017) 1702522.

30. S.K. Su, C.P. Chuu, M.Y. Li, C.C. Cheng, H.S.P. Wong, L.J. Li, Layered semiconducting 2D materials for future transistor applications. *Small Structures,* 2 (2021) 2000103.

31. M.S. Lundstrom (2017). *Fundamentals of Nanotransistors* (Vol. 6). World Scientific Publishing Company.

32. G. Hiblot, Q. Rafhay, F. Boeuf, G. Ghibaudo, Analytical relationship between subthreshold swing of thermionic and tunnelling currents. *Electron. Lett.,* 50 (2014) 1745–1747.

33. Y. Liu, X. Duan, H.J. Shin, S. Park, Y. Huang, X. Duan, Promises and prospects of two-dimensional transistors. *Nature,* 591 (2021) 43–53.

34. Z. Jin, X. Li, J.T. Mullen, K.W. Kim, Intrinsic transport properties of electrons and holes in monolayer transition-metal dichalcogenides. *Phys. Rev. B,* 90 (2014) 045422.

35. S. Manzeli, D. Ovchinnikov, D. Pasquier, O.V. Yazyev, A. Kis, 2D transition metal dichalcogenides. *Nat. Rev. Mater.,* 2 (2017) 1–15.

36. G.V. Resta, S. Sutar, Y. Balaji, D. Lin, P. Raghavan, I. Radu, F. Catthoor, A. Thean, P.E. Gaillardon, G. De Micheli, Polarity control in WSe2 double-gate transistors. *Sci. Rep.,* 6 (2016) 29448.

37. H. Oh, J. Park, W. Choi, H. Kim, Y. Tchoe, A. Agrawal, G.C. Yi, Vertical ZnO nanotube transistor on a graphene film for flexible inorganic electronics. *Small,* 14 (2018) 1800240.

38. C. He, X. Liu, The rise of halide perovskite semiconductors. *Light: Sci. App.,* 12 (2023) 15.

39. Z. Liu, J. Hu, H. Jiao, L. Li, G. Zheng, Y. Chen, Y. Huang, Q. Zhang, C. Shen, Q. Chen, H. Zhou, Chemical reduction of intrinsic defects in thicker heterojunction planar perovskite solar cells. *Adv. Mater.,* 29 (2017) 1606774.

40. J. Zhao, L. Zhao, Y. Deng, X. Xiao, Z. Ni, S. Xu, J. Huang, Perovskite-filled membranes for flexible and large-area direct-conversion x-ray detector arrays. *Nat. Photonics,* 14 (2020) 612–617.

41. H. Cho, S.H. Jeong, M.H. Park, Y.H. Kim, C. Wolf, C.L. Lee, J.H. Heo, A. Sadhanala, N. Myoung, S. Yoo, T.W. Lee, Overcoming the electroluminescence efficiency limitations of perovskite light-emitting diodes. *Science,* 350 (2015) 1222–1225.

42. A. Agrawal, J.Y. Park, P. Sen, G.C. Yi, Unraveling absorptive and refractive optical nonlinearities in CVD grown graphene layers transferred onto a foreign quartz substrate. *Appl. Surf. Sci.,* 505 (2020) 144392.

43. A. Agrawal, R.K. Saroj, T.A. Dar, P. Baraskar, P. Sen, S. Dhar, Insight into the effect of screw disloca-tions and oxygen vacancy defects on the optical nonlinear refraction response in chemically grown ZnO/Al_2O_3 films. *J. Appl. Phys.,* 122 (2017).

44. T.A. Dar, A. Agrawal, P. Sen, Pulsed laser deposited nickel doped zinc oxide thin films: Structural and optical investigations. *J. Nano Electron. Phys.,* 5 (2013) 02025–27.

45. P. Baraskar, R. Chouhan, A. Agrawal, R.J. Choudhary, P.K. Sen, P. Sen, Magnetic field induced changes in linear and nonlinear optical properties of Ti incorporated Cr2O3 nanostructured thin film. *Phys Lett. A,* 382 (2018) 860–864.

46. A. Agrawal, Y. Tchoe, Scaling study of molecular beam epitaxy grown InAs/Al2O3 films using atomic force microscopy. *Thin Solid Films,* 709 (2020) 138204.

47. A. Agrawal, C.M. Hussain (2023). Materials and technologies for flexible and wearable sensors. *Flexible and Wearable Sensors: Materials, Technologies, and Challenges* (Vol. 21). CRC Press.

7 Organic Semiconductor Devices
Materials and Technology

Periyasamy Angamuthu Praveen
and Thangavel Kanagasekaran

7.1 INTRODUCTION

Organic semiconductors (OSCs), as the name implies, are the organic or carbon version of semiconductors. Like the conventional inorganic semiconductors (ISCs), they possess band gaps in the range of 3.0–4.0 eV. However, the fundamental difference between ISCs and OSCs is their energy bands. ISCs possess continuous energy bands whereas their counterparts have discrete energy levels in the form of molecular orbitals. Due to this, molecular systems offer unique ways to tune and customize their charge-transferring capabilities and achieve specific properties of interest. These features majorly arise due to the π-bonding orbitals in the OSC and can be precisely modeled using quantum chemical techniques. This enables the possibility of molecular design and synthesizing the materials of interest with required highest occupied molecular orbitals (HOMOs–equivalent to the valence band in ISC) and lowest unoccupied molecular orbitals (LUMOs–equivalent to the conduction band) energy levels [1].

The first organic charge transfer complexes were reported during the 1940s. Since then, the realm has seen several important discoveries, ranging from crystalline theories to commercial success like active-matrix organic light emitting diodes (AMOLEDs) and a few Nobel prizes [2, 3]. Already the sister fields, such as organic field effect transistors (OFETs) and organic photovoltaic (OPV) devices, are matured and are reaching the performance of commercial systems [4]. Further, organic semiconductor devices (OSD) are expected to contribute in other important areas such as biosensors, photodetectors, and the organic lasers [5]. We hope a chapter on organic semiconductors would shed a ray of new possibilities in a book that discusses the key advances and technology of inorganic semiconductors.

One of the major advantages of OSCs is that they can be tailor-made, which results in variety of materials ranging from small molecules to dendrimers and then to polymers. Discussion of such a vast database of materials is cumbersome and not suitable for the objectives of this chapter. So we will restrict the discussions within the small molecule OSC, particularly of single crystals in nature. This is advantageous as this knowledge can be further extrapolated to other systems like polymers. The chapter is organized in the following way: Section 7.2 describes the basic materials aspects of OSCs. Section 7.3 introduces the theory of charge transfer in organic complexes followed by the discussion of major fabrication techniques in OSDs in Section 7.4. Section 7.5 briefly describes the key characterization techniques followed by the working principles of different OSDs. Section 7.6 describes the notable applications and Sections 7.7 and 7.8 summarize the chapter with a remark on key open problems.

7.2 STRUCTURAL PROPERTIES

In OSCs, molecular structure and their packing features determine their semiconducting behavior. With a proper molecular design and crystalline packing, the electrical and optical properties of the systems can be tuned. Considering the molecular structure, OSCs are usually π-conjugated systems, resulting in their conducting behavior. These π-orbitals overlap with the adjacent single

DOI: 10.1201/9781003450146-7

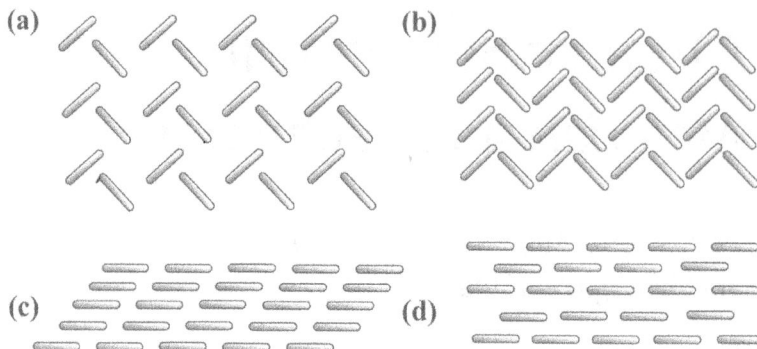

FIGURE 7.1 Different packing motifs in OSCs: (a) herringbone, (b) sandwiched herringbone, (c) lamellar packing with 1D π-π stacking, and (d) lamellar packing with 2D π-π stacking.

bonds leading to the delocalization of electrons along all the π-orbitals in the molecule. This overlapping is called sp²-hybridization and results in two bonds termed as π and σ. However, the energy gap between the bonding and antibonding orbitals is very high for σ bonds resulting an insulation behavior. On the other hand, a small energy difference in the case of π bonds results in the semiconducting property. If the OSC is a planar system, the π bonds become delocalized to form a π-system, which in turn reduces the energy gap and improves the charge transport. In the case of polymers, the π bonds are delocalized along the entire chain, forming a single transport channel. Another important parameter in OSCs that determines their charge transport properties is their molecular packing motifs [6]. Depending upon the molecular packing, the intermolecular interaction between the OSCs can be determined. There are four major molecular packings in the case of OSCs (Figure 7.1): (a) herringbone, (b) sandwiched herringbone, (c) lamellar packing with one-dimensional (1D) π-π stacking, and (d) lamellar packing with two-dimensional (2D) stacking.

Of these, π-π stacking is the best motif for organic electronics and optoelectronic devices. Since the charge transfer in OSCs is between the neighboring molecules, π-π stacking can stabilize the charged molecular states. This transfer process is called polaron hopping mechanism and is discussed in detail in the following section.

7.3 CHARGE TRANSPORT IN OSCs

The charge transport in OSC can be modeled as the hopping process between two molecules (called dimers) [7]. The charge transport between the molecules in a dimer can be interpreted as charge mobilities, a value indicates how fast an electron or hole moves between one molecule and another. In case the system possesses high hole mobility, it is termed as p-type, and if it has higher electron mobility, it is called n-type. If the OSC possesses both the hole and electron transport, it is termed ambipolar. The charge hopping between two identical molecules can be expressed as $M^{+/-}... \rightarrow M...M^{+/-}$. When the charge is injected to a molecule, it undergoes structural relaxation to obtain a stable structure in the charged state. Such distorted molecules are called polarons. The polaron falls back to neutral state by transferring its charge to the neighboring molecule. In modeling OSC charge transport, two parameters play a dominant role: (1) reorganization energy and (2) charge transfer integral. Using these two parameters, the charge transfer rate k can be described as:

$$k = \frac{4\pi^2}{h} \frac{J_e^{\pm 2}}{\sqrt{4\pi\lambda^{\pm}k_B T}} \exp\left(-\frac{\lambda^{\pm}}{4k_B T}\right) \tag{7.1}$$

Here $\lambda^{\pm}$ is the reorganization energy for the hole or electron, $J_e^{\pm}$ is the transfer integral for the hole or electron, h is the Planck's constant, T is the temperature, and k_B is the Boltzmann constant.

Of which the λ can be calculated as the energy difference between the neutral and charged states of a molecule. Whereas the transfer integrals are obtained by modeling the orbital splitting in dimer systems. According to Marcus theory, λ is the measure of energy variation in a molecule by adding or removing a charge, which in turn induces a structural change and is often expressed as inner and outer sphere reorganization energies. However, in most of the theoretical calculations, only inner sphere reorganization values are calculated owing to the high computational cost and its minimal contribution to the overall charge transport. The inner reorganization energy can be expressed as:

$$\lambda^{\pm} = \lambda_1^{\pm} + \lambda_2^{\pm} \tag{7.2}$$

Here $\pm$ is the hole or electron state of the molecule and the $\lambda_1^{\pm}$ and $\lambda_2^{\pm}$ values can be calculated from the difference in adiabatic potential energy as:

$$\lambda_1^{\pm} = E_2^{\pm} - E_3^{\pm} \tag{7.3}$$

$$\lambda_2^{\pm} = E_4^{\pm} - E_1^{\pm} \tag{7.4}$$

where E_1 is the neutral state in neutral geometry, E_2 is the charged state in neutral geometry, E_3 is the charged state in charged geometry, and E_4 is the neutral state in charged geometry. In a gas phase computational calculation (Figure 7.2a), E_1 is the energy of the optimized molecule. Then the charge state of the molecule is changed to either cation or anion, resulting in a variation in molecular structure, and the corresponding energy is E_2. Then this structure will be optimized to global minima to obtain the E_3 value. Finally, the charge of the molecule is changed to a neutral state to obtain the E_4 values.

The other parameter, J_e, indicates how strong the coupling between two molecules in a dimer is to exchange charge carriers between them (Figure 7.2b). Usually, the dimer consists of identical molecules. However, due to their molecular packing in crystalline state, the angle between them varies up to 90°. As a simple case, let us assume an identical dimer system with a co-facial geometry, and the distance between the two molecules is d in angstrom. In the dimer state, the HOMO and

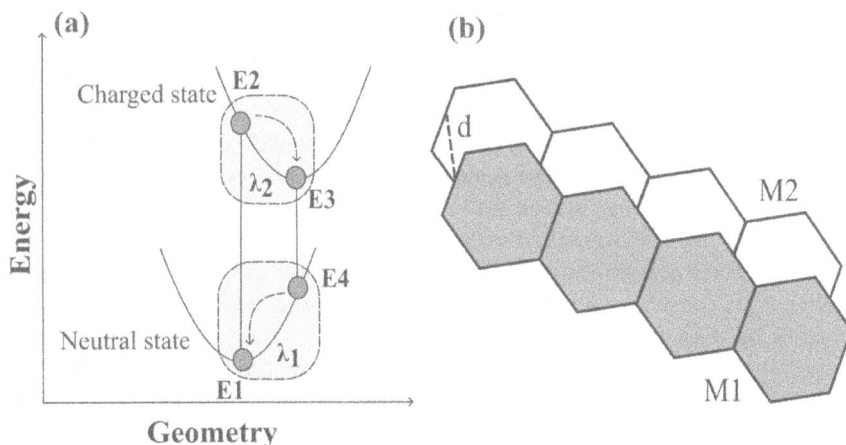

FIGURE 7.2 (a) Reorganization energy calculation from neutral and charged states. (b) Co-facial dimer separated at a distance d.

LUMO of the molecules are combined to form the HOMO and HOMO-1 (or LUMO and LUMO+1), inducing an energy splitting in the system, given as:

$$J_e^\pm = \left(\frac{E_{HOMO} - E_{HOMO-1}}{2} \right) \tag{7.5}$$

By substituting the resultant values, the charge transfer rate can be estimated. Further, by substituting the k values in the Einstein's relation, the charge transfer mobility can be obtained as,

$$\mu = \frac{ed^2k}{2k_BT} \tag{7.6}$$

Here e is the electronic charge. As noted earlier, $\lambda^\pm$ and $J_e^\pm$ influence the μ values and ideally higher $J_e^\pm$ values and minimal $\lambda^\pm$ values are required to get higher charge transport in an OSC.

7.4 FABRICATION OF OSC DEVICES

OSC devices are primarily fabricated in planar architecture using either a single crystal or a thin film. From the application point of view, thin films of either vacuum evaporated or solution processed such as spin casting is used [8]. However, many small molecule OSCs are insoluble in organic solvents, limiting their solution processability. In such cases, thermal evaporation of the material is preferred. On the other hand, the single crystalline form of the material is used in laboratory conditions to obtain high quality results as well as to understand the structure property relationship. Again, owing to their limited solubility, small molecule OSC single crystals are grown thermally using methods such as Bridgman, Chozcralski, or physical vapor transport (PVT). Of these, PVT is the widely adopted method for organic electronics due to its simplicity. The following sections briefly describe the crystal growth and metal contact deposition using PVT and thermal evaporation.

7.4.1 PHYSICAL VAPOR TRANSPORT

In OSC, the sublimation point is typically lower than their melting and boiling points. So, it is possible to convert them directly from solid state into gaseous state at lower temperatures. The sublimed materials can be, with proper temperature profiles, fabricated as high-quality single crystals. PVT is one of the finest sublimation point-based crystal growth techniques and can produce high-quality OSC single crystals [9]. Another interesting and useful aspect of utilizing sublimation is that different materials have different vapor pressures and deposits at different temperature ranges. So, minute impurities present in the material can be purified in this way. On the other hand, by utilizing the same principle, with materials of similar sublimation points and crystal lattices, doping can also be done [10]. A schematic of the PVT setup is shown in Figure 7.3. It consists of a tube furnace featuring two or three distinct temperature zones, and a quartz tube with open ends. The initial materials are positioned and vaporized within the high-temperature zone. For better transport, usually an inert carrier gas such as Ar, N_2, He, or H_2 is used. But often N_2 is considered less than the others due to the presence of oxygen impurities in nitrogen. Removing oxygen from nitrogen is technologically more challenging and may lead to the formation of oxygenated byproducts. The evaporated material is subsequently conveyed by the carrier gas to the low-temperature zone for crystallization. Both material and its impurities undergo crystallization; however, the impurities are deposited either preceding or succeeding the crystallization zone. There is also the possibility of forming co-crystals with the material.

The temperature of the high-temperature zone governs the sublimation rate and holds a pivotal role in crystal growth. Slightly surpassing the sublimation point leads to slow crystal

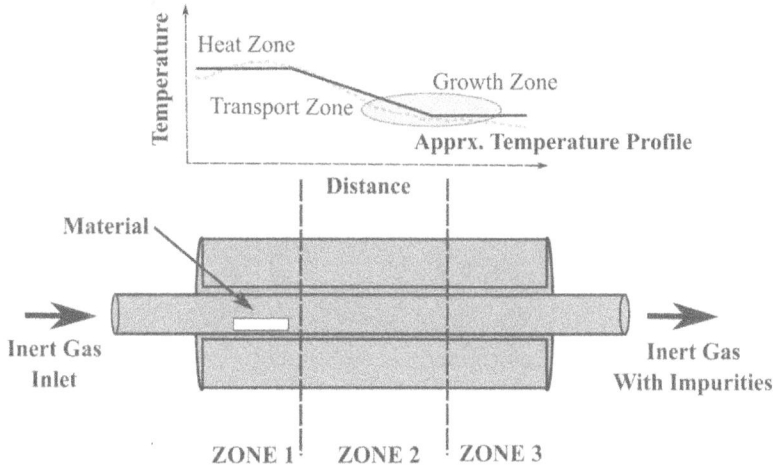

FIGURE 7.3 Approximate temperature profile with different growth zones in a physical vapor transport system.

growth generating large, high-quality crystals. Modulating this temperature can yield different crystal morphologies or molecular arrangements. This is likely because higher temperatures facilitate more molecular movement. In such cases, molecules tend to grow along orientations with stronger intermolecular interactions, resulting in crystals with an equilibrium shape. Moreover, both the quantity and purity of the initial materials play a significant role in subsequent crystal growth. Larger amounts of starting materials result in a higher evaporation rate, leading to faster crystallization.

7.4.2 Substrate Preparation

For OSC device fabrication, a Si^{++} wafer with a SiO_2 dielectric layer of thickness 100 to 500 nm often serves as the substrate. Indium tin oxide (ITO) on a fused silica is another widely used substrate. However, in the case of Si wafer, in addition to SiO_2, polymethyl methacrylate (PMMA) or polystyrene (PS) dielectric layers have also been used to improve the carrier transport and to reduce the trap states [11]. Furthermore, when the single crystals are laminated on these substrates, these extra layers provide better lamination through electrostatic force. In addition to polymers, inorganic materials like Al_2O_3 and monolayers of hexamethyldisilazane are also used as dielectric layers. A combination of these layers along with polymers was also reported. Ideally, the higher the dielectric values the better the charge transport in the system. The thickness of this additional layer can be tailored based on the polymer weight percentage if it is solution-processed. Before the deposition of the polymer layer, the substrate usually undergoes a rigorous cleaning process involving treatment with piranha solution, rinsing in IPA, and an ozone plasma treatment [12]. Subsequently, a polymer solution is applied to form a thin layer on the prepared substrate, which is then annealed overnight in an inert atmosphere. Finally, PVT-grown crystals are electrostatically transferred and laminated onto the substrate. Alternatively, OSC films are deposited by means of thermal evaporation or solution processing.

7.4.3 Contacts Deposition

In OSC devices, fabrication of contacts plays a crucial role in the efficiency and overall performance of the device. Usually, two types of device architectures are possible in this realm. One is based on diodes and another one is based on transistors. Organic light-emitting diodes (OLEDs), OPVs are

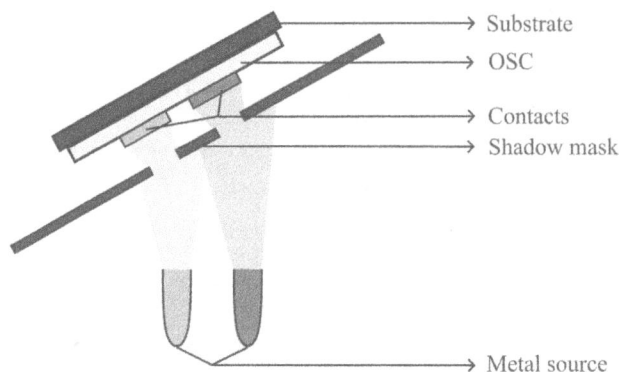

FIGURE 7.4 Deposition of two different metal sources using a shadow mask.

fabricated in diode architecture, with additional layers like electron and hole injection layers, etc. Whereas OFETs are the base architecture for organic photodetectors (OPDs) and organic biosensors (OBSs). The devices are often fabricated as a stacked structure where more than one layer is deposited using a solution or evaporation process. On top of these stacked layers, often two contacts, say, source and drain, are deposited to inject and collect the charges through the device. Depending upon the HOMO and LUMO values of the molecule, the type of metal should be chosen. For example, if a molecule has a HOMO value of 5 eV, then metal with a work function near to 5 eV, like Au, can be used as the contact. Similarly, if the material has a LUMO value of 3 eV, Ca with work function of 2.7 eV could be a better choice. A mismatch between the orbital energy values and the metal work function would significantly affect the charge injection and extraction from the device. Usually, materials like WO_3 or MoO_3 for hole injection and V_2O_5 or CsF for electron injection are used to address this issue [13]. Other combinations like introducing tetratetraconate as an interlayer have also proven to be very efficient [14].

Both contacts and interlayers are often fabricated using physical methods such as thermal evaporation in the ultrahigh vacuum of around 10^{-6} Torr. When air-sensitive materials like Ca are used for the fabrication, entire device fabrication as well as the characterization will be done inside the inert atmosphere. If more than one material is involved in the fabrication of contacts/interlayers, a shadow mask approach is utilized to avoid multiple exposures to vacuum and inert atmosphere. Figure 7.4 shows the schematic of a shadow mask-based contact fabrication. The sources and substrates are aligned precisely so that the source and drain are fabricated with different materials. The quality of the contacts and overall device performance depend on the material purity as well as the deposition parameters such as rate and thickness. Post-deposition processes can also be employed to further optimize the device. This could involve controlled annealing at specific temperatures to enhance the interfaces between metal and semiconductor materials, as well as to alleviate any residual stresses.

7.5 CHARACTERIZATION OF OSC DEVICES

7.5.1 DIODE CHARACTERISTICS

OLEDs and OPVs are the two major devices fabricated as diodes, and for the analysis of their performance and efficiency, many different characterization tools are employed. However, their I-V characteristics can be used as a foundation for assessing their efficiency, stability, and potential for various applications in optoelectronics as well as in energy harvesting.

In the case of OLEDs, when they are operating in the forward bias region (Figure 7.5a), an exponential increase in current with respect to voltage can be observed [15]. This is due to the charge

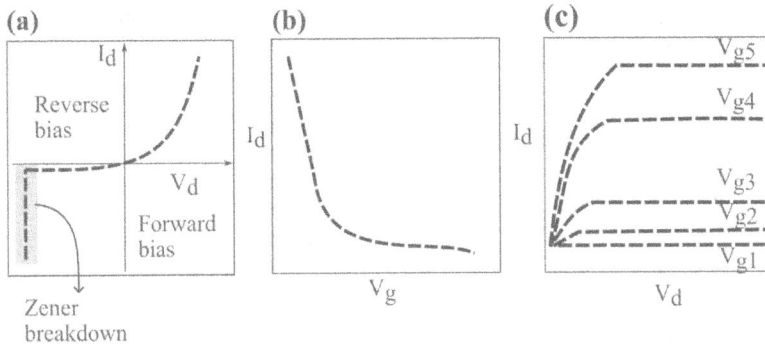

FIGURE 7.5 (a) I-V characteristics of a diode, (b) transfer, and (c) output characteristics of a OFET.

carriers are injected from the electrodes into the active medium. The subsequent charge recombination leads to light emission from the device. By using the ideality factor from diode characteristics, the quality of charge injection and recombination in the device can be determined. In the reverse bias region, initially a sharp increase in current with respect to the voltage can be observed, indicating the onset of device breakdown. This is due to the injection of carriers from the opposite electrodes, resulting undesirable effects such as emission quenching. Usually this would potentially damage the organic layer and device; however, understanding and characterizing the reverse bias behavior is crucial for ensuring the stability and longevity of the OLED devices.

For OPVs, I-V characteristics can be used to evaluate the power conversion efficiency in the devices [16]. Like OLEDs, OPVs operate like a diode in the forward bias region, allowing the flow of current and generating power. Parameters such as short-circuit current (I_{sc}) – maximum current obtained from the device with zero voltage – and open-circuit voltage (V_{oc}) – the voltage across the device when no current is flowing – can be obtained from this region. The shape of an OPV I-V curve can also provide information about the fill factor (FF) of the device, which quantifies the efficiency of power extraction from the cell and the power conversion efficiency (PCE) of devices, a critical parameter for evaluating the performance in solar cell applications. It is worth noting that these parameters might be influenced by a variety of factors, including the material properties, device architecture, and fabrication techniques used.

7.5.2 TRANSISTOR CHARACTERISTICS

Transistor characteristics are majorly analyzed in the form of their transfer and output characteristics and serve as a fundamental tool to optimize the performance of OFETs and related devices.

7.5.2.1 Transfer Characteristics

The transfer characteristics of an OFET can be depicted (Figure 7.5b) as a relationship between input gate voltage (V_g) and the resulting drain current (I_d) while maintaining a constant drain-source voltage (V_{ds}) [17]. The sub-threshold region, followed by the threshold voltage (V_{th}), indicates the role of V_g on the operation of a transistor and its switching behavior. Transfer curves can be influenced by factors such as trap states, interface charges, and the semiconductor-dielectric interface quality. Above the threshold region, the drain current saturates and it is no longer affected by the gate voltage. The sub-threshold slope (S) shows the rate at which the logarithm of the drain current changes with respect to the gate voltage. A lower sub-threshold slope indicates a better control efficiency of the gate. Further, the threshold voltage (V_{th}) obtained from the slope of the transfer curve indicates the required gate voltage to initiate a current flow in the channel. Finally, the field-effect mobility (μ), extracted from the transfer curve, serves as a vital measure to estimate charge carrier mobility in the device. Additionally, the on-off ratio (I_{on}/I_{off}), a crucial factor for digital applications, can also be extracted and used to analyze the switching ability of OFETs.

7.5.2.2 Output Characteristics

The output characteristics (Figure 7.5c) are plotted between the drain current (I_d) against the drain-source voltage (V_{ds}) with varying gate-source voltages (V_{gs}) [18]. It contains a linear and saturation regime. In the linear region, the drain current increases linearly with V_{ds}, and is termed as a resistance-controlled regime. On the other hand, the saturation region indicates that the channel is fully conducting and operating at its maximum current-carrying capacity. Basically, output characteristics are used to obtain information about the saturation current (I_{d_sat}), and it represents the maximum current that OFET can carry under specific operating conditions, a parameter essential to estimate efficiency for high-current applications. Further, the output conductance (g_{ds}) indicates variation in drain current with respect to the changes in V_{ds}. A lower g_{ds} indicates a better channel pinch-off and higher output resistance.

7.6 APPLICATIONS

7.6.1 Organic Light Emitting Diodes

OLEDs are well-known candidates for light-emitting devices and have gained significant attention and application in the field of display technology. Current commercial OLEDs offer potential advantages such as high contrast, wide viewing angles, and energy efficiency compared to their inorganic counterparts [19–20]. In OLEDs, light is emitted in the form of electroluminescence due to the recombination of electrons and holes within the active medium. Usually, OLEDs are made of multiple layers of organic materials sandwiched between two electrodes termed a cathode and an anode. When the voltage is applied across the electrodes, electrons are injected from the cathode and holes from the anode into the OSC layer, usually through an electron transmitting layer (ETL) and hole transmitting layers (HTLs). These layers enhance the chances for more charges to be transferred to the active medium and improve the charge recombination rate. As the emission color depends on the molecular energy levels of the active medium, different colors can be achieved by using different organic systems. Alternatively, molecules can be tailor made to achieve a particular emission wavelength. In the case of commercial displays, full-color emission is achieved by the matrix of different organic emissive layers. Even though almost all commercial displays are made on a rigid flat panel of glass, due to their processability, OLEDs can also be fabricated on flexible substrates like plastics, and few flexible commercial products have already appeared in the market. Yet, issues like durability are still a major concern in this arena, and often OLEDs are encapsulated with polymeric protective layers such as CYTOP [21]. Another notable challenge lies in achieving high-performance OLEDs, and it is expected that designing and synthesizing advanced molecular emitting materials with efficient energy conversion from singlet and triplet excitons can address this issue. Solutions like thermally activated delayed fluorescence (TADF) are already proven to be efficient in this regard [22].

7.6.2 Organic Light Emitting Transistors

Organic light emitting transistors (OLETs) are a combination of two well-known components in modern electronics such as the OFET and OLED. Unlike the OLED, which requires a back panel transistor support, OLETs can control both electrical and optical properties in a single, compact unit [23]. As a transistor, OLET is essentially a three-terminal device containing a source, drain, and gate. The crucial difference is that the OSC acts as the channel for both charge transport as well as for light emission. Though there are reports with ETL and HTL as in OLEDs, due to the ambipolar nature of its active medium, most of the OLETs are reported as a single-layer device.

So, this architecture holds great promise for the investigation of the structure-property relationship in OSCs [24]. On the other hand, materials for OLETs should have both high mobility as well as high photoluminescence quantum yield, which is very scarce among OSCs.

The operation of the OLET is the same as that of the OFET. A field is applied between the source and drain electrodes and is manipulated by means of applied gate voltage. With a suitable V_g the charges recombine in the form of photons to emit light. Like an OLED, the emission color depends on the molecular energy levels and can be tuned by altering the structure, planarity, or crystal packing [25].

Though the field is in active research for more than a decade, still many challenges must be addressed for the successful commercialization of the devices. Improvements like luminescence quenching due to contacts, high threshold voltages, device stability, and new active mediums that cover the entire visible spectrum are required. Already solutions like transparent contacts and high k-dielectrics like Al_2O_3 are reported as possible and potential solutions to overcome some of the issues [26].

7.6.3 ORGANIC SEMICONDUCTOR LASERS

Realization of an electrically pumped organic semiconductor laser (OSL) is one of the holy grails in the arena of organic optoelectronics. To employ organics for lasing, two essential properties are needed: high gain by the effective conversion of excitons into light emission and low threshold with minimal electrical or optical energy as input for pumping states [27]. A practical OSL consists of three essential components such as the organic medium, a microresonator, and metal contacts [28, 29]. Applying potential through the contacts induces charge carriers in the organic medium and is responsible for light emission. The generated light is called spontaneous emission, which is then amplified by the built-in resonators, and the resulting edge emission contains the output laser beam. To achieve the lasing, organics often function as a four-level system (Figure 7.6a). Initially, electrons are excited from the ground state to a higher energy state using electrical or optical pumping. Within a picosecond time frame, these excited species transit to a lower, stable excited state (referred to as S_1) via vibrational relaxation. If there are more excited species than ground-state species, this is known as population inversion. Then these excited electrons fall back to the ground state through the photon emission and ground state thermalization. As Figure 7.6b depicts, a sudden increment in output intensity with respect to the reduction of full-width half-maximum (FWHM), possibly approaching the resolution limit of the detectors, is the vital sign to confirm the lasing from a system. However, this is specific to singlets; in the presence of triplets, intersystem crossing hinders the population inversion and consequently, the output. Additionally, polarons and other device-related issues can significantly impact the device and final output [30, 31].

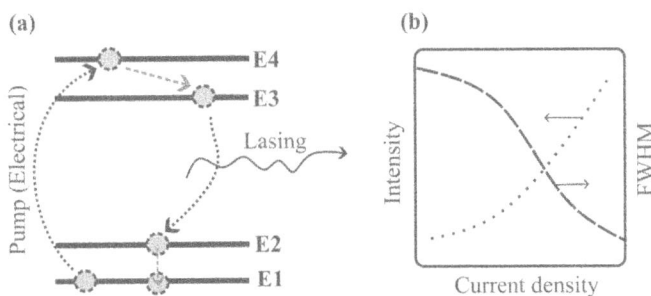

FIGURE 7.6 (a) Energy level diagram for a lasing medium. (b) Output profile to recognize lasing from an organic system.

The quest for new gain materials is imperative in this arena. As mentioned earlier, in organic emissive systems there is a trade-off between mobility and photoluminescence quantum yield (PLQY). High mobility materials tend to have low PLQY and vice versa due to the vibrational coupling effects in the organic medium. Achieving a balance between moderate mobility and moderate PLQY is crucial for lasing, yet such materials are currently scarce. Theoretical design and engineering of material properties hold great promise for synthesizing new materials to meet these criteria.

7.6.4 ORGANIC PHOTODETECTORS

OPDs are a class of devices that can efficiently convert light into an electrical signal and have potential applications in imaging, sensing, and optical communication [32]. OFET-based OPDs operate on the principle of photoconduction, i.e., the conductivity of the device changes with respect to the incident light, allowing it to detect and convert light signals into electrical signals. Like most of its counterparts, OPDs have an OSC as an active layer, and this layer is responsible for the charge transport in response to light exposure. There can be a few other layers to effectively inject photons and extract electrons. When a photon of appropriate energy is incident on the OSC layer, the electrons in the HOMO level are excited to the LUMO level, creating electron-hole pairs. The number of electron-hole pairs generated is directly proportional to the intensity of the incident light. The generated charges are then moved through the active layer under the influence of an electric field applied through the drain and source electrodes. In the case of an OFET architecture, gate voltage can be applied to further control the flow of charges and thus enhance the sensitivity and response of the photodetector. The change in conductivity due to incident light can be read out as an electrical signal and can be further amplified and processed to extract useful information about the incident light, such as intensity, wavelength, or spatial distribution [33].

Despite the numerous advantages of OPDs, their practical applications have not materialized yet. Still the performance of OPDs is inferior to their inorganic counterparts [40]. While the OSC offers benefits like flexibility, large area, and low cost, these advantages are not fully incorporated into OPDs. Most research on OPDs is currently at laboratory level with limited participation from companies. Like their sister fields, OPDs are also susceptible to degradation by means of atmospheric contamination, and this limits the device's lifetime and the practical applications [34].

7.6.5 ORGANIC BIOSENSORS

OBSs are often constructed as an extension of OFETs and are used to detect active chemicals and biological species. Like all OSC devices, OBSs are advantageous in the sense of flexibility and low-cost fabrication. But the major advantage lies in their compatibility with biological systems. OBSs are working on the principle that the interaction between OSCs and target molecules such as proteins or DNA strands would alter the conductivity of the device and act as the sensing element [35]. OBSs can be designed in a way that either the active medium or the extended gate acts as the target site. The extended gate mechanism is preferable in the case of reusable sensors, and usually a metal-organic framework or functionalized proteins are deposited on the gate electrodes to ensure the proper interaction. Various mechanisms can be accounted for the interaction between OBS and target molecules, including charge transfer, electrostatic interactions, or chemical binding. For example, in DNA sensors, the DNA strings hybridize with the complementary sequences on the OSC surface, leading to a change in charge distribution and consequently altering the electronic properties. Such alterations can then be transduced into an electrical signal.

OBSs have a wide range of applications in fields such as healthcare. For example, OBS-based flexible glucose sensors are demonstrated for continuous monitoring of sugar levels in diabetic patients using their sweat. Additionally, the active OSCs in OBSs can be tailor-made to detect various proteins, viruses, and genetic mutations, aiding in the early detection of diseases. OBSs also have significant potential in environmental monitoring and can be employed to detect pollutants, toxins, and pathogens in air, water, and soil [36].

7.7 CHALLENGES

One of the primary challenges in the field of OSC materials and devices is to achieve higher charge carrier mobilities [37]. The OSC charge carrier mobility is often much lower than the ISC and this is a critical factor for applications like detectors and sensors. Developing strategies to enhance charge transport properties while maintaining the advantages of organic materials remains a significant research focus. Another challenge lies in achieving the long-term stability and reliability of OSDs [38]. Most of the OSCs are susceptible to degradation over time, particularly when exposed to environmental factors such as moisture, oxygen, and light. This is an important concern for devices to operate over extended periods, such as in displays or medical implants. Strategies to encapsulate and protect OSDs from environmental factors are actively being explored. The development of efficient and stable interfaces between OSCs and other materials, such as electrodes or dielectrics, is also a crucial challenge [39]. Achieving good charge injection, extraction, and transport at these interfaces is necessary for optimizing the device performance. As we have seen in Section 7.2, the arrangement of molecules within the active layer of an organic device strongly influences its electronic properties [40]. Achieving uniform and well-ordered film morphologies is critical for reproducibility and device optimization. Methods to precisely control film deposition, such as advanced processing methods and molecular alignment techniques, are actively being explored. While OSC offers advantages such as low cost and flexibility, achieving high throughput and yield in large-scale production remains a complex problem [41]. Techniques for high-speed printing, roll-to-roll processing, and other scalable fabrication methods are continuously being developed to address this challenge. Lastly, variations in material properties, fabrication processes, and environmental conditions can lead to inconsistencies in device performance. Developing techniques for accurate characterization and quality control, as well as designing robust device architectures, are essential for ensuring the reliable and consistent operation of OSDs in practical applications.

7.8 CONCLUSION

OSCs are a class of materials that have potential applications in the arena of optoelectronics. Unlike their inorganic counterparts, organic materials have discrete bands termed molecular orbitals, and their properties depend entirely on the molecular structure and its packing. Of the four major motif packings, π-π stacking is the most favorable for optoelectronic applications. These materials can be processed both in solution and in thermal evaporation process. However, owing to their limited solubility, small molecular systems are often fabricated using thermal methods. Physical vapor transport is one of the widely used techniques for the growth of OSCs, and depending upon their HOMO and LUMO, a choice of metal contacts can be fabricated. Once fabricated, most of the devices are characterized for current-voltage characteristics, and based on that, their performance and efficiency can be estimated. OLEDs are among the widely recognized applications of OSCs, and their sister fields, such as photovoltaic cells, sensors, and detectors, are maturing and approaching commercial device limits. To fully recognize the potential of OSCs, their mobility and overall stability still have to be improved.

ACKNOWLEDGMENTS

The authors acknowledge the financial support from SERB CRG in the form of a research grant by the Government of India (CRG/2022/006100). One of the author PAP thanks the Indian Institute of Science Education and Research – Tirupati for the financial support.

REFERENCES

1. M. Gregor, T. Grasser, eds. Organic electronics. Springer Berlin, Heidelberg (2009).
2. W. Wei, L. Luo, P. Sheng, J. Zhang, Q. Zhang, Multifunctional features of organic charge-transfer complexes: Advances and perspectives, Chemistry–A European Journal 27(2) (2021) 464–490.
3. F. Stephen R., M. E. Thompson, Introduction: Organic electronics and optoelectronics, Chemical Reviews 107(4) (2007) 923–925.
4. W. Cong, X. Zhang, H. Dong, X. Chen, W. Hu, Challenges and emerging opportunities in high-mobility and low-energy-consumption organic field-effect transistors, Advanced Energy Materials 10(29) (2020) 2000955.
5. O. Toshihiro, C. P. Yu, C. Mitsui, M. Yamagishi, H. Ishii, J. Takeya, Bent-shaped p-type small-molecule organic semiconductors: A molecular design strategy for next-generation practical applications, Journal of the American Chemical Society 142(20) (2020) 9083–9096.
6. H. Guangchao, Y. Yi, Z. Shuai, From molecular packing structures to electronic processes: Theoretical simulations for organic solar cells, Advanced Energy Materials 8(28) (2018) 1702743.
7. B. Arka, P. A. Praveen, S. V. Bhat, S. Dhanapal, A. Kandhasamy, T. Kanagasekaran, Theoretical insights on pyrene end-capped thiophenes/furans and their suitability towards optoelectronic applications, Computational and Theoretical Chemistry 1225 (2023) 114135.
8. A. K. Chauhan, J. Purushottam, D. K. Aswal, J. V. Yakhmi, Organic devices: Fabrication, applications, and challenges, Journal of Electronic Materials 51(2) (2022) 1–39.
9. J. Hui, C. Kloc, Single-crystal growth of organic semiconductors, MRS Bulletin 38(1) (2013) 28–33.
10. Z. Xiaotao, H. Dong, W. Hu, Organic semiconductor single crystals for electronics and photonics, Advanced Materials 30(44) (2018) 1801048.
11. B. Arka, P. A. Praveen, T. Kanagasekaran, A combined theoretical and experimental approach to deduce the role of dielectric layer on interface trap density in single crystal organic field-effect transistors, Crystal Research and Technology 58(7) (2023) 2200263.
12. H. H. Richter, A. Wolff, K. Blum, K. Hoeppner, D. Krüger, R. Sorge, Damage to Si substrates during SiO_2 etching: Opportunities of subsequent removal by optimized cleaning procedures, Vacuum 47(5) (1996) 437–443.
13. H. Szuheng, S. Liu, Y. Chen, F. So, Review of recent progress in multilayer solution-processed organic light-emitting diodes, Journal of Photonics for Energy 5(1) (2015) 057611–057611.
14. K. Thangavel, H. Shimotani, R. Shimizu, T. Hitosugi, K. Tanigaki, A new electrode design for ambipolar injection in organic semiconductors, Nature Communications 8(1) (2017) 999.
15. V. David, OLED characterization, OLED microdisplays: Technology and applications, Wiley (2014) 53–94.
16. S. Vishal, G. Li, Y. Yao, T. Moriarty, K. Emery, Y. Yang, Accurate measurement and characterization of organic solar cells, Advanced Functional Materials 16(15) (2006) 2016–2023.
17. K. Ioannis, Organic field effect transistors: Theory, fabrication and characterization, Springer Science & Business Media (2008).
18. W. Yugeng, Y. Liu, Y. Guo, G. Yu, W. Hu, Experimental techniques for the fabrication and characterization of organic thin films for field-effect transistors, Chemical Reviews 111(5) (2011) 3358–3406.
19. K. N. Thejo, S. J. Dhoble, Organic light emitting diodes: Energy saving lighting technology—A review, Renewable and Sustainable Energy Reviews 16(5) (2012) 2696–2723.
20. L. Yuchao, C. Li, Z. Ren, S. Yan, M. R. Bryce, All-organic thermally activated delayed fluorescence materials for organic light-emitting diodes, Nature Reviews Materials 3(4) (2018) 1–20.
21. J. Granstrom, J. S. Swensen, J. S. Moon, G. Rowell, J. Yuen, A. J. Heeger, Encapsulation of organic light-emitting devices using a perfluorinated polymer, Applied Physics Letters 93(19) (2008).
22. S. Jinouk, H. Lee, E. G. Jeong, K. C. Choi, S. Yoo, Organic light-emitting diodes: Pushing toward the limits and beyond, Advanced Materials 32(35) (2020) 1907539.
23. Q. Zhengsheng, H. Gao, H. Dong, W. Hu, Organic light-emitting transistors entering a new development stage, Advanced Materials 33(31) (2021) 2007149.
24. Z. Congcong, P. Chen, W. Hu, Organic light-emitting transistors: Materials, device configurations, and operations, Small 12(10) (2016) 1252–1294.
25. Y. Dafei, V. Sharapov, X. Liu, L. Yu, Design of high-performance organic light-emitting transistors, ACS Omega 5(1) (2019) 68–74.
26. S. Caterina, Engineering dielectric materials for high-performance organic light emitting transistors (OLETs, Materials 14(13) (2021) 3756.

27. K. Alexander JC, M. C. Gather, Organic lasers: Recent developments on materials, device geometries, and fabrication techniques, Chemical Reviews 116(21) (2016) 12823–12864.

28. J. Yi, Y.-Y. Liu, X. Liu, H. Lin, K. Gao, W.-Y. Lai, W. Huang, Organic solid-state lasers: A materials view and future development, Chemical Society Reviews 49(16) (2020) 5885–5944.

29. W. Jun-Jie, X.-D. Wang, L. S. Liao, Advances in energy-level systems of organic lasers, Laser & Photonics Reviews 16(12) (2022) 2200366.

30. A. S. D. Sandanayaka, T. Matsushima, F. Bencheikh, S. Terakawa, W. J. Potscavage, C. Qin, T. Fujihara, K. Goushi, J.-C. Ribierre, C. Adachi, Indication of current-injection lasing from an organic semiconductor, Applied Physics Express 12(6) (2019) 061010.

31. K. Thangavel, H. Shimotani, K. Kasai, S. Onuki, R. D. Kavthe, R. Kumashiro, N. Hiroshiba, T. Jin, N. Asao, K. Tanigaki, Towards electrically driven organic semiconductor laser with field-effective transistor structure, arXiv preprint arXiv:1903.08869 (2019).

32. R. Hao, J. D. Chen, Y. Li, J. Tang, Recent progress in organic photodetectors and their applications, Advanced Science 8(1) (2021) 2002418.

33. C. Philip, T. Someya, Organic photodetectors for next-generation wearable electronics, Advanced Materials 32(15) (2020) 1902045.

34. Y. Dezhi, D. Ma, Development of organic semiconductor photodetectors: From mechanism to applications, Advanced Optical Materials 7(1) (2019) 1800522.

35. W. Juan, D. Ye, Q. Meng, C. Di, D. Zhu, Advances in organic transistor-based biosensors, Advanced Materials Technologies 5(7) (2020) 2000218.

36. S. Chenfang, X. Wang, M. A. Auwalu, S. Cheng, W. Hu, Organic thin film transistors-based biosensors, EcoMat 3(2) (2021) e12094.

37. H. Hamna F, A. M. Zeidell, O. D. Jurchescu, Charge carrier traps in organic semiconductors: A review on the underlying physics and impact on electronic devices, Journal of Materials Chemistry C 8(3) (2020) 759–787.

38. A. Seung-Hee, J. Jeong, S. June Kim, Emerging encapsulation technologies for long-term reliability of microfabricated implantable devices, Micromachines 10(8) (2019) 508.

39. C. Hongliang, W. Zhang, M. Li, G. He, X. Guo, Interface engineering in organic field-effect transistors: Principles, applications, and perspectives, Chemical Reviews 120(5) (2020) 2879–2949.

40. Z. Fuwen, C. Wang, X. Zhan, Morphology control in organic solar cells, Advanced Energy Materials 8(28) (2018) 1703147.

41. P. Sungmin, T. Kim, S. Yoon, C. W. Koh, H. Y. Woo, H. J. Son, Progress in materials, solution processes, and long-term stability for large-area organic photovoltaics, Advanced Materials 32(51) (2020) 2002217.

8 Organic Semiconductors for Electrochemical Energy Applications

Jeffery Horinek, Allen Davis, and Ram K. Gupta

8.1 INTRODUCTION

As global energy requirements continue to rise, the need for sustainable materials has never been more apparent. One field that stands to benefit the most from stability is that of energy. In its current state, the major applications for energy materials can be broken into three distinct categories. The first of these categories is that of materials for energy production. This group usually encompasses materials such as petroleum products, natural gas, and coal [1]. In terms of stainability, renewable sources such as wind, hydroelectric, and solar have been explored. The second category of materials are those involved in energy transportation. This section of materials typically encompasses conductive materials such as copper, steel, and aluminum [2]. Sustainability in this field primarily involves metal reclamation and recycling, as well as research towards cleaner source extraction. The final collection of materials is those involved in energy storage. This class of materials includes alkali/alkali earth metals, transition metal oxides, and some carbon-based materials. These base materials are then applied to energy storage devices such as batteries, supercapacitors, and fuel cells. Sustainability in energy storage is of extreme interest in the scientific community, as there are many different paths to consider. Lithium is a powerful option for batteries but is limited in availability. Meanwhile, precious metals such as platinum and iridium are efficient fuel cell electrocatalysts but suffer from cost, scarcity, and stability. These examples express the immense need for more varied and efficient materials, of which nanoparticles are a potential solution.

Nanoparticles are a unique class of material that receive their name from their extremely small size. Nanoparticles consist of solid colloidal particles that range in size from as small as 10 nm up to 1000 nm [3]. In conjunction with their small size, nanoparticles exhibit a wide range of shapes. These shapes range from simple spheres, cubes, rods, and plates to more complex forms such as ribbons, flowers, and crystals [4].

Due to their small size and plethora of shapes, nanoparticles can be used in a variety of applications. These applications include drug delivery, chemical sensing, catalysis, water treatment, and many more [5]. As for their application in energy storage, there are several types of nanoparticles that can be used, each with their own unique advantages. One of the more common varieties are metal nanoparticles. This class of nanoparticles includes noble metals such as gold, silver, and copper, as well as more common transition metals like iron and nickel. The advantages of metal nanoparticles are high thermal stability, low cost, ecological compatibility, and high conductivity. Metal oxide nanoparticles exist as a derivative of transition metal-based nanoparticles. Despite their initial similarities, metal oxide nanoparticles can exhibit vastly different properties. Rapid ion transport, high electronic conductivity, high surface area, high porosity, and high theoretical capacitance are all factors seen in these materials [6, 7]. Another commonly used nanoparticle class is that of carbon-based nanoparticles. This family of nanoparticles focuses on allotropes of carbon, such as fullerenes, carbon nanotubes, nanodiamonds, and graphene. Examples of these nanoparticles can be seen in Figure 8.1.

DOI: 10.1201/9781003450146-8

FIGURE 8.1 Examples of various carbon-based nanoparticles. (Adapted with permission [8]. Copyright 2021, American Chemical Society.)

These nanoparticles offer a plethora of desirable characteristics such as a high surface area, tunable morphologies, excellent thermal/chemical stability, and high mechanical strength [9]. The final class of nanoparticles are known as organic nanoparticles. This group of materials encompasses metal-organic frameworks (MOFs), nanopolymers, dendrimers, and certain biological macromolecules such as micelles and proteins [10, 11]. This class of nanoparticles offers advantages like increased ion transport, low cost, recyclability, and increased redox potential when applied to energy applications [12]. Despite the wide range of options available, few organic materials are compatible with the energy field, Of the materials that demonstrate compatibility, organic semiconductors are the most promising. Organic semiconductors can be divided between two distinct types. These are polymer-based semiconductors and small molecule-based semiconductors. Polymeric semiconductors are comprised of long polymer chains and tend to be semicrystalline in nature. Opposingly, small molecule semiconductors tend to have discrete, well-defined shapes, with distinct crystalline features [13]. The semiconductive nature of these materials is defined by the size of their band gap. Like insulators, they feature a noticeable band gap that inhibits electron flow. Unlike insulators, this gap is greatly reduced to a size of ~1–3 electron volts. Through doping, this gap can be further reduced, endowing conductivity to the material. This in turn allows for the material to switch between a conductor and an insulator depending on the need.

8.2 SYNTHESIS AND CHARACTERIZATION OF ORGANIC SEMICONDUCTORS

As stated earlier there are two types of organic semiconductors, polymeric semiconductors and small molecule semiconductors, and each of these materials have their own unique properties. Additionally, they may vary in their synthesis methods. In the case of polymeric semiconductors, they are strictly limited to solution synthesis, whereas small-molecule semiconductors can be synthesized by either solution or vacuum deposition methods [14, 15]. Furthermore, these two conductor types also differ in the advantages they possess. In the case of polymer organic semiconductors, they feature easy processability, good bandgap adjustability, excellent mechanical properties, and low-temperature processing. This type of semiconductor can be processed at low temperatures as allowed by the mechanics of solution synthesis, since high temperatures are not required [16]. As for small molecule organic semiconductors, they offer a well-defined chemical structure, definite molecular weight, and low batch-to-batch variation. The limited variation allows the semiconductors produced by vacuum deposition to demonstrate remarkable consistency, leading to more consistent products [17].

An example of polymer organic semiconductor synthesis was presented in a study conducted by Jarrahi et al., wherein polyaniline was combined with cadmium sulfide in a one-pot solution synthesis to fabricate an organic polymer semiconductor nanocomposite for photovoltaic applications [18]. This process was accomplished by synthesizing three solutions. To start, solution 1 was synthesized by dissolving 15 mmol of ammonium persulfate in 50 mL of distilled water. Solution 2 was then prepared by dissolving sodium sulfide in 50 mL of distilled water. Finally, solution 3 was created by dissolving 17 mmol paratoluene sulfonic acid in 50 mL of distilled water. Aniline and cadmium chloride were added to this solution, then sonicated and stirred simultaneously for one hour. Following this, solutions 1 and 2 were added to solution 3 dropwise at a rate of one drop every 20 seconds. This step was done under a neutral atmosphere at 0°C. The solution was then stirred under the same conditions for 24 hours. Afterward, the solution was washed several times with distilled water and ethanol to remove any unreacted materials. The solution was then filtered to retrieve the nanocomposite materials, which were then subsequently dried in an oven for 24 hours at 60°C. A scheme depicting this reaction process can be seen in Figure 8.2 [18].

Small molecule semiconductors made via solution synthesis typically follow procedures like their polymeric counterparts. Vacuum deposition synthesis, on the other hand, is a completely different process. An example of this process was shown in a study by Kim et al., where N,N′-bis(2-ethylhexyl)-1,7-dicyanoperylene-3,4:9,10-bis(dicarboxyimide) (PDI-CN2) semiconductors were synthesized for electronic applications [19]. This was accomplished by first coating a mesh wafer in silicon oxide. After coating, this wafer was vigorously washed with distilled water. Following this, the wafer was coated with the organic coupling agent octadecyltrichlorosilane via dip coating

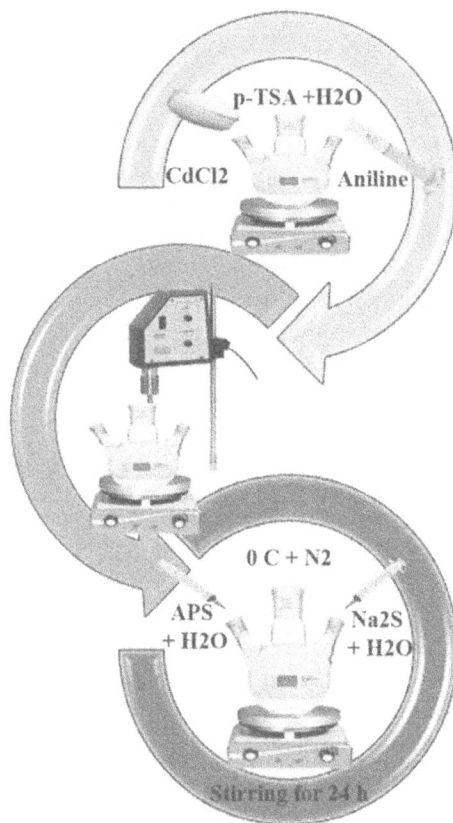

FIGURE 8.2 Example of the synthesis scheme for the polyaniline-based organic semiconductor via solution synthesis. (Adapted with permission [18]. Copyright 2023, American Chemical Society.)

to create a dielectric substrate. After dip coating, the substrates were baked at 120°C for 20 minutes under a nitrogen atmosphere. Finally, the PDI-CN2 was deposited on the substrate using organic molecular beam deposition. This process was completed in a vacuum with a pressure of 10^{-7} Torr.

Regardless of the semiconductor type, the characterization methods used are the same. To analyze the functional groups of these semiconductors Fourier-transform infrared (FTIR) spectroscopy is used. To analyze the crystallinity and crystal phase of these semiconductors, X-ray diffraction (XRD) is performed. To analyze the surface of these semiconductors, scanning electron microscopy (SEM) is used. These methods help to confirm pore size, surface area, and nanoparticle shape. One final characterization method is UV-Vis spectroscopy. This method is used to measure the transparency of the materials. More importantly, this method helps to confirm the conjugated pi bonds and band gaps that make up organic semiconductors [20].

8.3 ELECTROCHEMICAL ENERGY APPLICATIONS OF ORGANIC SEMICONDUCTORS

8.3.1 ORGANIC SEMICONDUCTORS FOR SUPERCAPACITORS

Supercapacitors are a class of energy storage devices known for possessing high power density at the cost of a low energy density. This factor makes them popular for applications that require a high volume of electricity to be delivered quickly. Along with this characteristic, supercapacitors also feature stable cycling performance, long use life, and relatively low cost [22]. The method of energy storage used in supercapacitors depends on the design of their electrodes. The first type of supercapacitor is the electrical double-layer capacitor (EDLC). This device functions by forming a Helmholtz layer at the interface of the electrode and the electrolyte, with charge being stored on the surface of the electrode [23]. EDLCs typically use carbonaceous materials such as carbon nanotubes and graphene due to their high porosity and surface area. However, the limitations inherent to surface-based energy storage causes EDLCs to suffer from limited energy density [24]. Pseudocapacitors are another common supercapacitor design that functions due to fast and reversible redox reactions on and within the surface of the electrodes. These capacitors typically rely on redox active materials such as metal oxides and sulfides. Unlike EDLCs, pseudocapacitors store much of their energy chemically. This provides pseudocapacitors with higher energy density, specific capacitance, and charge storage. Consequently, deformations in the structure of the material may occur due to continuous charge-discharge cycling, greatly reducing their stability [25]. The final family of supercapacitors are hybrid supercapacitors, which try to marry the effects of the prior two. These supercapacitors operate by combining the characteristics of pseudocapacitors with those of EDLC devices. As such, hybrid supercapacitors often rely on a combination of the prior materials in their construction [26].

Despite the vast quantity of materials that can be used in supercapacitor design, researchers are still searching for materials that can provide higher energy density, power density, and capacitance. In terms of supercapacitor application, organic semiconductors are most suited to electrode construction. This concept was illustrated in a study by Wang et al., wherein uniform polypyrrole nanowires were synthesized for use as supercapacitor electrodes [27]. An image detailing the synthesis of the polypyrrole nanowires can be seen in Figure 8.3 [27].

The nanowires were then suspended in ethanol and combined with acetylene black, polyvinylidene difluoride, and N-methylpyrrolidone to make a slurry. This slurry was then coated onto carbon cloth to fabricate the working electrode. Using the same methods, two composite electrodes were also fabricated. One of these composite electrodes had gemini surfactant added to it while the other had CTAB added. These electrodes had the material added during the first step of the nanowire synthesis and had the following molar ratios (0.2,0.4,0.6,0.8). This in turn created PPy-GS and PPy-CTAB electrodes. These three electrodes were tested using a three-electrode system in 1.0 mol/L

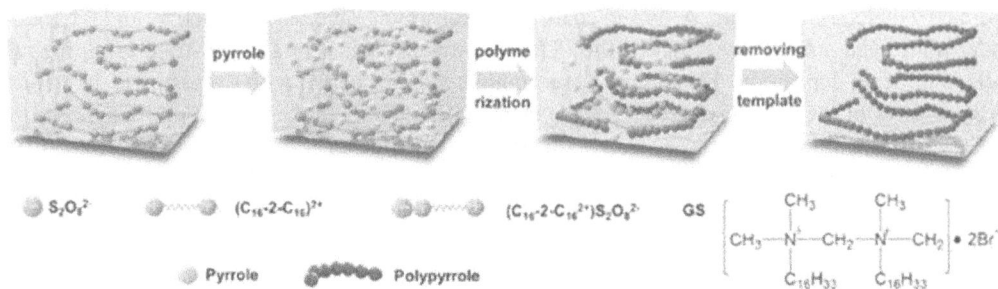

FIGURE 8.3 Depiction of the polypyrrole nanowire synthesis. (Adapted with permission [27]. Copyright 2023, American Chemical Society.)

H_2SO_4 aqueous electrolyte, and out of all the electrodes, the PPy-GS with a GS molar ratio 0.4 was found to have the superior electrochemical properties. The electrode featured a specific capacitance of about 556 F/g at 1 A/g. As for the energy and power density of these electrodes, they also fared the best as those values came in at about 49.4 Wh/kg and 400 W/kg. Additionally, this electrode featured a capacity retention of about 85.4% after 2000 cycles. The reason that this electrode was able to perform better than the others was attributed to the GS, as it provided a better distribution of the polypyrrole nanowires throughout the electrode.

Aside from being used as the nanoparticle in the electrode, organic semiconducting materials can be used to entrap nanoparticles to help improve their electrochemical properties. This concept was illustrated in a study by Baby et al. wherein the organic semi-conducting polymer, polyaniline, was combined with graphene oxide to fabricate nanocomposite electrodes with excellent electrochemical properties [28]. To begin, graphene oxide was synthesized using a modified Hummers method. The graphene was then modified using an azopyridine derivative to create GO-Azo. Following this, the polyaniline GO-Azo nanocomposite was synthesized via solution synthesis. The electrodes were then prepared by coating a slurry of active material (85 wt %), polyvinylidene fluoride (10 wt %), and carbon black (5 wt %) onto prepared carbon cloth pieces. In total, two working electrodes were prepared, one using the GO-Azo compound and the other using the polyaniline GO-Azo nanocomposite (PANI/GO-Azo). These electrodes then underwent extensive electrochemical analysis using a three-electrode testing system to examine the EIS, CV, and GCD of each electrode. A figure showing the results of these tests can be seen in Figure 8.4 [28].

As seen above, the addition of PANI to GO-Azo outperformed the standard GO-Azo handily. At a current density of 0.25 A/g, this electrode recorded the highest gravimetric capacitance and areal-specific capacitance with a value of about 426.25 F/g and 622 mF/cm^2, respectively. This electrode also demonstrated excellent columbic efficiency as after 5,000 cycles at a current density of 7 A/g the columbic efficiency was found to be 98.5%. Furthermore, this electrode featured a high electrochemical durability as after 5,000 cycles a charge retention rate of 89% was found.

Another way supercapacitor electrodes can be fabricated is by bonding an organic thin film to a substrate. This concept was shown in a study by Han et al., where poly(3,4-ethylenedioxythiophene) (PEDOT) nanowires were used to fabricate PEDOT thin films for supercapacitor electrodes [29]. To start with, PEDOT nanowires were synthesized via solution synthesis and then suspended in a methanol solution during polymerization. The thin films were polymerized for different lengths of time in order to produce different films. After polymerization, the films were retrieved via vacuum filtration. The thin films were then cut into circular shapes with identical geometric areas of 1.13 cm^2. These electrodes were then tested using the three-electrode testing system. After testing, it was discovered that the PEDOT film that polymerized for six hours yielded the best results. This electrode featured an energy density of 1.69 mWh cm^{-3}. Along with this energy density, a power

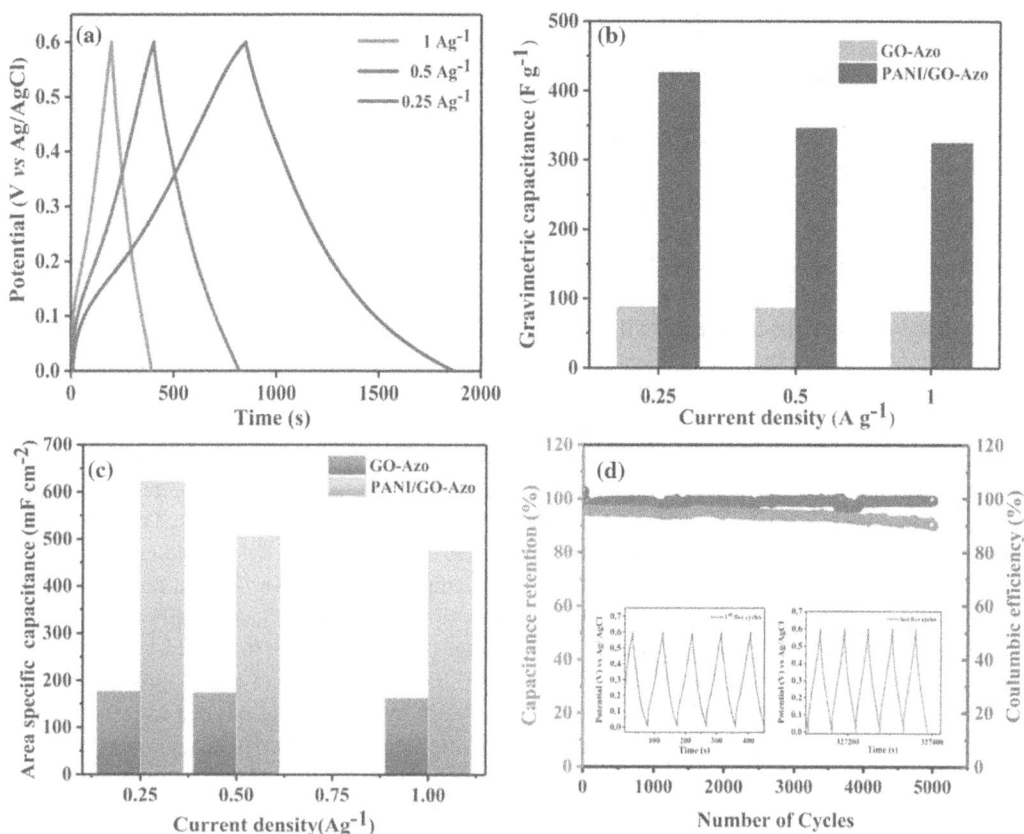

FIGURE 8.4 (a) GCD at different current densities, (b) gravimetric specific capacitance at various current densities, (c) areal specific capacitance at different current densities, and (d) capacitance retention vs. coulombic efficiency. (Adapted with permission [28]. Copyright 2023, American Chemical Society.)

density of 30.5 mW cm^{-3}. Additionally, at a current density of 0.3 A/g, a specific capacitance of 117 F/g was achieved. Along with these values, a coulombic efficiency of 100% was found after 10,000 charge-discharge cycles. This shows the PEDOT films that were synthesized exhibit excellent electrochemical stability. Also, after 10,000 charge-discharge cycles, a capacitance retention of 98% was found.

Even though organic semiconductors perform very well in supercapacitor applications on their own, researchers recently have begun combining organic semiconductors with MOFs. This was presented in a study by Yue et al. wherein a porous polypyrrole scaffold was combined with MOFs to fabricate flexible solid-state supercapacitors. This was accomplished by first using solution synthesis to produce a porous polypyrrole scaffold. This scaffold was then immersed in a solution of copper acetate monohydrate and 2, 3, 6, 7, 10, 11-hexahydroxytriphenylene at 85°C for 6 hours. After this, the product was removed and washed with distilled water and dimethylformamide (DMF) before being soaked in acetone at 80°C for two days. The product was then combined with polyvinylidene fluoride and carbon black to make a slurry. This slurry was then coated onto prepared carbon cloth sheets to fabricate a working electrode. This electrode then underwent testing using a three-electrode system. This electrode featured a high specific capacitance of about 233 mF/cm at a current density of 0.5 mA/cm^2. Additionally, the electrode had an energy density of µWh/cm^2 and a power density of about 1.5mW/cm^2. Along with these values the electrode also had an incredibly high specific surface area of about 107 m^2/g.

8.3.2 Organic Semiconductors for Batteries

As an energy storage device, batteries occupy a happy medium between energy and power density. The most common batteries function using a relatively simple electrochemical redox process. The process starts at the anode (negative terminal) where electrons are produced through an oxidation reaction. These electrons are then accepted by the cathode (positive terminal) and used in a reduction reaction. Between these two reaction sites exists an electrolyte that is used to help ferry ions between the two reactions. This reaction cycle is what helps batteries to have a higher energy density when compared to supercapacitors. More advanced batteries rely on intercalation to function. This process is the reversable insertion of a molecule into a layered material. Through this process the ions can flow between the terminals of the battery and be stored to a much greater degree. One of the most common batteries that uses intercalation is the lithium-ion battery [30]. As stated in the name, this battery relies on lithium as the redox agent, where it offers several desirable characteristics such as excellent reversible capacity, high energy density, long service life, and wide operating temperature [31]. Despite these excellent characteristics, lithium suffers from extremely limited stores and detrimental environmental impact [32]. Another common battery is the zinc ion battery. Similarly, to lithium ion, these batteries function via intercalation of zinc ions in the battery. These batteries feature advantages like high gravimetric and volumetric capacities, high safety and environmental friendliness, the abundance on earth, and ease of manufacturing [33]. Another battery type is the alkaline battery. This battery functions via a redox reaction. These batteries typically use materials like zinc, manganese oxide, and graphite. The advantages associated with this battery include cost, safety, ability to deal with heavy-duty discharge, long storage life, and being leak proof [34]. One final battery is the metal air battery. These batteries work by first letting oxygen into the battery at the cathode. This oxygen is then reduced to form hydroxide ions. These hydroxide ions travel to the anode through the electrolyte where they participate in the oxidation of the metal in the anode. The oxidation of this metal produces free electrons and metal cations. These batteries are typically made of materials like zinc, aluminum, magnesium, and lithium. Additionally they offer advantages like cost, energy efficiency, high energy density, and environmental friendliness [34, 35]. Although these batteries have excellent properties, researchers are continuing to dive deeper into the inclusion of organic semiconductors into everyday batteries. As it stands now organic semiconductors offer several advantages that other materials cannot, such as low cost, high flexibility, large area fabrication, easy processability, light weight, and low toxicity [36].

With the push for flexible electronics garnering evermore support, extremely flexible organic semiconductors have gained increased attention. This fact was demonstrated in a study by Cang et al. [37], wherein poly(3,4,9,10-perylentetracarboxylic dimide) (PPTCDI), an organic semiconducting material, was used to fabricate the anode in a flexible zinc ion battery. Firstly, PPTCDI particles were prepared via solution synthesis, then the PPTCDI, polyvinylidene difluoride, acetylene black, and 1-methyl-2-pyrrolidone was mixed to prepare a slurry. This slurry was then coated on carbon fiber cloth in order to fabricate a working anode terminal. This anode was combined with a zinc electrolyte and activated carbon cathode to fabricate the battery. Said battery then underwent electrochemical analysis using a LAND battery testing system [37].

Using an organic semiconducting polymer as an anode, the zinc batteries achieved an excellent reversible capacity of 199.1 mAh/g at 100 mA/g. Along with the reversible capacity, a current density of 100 mA/g between 0 and 1.8 V was found, alongside an energy density of 174.3 Wh/kg. Finally, a coulombic efficiency of nearly 100% was found after 5,000 charge-discharge cycles, with a capacitance retention of 98%.

While the hallmark flexibility of some organic semiconductors pushes them into the realm of flexible batteries, they still demonstrate usefulness in the fabrication of more traditional batteries. This is due to the ability of organic semiconductors to provide the battery with increased functionality in various climates. This was seen in a study by Wu et al. wherein the organic semiconductor 5,7,12,14-pentacenetetrone (PT) was used as a cathode material in aqueous zinc-organic batteries [38]. This was accomplished by fabricating a cathode electrode using 50 wt % PT, 40 wt % conductive carbon, and 10 wt % polytetrafluoroethylene binder on a 400-mesh stainless steel mesh. This electrode was then combined with a zinc anode and a trimethyl phosphate water hybrid electrolyte to fabricate the battery. After undergoing electrochemical analysis, it was discovered that these batteries function well in a range of temperatures. At room temperature, these batteries featured a specific capacitance of 229.4 mA h/g at a current density of 0.1 A/g. This was complemented by an excellent coulombic efficiency of 99.99% after 9,600 cycles, showing these batteries are extremely stable. Additionally, the battery was found to have a capacity retention of 84.2%. This equates to a retention loss of 0.16% per 100 cycles. Following this, the batteries were tested at three other temperatures (–20, 45, and 60°C). At –20°C a specific capacitance of 83 mA h g^{-1} at a current density of 1 A/g was found after 3,000 cycles. As the temperature was increased to 45°C, the specific capacitance saw a drastic increase up to 111.8 mAh/g at a current density of 1 A/g after 1,200 cycles. Finally, the batteries were tested at 65°C. At this temperature, the specific capacitance decreased slightly but was still high at about 103.2 mAh/g after 1,200 cycles. Additionally, at all these temperatures the coulombic efficiency was 99.99%. The graphs produced from cycling analysis at different temperatures demonstrate that these batteries featured extreme electrochemical stability at all observed temperatures.

Another reason organic semiconductors are used in the fabrication of batteries is because of their ability to be readily modified. This was presented by a study by Luo et al. where the organic semiconductor perylenetetracarboxylic dianhydride (PTCDA) was sulfurized and then used to fabricate the cathode in an aluminum organic battery [39]. This was accomplished by taking a novel PTCDA powder and combining it with sublimed sulfur powder at a ratio of 1:1.3. This mixture was then heat-treated under an argon atmosphere at 500°C for 5 hours. After heating, this mixture was washed in a carbon disulfide solvent to remove unreacted sulfur. Finally, the mixture was washed with water and ethanol, then dried overnight at 70°C. To confirm successful sulfurization, an XPS spectrum was run on the powder. During the analysis of the spectrum at peak value of approximately 284.5 eV, it indicated the presence of a C-S-C bond confirming the sulfurization. Once sulfurization was confirmed, the powder was used to fabricate the cathode for an aluminum organic battery. This battery was then tested and compared to a battery using a pristine cathode. During early testing both batteries performed very similarly, with CV curves for the pristine and sulfurized electrodes at 1.38/0.98 V and 1.32/1.04 V, respectively, demonstrating that both batteries had excellent reversible redox potential. Additionally, after the first charge discharge cycle, both electrodes featured similar specific capacities of about 264 and 252 mAh/g, respectively. However, when looking at the specific capacities of the reverse reaction, the sulfurized cathode had a value of 113 mAh/g after 200 cycles at a current density of 1,000 mA/g, whereas the pristine electrode only had a value of about 83 mAh/g.

8.3.3 Organic Semiconductors for Fuel Cells

Fuel cells are a type of device that is nearly as old as batteries. Despite this, their widespread use has been rather limited due to a number of issues. These issues include costly materials, volatile fuel sources, and short lifespans. With the growing need for renewable energy, however, these devices are rapidly growing in popularity. One of the most common fuel cells used are proton-exchange membrane (PEM) fuel cells. PEMs function by pumping hydrogen into the anode. These hydrogen atoms then undergo an oxidation reaction to liberate electrons and hydrogen cations. Simultaneously at the cathode, O_2 is pumped into the cell, where the oxygen undergoes a reduction reaction to produce oxygen anions. To complete these reactions, the hydrogen cations travel across the proton exchange membrane to the anode, where they bond with the oxygen anions to produce water as a byproduct [40]. A diagram of this cell can be seen in Figure 8.5 [41].

Another common fuel cell type is the alkaline fuel cell. Alkaline fuel cells function similarly to PEMs; however, there is one key difference. In this cell, hydrogen cations are not transferred to the

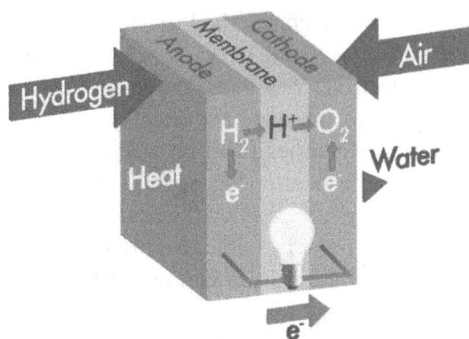

FIGURE 8.5   Diagram of how a PEM fuel cell functions. (Adapted with permission [41]. Copyright 2014, American Chemical Society.)

cathode; rather, the oxygen anions are transferred to the anode through the anion exchange membrane. One final variety of fuel cell type is the direct methanol fuel cell. Methanol fuel cells work by pumping a dilute methanol solution into the anode to produce CO_2, electrons, and hydrogen cations. After this point the reactions that take place are like PEMs.

Even though these examples function via different processes, they typically use similar materials. One of the most common materials shared by these cells is platinum. In the cell, platinum is often used in the electrodes as an efficient electrocatalyst for redox reactions. Other rare earth metals like ruthenium and iridium are also popular for this same reason. [42]. Aside from these rare metals, there are several other materials used as catalysts. These materials include metal oxides and a plethora of carbon-based materials [43, 44]. As for organic semiconductors, these are typically used as proton and anion exchange membranes, with limited research being done on their effectiveness as a cathode material.

One of the primary ways that organic semiconductors can be used in fuel cells is as a component of the proton exchange membrane. These membranes help conduct protons from the anode to the cathode while insulating against the current and preventing the crossover of the reagent gases. The advantages of organic semiconductors for this purpose are small size, low weight, efficient energy conversion, high current, power densities, and ideal conductive properties [45]. Several of these properties can been seen in a study by Wang et al. wherein the organic semiconducting polymer, polyethyleneimine, was used to fabricate a free-standing covalent organic framework membrane with high proton conductivity [46]. First, polyethylenimine was synthesized via the solution synthesis method. The resulting product was then added to a solution of 2,5-diaminobenzenesulfonic acid and 1,3,5-trihydroxyhomobenzaldehyde. The amount of polyethylenimine added to the preceding solution varied from 0 to 0.1 g. Finally, p-toluenesulfonic acid and water were added to the solution, after which films were cast onto glass to form a membrane. After the membranes were fabricated, they underwent extensive mechanical and electrochemical analysis. After the completion of all the testing, the membrane made from 0.02 g of polyethylenimine stood out from the other electrodes. This material had the highest tensile strength of ~45.4 MPa. Additionally, this membrane featured a swelling ratio of less than 2%. As for the electrochemical properties, this membrane had a high conductivity at about 223 mS/cm. Along with this high conductivity, extraordinary power and energy densities were found. These values came out to be 111.1 mW/cm and 664 mA/cm, respectively.

Another application for organic semiconductors in fuel cells is as an anion exchange membrane. This fact was illustrated in a study by Najafi et al. wherein the organic semiconductor poly(aryleneimidazolium) was used to fabricate anion exchange membranes with high conductivity [47]. This was achieved by first synthesizing two aryl imidazolium monomers via solution synthesis. These monomers were then taken and used to fabricate two batches of poly(arylene-imidazolium).

Following this, a portion of each of the batches of poly(arylene-imidazolium) were mixed. The mixture was then combined with N-methyl-2-pyrrolidone and poly(vinylpyrrolidone) (PVP). This solution was then heated in a petri dish to make a blended membrane of batch one and two PAIm1,2/PVP. Another membrane was made just using batch one of the poly(arylene-imidazolium); however, this batch was crosslinked with the PVP to fabricate the cross-PAIm1-PVP membrane. These membranes underwent testing for their electrochemical properties, and between the two, the Cross-PAIm1-PVP membrane performed the best. The membrane featured a conductivity of about 356.1 mS/cm. The membrane also showed excellent power density as peak power densities of 17.7, 11.1, and 9.1 mW/cm were recorded.

Aside from being used as exchange membranes in fuel cells, organic semiconductors also have the potential to be used as an electrode material. This was observed in a study by Miglbauer et al. wherein hydrogen peroxide fuel cells were fabricated using the organic semiconductor poly(3,4-ethylenedioxythiophene) as a cathode [48]. This was done by first synthesizing PEDOT via solution synthesis. This solution was then mixed with dimethylsulfoxide and (3-glycidyloxypropyl) trimethoxysilane. This mixture was then poured in a petri dish and dried overnight to obtain a, poly(3,4-ethylenedioxythiophene) polystyrene sulfonate (PEDOT:PSS) foil. This foil then had its electrochemical properties analyzed in conjunction with a nickel mesh anode. This testing revealed that the foil had an open circuit voltage of 0.56V, in conjunction with a short current density of 2.0 mA/cm. Additionally, the decomposition of peroxide was analyzed. It was discovered that a 0.1 M H_2O_2 in 0.05 M HCl decomposes at a rate of 0.52 mM/h. When a nickel anode was added to this, decomposition reached 2.6 mM/h. Then when the PEDOT:PSS cathode was added, the decomposition was stabilized to 0.34 mM/h. This shows how the PEDOT:PSS cathode not only stabilizes the decomposition of hydrogen peroxide, but also induces it. With more work the PEDOT:PSS foil could be used as both a cathode and anode in large-scale hydrogen fuel cells. Outside of fuel cells, organic semiconductors have found their way into another high energy density application, and that application is the solar cell. This is due to the ability of these semiconductors to be fabricated as thin films with extremely low transparency.

8.3.4 Organic Semiconductors for Solar Cells

Like fuel cells, solar cells, also known as photovoltaic cells, are used in the generation of energy. To generate energy, solar cells rely on the absorption of light, often from the sun. First photons are absorbed by a semiconducting material embedded in the cell. These photons provide the semiconducting material with energy to dislodge its electrons. When the electrons dislodge, they create holes. These holes facilitate the flow of electrons across the material where they can be siphoned off into a circuit as a usable form of energy [49]. Due to most solar cells requiring a semiconducting material, organic semiconductors have garnered increased attention for this role due to a plethora of reasons which include tunable properties, manufactured using non-toxic raw materials, use manufacturing technologies, capable of coating large areas inexpensively and fast, synthesis uses little material consumption, low-temperature processing, and excellent compatibility with flexible substrates [13].

As the push for flexible solar cells becomes increasingly popular, the ability of organic semiconductors to bind to flexible substrates shines. This idea was shown in a study by Kwon et al., where the organic semiconducting polymer, PEDOT:PSS, was used to fabricate a stretchable electrode [50]. To begin, a wavy silver nanowire network was deposited on a pre-strained substrate. This substrate was then spin coated with a PEDOT:PSS solution to form a nanowire organic conducting polymer composite. The composite was then run through spin coating again to coat it in an ionic solution to increase the flexibility and conductivity of the PEDOT:PSS. Then, using thermal annealing, the ionic solution was permeated into the composite. Then after washing and drying, a stretchable, transparent electrode was retrieved. A depiction of the electrode fabrication can be seen in Figure 8.6 [50].

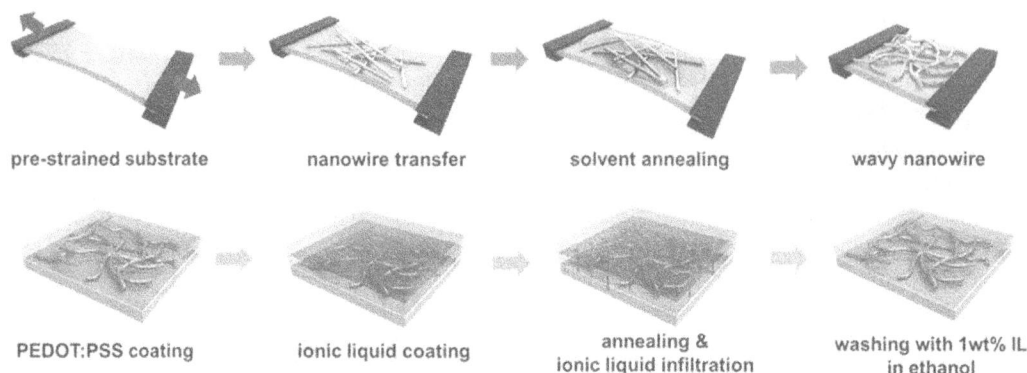

FIGURE 8.6 The process of making the sequentially introduced wavy nanowire network/PEDOT:PSS/ionic liquid (S–NPI) electrode, starting with the pre-strained substrate and finishing with the fabricated electrode. (Adapted with permission [50]. Copyright 2023, American Chemical Society.)

Using the aforementioned process, three separate electrodes were synthesized. These three electrodes were put through electrochemical analysis and tensile testing to analyze their properties. Of all the electrodes tested, the S-NPI composite electrode performed the best in all these tests. To begin, the conductivity of the electrodes was tested using conductive atomic force microscopy, where it was revealed that the S-NPI electrode had the highest current flowing through. This showed that the electrode had the most uniform dispersion of nanowires, providing it with the highest conductivity of all the electrodes. The tensile testing of this electrode revealed this electrode was also the most flexible, as it had a tensile modulus of ~2.7 GPa. This was significantly lower than the other two electrodes and explained why this was the only material to not experience structural defects in its surface during the testing. This electrode also exhibited the highest power conversion efficiency at 11.3%, open circuit voltage at 0.781V, short circuit current at 22.74 mA/cm, and fill factor at 0.63. These factors are mainly attributed to the electrode's low surface roughness of 5.1 nm and sheet resistance of 33.5 Ω.

When looking at the fabrication of solar devices, organic semiconductors have one specific advantage that allows them to outpace inorganic semiconductors, which is the ability to have their colors modified without sacrificing performance. This ability allows these solar cells to be used in more diverse and/or aestetic applications. [51]. In a study by Li et al., organic polymer solar cells featuring vivid colors were fabricated [52]. This was accomplished by taking prepared indium tin oxide glass substrates and spin coating them in the organic semiconducting polymer PEDOT:PSS. Then in a nitrogen filled glove box, PM6 and Y6, common industrial compounds that facilitate the transfer of electrons, were spin coated onto the electrodes as an active layer. Finally, to fabricate colorful electrodes, silver and tellurium oxide were spin coated on at varying thicknesses [52].

After all the electrodes underwent testing for their photovoltaic capabilities, it was discovered the R-3 electrode performed the best out of all the colored electrodes. This electrode featured an open circuit voltage of about 0.85 V. Along with that, it featured a short circuit current of about 24.1 mA cm^{-2}. Additionally, this electrode featured a fill factor of 0.76. Finally, and most importantly, a power conversion efficiency of about 15.57% was displayed. This power conversion was only 0.37% off the recorded value of a transparent electrode. This fact proves that colored solar cells fabricated using organic semiconductors have the potential to be as efficient as transparent solar cells, potentially guiding solar cells to be used in new and diverse applications.

Speaking more specifically on diverse applications there is one fabrication method that is slowly catching up to more traditional fabrication methods, and that is the 3D printing of solar cell devices. The 3D printing of solar cells relies heavily on the processing of solutions [53]. This is where polymer organic semiconductors shine as they rely solely on solution synthesis. Furthermore, polymer organic semiconductors can be used as a matrix to fabricate composite solar cell electrodes. These two concepts were illustrated in a study conducted by Kim et al., where PEDOT:PSS thylakoid composite electrodes were fabricated. This was achieved by first lyophilizing a prepared PEDOT:PSS solution. This solution was then dissolved in a 15% DMSO solvent to produce a 6% 3D-printable conductive ink. To this ink, a crude TM pellet made from spinach leaves was dissolved. This ink was then loaded into an air pressure 3D printer to fabricate an electrode. Three different electrodes were fabricated. The difference between each electrode was the number of layers varying from 1 to 3 layers. After extensive electrochemical testing, it was discovered the three-layer electrode performed the best. This electrode had an excellent open circuit voltage of 689 mV when illuminated. While unilluminated, the open circuit voltage only measured 168 mV. Along with these values the short circuit current density was also found. Under illumination it was found to be 0.57 mA/cm^2. Without illumination the short current density was only 0.27 mA/cm^2. Finally, the power density of this electrode was found. Under illumination it was found to be 101.7 µW/cm^2. Without the illumination the power density was significantly lower at about 37.6 µW/cm^2. As seen above, all the recorded electrochemical values increased when the electrode was tested under illumination. This fact shows how these electrodes are perfect for solar cell application as light increased the value of every category. Thus, 3D electrodes fabricated using organic semiconducting polymers may have a place in the solar cell industry. This area of energy storage uses more organic semiconductors than any of the other areas. This is due to the fact that most solar cells require thin semiconductors to properly set under the glass and to keep the size of the cell down. With this area containing the largest share of organic semiconductors, it is often looked at as the most promising future of organic semiconductors; as of now, organic solar cells are one of the most popular research areas when it comes to solar cells [54].

8.4   CONCLUSION AND OUTLOOK

With the world's global energy needs constantly increasing, it is no surprise that researchers are working around the clock to find solutions. Green sourcing, sustainability, cost effectiveness and general efficiency are all factors that are explored to conquer these growing energy needs. Due to these factors, organic semiconductors have positioned themselves as a promising answer. Organic semiconductors attribute their low cost to simple synthesis methods and resource availability. Solution synthesis offers a facile method for nanoparticle production, while the various polymerization methods are conducive for batch processing. Meanwhile, the wide availability of different organic compounds leads to lower overhead costs when compared to purified silicon wafers. These properties translate into efficiency gains, even before factors like energy efficiency and reliability are involved. For batteries and supercapacitors, organic semiconductors imbue both physical and chemical flexibility. Energy storage devices that can flex and bend are greatly desired for their applications in wearable electronics. Medical devices designed to monitor vital signs and general consumer electronics stand to benefit most from these advances. For energy cell applications, organic semiconductors demonstrated their applicability towards different processing methods and uses. With the constant advances observed with material processing technology, new materials need to be developed to match. Conducting polymers and inks rose to this occasion, becoming particularly popular in the field of 3D printing. Meanwhile, the variety of properties present in these materials allows for their application to different parts of the apparatus, further expanding their use. This is all without mentioning the electrochemical properties of organic semiconductors, which have moreover proven their competitiveness with traditional mediums.

With the diversity of advantages seen in organic semiconductor applications, it comes with little shock that their outlook is so promising. By the year 2030, the global market for organic supercapacitors is expected to reach $566.38 [55]. A majority of this will be centered in the energy sector, as that is where the largest share sits currently. As stated earlier, one of the most promising futures for organic semiconductors is in organic solar cells. However, this is not the only area. As more structurally complex organic semiconducting materials are developed, they will experience increased usage in the various applications that they can be found in.

REFERENCES

1. M. Fichtner, Materials for sustainable energy production, storage, and conversion, Beilstein J. Nanotechnol. 6 (2015) 1601–1602.
2. Transmission Lines, in: Electr. Power Syst. Basics Nonelectrical Prof., 2017: pp. 43–52. https://ieeexplore.ieee.org/document/779420210.1002/9781119180227.ch3
3. S.A.A. Rizvi, A.M. Saleh, Applications of nanoparticle systems in drug delivery technology, Saudi Pharm. J. 26 (2018) 64–70.
4. G. Adamo, S. Campora, G. Ghersi, Chapter 3 - Functionalization of nanoparticles in specific targeting and mechanism release, in: D. Ficai and A.M. Grumezescu (Eds.), Micro Nano Technol, Elsevier, 2017: pp. 57–80.
5. S. Moeinzadeh, E. Jabbari, Nanoparticles and their applications BT, in: B. Bhushan (Ed.), Springer Handbook of Nanotechnology, Springer Berlin Heidelberg, Berlin, Heidelberg, 2017: pp. 335–361.
6. A.A. Yaqoob, A. Ahmad, M.N.M. Ibrahim, R.R. Karri, M. Rashid, Z. Ahamd, Chapter 23 – Synthesis of metal oxide–based nanocomposites for energy storage application, in: J.R. Koduru, R.R. Karri, N.M. Mubarak, and E.R. Bandala (Eds.), Sustainable Nanotechnology for Environmental Remediation, Elsevier, 2022: pp. 611–635.
7. P. Kamaraj, R. Vennila, M. Sridharan, P.A. Vivekanand, Super capacitance of metal oxide nanoparticles BT, in: O.V. Kharissova, L.M.T. Martínez, B.I. Kharisov (Eds.), Handbook of Nanomaterials and Nanocomposites for Energy and Environmental Applications, Springer International Publishing, Cham, 2020: pp. 1–14. https://doi.org/10.1007/978-3-030-11155-7_120-1
8. S. Wickramasinghe, J. Wang, B. Morsi, B. Li, Carbon dioxide conversion to nanomaterials: Methods, applications, and challenges, Energy & Fuels. 35 (2021) 11820–11834.
9. G. Kothandam, G. Singh, X. Guan, J.M. Lee, K. Ramadass, S. Joseph, M. Benzigar, A. Karakoti, J. Yi, P. Kumar, A. Vinu, Recent advances in carbon-based electrodes for energy storage and conversion, Adv. Sci. 10 (2023) 2301045.
10. K.R. Reddy, B. Hemavathi, G.R. Balakrishna, A.V. Raghu, S. Naveen, M.V. Shankar, 11 – Organic conjugated polymer-based functional nanohybrids: Synthesis methods, mechanisms and its applications in electrochemical energy storage supercapacitors and solar cells, in: K. Pielichowski and T.M. Majka (Eds.), Polymer Composites with Functionalized Nanoparticles: Synthesis, Properties, and Applications, Elsevier, 2019: pp. 357–379.
11. J. Ding, W. Xu, X. Zhu, Z. Liu, Y. Zhang, Z. Jiang, All-organic nanocomposite dielectrics contained with polymer dots for high-temperature capacitive energy storage, Nano Res. 16 (2023) 10183–10190.
12. C.N. Gannett, L. Melecio-Zambrano, M.J. Theibault, B.M. Peterson, B.P. Fors, H.D. Abruña, Organic electrode materials for fast-rate, high-power battery applications, Mater. Reports Energy. 1 (2021) 100008.
13. M. Riede, D. Spoltore, K. Leo, Organic solar cells—The path to commercial success, Adv Energy Mater. 11 (2021) 2002653.
14. T.J. Dawidczyk, H. Kong, H.E. Katz, 20 - Organic semiconductors (OSCs) for electronic chemical sensors, in: O. Ostroverkhova (Ed.), Handbook of Organic Materials for Optical and (Opto)electronic Devices: Properties and Applications, Woodhead Publishing, 2013: pp. 577–596.
15. M. Sawatzki-Park, S.-J. Wang, H. Kleemann, K. Leo, Highly ordered small molecule organic semiconductor thin-films enabling complex, high-performance multi-junction devices, Chem. Rev. 123 (2023) 8232–8250.
16. Z. Zhou, N. Luo, X. Shao, H.-L. Zhang, Z. Liu, Hyperbranched polymers for organic semiconductors, Chempluschem. 88 (2023) e202300261.
17. A. Mishra, P. Bäuerle, Small molecule organic semiconductors on the move: Promises for future solar energy technology, Angew, Chemie Int. Ed. 51 (2012) 2020–2067.
18. Z. Jarrahi, G. Farzi, A. Fischer, Synthesis of polyaniline/cadmium sulfide hybrid nanocomposites via a one-pot process: Morphology control for improvement of photovoltaic performance, ACS Appl. Electron. Mater. 5 (2023) 1156–1163.
19. J.-H. Kim, S. Han, H. Jeong, H. Jang, S. Back, J. Hu, M. Lee, B. Choi, H.S. Lee, Thermal gradient during vacuum-deposition dramatically enhances charge transport in organic semiconductors: Toward high-performance n-type organic field-effect transistors, ACS Appl. Mater. Interfaces. 9 (2017) 9910–9917.

20. V.J. Rao, M. Matthiesen, K.P. Goetz, C. Huck, C. Yim, R. Siris, J. Han, S. Hahn, U.H.F. Bunz, A. Dreuw, G.S. Duesberg, A. Pucci, J. Zaumseil, AFM-IR and IR-SNOM for the characterization of small molecule organic semiconductors, J. Phys. Chem. C. 124 (2020) 5331–5344.
21. D. Zhu, J. Hui, N. Rowell, Y. Liu, Q.Y. Chen, T. Steegemans, H. Fan, M. Zhang, K. Yu, Interpreting the ultraviolet absorption in the spectrum of 415 nm-bandgap CdSe magic-size clusters, J. Phys. Chem. Lett. 9 (2018) 2818–2824.
22. X. Liu, X. Yi, J. Zhang, X. Zhao, S. Liu, T. Wang, S. Cui, Synthetic strategy for MnO2 Nanoparticle/ Carbon aerogel heterostructures for improved supercapacitor performance, ACS Appl. Nano Mater. 6 (2023) 14127–14135.
23. S. Cho, J. Lim, Y. Seo, Flexible solid supercapacitors of novel nanostructured electrodes outperform most supercapacitors, ACS Omega. 7 (2022) 37825–37833.
24. Y. Abdul Wahab, M.N. Naseer, A.A. Zaidi, T. Umair, H. Khan, M.M. Siddiqi, M.S. Javed, Super capacitors in various dimensionalities: Applications and recent advancements, in: E. Cabeza (Ed.), Super Capacitors in Various Dimensionalities: Applications and Recent Advancements Elsevier, Oxford, 2022: pp. 682–691.
25. L. Zhou, C. Li, X. Liu, Y. Zhu, Y. Wu, T. van Ree, 7 - Metal oxides in supercapacitors, in:Y. Wu (Ed.), Metal Oxides in Energy Technologies. A Volume in Metal Oxides, Elsevier, 2019: pp. 169–203.
26. A. Afif, S.M.H. Rahman, A. Tasfiah Azad, J. Zaini, M.A. Islan, A.K. Azad, Advanced materials and technologies for hybrid supercapacitors for energy storage – A review, J. Energy Storage. 25(2019) 100852.
27. Y. Wang, H. Wang, W. Zhang, G. Fei, K. Shu, L. Sun, S. Tian, H. Niu, M. Wang, G. Hu, Y. Duan, A simple route to fabricate ultralong and uniform polypyrrole nanowires with high electrochemical capacitance for supercapacitor electrodes, ACS Appl. Polym. Mater. 5 (2023) 1254–1263.
28. A. Baby, S. Sunny, J. Vigneshwaran, S. Abraham, S.P. Jose, W.S. Saeed, M.R. Pallavolu, J. Cherusseri, S. Puthenveetil Balakrishnan, Azopyridine as a linker molecule in polyaniline-grafted graphene oxide nanocomposite electrodes for asymmetric supercapacitors, ACS Appl. Energy Mater. 6 (2023) 10442–10456.
29. X. Han, J. Sun, Q. Li, X. He, L. Dang, Z. Liu, Z. Lei, Highly flexible PEDOT film assembled with solution-processed nanowires for high-rate and long-life solid-state supercapacitors, ACS Sustain. Chem. Eng. 11 (2023) 2938–2948.
30. R.C. Massé, C. Liu, Y. Li, L. Mai, G. Cao, Energy storage through intercalation reactions: Electrodes for rechargeable batteries, Natl. Sci. Rev. 4 (2017) 26–53.
31. H. Liu, X. Liu, H. Wang, Y. Zheng, H. Zhang, J. Shi, W. Liu, M. Huang, J. Kan, X. Zhao, D. Li, High-performance sodium-ion capacitor constructed by well-matched dual-carbon electrodes from a single biomass, ACS Sustain. Chem. Eng. 7 (2019) 12188–12199.
32. X. Jia, Y. Ge, L. Shao, C. Wang, G.G. Wallace, Tunable conducting polymers: Toward sustainable and versatile batteries, ACS Sustain. Chem. Eng. 7 (2019) 14321–14340.
33. P. Samanta, S. Ghosh, A. Kundu, P. Samanta, N.C. Murmu, T. Kuila, Recent progress on the performance of Zn-ion battery using various electrolyte salt and solvent concentrations, ACS Appl. Electron. Mater. 5 (2023) 100–116.
34. T. Takamura, Primary batteries – Aqueous systems | Alkaline manganese–zinc. Encyclopedia of Electrochemical Power Sources. (2009) 28–42. https://www.sciencedirect.com/referencework/978044 4527455/encyclopedia-of-electrochemical-power-sources
35. M. Salado, E. Lizundia, Advances, challenges, and environmental impacts in metal–air battery electrolytes, Mater. Today Energy. 28 (2022) 101064.
36. N. Goujon, N. Casado, N. Patil, R. Marcilla, D. Mecerreyes, Organic batteries based on just redox polymers, Prog. Polym. Sci. 122 (2021) 101449.
37. R. Cang, K. Ye, K. Zhu, D. Cao, Environmentally friendly and flexible aqueous zinc ion batteries using an organic anode and activated carbon as the cathode, ACS Sustain. Chem. Eng. 11 (2023) 5065–5071.
38. M. Wu, W. Su, X. Wang, Z. Liu, F. Zhang, Z. Luo, A. Yang, P. Yeleken, Z. Miao, Y. Huang, Long-life aqueous Zinc–Organic batteries with a trimethyl phosphate electrolyte and organic cathode, ACS Sustain. Chem. Eng. 11 (2023) 957–964.
39. W. Luo, Y. Liu, Z. Zhang, F. Li, Z. Chao, J. Fan, Rational molecular design strategy of a carbonyl cathode for better aluminum organic batteries, ACS Sustain. Chem. Eng. 11 (2023) 11406–11414.
40. L. Fan, Z. Tu, S.H. Chan, Recent development of hydrogen and fuel cell technologies: A review, Energy Reports. 7 (2021) 8421–8446.
41. A. Bailey, L. Andrews, A. Khot, L. Rubin, J. Young, T.D. Allston, G.A. Takacs, Hydrogen storage experiments for an undergraduate laboratory course—Clean energy: Hydrogen/Fuel cells, J. Chem. Educ. 92 (2015) 688–692.

42. X. Hu, B. Yang, S. Ke, Y. Liu, M. Fang, Z. Huang, X. Min, Review and perspectives of carbon-supported platinum-based catalysts for proton exchange membrane fuel cells, Energy & Fuels. 37, 16 (2023) 11532–11566.

43. M. Gasik, 1 - Introduction: Materials challenges in fuel cells, in: M. Gasik (Ed.), Materials for Fuel Cells. A Volume in Woodhead Publishing Series in Electronic and Optical Materials, Woodhead Publishing, 2008: pp. 1–5.

44. K. Föger, 2 - Materials basics for fuel cells, in: M. Gasik (Ed.), Materials for Fuel Cells. A Volume in Woodhead Publishing Series in Electronic and Optical Materials, Woodhead Publishing, 2008: pp. 6–63.

45. H.A. Elwan, M. Mamlouk, K. Scott, A review of proton exchange membranes based on protic ionic liquid/polymer blends for polymer electrolyte membrane fuel cells, J. Power Sources. 484 (2021) 229197.

46. L. Wang, C. Wang, Y. Ren, Z. Yang, Y. Zheng, Q. Zhang, W. Wu, J. Wang, Free-standing polymer covalent organic framework membrane with high proton conductivity and structure stability, ACS Appl. Polym. Mater. 5 (2023) 7562–7570.

47. H.N. Fath Dehghan, A. Abdolmaleki, M. Zhiani, Highly conducting poly(arylene-imidazolium) anion exchange membranes containing thin-span channels constructed by Cation–Dipole interaction, ACS Appl. Polym. Mater. 5 (2023) 1965–1976.

48. E. Miglbauer, P.J. Wójcik, E.D. Głowacki, Single-compartment hydrogen peroxide fuel cells with poly(3,4-ethylenedioxythiophene) cathodes, Chem. Commun. 54 (2018) 11873–11876.

49. S. Battersby, News feature: The solar cell of the future, Proc. Natl. Acad. Sci. U. S. A. 116 (2019) 7–10.

50. H.J. Kwon, G.-U. Kim, C. Lim, J.K. Kim, S.-S. Lee, J. Cho, H.-J. Koo, B.J. Kim, K. Char, J.G. Son, Sequentially coated wavy nanowire composite transparent electrode for stretchable solar cells, ACS Appl. Mater. Interfaces. 15 (2023) 13656–13667.

51. J. Kong, M. Mohadjer Beromi, M. Mariano, T. Goh, F. Antonio, N. Hazari, A.D. Taylor, Colorful polymer solar cells employing an energy transfer dye molecule, Nano Energy. 38 (2017) 36–42.

52. X. Li, R. Xia, K. Yan, J. Ren, H.-L. Yip, C.-Z. Li, H. Chen, Semitransparent organic solar cells with vivid colors, ACS Energy Lett. 5 (2020) 3115–3123.

53. B.R. Hunde, A.D. Woldeyohannes, 3D printing and solar cell fabrication methods: A review of challenges, opportunities, and future prospects, Results Opt. 11 (2023) 100385.

54. Q. Zhang, W. Hu, H. Sirringhaus, K. Müllen, Recent progress in emerging organic semiconductors, Adv. Mater. 34 (2022) 2108701.

9 Nanotechnology in Semiconductors
Role of Nano-Dimensions and Thin Film Structure

*Nasrin Babazadeh, Amir Ershad-Langroudi,
Seyed Mehdi Mousaei, and Farhad Alizadegan*

9.1 AN INTRODUCTION TO NANOTECHNOLOGY AND SEMICONDUCTORS

9.1.1 BASIC CONCEPTS OF NANOTECHNOLOGY

Nanotechnology can be defined as a branch of technology that pertains to the manipulation and utilization of objects and systems at the nanoscale, or on a minuscule level [1]. However, the term "nano" carries an additional connotation within scientific discourse, denoting a unit of measurement equivalent to one billionth of a meter, commonly known as a "nanometer" [2, 3]. In light of this, nanotechnology is associated with technologies functioning at the scale of nanometers. Nanotechnology encompasses materials and systems characterized by altered chemical, biological, and physical characteristics resulting from their dimensions at the nanoscale. Materials composed of structures with dimensions ranging from approximately 1–100 nm exhibit significant alterations in their properties when compared to isolated molecules (1 nm) or bulk materials [4].

Nanotechnology has emerged as a prominent concept in the current millennium, contributing greatly to the improvement of living standards. Consequently, this era might be aptly characterized as the age of nanomaterials. This article presents a comprehensive overview of the fundamental elements of nanotechnology, introducing the expansive subject of transition metal nanoparticles. It delves into their properties, as well as the techniques employed for their production and characterization [1].

9.1.2 PROPERTIES OF NANO-DIMENSIONS IN SEMICONDUCTORS

The characteristics associated with nanoscale dimensions in semiconductors present a wide range of remarkable capacities. Semiconductors demonstrate distinct properties at the nanoscale, distinguishing them from their larger-scale counterparts. The phenomenon of quantum confinement gives rise to size-dependent fluctuations in the electrical, optical, and thermal characteristics of materials, thereby facilitating meticulous manipulation of their behavior. Nanostructured semiconductors frequently exhibit heightened electrical conductivity, higher carrier mobility, and adjustable bandgaps, rendering them very suitable for the development of high-performance electronic and optoelectronic devices. The discrete energy levels observed in nanostructures, such as quantum dots, offer customized light emission, thereby opening up possibilities for enhanced applications in the fields of displays and imaging technology. The utilization of these qualities enables the semiconductor industry to drive innovation and bring forth revolutionary advancements in diverse technical domains [5].

DOI: 10.1201/9781003450146-9

Furthermore, the characteristics associated with nanoscale dimensions in semiconductors transcend the domain of electronics. These distinct materials are utilized in several domains, including as medicine, energy, and environmental monitoring. The utilization of nanostructured semiconductors is observed in targeted drug delivery systems, wherein their small size enables accurate administration of medicinal drugs to certain cells, hence reducing adverse effects and optimizing treatment effectiveness.

Nanostructured semiconductors have been found to significantly enhance the absorption of light, hence leading to a notable increase in the overall efficiency of solar panels. Furthermore, they enhance the efficiency of batteries and supercapacitors by the provision of an increased surface area for the storage of charges. This facilitates expedited charging processes and enables the attainment of prolonged energy storage capabilities [6].

Nanotechnology also serves a crucial function in the realm of environmental monitoring and pollution detection. Sensors that are capable of detecting and analyzing contaminants in air and water utilize semiconductors with nanoscale dimensions. They enable real-time monitoring because of their both sensitivity and selectivity, which aids in our ability to comprehend and manage environmental issues [7].

The properties of semiconductor dimensions at the nanoscale present novel prospects. Engineered materials play a significant role in shaping technology and fostering global sustainability and connectivity through their transformative impact on the fields of electronics, medicine, and the environment. As the scientific community continues to explore and harness the capabilities of semiconductor nanotechnology, it is anticipated that there will be further notable advancements in this field. Nano-sized structures in semiconductors demonstrate distinct characteristics compared to their larger counterparts, mostly attributed to quantum phenomena and an enhanced ratio of surface area to volume.

9.1.2.1 Band Gap Engineering

This is feasible to achieve nanoscale modification of the band gap of a semiconductor, which refers to the energy range devoid of electron states. The ability to modify the nanostructure size and composition allows researchers to tailor the band gap to certain optical and electrical characteristics [8]. These findings have significant effects on optoelectronic devices such as lasers and photodetectors.

Abnormal electrical transport and optical effects have been observed in band gap-designed quantum devices, including lasers and heterojunction bipolar transistors. The emergence of synthetic materials, including carbon fullerenes, carbon nanotubes, and organized mesoporous materials, has significantly contributed to the advancement of nanotechnology and nanomaterials research [5].

9.1.2.2 Enhanced Surface Effects

Nanoscale semiconductors exhibit a greater surface area in proportion to their volume, resulting in enhanced surface phenomena. The influence of surface states and defects on the properties of materials is considerable, as they impact various aspects such as charge carrier dynamics, recombination rates, and chemical reactivity [9].

The nanocrystals have a comparatively greater surface area in relation to an equivalent amount of material composed of larger particles [10]. The augmented surface area provides a greater number of atoms, hence enhancing chemical and physical processes such as catalysis and detecting reactions [11].

9.1.2.3 Size-Dependent Optical Properties

The advantages of semiconductors are contingent upon their ability to manipulate material structures at a much-reduced scale in order to get the desired properties [10]. In the world of nanoscale semiconductors, two primary phenomena that govern their behavior are the size and the surface effects. Semiconductor nanocrystals possess distinctive physical and chemical characteristics that render them well-suited for implementation in several new technologies. These applications encompass catalysis, nanoelectronics, nanophotonics, energy conversion, non-linear optics, solar cells, and detectors, among others [12]. The electrical conductivity of semiconductor nanocrystals, as well as their optical properties, including absorption and fluorescence spectra, refractive index, and absorption coefficient, can be substantially modified by manipulating the crystalline sizes [10].

The size-dependent tuning of optical characteristics in nanoscale semiconductors is a notable characteristic. Quantum dots, for example, demonstrate emission spectra that rely on their size. The aforementioned characteristic is used in many applications such as displays, sensors, and biological imaging [13].

9.1.2.4 Thermal Properties

The thermal conductivity of nanostructures may be modified as a result of photon scattering occurring at the interfaces. The impacts of this feature are relevant to the fields of nanoelectronics and thermoelectric applications. The scientific community has shown significant interest in the thermal characteristics of wide band gap (WBG) nanowires, since these qualities play a crucial role in environmental applications, namely in the fields of energy conversion and thermoelectric devices. Recent studies have shown that the thermal characteristics of thin-film WBG materials have been effectively manipulated by several creative methods, including strain [14, 15] or buffer layer engineering [16].

9.1.3 Applications of Nanotechnology in the Semiconductor Industry

The advent of nanotechnology has brought about a paradigm shift in the semiconductor industry. The use of nanoscale materials and technologies has facilitated the development of electronic components that are characterized by reduced size, enhanced speed, and improved energy efficiency. The strict control of nanomaterials has facilitated the development of state-of-the-art semiconductor technology [5]. The semiconductor industry has seen many uses of nanotechnology, which are described below.

9.1.3.1 Smaller Transistors

The field of nanotechnology has facilitated the development of transistors that are both smaller in size and more efficient in their performance. As the size of transistors decreases to nanoscale levels, their operational speed increases, power consumption decreases, and the potential for more device integration on a single chip is enabled. The aforementioned progress has propelled the evolution of electronic gadgets that are characterized by enhanced speed and increased power [17].

The transparency of applications is contingent upon the quality of materials used and the functioning of the device. Therefore, it is important to thoroughly analyze the circuit, system, and manufacturing obstacles in order to conduct a realistic assessment of transparent nanotechnology [18]. Hoffman et al. [18] examined the progression of a distinctive nanostructure, the characteristics of the materials used, and the efficacy of nanoelectronics. Thin-film transistors (TFTs) are produced via optical transmission and are composed of very transparent ZnO-based materials. These TFTs operate in the enhancement mode and exhibit a transmission efficiency of 75% within the visible range of the electromagnetic spectrum.

The field of nanotechnology has significantly contributed to the development of cutting-edge gadgets and computers, characterized by their enhanced speed, reduced size, portability, and substantial data storage capabilities. The manufacturing process of transistors is seeing a reduction in size via the use of nanotechnology. Companies including Intel, IBM, and Samsung have made significant advancements in the development of nanoscale transistors [19].

9.1.3.2 Quantum Dots (QDs) for Displays

QDs refer to semiconductors at the nanoscale that exhibit size-dependent emission of distinct hues of light. Display technologies often use these components to improve the precision of color reproduction and optimize energy usage. Qd-enhanced displays have been shown to provide heightened brightness and increased color vibrancy, hence enhancing the visual perception and quality of devices like TVs, desktop computers, and smartphones [20].

García de Arquer et al. [21] published a review with the title of "Semiconductor quantum dots: Technological progress and future challenges." The authors provided a comprehensive analysis of recent progress in the synthesis and comprehension of QD nanomaterials, with particular emphasis on colloidal QDs. They also examined the potential applications of these materials in many technologies, including displays and lighting, lasers, sensors, electronics, solar energy conversion, photocatalysis, and quantum information [21].

9.1.3.3 Nanowires for Sensors

Nanowires, characterized by their nanometer scale, are used in a multitude of applications as sensors. Nanowires may consist of conducting materials, such as aluminum, platinum, and gold, as well as semiconducting materials, such as silicon, germanium, and gallium nitride. Additionally, nanowires can be composed of insulating materials, including SiO_2, TiO_2, H_fO_2, and so on [22].

The distinctive characteristics shown by nanowires, such as their superior electron transport capabilities, exceptional mechanical durability, expansive surface area, and capacity for intrinsic property manipulation, facilitate the emergence of new categories of nanoelectromechanical systems (NEMS) [23].

These entities possess the capability to discern variations in temperature, pressure, or chemical composition with a notable degree of sensitivity. Nanowire-based sensors have been used in several fields such as healthcare, environmental monitoring, and industrial operations [5].

The distinctive electrical characteristics shown by semiconductor nanowires, namely silicon nanowires (SiNWs), render them very appealing for the detection of diverse gases in a label-free, real-time, and highly sensitive manner. Consequently, significant endeavors have been undertaken in the last twenty years to investigate the gas-sensing capabilities of nanowires. Akbari-Saatlu et al. [24] in their review paper, provided an overview of the current advancements in the use of SiNWs for gas sensing purposes. The material started with introducing the two fundamental synthesis methodologies, namely top-down and bottom-up methods, followed by a comprehensive examination of the merits and drawbacks associated with each technique. Then, a concise overview was provided on the fundamental sensing mechanism of SiNWs in both resistor and field effect transistor configurations. Additionally, the sensitivity and selectivity of SiNWs towards various gases following distinct functionalization techniques were then elucidated.

9.1.3.4 Nanoelectromechanical Systems (NEMS)

NEMS integrates nanotechnology with microelectromechanical systems (MEMS) to construct very miniature mechanical systems. These devices possess many applications in the domains of sensing, actuation, and signal processing. Accelerometers, resonators, and gyroscopes find use in many electrical devices. Nanowires composed of WBG semiconductors have been a prominent area of study due to their significant potential in the field of NEMS. Specifically, these nanowires hold great promise for applications such as environmental monitoring and energy harvesting [23]. Pham et al. [23] provided a thorough examination of the latest advancements in the development, characteristics, and uses of silicon carbide (SiC), group III-nitrides, and diamond nanowires as the preferred materials for NEMS.

9.1.3.5 Photolithography Advances

Nanotechnology has made significant advancements in the field of semiconductor photolithography, which is widely recognized as a crucial manufacturing process. Enhanced nanolithography technologies enable the fabrication of high-density integrated circuits, characterized by reduced feature sizes and improved patterning accuracy [25].

Silicon wafers are frequently employed in many fabrication methods, including photolithography patterning, doping and ion implantation, thin-film deposition, as well as etching. Gopinath et al. [26], conducted a comprehensive review that highlights the significance of semiconductor materials in combination with biosensors for the purpose of health monitoring and diagnosis. The process of wet thermal oxidation was used to oxidize a silicon wafer at a temperature of 1000°C, resulting in the formation of a thin layer of silicon oxide. Silica can be employed in the process of photolithography production to generate two distinct visible epilayers when seen from a top-down perspective. Al metal is then used as an intermediate metal, where it is put onto a silica substrate and subsequently subjected to etching using the photolithography process.

9.1.3.6 Carbon Nanotubes and Graphene

Nano-sized carbon-based materials such as carbon nanotubes and graphene exhibit distinctive electrical and mechanical characteristics. The extraordinary conductivity, strength, and flexibility of these materials have led to their investigation for potential applications in transistors, interconnects, and flexible electronics [27]. In 2022, Ramalingam et al. [28], published a review with the title of "A review of graphene-based semiconductors for photocatalytic degradation of pollutants in wastewater." Graphene-based photocatalysts have garnered significant attention in the field of waste water treatment owing to their exceptional physical, chemical, and mechanical characteristics. These photocatalysts exhibit remarkable electron conductivity, wide light absorption spectrum, extensive surface area, and substantial adsorption capacity. The incorporation of these substances into various materials such as metals, metal-containing nanocomposites, semiconductor nanocomposites, polymers, MXene, and other compounds has been shown to significantly enhance the photocatalytic activity for the purpose of photodegradation of pollutants.

Graphene finds use in single-electron transistors and high-frequency graphene field-effect transistors owing to its electrical properties. The electron mobility in graphene has a much higher speed, ranging from 1000 to 10,000 times quicker than that seen in silicon [29]. Graphene electrodes find use in rechargeable batteries and the construction of electrochemical double capacitors [30]. The efficiency of batteries is enhanced via accelerated charging, increased energy production, extended cycle periods, and prolonged lives.

9.1.3.7 Energy Harvesting

The combination of conducting polymers with inorganic hybrid and organic nanomaterials has led to the development of multifunctional hybrid nanocomposites that exhibit enhanced performance in many applications, such as sensors, energy storage, energy harvesting, and defense devices. The use of nanoscale materials and structures has the potential to significantly improve the operational efficiency of several energy-related technologies, such as solar cells, thermoelectric generators, and energy storage devices. Due to its capacity to improve polymers' thermal, morphological, electrical, and mechanical characteristics, conducting polymers with carbon nanomaterial composites has become popular for chemical sensors and energy harvesting. Table 9.1 presents a comprehensive compilation of multiple applications of conducting polymers in the field of energy harvesting [31].

9.1.3.8 Nanocomposites for Packaging

Semiconductor packaging materials serve the dual purpose of safeguarding chips and facilitating electrical connectivity. Nanocomposite materials have been shown to enhance thermal conductivity, mechanical strength, and electromagnetic interference insulation, hence enhancing the reliability and performance of electronic devices [32].

TABLE 9.1

Some Energy Harvesting Applications for Conducting Polymers

Polymer	Possible Applications
Polypyrrole	Rechargeable batteries, printed electronics, circuit boards, condensers, chemical sensors, electroplating, electroacoustic devices, adhesives, electromagnetic shielding, transparent coating, electro-photochemical cells, photocatalysts, physiological implantations, field effect transistors
Polyaniline	Rechargeable batteries, chemical and biosensors, indicator devices, electrochromic devices, textiles
Poly (p-phenylene)	Rechargeable batteries, photocatalysts, fillers
Polythiophene	Rechargeable batteries, gas and chemical sensors, optoelectronics, display devices, fillers, field effect transistors, photocatalysts

The rapid advancement of personal portable electronic devices necessitates the investigation of novel die attach film (DAF) materials within the constraints of restricted mounting area and height. This exploration is crucial to fulfill the demands of achieving high package density and operating speed. In this regard, Sun et al. [33] in 2022, presented melamine–graphene epoxy nanocomposite-based DAFs for advanced three-dimensional semiconductor packing applications. Melamine functionalized graphene was synthesized by a nondestructive ball milling technique and then incorporated into an epoxy matrix at different weight percentages, namely 1, 4, 7, and 10 wt%. The melamine molecule has a sp^2 hybridized conjugated structure, allowing for π–π stacking interactions with graphene nanosheets. This contact leads to improved dispersion of melamine on the graphene surface, while avoiding the formation of any surface defects. The dependability of DAFs is influenced by the adsorption of moisture since the presence of adsorbed moisture may lead to vaporization during high-temperature processing. The resulting high-pressure water vapor can induce package breaking or interface delamination, which is popularly referred to as popcorn failure in semiconductor packaging. Furthermore, moisture serves as a plasticizer in order to decrease the modulus and T_g of the cured epoxy composites. The moisture adsorption of neat epoxy resins is comparatively greater than that of M-G epoxy nanocomposites due to the absence of moisture absorption by inorganic graphene fillers. Furthermore, melamine modifications result in a reduction in the mobility of polymer chains and an increase in cross-linking, hence restricting the rotational and alignment movements of the polar groups in both melamine and epoxy. The polarity of the epoxy composite is decreased as a result of this. Nevertheless, the presence of melamine results in an increased number of amine polar groups being present on the surface of graphene. When the filler loading is increased from 4 wt% to 7 wt%, the impact of these two conflicting effects on moisture adsorption becomes almost equal [33].

The aforementioned applications exemplify the profound influence of nanotechnology on the semiconductor sector, resulting in the development of electronic devices that are characterized by enhanced power, efficiency, and versatility.

9.2   SEMICONDUCTOR NANOSTRUCTURES

9.2.1   Methods of Making Nanostructures in Semiconductors

Various techniques are utilized to fabricate nanostructures in semiconductors. Methods such as chemical vapor deposition, lithography, and self-assembly facilitate the attainment of meticulous control in the process of manufacturing. Etching techniques and electrochemical deposition play a significant role in the fabrication of complex nanostructures, which are crucial for a wide range of semiconductor applications [5, 34].

9.2.1.1 Lithography

The process of lithography entails the generation of patterns on a substrate via the use of masks and exposure to light. Various techniques, including as optical lithography, electron beam lithography, and extreme ultraviolet (EUV) lithography, enable the accurate formation of nanostructures on semiconductor surfaces [35].

Lithography, also known as photoengraving, involves the transfer of a pattern onto a reactive polymer film known as resist. This pattern is then replicated onto a thin film or substrate situated underneath the resist. Over the last fifty years, several lithography methods have been developed, using different lens systems and exposure radiation sources such as photons, X-rays, electrons, ions, and neutral atoms [5].

9.2.1.2 Chemical Vapor Deposition (CVD)

CVD involves the chemical reaction of a volatile component of a specific material with other gases. This reaction results in the formation of a nonvolatile solid, which is then deposited atomistically onto a substrate that has been appropriately positioned. This is a widely used technique in which thin films of various materials are deposited onto a substrate. The growth of nanowires, nanotubes, or thin films with nanoscale characteristics may be achieved depending on the specific process circumstances. Metal-organic chemical vapor deposition (MOCVD) and plasma-enhanced chemical vapor deposition (PECVD) are widely used techniques in the field of semiconductor manufacture [36].

Due to the various features of CVD, its chemistry exhibits a high degree of complexity, including a diverse array of chemical reactions. The significance of gas phase reactions (homogeneous) increases as the temperature and relative pressure of the reactants rise. Gas phase reactions become prominent, and homogenous nucleation occurs when there is an exceptionally high concentration of reactants. In order to achieve the deposition of high-quality films, it is essential to prevent the occurrence of homogenous nucleation. Chemical reactions include a diverse range of processes, which may be categorized into several groups based on the precursors used and the circumstances under which deposition occurs. These groups include pyrolysis, reduction, oxidation, compound synthesis, disproportionation, and reversible transfer [5].

9.2.1.3 Molecular Beam Epitaxy (MBE)

The process of MBE entails the precise deposition of crystalline layers of semiconductor materials at the atomic level. The mentioned technique is well-suited for the fabrication of nanostructures of superior quality, allowing for meticulous regulation of both layer thickness and composition. In the study of Nguyen et al. [37] in 2023, they utilized a two-step co-deposition technique and molecular beam epitaxy on an Al_2O_3 (0001) substrate at temperatures below 250°C, to form high-quality ordered L10-FeNi films. Reflection high energy electron diffraction and X-ray diffraction patterns verified that the films were epitaxially formed in the (111) orientation. To design the next generation of rare-earth-free constant magnets, they presented an efficient molecular beam epitaxy growth technique of high-quality ordered $L1_0$-FeNi (111) film.

9.2.1.4 Self-Assembly

Self-assembly is a broad word used to denote a phenomenon whereby the spontaneous organization of molecules and microscopic components, such as particles, takes place in an organized manner due to the effect of many factors, including chemical reactions, electrostatic attraction, and capillary forces [38]. The process of self-assembly is predicated upon the inherent characteristics of materials, which enable them to autonomously organize into certain nanostructures. The aforementioned technique demonstrates use in the generation of surface patterns and the organized arrangement of nanoparticles [5].

The utilization of block-copolymer self-assembly exhibits significant potential for extensive nano-manufacturing applications in both bulk and thin film types. The resulting nanopatterns possess the capability to serve as templates for the self-assembly of functional nanomaterials, thereby facilitating the development of energy-harvesting devices, photonic metasurfaces, nanofiltration membranes, and antibacterial coatings [39, 40].

In 2023, Walker et al. [41] examined the feasibility of using spin-coating techniques to facilitate the self-assembly process of silica nanocolloids with a diameter of 150 nm into extensive crystal formations on mica. The researchers explored various colloidal concentrations, accelerations, and rotational speeds as part of their investigation. The ordering arrangement was influenced by the low concentration of colloidal particles, which exhibited a size-dependent behavior. The nanocolloidal particles of the greatest size exhibited crystalline close-packed arrangements, whereas progressively smaller nanocolloids displayed configurations characterized by polycrystalline or amorphous structures.

9.2.1.5 Etching Techniques

Etching techniques, such as deep reactive ion etching (DRIE) and reactive ion etching (RIE) are used for the purpose of selectively eliminating material from a substrate, hence generating precisely defined nanostructures. Techniques such as atomic layer etching (ALE) enable meticulous manipulation of etching processes at the atomic scale.

In 2023, Ratha et al. [42], focused on optimizing the circumstances of reactive ion etching to minimize damage in BiFeO3-based thin films with Eu/Co substitution. The objective was to enhance the functionality of magnetic nanodevices in these films. In their work, they tested the ability of thin films to endure etching, a fundamental step in microfabrication. They claimed that reactive ion etching might end up in an alteration in crystallography that could alter the ferromagnetic and ferroelectric phases of multi-ferroic thin films. Consequently, the researchers undertake an investigation into the potential morphological and magnetic impairments resulting from the process of reactive ion etching on thin films based on BFO ((Bi,Eu)(Fe,Co)O$_3$). In addition, the researchers examined an optimal etching condition in order to minimize magnetic damage in the thin film of (Bi,Eu)(Fe,Co)O$_3$.

9.2.1.6 Sol-Gel Method

Sol-gel is a process that is a commonly employed technique for the synthesis of both inorganic and organic-inorganic hybrid materials. This versatile method enables the production of various structures such as nanoparticles, nanorods, thin films, and monoliths. The sol-gel process includes the transformation of precursor solutions into solid materials via chemical reactions. It is particularly utilized for the fabrication of nanoparticles, thin films, and coatings that possess nanoscale characteristics [5]. Balakrishnan and John [43], synthesized the multiphase TiO$_2$ (TAB)-ZnO (ZW) nanostructure through sol-gel technique. Sol-gel-produced TiO$_2$-ZnO nanostructure was studied for its structure, electrical properties, and photocatalytic performance.

These techniques allow for the precise manufacture of semiconductor nanostructures, giving scientists and engineers unprecedented control over the characteristics of the final goods for use in electronics, photonics, and other areas.

9.2.2 Electronic and Optical Properties of Semiconductor Nanostructures

The size, shape, and composition of semiconductor nanostructures have a significant impact on their electronic and optical characteristics. The electronic band structure is modified as a result of the discrete energy levels caused by quantum confinement phenomena. Band gap energies may be manipulated via the use of nanostructures to create tailored optoelectronics. Surface effects, which affect charge carrier dynamics and recombination rates, are modified by increased surface-to-volume ratios. Quantum dots, with their size-dependent optical features such as emission spectra, are used in displays and sensors [12].

9.3 NANOTECHNOLOGY CHALLENGES IN SEMICONDUCTORS

9.3.1 PROBLEMS OF MANUFACTURING AND CONTROLLING NANOSTRUCTURES AND THIN FILMS

The semiconductor business has been profoundly impacted by nanotechnology, which has allowed for the creation of smaller, quicker, and more efficient electronic devices. It raises a number of difficulties, however, that must be overcome. Nanostructures and thin films have unique difficulties in manufacturing and regulating. Some typical issues that arise during these procedures are as follows:

Nanostructures and thin films may be difficult to fabricate with a high degree of uniformity. Film thickness, content, and shape might vary depending on the deposition rate, substrate qualities, and other process factors. Particularly challenging is maintaining uniformity over wide regions or complicated structures, since this calls for careful regulation of deposition methods and process conditions. Soltani-kordshuli et al. [44] documented the first achievement in producing graphene-doped PEDOT:PSS composite thin films that had both high conductivity and transparency. This was accomplished by the use of conventional spray coating and substrate vibration-assisted ultrasonic spray coating (SVASC) techniques. In order to mitigate the aggregation of dopant particles, enhance conductivity, and enhance film homogeneity and nanostructure, composite PEDOT: PSS films were subjected to concurrent vertical and lateral ultrasonic vibration.

The reliability and efficacy of nanostructures and thin films may be greatly affected by the presence of contamination and flaws. Contaminants, including impurities, particles, and residual gases, have the potential to be introduced throughout the manufacturing process, which may result in adverse impacts on the characteristics of materials and the performance of devices. The presence of various imperfections, including as vacancies, dislocations, and grain boundaries, may have significant implications for the structural integrity, electrical characteristics, and mechanical stability of materials. It is of utmost importance to minimize the occurrence of contamination and defects in order to guarantee the intended attributes and performance of nanostructures and thin films [45].

The attainment of the intended surface roughness and morphology of nanostructures and thin films is essential in order to optimize their functional and optical characteristics. The presence of uncontrolled surface roughness has the potential to significantly impact the performance of devices, as well as influence light scattering and adhesion qualities. In order to get the appropriate surface roughness and morphology, it is essential to have control over the growth or deposition parameters, including temperature, pressure, and precursor flow [46].

The adhesion between thin films and the substrate is of utmost importance in ensuring their mechanical stability and long-term functionality. Insufficient adhesion has the potential to result in the separation of the film, hence compromising the structural integrity and performance of the system. To provide a robust bond between the film and substrate, it is essential to undertake appropriate surface preparation techniques, engage in interfacial engineering practices, and optimize the deposition parameters [47].

The precise characterization of nanostructures and thin films plays a vital role in comprehending their characteristics and evaluating their performance. Nevertheless, the characterization of nanoscale features and thin film characteristics may pose significant challenges as a result of the restrictions inherent in existing methodologies and equipment. The current field of study involves the development of proficient characterization methods that have the capability to investigate nanoscale characteristics and precisely quantify aspects of thin films [48].

9.3.2 THE EFFECT OF ENVIRONMENT AND EXTERNAL CONDITIONS ON THE BEHAVIOR OF NANOSTRUCTURES AND THIN FILMS

The performance of nanostructures and thin films may be significantly impacted by their surrounding environment and external factors. These minuscule substances, characterized by their sometimes nanoscale size, have exceptional characteristics that may either thrive or decline based on a range of variables. This part aims to examine the intricate relationship between nanostructures, thin films, and their surrounding environment, elucidating the significant impact of external factors on their performance.

The efficiency and durability of these materials may be considerably influenced by several environmental factors, such as ambient temperature, humidity, mechanical stress, and chemical interactions. Furthermore, it is crucial to comprehend and regulate variables such as radiation exposure and atmospheric gases, since they have the potential to induce transformational alterations. This understanding and control are essential for a wide range of applications.

The behavior of nanostructures and thin films is significantly influenced by temperature. The phenomenon of thermal expansion and contraction has the ability to induce mechanical stress and strain inside materials, which may result in various forms of deformation, such as buckling or delamination. Furthermore, changes in temperature have the potential to impact the crystal structure, phase transitions, and electrical characteristics of certain materials [49]. The maintenance of stability and performance in nanostructures and thin films is heavily reliant on effective thermal management. Gahtar et al. [50] employed the spray pyrolysis process to fabricate novel nickel sulfide thin films. These materials were intended to serve as potential co-catalysts, with the aim of enhancing photocatalytic efficiency or superconductivity. Subsequently, an investigation was conducted to examine the impact of deposition temperatures of 523, 573, and 623 K on the optical, structural, and electrical features. The thin films that were fabricated at temperatures of 523 K and 573 K have a favorable transmittance level of around 20%. The Ni_1S_2 thin film was characterized by determining its lowest optical band gap and Urbach energy at a temperature of 623 K. At 573K, and the compound $Ni_{17}S_{18}$ exhibited a maximum electrical conductivity of 4.29×10^5 $(\Omega.cm)^{-1}$. The nickel sulfide thin films that were deposited at 573 K exhibited favorable characteristics in terms of their optical, structural, and electrical properties.

The presence of humidity and moisture may adversely impact the qualities and performance of nanostructures and thin films. These factors have the potential to initiate corrosion, oxidation, or chemical reactions, resulting in the deterioration of material characteristics, changes in surface morphology, and diminished performance of the device. Thin films are susceptible to swelling and delamination when exposed to moisture. Protection against moisture and humidity requires proper encapsulation, sealing, or the application of coatings [51].

Nanostructures and thin films may be profoundly affected by the presence of certain gases. Thin film growth dynamics, surface composition, and the induction of chemical reactions are all influenced by the presence of reactive gases. On the other hand, inert gases may provide a safe space free from any reactions. Maintaining the intended characteristics and functionalities of nanostructures and thin films requires strict control over gas composition, pressure, and purity throughout manufacturing, processing, and operation. The effects of NO_2 and reducing gases including H_2 and NH_3, as well as relative humidity, on the physicochemical characteristics of ZnO nanostructures were investigated by Procek et al. [52]. Extensive experiments were conducted at room temperature and 200°C to determine the impact of the gases under study on the resistivity of the ZnO nanostructures. These experiments demonstrated a remarkable sensitivity of ZnO nanostructures to even low NO_2 concentrations. Approximately 600% in N_2/230% in the air at 200°C in the dark and 430% in N_2/340% in the air at RT accompanied by UV excitation were the structural responses to 1 ppm of NO_2. The structural reaction to the impact of NO_2 at a temperature of 200°C has a magnitude that surpasses the response to NH_3 by a factor exceeding 105 and exceeds the response to H_2 by a factor exceeding 106, both in relation to a concentration of 1 ppm.

The behavior of nanostructures and thin films may be influenced by light and radiation via a variety of mechanisms. The existence of light may impact optical qualities, including absorption, reflection, and transmission. Nanostructures and thin films are susceptible to photon-induced processes such as photoluminescence, photoexcitation, and photochemical reactions. Furthermore, ionizing radiation exposure may cause radiation-induced flaws, changes in material features, and device performance deterioration [53].

REFERENCES

1. P. Szczyglewska, A. Feliczak-Guzik, I. Nowak, Nanotechnology–General aspects: A chemical reduction approach to the synthesis of nanoparticles, Molecules. 28 (2023) 4932.
2. R.W. Whatmore, Nanotechnology—What is it? Should we be worried? Occup. Med. 56 (2006) 295–299.
3. S. Bayda, M. Adeel, T. Tuccinardi, M. Cordani, F. Rizzolio, The history of nanoscience and nanotechnology: From chemical–physical applications to nanomedicine, Molecules. 25 (2019) 112.
4. A. Selmani, D. Kovačević, K. Bohinc, Nanoparticles: From synthesis to applications and beyond, Adv. Colloid Interface Sci. 303 (2022) 102640.
5. G. Cao, Nanostructures & Nanomaterials: Synthesis, Properties & Applications, Imperial College Press (2004).
6. A. Nayak, B. Bhushan, S. Kotnala, N. Kukretee, P. Chaudhary, A.R. Tripathy, K. Ghai, S.L. Mudliar, Nanomaterials for supercapacitors as energy storage application: Focus on its characteristics and limitations, Mater. Today Proc. 73 (2023) 227–232.
7. M. Ikram, U. Qumar, S. Ali, A. Ul-Hamid, Impact of metal oxide nanoparticles on adsorptive and photocatalytic schemes: Fundamentals to Applications, in: Chapter 3 of Advanced Materials for Wastewater Treatment and Desalination Editors: A.F. Ismail, P.S. Goh, H. Hasbullah, and F. Aziz, CRC Press (2022).
8. Z. Liu, S. Fu, X. Liu, A. Narita, P. Samorì, M. Bonn, H.I. Wang, Small size, big impact: Recent progress in bottom-up synthesized nanographenes for optoelectronic and energy applications, Adv. Sci. 9 (2022) 2106055.
9. A. Niyati, M. Haghighi, M. Shabani, Solar-assisted photocatalytic elimination of Azo dye effluent using plasmonic AgCl anchored flower-like Bi4O5I2 as staggered nano-sized photocatalyst designed via sono-precipitation method, J. Taiwan Inst. Chem. Eng. 115 (2020) 144–159.
10. A.I. Onyia, H.I. Ikeri, A.I. Chima, Surface and quantum effects in nanosized semiconductor, Am. J. Nano Res. Appl. 8 (2020) 35–41.
11. P.E. Sheehan, L.J. Whitman, Detection limits for nanoscale biosensors, Nano Lett. 5 (2005) 803–807.
12. L.E. Brus, Electron–electron and electron-hole interactions in small semiconductor crystallites: The size dependence of the lowest excited electronic state, J. Chem. Phys. 80 (1984) 4403–4409.
13. S. Kargozar, S.J. Hoseini, P.B. Milan, S. Hooshmand, H. Kim, M. Mozafari, Quantum dots: A review from concept to clinic, Biotechnol. J. 15 (2020) 2000117.
14. X. Tang, J. Wang, L. Zhu, W. Yin, Simulating stress-tunable phonon and thermal properties in heterostructured AlN/GaN/AlN-nanofilms, Mater. Res. Express. 6 (2018) 15018.
15. S. Zhang, X. Tang, H. Ruan, L. Zhu, Effects of surface/interface stress on phonon properties and thermal conductivity in AlN/GaN/AlN heterostructural nanofilms, Appl. Phys. A. 125 (2019) 1–14.
16. A.S. Yalamarthy, H. So, M. Muñoz Rojo, A.J. Suria, X. Xu, E. Pop, D.G. Senesky, Tuning electrical and thermal transport in AlGaN/GaN heterostructures via buffer layer engineering, Adv. Funct. Mater. 28 (2018) 1705823.
17. E.K. Hamza, S.N. Jaafar, Nanotechnology application for wireless communication system, in: Chapter 6 of Nanotechnology for Electronic Applications, Editors: N.M. Mubarak, S. Gopi, P. Balakrishnan, Springer Nature Singapore (2022), pp. 115–130.
18. R.L. Hoffman, B.J. Norris, J.F. Wager, ZnO-based transparent thin-film transistors, Appl. Phys. Lett. 82 (2003) 733–735.
19. B. Shrestha, Nanotechnology for biosensor applications, in: Chapter 20 of Sustainable Nanotechnology for Environmental Remediation, Elsevier (2022), pp. 513–531.
20. Z. Liu, C.-H. Lin, B.-R. Hyun, C.-W. Sher, Z. Lv, B. Luo, F. Jiang, T. Wu, C.-H. Ho, H.-C. Kuo, Micro-light-emitting diodes with quantum dots in display technology, Light Sci. Appl. 9 (2020) 83.
21. F.P. García de Arquer, D.V. Talapin, V.I. Klimov, Y. Arakawa, M. Bayer, E.H. Sargent, Semiconductor quantum dots: Technological progress and future challenges, Science. 373 (2021) eaaz8541.
22. S.K. Sharma, P. Kumar, B. Raj, Introduction to nanowires: Types, proprieties, and application of nanowires, in: Chapter 1 of Innovative Applications of Nanowires for Circuit Design, Editor: B. Raj, IGI Global (2021), pp. 1–15.
23. T. Pham, A. Qamar, T. Dinh, M.K. Masud, M. Rais-Zadeh, D.G. Senesky, Y. Yamauchi, N. Nguyen, H. Phan, Nanoarchitectonics for wide bandgap semiconductor nanowires: Toward the next generation of nanoelectromechanical systems for environmental monitoring, Adv. Sci. 7 (2020) 2001294.
24. M. Akbari-Saatlu, M. Procek, C. Mattsson, G. Thungström, H.-E. Nilsson, W. Xiong, B. Xu, Y. Li, H.H. Radamson, Silicon nanowires for gas sensing: A review, Nanomaterials. 10 (2020) 2215.
25. K. Okamoto, M. Sugiyama, S. Mabu, Importance of advanced metrology in semiconductor industry and value-added creation using AI/ML, E-Journal Surf. Sci. Nanotechnol. 18 (2020) 214–222.

26. C.B. Gopinath, S. Ramanathan, M.N.M. Yasin, M.I. Shapiai, Z.H. Ismail, S. Subramaniam, Essential semiconductor films in micro-/nano-biosensors: Current scenarios, J. Taiwan Inst. Chem. Eng. 127 (2021) 302–311.

27. G. Boschetto, S. Carapezzi, A. Todri-Sanial, Graphene and carbon nanotubes for electronics nanopackaging, IEEE Open J. Nanotechnol. 2 (2021) 120–128.

28. G. Ramalingam, N. Perumal, A.K. Priya, S. Rajendran, A review of graphene-based semiconductors for photocatalytic degradation of pollutants in wastewater, Chemosphere. 300 (2022) 134391.

29. S. Trivedi, K. Lobo, H.S.S.R. Matte, Synthesis, properties, and applications of graphene, in: Chapter 3 of Fundamentals and Sensing Applications of 2D Materials, Editors: M. Hywel, C. S. Rout, and D.J. Late, Elsevier (2019), pp. 25–90.

30. M.B. Tahir, T. Iqbal, M. Rafique, M.S. Rafique, T. Nawaz, M. Sagir, Nanomaterials for photocatalysis, in: Nanotechnol. Photocatal. Environ. Appl., Elsevier, 2020: pp. 65–76.

31. S.K. Verma, S. Samanta, A.K. Srivastava, S. Biswas, R.M. Alsharabi, S. Rajput, Conducting polymer nanocomposite for energy storage and energy harvesting systems, Adv. Mater. Sci. Eng. 2022 (2022).

32. T. Li, P. Li, R. Sun, S. Yu, Polymer-based nanocomposites in semiconductor packaging, IET Nanodielectrics. 6 (2023) 147–158.

33. Z. Sun, R. Wong, Y. Liu, M. Yu, J. Li, D. Spence, M. Zhang, M. Kathaperumal, C.-P. Wong, Melamine–graphene epoxy nanocomposite based die attach films for advanced 3D semiconductor packaging applications, Nanoscale. 14 (2022) 15193–15202.

34. S. Suresh, Semiconductor nanomaterials, methods and applications: A review, Nanosci. Nanotechnol. 3 (2013) 62–74.

35. C.D. Dieleman, W. Ding, L. Wu, N. Thakur, I. Bespalov, B. Daiber, Y. Ekinci, S. Castellanos, B. Ehrler, Universal direct patterning of colloidal quantum dots by (extreme) ultraviolet and electron beam lithography, Nanoscale. 12 (2020) 11306–11316.

36. M.K. Patil, S. Shaikh, I. Ganesh, Recent advances on TiO2 thin film based photocatalytic applications (a review), Curr. Nanosci. 11 (2015) 271–285.

37. B.-H. Jun, Y.-B. Chun, J.H. Lee, Ordered L10-FeNi (111) epitaxial thin film on Al2O3 (0001) substrate: Molecular beam epitaxy growth and characterizations, Thin Solid Films. 780 (2023) 139962.

38. A.A. Kulkarni, G.S. Doerk, Thin film block copolymer self-assembly for nanophotonics, Nanotechnology. 33 (2022) 292001.

39. C. Cummins, R. Lundy, J.J. Walsh, V. Ponsinet, G. Fleury, M.A. Morris, Enabling future nanomanufacturing through block copolymer self-assembly: A review, Nano Today. 35 (2020) 100936.

40. J.H. Kim, H.M. Jin, G.G. Yang, K.H. Han, T. Yun, J.Y. Shin, S. Jeong, S.O. Kim, Smart nanostructured materials based on self-assembly of block copolymers, Adv. Funct. Mater. 30 (2020) 1902049.

41. J. Walker, V. Koutsos, Spin coating of silica nanocolloids on mica: Self-assembly of two-dimensional colloid crystal structures and thin films, Coatings. 13 (2023) 1488.

42. S. Ratha, G. Egawa, S. Yoshimura, Optimizing the Reactive Ion Etching Conditions with Minimal Damage for High Functional Magnetic Nano Device Application in Bifeo3-Based Thin Film by Eu/Co Substitution, Co Substit. (n.d.).

43. M. Balakrishnan, R. John, Properties of sol-gel synthesized multiphase TiO2 (AB)-ZnO (ZW) semiconductor nanostructure: An effective catalyst for methylene blue dye degradation, Iran. J. Catal. 10 (2020) 1–16.

44. F. Soltani-kordshuli, F. Zabihi, M. Eslamian, Graphene-doped PEDOT: PSS nanocomposite thin films fabricated by conventional and substrate vibration-assisted spray coating (SVASC, Eng. Sci. Technol. an Int. J. 19 (2016) 1216–1223.

45. X. Wu, X. Chen, Q.M. Zhang, D.Q. Tan, Advanced dielectric polymers for energy storage, Energy Storage Mater. 44 (2022) 29–47.

46. R. Kesarwani, P.P. Dey, A. Khare, Correlation between surface scaling behavior and surface plasmon resonance properties of semitransparent nanostructured Cu thin films deposited via PLD, RSC Adv. 9 (2019) 7967–7974.

47. J.-H. Kim, H.-J. Kil, S. Lee, J. Park, J.-W. Park, Interfacial delamination at multilayer thin films in semiconductor devices, ACS Omega. 7 (2022) 25219–25228.

48. M.R. Baklanov, K.P. Mogilnikov, A.S. Vishnevskiy, Challenges in porosity characterization of thin films: Cross-evaluation of different techniques, J. Vac. Sci. Technol. A. 41 (2023).

49. M. Huff, Residual stresses in deposited thin-film material layers for micro-and nano-systems manufacturing, Micromachines. 13 (2022) 2084.

50. A. Gahtar, S. Benramache, C. Zaouche, A. Boukacham, A. Sayah, Effect of temperature on the properties of nickel sulfide films performed by spray pyrolysis technique, Adv. Mater. Sci. 20 (2020) 36–51.

51. C. Han, Analysis of moisture-induced degradation of thin-film photovoltaic module, Sol. Energy Mater. Sol. Cells. 210 (2020) 110488.
52. M. Procek, T. Pustelny, A. Stolarczyk, Influence of external gaseous environments on the electrical properties of ZnO nanostructures obtained by a hydrothermal method, Nanomaterials. 6 (2016) 227.
53. A. Peter Amalathas, M.M. Alkaisi, Nanostructures for light trapping in thin film solar cells, Micromachines. 10 (2019) 619.

10 The Development and Processing of Advanced Low-Dimensional Semiconductors

Xinghui Liu, Shiheng Xin, Fuchun Zhang, and Chunyi Zhi

10.1 INTRODUCTION

The development of materials has been closely intertwined with the progress of human civilization. With the advent of the "Silicon Age" at the end of the 20th century, research on new materials has gained significant attention, marking a pivotal era in the 21st century. Traditional semiconductors have played a prominent role in this historical period, accompanied by advancements in fabrication processes. However, in recent decades, the rapid development of low-dimensional (LD) semiconductors, such as carbon-based allotropes, phosphorus allotropes, and transition-metal dichalcogenides (TMDs), has garnered immense interest as potential key components for electronic devices.

This chapter aims to introduce a representative traditional semiconductor and delve into the associated eight fabrication processes. By comprehensively understanding the intricacies of semiconductor development, we can unlock new possibilities and drive innovation in various fields, ultimately shaping the trajectory of modern civilization in the 21st century and beyond.

10.2 CONVENTIONAL SEMICONDUCTOR MATERIALS AND PROCESSES

Semiconductors are used in every aspect of our lives: From smartphones and laptops to credit cards and subways, semiconductors are used in many items we rely on daily. This section will introduce common and well-established semiconductor materials and how they are processed.

10.2.1 SILICON (SI)

Silicon (Si), the well-established and long-applied semiconductor material, has contributed indelibly to the world's development. Silicon exists in large quantities on the earth, second only to oxygen in the earth's crust, accounting for about 25.8% of the mass of the earth's crust; the silicon content in sand can be up to 90% [1–3].

The semiconductor processing of silicon-type, including germanium, has become so sophisticated today that scientists have researched whether other semiconductor materials can also replace the high-cost silicon-type materials [4–6].

10.2.2 OVERVIEW OF THE DEVELOPMENT OF OTHER SEMICONDUCTOR MATERIALS

10.2.2.1 Gallium Arsenide (GaAs), Indium Phosphide (InP), and Gallium Nitride (GaN)

GaAs, InP, and GaN materials have high electron mobility due to a small, effective mass of electrons, making them suitable for high-frequency and high-speed electronic devices such as transistors and high-frequency power amplifiers. They have good optoelectronic properties for laser

DOI: 10.1201/9781003450146-10

diodes (LEDs) and photodetectors because of their excellent light emission and detection properties. Through years of research, GaAs, InP, and GaN materials have been widely used in fiberoptic, radar, and wireless communications.

However, the cost of producing GaAs, InP, and GaN manufacturing is relatively high because of the preparation and processing technology requirements; the process is pretty complex and requires unique technology and equipment. They are susceptible to oxidation and moisture, requiring special handling and packaging to ensure device stability. Devices made with GaAs materials typically have relatively high power consumption, which may not be applicable in specific low-power applications. Moreover, InP materials are sensitive to oxidation and moisture and require special treatment. Choosing the suitable substrate is critical and a challenge to the growth and preparation process of InP materials [7–10].

10.2.2.2   Silicon Carbide (SiC)

SiC has high thermal stability and can be applied to high-temperature electronic devices, which work stably in high-temperature environments. It has high electron mobility, which can be used for high-frequency and high-power devices. Simultaneously, it has good radiation resistance, making it suitable for nuclear energy and space applications. However, SiC is relatively expensive to manufacture since making the quality and homogeneity of SiC materials is a challenge [11].

10.2.2.3   Diamond

Diamond materials have ultra-high thermal stability and can be used in high-temperature, high-power, high-frequency electronic devices. Their harmful effects have excellent thermal conductivity and can be used for heat dissipation in high-power devices. However, diamond is costly and requires high preparation and processing technology [12].

10.2.2.4   Gallium Oxide (GaO)

GaO has high electron mobility, small electron effective mass, and excellent insulating properties, which helps to improve e-motion performance and makes it stand out from many semiconductor materials. Furthermore, GaO preparation and processing technologies are being explored step by step, with numerous production and cost challenges [13–16].

10.2.3   PROCESSING FLOW OF THE TRADITIONAL SEMICONDUCTOR

Over the years, the traditional semiconductor processing described above has matured into industrialization and commercialization. The eight semiconductor processing steps in the industrialization process will be described here.

10.2.3.1   Wafer Processing

All semiconductor processes begin with a grain of sand because the silicon contained in the sand is the raw material needed to produce wafers. A wafer is a round sheet formed by cutting a single crystal column made of silicon. To extract high-purity silicon materials, we need to use silica sand, a silica content of up to 95% of the unique materials, but the production of wafers is the primary raw material.

- **Ingots**
  The sand is first heated to separate the carbon monoxide from the silicon, and the process is repeated until ultra-high purity electronic-grade silicon (EG-Si) is obtained. The first step in semiconductor manufacturing is melting high-purity silicon into a liquid, then solidified into a single-crystal solid form called an ingot. Silicon ingots (pillars) are made with a high degree of precision down to the nanometer level, and the widely used manufacturing method is the lifting and pulling method.

- **Spindle Cutting**
  Once the previous step is completed, the ends of the ingot must be cut off with a diamond saw and then cut into thin slices of a certain thickness. Larger and thinner wafers can be divided into more usable units, which helps to reduce production costs. After cutting the ingots, the wafers are marked with flat zones or dents, which can be used as a standard to set the processing direction in subsequent steps.
- **Wafer Surface Polishing**
  The thin wafers obtained through the above cutting process are called bare wafers, which are unprocessed raw wafers. The surface of a bare wafer is so uneven that it is impossible to print circuit graphics on it directly. Therefore, surface imperfections are removed by grinding and chemical etching, then polished to create a clean surface, and then cleaned to remove residual contaminants, resulting in a finished wafer with a clean surface [3, 17–20].

10.2.3.2 Oxidation

The function of the oxidation process is to form a protective film on the surface of the wafer. It protects the wafer from chemical impurities, prevents leakage currents from entering the circuit, prevents diffusion during ion implantation, and prevents the wafer from slipping during etching.

The first step in the oxidation process is the removal of impurities and contaminants, which requires four steps to remove impurities such as organics, metals, and evaporated water residues. Once cleaned, the wafers can be placed in a high-temperature environment of 800–1200°C through the flow of oxygen or vapor on the surface of the wafer to form a layer of silicon dioxide (i.e., oxide).

- **Dry and wet oxidation**
  Depending on the oxidizing agent in the oxidation reaction, the thermal oxidation process can be divided into dry oxidation, which uses pure oxygen to produce a silica layer at a slow rate but with a thin and dense oxide layer, and wet oxidation, which requires the simultaneous use of oxygen and highly soluble water vapor and is characterized by a fast growth rate but with a relatively thick and low-density protective layer.

In addition to the oxidizer, other variables affect the thickness of the silica layer. First, the wafer structure, surface defects, and internal doping concentration affect the oxide layer's generation rate. In addition, the higher the pressure and temperature generated by the oxidizing equipment, the faster the oxide layer is generated [13, 14, 21, 22].

10.2.3.3 Photolithography

Photolithography is the printing of circuit patterns onto wafers by means of light, which we can understand as drawing a planar design on the surface of a wafer necessary for semiconductor manufacturing. The higher the fineness of the circuit pattern, the higher the integration of the finished chip, which must be realized through advanced photolithography. Specifically, lithography can be divided into three steps: coating photoresist, exposure, and development.

- **Coated photoresist**
  The first step in drawing a circuit on a wafer is to coat the oxide layer with photoresist. The photoresist turns the wafer into photo paper by changing its chemical properties. The thinner the layer of photoresist on the surface of the wafer and the more uniformly it is coated, the finer the pattern that can be printed.
- **Exposure**
  We can selectively pass light through an exposure device, and when the light passes through the mask containing the circuit pattern, the circuit is printed onto the wafer with the photoresist film underneath.

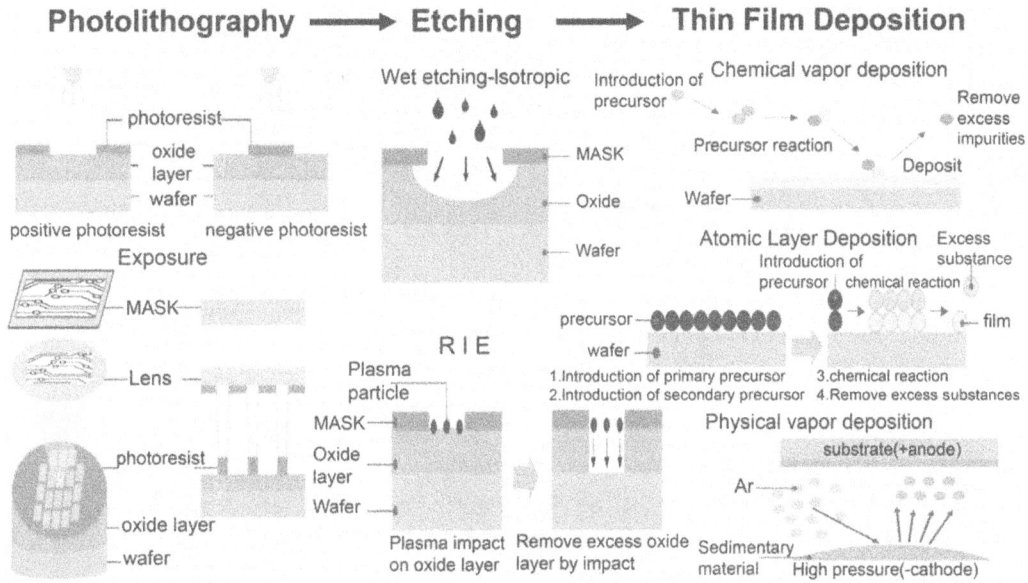

FIGURE 10.1 Photolithography, etching, and thin film deposition.

The finer the print pattern during the exposure process, the more components the final chip will be able to accommodate, which helps to increase production efficiency and reduce the cost of individual components. A new technology currently receiving much attention in this area is extreme ultraviolet (EUV) lithography.

- **Expose (a photographic plate)**
  The step after exposure is to spray a developer on the wafer to remove the photoresist from the uncovered areas of the graphic, thus allowing the printed circuit pattern to appear. After the development is completed, it needs to be checked by various measuring devices and optical microscopes to ensure the quality of the circuit drawing [6, 22–24]. The process of photolithography, etching, and thin film deposition is shown in Figure 10.1.

10.2.3.4 Etching

After the circuit diagram is photolithographed on the wafer, an etching process removes any excess oxide film and leaves only the semiconductor circuit diagram. This is accomplished using a liquid, gas, or plasma to remove the selected excess.

There are two main types of etching methods, depending on the substance used: wet etching, which uses a chemical reaction with a specific chemical solution to remove the oxide film, and dry etching, which uses gas or plasma.

- **Wet etching**
  Wet etching using a chemical solution to remove the oxide film has the advantage of low cost, high etching speed, and high productivity. However, wet etching is isotropic, i.e., its velocity is the same in any direction. This results in the mask (or sensitive film) not being perfectly aligned with the etched oxide film, making it challenging to process excellent circuit diagrams.
- **Dry etching**
  Dry etching can be categorized into three different types. The first is chemical etching, which uses an etching gas (mainly hydrogen fluoride). Like wet etching, this method is isotropic, which means it is also unsuitable for fine etching.

The second method is physical sputtering, where ions in a plasma are used to impact and remove the excess oxide layer. As an anisotropic etching method, sputter etching has different etching speeds in the horizontal and vertical directions, and therefore, it is finer than chemical etching. However, the disadvantage of this method is that the etching rate is slower, because it relies entirely on the physical reaction caused by ion collisions.

The final third method is reactive ion etching (RIE), which combines the first two methods, i.e., ionized physical etching by plasma and chemical etching by free radicals generated by plasma activation. In addition to the etching speed exceeding that of the previous two methods, RIE can utilize the anisotropic properties of ions to etch highly detailed patterns.

Dry etching is widely used today to improve the yield of exemplary semiconductor circuits. Maintaining whole wafer etch uniformity and increasing etch speeds is critical, and today's state-of-the-art dry etch equipment supports the production of the most advanced logic and memory chips at higher performance [2, 16, 22, 24–26].

10.2.3.5   Thin Film Deposition

To create the miniature devices inside the chip, we need to continuously deposit layers of film, remove the excess by etching, and add materials to separate the different devices. Each transistor or memory cell is built step by step through this process. By thin film, we mean a film that is less than one micron (μm, one-millionth of a meter) thick and cannot be made by ordinary mechanical processes. Placing a thin film containing the desired molecular or atomic units onto a wafer is called deposition.

To form multilayer semiconductor structures, we need to create device stacks, which are alternating layers of thin metallic (conductive) and dielectric (insulating) films on the wafer surface, followed by a repetitive etching process that removes the excess and creates a three-dimensional structure. Techniques that can be used for the deposition process include chemical vapor deposition (CVD), atomic layer deposition (ALD), and physical vapor deposition (PVD), and the methods used for these techniques can be categorized into dry and wet deposition.

- **Chemical vapor deposition**
  In CVD, the precursor gas reacts chemically in the reaction chamber, producing a thin film that adheres to the wafer's surface and byproducts pumped out of the chamber.
  Plasma-enhanced CVD requires plasma to generate a reaction gas. This method reduces the reaction temperature, making it ideal for temperature-sensitive structures. Plasma also reduces the number of deposit passes, often resulting in higher-quality films.
- **Atomic Layer Deposition**
  ALD forms thin films by depositing only a few atomic layers at a time. The key to the method is to cycle through the separate steps in a specific order and to maintain reasonable control. Coating the wafer surface with a precursor is the first step, after which different gases are introduced to react with the precursor to form the desired substance on the wafer surface.
- **Physical vapor deposition**
  As the name implies, PVD refers to forming thin films by physical means. Sputtering is a PVD method based on the principle that atoms of the target material are sputtered by argon plasma bombardment and deposited on the wafer surface to form a thin film.

In some cases, the deposited film can be treated, and its properties are improved by ultraviolet thermal treatment (UVTP) [6, 24].

10.2.3.6   Interconnection

Semiconductors conduct electricity between conductors and non-conductors (i.e., insulators), a property that gives us complete control over electric currents. Components such as transistors can be constructed through wafer-based lithography, etching, and deposition processes, but they also need to be connected to send and receive power and signals.

Metals are used for circuit interconnections because of their electrical conductivity. Metals used in semiconductors need to fulfill the following conditions:

- *Low resistivity*: Since metal circuits are required to transmit current, the metal in them should have a low resistance.
- *Thermo-chemical stability*: The properties of the metal material must remain unchanged during the metal interconnection process.
- *High reliability*: As IC technology evolves, even small amounts of metallic interconnect materials must be durable enough.
- *Manufacturing cost*: Even if the previous three conditions have been met, if the material cost is too high, it will not be able to meet the needs of mass production.

The interconnect process uses two main substances, aluminum and copper [1, 3, 17, 18, 20, 22].

10.2.3.7   Testing
The main goal of testing is to verify that the quality of semiconductor chips meets specific standards, thereby eliminating defective products and improving chip reliability. Additionally, products tested for defects do not make it to the packaging step, helping to save costs and time. Electronic core sorting (EDS) is one such test method for wafers. EDS is a process that verifies the electrical characteristics of each chip in the wafer state, thus improving semiconductor yields [1, 21].

10.2.3.8   Encapsulation
After the previous processes, square chips of equal size (or individual wafers) are formed on the wafer. The next step is to obtain individual chips by dicing. Freshly diced chips are fragile, cannot exchange electrical signals, and must be handled separately. This process is called encapsulation, which involves forming a protective shell around the semiconductor chips and allowing them to exchange electrical signals with the outside world.

Through hundreds of processes and eight significant steps, the standard "sand" of our daily life is processed into various semiconductor devices and chips that serve all aspects of human life [1, 21].

10.3   LOW-DIMENSIONAL SEMICONDUCTORS AND PROCESSES

As Moore's law ends, the future design of chips is beginning to face various difficulties. As the feature size of functional devices continues to decrease, the quantum size effect, short channel effect, quantum tunneling effect, and thermal effect in the devices will lead to the degradation of device performance or even failure. Silicon-based functional devices based on conventional semiconductor materials have reached their limits. In that case, the effect is always limited, so it is undoubtedly an excellent direction to look for new semiconductor materials from the new semiconductor materials themselves.

Low-dimensional materials refer to the three-dimensional space in at least one dimension in the nanoscale range or as the basic unit of the material according to the number of sizes. These nanostructures are distinguished from bulk materials and are called low-dimensional nanostructures.

10.3.1   Basic Properties of LD Semiconductor Material

10.3.1.1   Quantum Size Effect
The quantum size effect is a phenomenon in which the electronic energy levels near the Fermi energy level change from quasi-continuous to discrete energy levels, also known as energy level

splitting or energy gap widening, when the particle size decreases to a particular value. When the degree of change in the energy levels is greater than the change in thermal, optical, and electromagnetic energy, it results in nanoparticle magnetic, optical, acoustic, thermal, electrical, and superconducting properties significantly different from those of conventional materials [27–29].

10.3.1.2 Short Channel Effect

These are some effects in semiconductor field effect transistors when the conductive channel length is reduced to a few nanometers. These effects include, among others, a decrease in threshold voltage with channel length, a reduction in the leakage barrier, carrier surface scattering, velocity saturation, ionization, and hot-electron effects. This is particularly noticeable in short channels and can lead to severe source-drain-through device failure [27, 28, 30].

10.3.1.3 Small Size Effect

The small size effect is a phenomenon in which electron transport and device performance are significantly affected when the critical dimension of a semiconductor device is reduced to the nanometer level. This includes quantum size and short channel effects but can consist of other factors such as charge quantization and electron tunneling effects [27, 28].

10.3.1.4 Surface Effects

Surface effects include the appearance of surface states and electron transport between surface and bulk states. These effects can affect the electrical conductivity, chemical reactivity, and optical properties of the material [27, 28, 31].

10.3.1.5 Cooren Blockage Effect

As electrons move through a semiconductor, they are subjected to coulombic interaction forces from other electrons. The coulomb blocking effect refers to the fact that this interaction causes the movement of electrons to be hindered, especially at high electron concentrations, and it can reduce electron mobility and conductivity [27, 28, 32].

10.3.1.6 Quantum Interference Effects

When electrons propagate as waves, they can undergo interference effects, similar to optical interference. In semiconductor nanostructures, this effect can affect the wave function distribution of electrons, leading to changes in the energy band structure and transport properties of electrons in the crystal [27, 28].

10.3.1.7 Two-dimensional Electron Gas and the Quantum Hall Effect

- *Two-dimensional electron gas*: In a two-dimensional material or two-dimensional electronic system, electrons are confined in a plane and are free to move in a direction perpendicular to that plane [27, 28, 33].
- *Quantum hall effect*: This is a quantum effect that occurs in two-dimensional electron gases, where electrons form discrete energy levels perpendicular to the direction of the magnetic field and exhibit a high degree of conductivity when subjected to a strong magnetic field [27, 28, 34].

10.3.1.8 Quantum Tunneling Effects

It is assumed that a particle with a specific energy moves from the left side of the potential barrier to the right side. In classical mechanics, only particles with energy more incredible than the barrier can cross the barrier and move to the right of the barrier, while particles with energy less than the barrier are reflected and cannot pass through the wall. The quantum tunneling effect is the phenomenon that particles with energy less than the height of the potential barrier can still penetrate the barrier. The probability of quantum tunneling is related to the height and thickness of the potential barrier and the effective mass of the particle; in resonance tunneling, it is also associated with the width of the potential well and the energy band structure of the material [27, 28].

10.3.2 LOW-DIMENSIONAL SEMICONDUCTORS

LD materials have attracted everyone's attention due to their unique characteristics and structures [27, 28]. Among them, carbon-based and phosphorus-based materials are representative LD material, which is demonstrated as follows (Figure 10.2).

10.3.2.1 Carbon-based Materials

Carbon-based semiconductor materials include carbon nanotubes (CNT), graphene-based nanoribbons, and graphdiyne. Carbon-based two-dimensional materials may continue the electronic Moore's law. LD materials can be constructed as a few or even a single atomic layer, thus providing fragile channel areas and eliminating the need to worry about short-channel effects [35, 36].

10.3.2.1.1   Carbon Nanotubes (CNTs)

CNT materials have high electron and hole mobility, allowing them to excel in high-speed electronic devices such as transistors and radio frequency (RF) switches. Their structural versatility allows for selection of different types of CNTs, including semiconductors, metals, and insulators, to

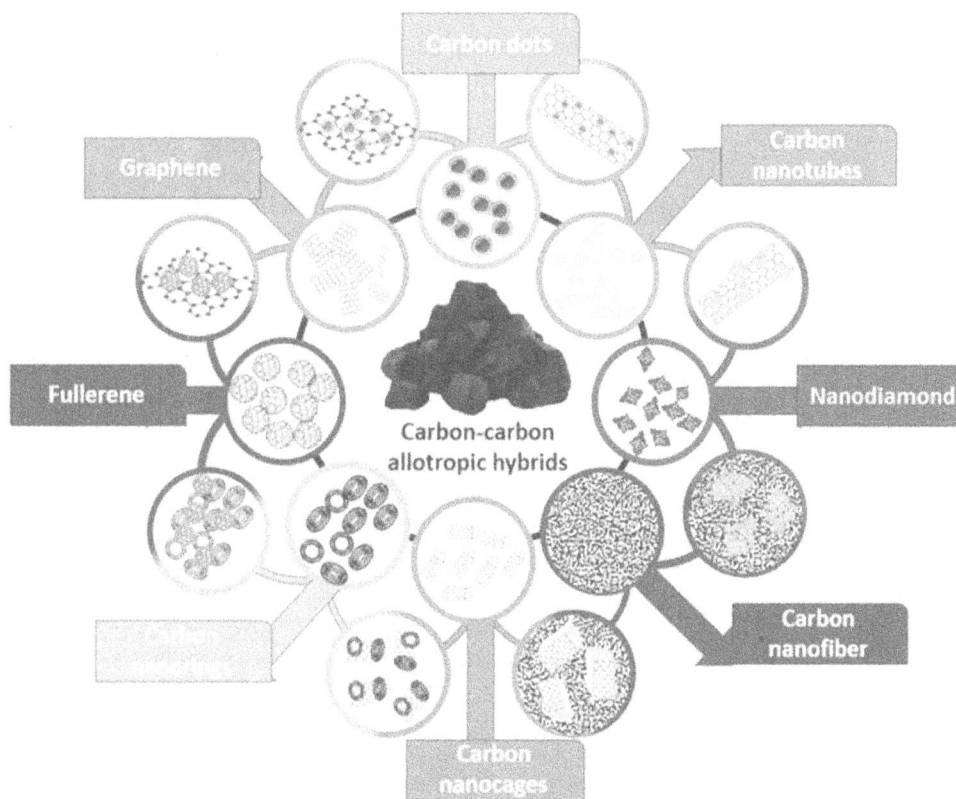

FIGURE 10.2   Carbon-based materials. (Adapted with permission [35]. Copyright [2019], ACS publications.)

meet the needs of varying semiconductor applications. CNT materials also have excellent thermal conductivity, allowing them to be used in high-performance heat sinks and thermal management devices. The electrical properties of CNTs are unstable due to their structure and defects, which leads to inconsistent performance. Producing high-quality CNT is costly, which may limit its commercialization in some applications [35, 37].

10.3.2.1.2 Graphene-based Materials

Graphene-based materials have high electron and hole mobility and are suitable for high-frequency and high-speed electronic devices such as transistors and photodetectors. However, graphene is typically a zero-bandgap material and lacks a bandgap, which limits its use in conventional semiconductor applications. It is structurally unstable and susceptible to structural defects and oxidation, which affects its performance and stability. Difficult to integrate and prepare, integrating graphene into existing semiconductor processes can be challenging, and complex techniques are required to prepare high-quality graphene [27, 35, 38–40].

Graphene nanoribbons (GNRs) materials have a tunable bandgap that can be precisely controlled by width and edge structure adjustments, extending their use in semiconductors. GNRs have high electron mobility, making them suitable for high-performance semiconductor devices. However, preparing narrow and long GNRs is usually difficult and requires sophisticated preparation processes. The edge structure of GNRs is unstable and prone to defects, which may affect their performance. Moreover, their integration is difficult, and integrating GNRs into electronic devices is complex and challenging [35, 37].

10.3.2.1.3 Graphdiyne

Compared with two-dimensional graphene-based semiconductor materials, the emerging graphdiyne (graphyne) has attracted much attention for its theoretically higher mobility and narrower forbidden bandwidth than traditional two-dimensional materials. Graphdiyne has a tunable band gap and high electron mobility, making it suitable for high-performance semiconductor device applications. However, preparing high-purity graphdiyne is usually complex and requires sophisticated preparation processes that require further research and development [27, 35, 41, 42].

10.3.2.2 Phosphorus-based Materials (P)

10.3.2.2.1 Phosphorus-based Materials and Their Isomers

Among the low-dimensional semiconductors, the polycrystalline nature of phosphorus-based materials has attracted significant research interest, and their variable chemical bonding structure gives rise to a wide variety of micro- and nanostructures. The properties of monolithic phosphorus materials are closely related to the geometrical arrangement of phosphorus atoms. Among the monolithic phosphorus materials are black phosphorus (a bipolar semiconductor in which holes are slightly dominant), violet phosphorus (a p-type semiconductor), and fiber phosphorus (an n-type semiconductor). Most monolithic phosphorus semiconductors have in-plane anisotropy, making them suitable for constructing polarization-sensitive optoelectronic components. In addition to the energy band structure, the mobility of charge carriers plays an essential role in determining the electronic properties of semiconductors. Monolayer phosphenes tend to have high carrier mobility.

Monomorphic phosphorus and phosphorus isomers, black phosphorus, white phosphorus, red phosphorus, green phosphorus, blue phosphorus, violet phosphorus, and fibrous phosphorus are different forms of phosphorus isomers, which have their advantages and disadvantages in semiconductor applications and prospects. The following is a detailed description of the properties of each phosphorus heterogeneous body and their application prospects in the semiconductor field [43]. Atomic structures of white, violet, fibrous, black, blue, and green phosphorus are shown in Figure 10.3.

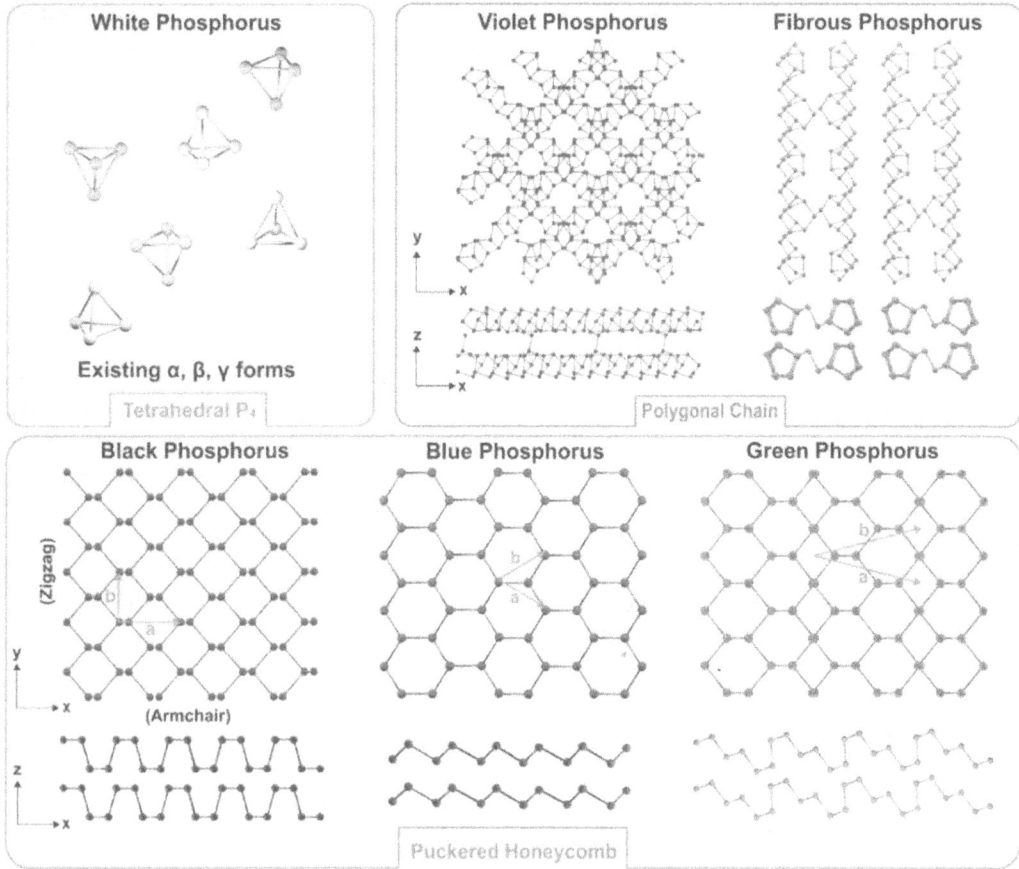

FIGURE 10.3 Atomic structures of white, violet, fibrous, black, blue, and green phosphorus. (Adapted with permission [43]. Copyright [2023], Chemical Society Reviews.)

10.3.2.2.2 Black Phosphorus

Black phosphorus exhibits high carrier mobility in a two-dimensional lattice structure, making it a candidate for high-performance field-effect transistor materials. The size of its tunable band gap can be regulated by changes in the number of layers, allowing it to be engineered in different applications. However, black phosphorus oxidizes quickly and loses its properties rapidly in air, requiring special treatment and encapsulation.

The material properties of black phosphorus make it promising for use in two-dimensional materials for high-performance transistors, nanoelectronic devices, and optoelectronic devices. However, further research is still needed to improve its stability and promote its application in the semiconductor industry [43–45].

10.3.2.2.3 White and Red Phosphorus

White phosphorus is widely found in nature, is easy to extract and process, and has a low production cost, making it economically beneficial. However, white phosphorus is toxic to human beings and requires special treatment, which brings many limitations to processing, and in semiconductor applications, it is unsuitable for high-performance electronic devices due to its low electron mobility. White phosphorus is more likely to be mainly used in gunpowder and military applications and is unsuitable for a wide range of applications in semiconductors [43, 46].

Red phosphorus can be found as widely in nature as white phosphorus and is relatively simple to prepare. Moreover, it is stable at room temperature, not readily oxidized, and has considerable stability characteristics. Red phosphorus may be used in flexible electronic devices, energy storage, and energy production in the future, but its main advantage lies in its stability rather than high performance [46].

The research on these phosphorus (green phosphorus, blue phosphorus, and purple phosphorus) heterostructures is more limited and has not yet reached the stage of industrial application. More research is needed to determine their potential and limitations in semiconductor applications [43–45, 47].

10.3.2.2.4 Fibrous Phosphorus

Fibrous phosphorus has a one-dimensional structure, and the structural properties of the one-dimensional system are expected to enable its use in the preparation of nanowires and nanotubes. However, the challenges in preparing and handling fibrous phosphorus have not yet been fully resolved, and the research is still in the preliminary stage. Fibrous phosphorus is promising for nanoelectronic devices, sensors, and nano-optoelectronic devices to play their role in semiconductor systems, but further research and engineering improvements are still needed [43].

The presence of intractable layers and structural defects in the preparation of monolithic phosphorus and its allotropes and interfacial resistance during detection are attributed to these differences. The different forms of phosphorus isomers have distinct advantages, disadvantages, and future perspectives for semiconductor applications. In addition to continuing to search for phosphorus materials or their micro- and nanostructures with superior properties, future research should fully exploit their potential to promote innovation and development in the semiconductor field [43].

10.3.2.3 Transition Metal Dichalcogenides (TMDs)

TMDs metal dichalcogenides are graphene-like, two-dimensional materials due to their similar structure, which is usually composed of metallic elements (e.g., W, Mo, and Ti) and sulfur elements (S, Se, and Te). For semiconductor applications, TMDs have the following advantages:

Band gap: The band gap of TMDs can be adjusted by changing the ratio of metal and sulfur elements, which gives them a wide range of applications in optoelectronic devices, solar cells, and other fields.

Variable carrier types: The carrier types of TMDs can be either electrons or holes, giving them the potential to make different semiconductor devices.

Larger atomic number: The larger atomic number of TMD makes its electronic structure more stable, with a higher melting point and hardness, which is conducive to improving the stability of the device.

However, there are some drawbacks, as below:

Difficult synthesis: The synthesis of TMD requires high-temperature and high-pressure conditions, and there are more side reactions during the preparation process, leading to synthesis difficulty.

Poor stability: TMD is easily oxidized in air, and moisture is easily adsorbed on the surface, affecting its performance and application.

Low carrier mobility: Although TMDs have variable types of carriers, they generally have low mobility, which will limit their application in high-speed electronics [15, 16, 48, 49].

10.3.3 Controlled Synthesis Technologies for Semiconductor Materials

Faced with how to continue Moore's Law, two-dimensional materials are potent seeds. However, at present, the way to industrialization of two-dimensional materials are challenging and a problem that needs a breakthrough. Today's preparation and synthesis methods for low-dimensional semiconductors focus on two directions (top-down and bottom-up) [36, 50].

10.3.3.1 Bottom-up Approach

The bottom-up approach is a process of molecular self-assembly to prepare materials based on molecules. It mainly includes self-assembly, solution method, and the vapor phase deposition method. Self-assembly refers to the self-assembly of macromolecules into a new structure, which can prepare two-dimensional materials, such as graphene and molybdenum disulfide. The solution method involves dissolving the synthesized molecules in an organic solvent and depositing them onto a substrate, followed by a process such as temperature or chemical reaction, which causes the molecules in the solution to form a suitable crystal structure. Vapor phase deposition is a method of preparing materials by compressing a gas or atmosphere into high temperature and pressure and is ideal for preparing low-dimensional materials such as nanowires and quantum dots [27, 28, 36, 50]. Figure 10.4 shows top-down and bottom-up approaches.

FIGURE 10.4 Top-down and bottom-up approach. (Adapted with permission [50]. Copyright [2019], ACS Publications.)

10.3.3.2 Top-down Approach

The top-down method is to prepare a large piece of material first and then gradually reduce its size employing cutting or chemical erosion. It mainly includes mechanical chipping, electron beam etching, and chemical vapor deposition methods. Mechanical chipping is the earliest preparation method to prepare high-quality monolayer graphene films and photonic crystals. The e-beam etching method is to focus and heat the electron beam directionally and then etch under the etching agent's action to prepare materials with nanofeatures [50].

Different nanostructures correspond to different preparation routes and synthesis methods. Nanomaterials are synthesized in top-down and bottom-up ways. The top-down approach is to etch the whole silicon wafer by photolithography and mask technology, adjust the photolithography size by changing the wavelength of the light beam and the size of the mask, and adjust the thickness by adjusting the evaporation parameters, which is suitable for the preparation of planar two-dimensional structures. The bottom-up method is through molecules, atoms, or ions by changing the external environment or chemical reaction so that the solute from the supersaturated dispersed phase (gas or liquid phase) forms a solid. It can regulate the composition, structure, size, and morphology of substances at the nanoscale, realizing the control of physical, chemical, biological, and mechanical properties [27, 28, 36, 50].

10.4 CONCLUSION

It is undoubtedly a highly disruptive process to enable the entire semiconductor industry to adopt new materials in industrial production. That's why LD materials are starting to become the focus of the industry's attention. However, LD materials today are only produced in small quantities in labs to support academic research. The process of inheriting LD materials and scaling them up for industrialization faces several issues, including changes in design tools, material growth, material transfer, and integration of production lines. Each step of the process requires specially designed and customized specialized production tools.

LD semiconductors have a unique atomic arrangement and excellent optoelectronic properties, such as high absorbance, mechanical properties, and the absence of tilted bonds on the surface. However, the applications of LD materials are limited by their properties, such as size constraints, considerable dark current, high noise, low quantum efficiency, and slow response speed. Here, we need to invest in the innovative development of science and technology to fully explore the potential of various semiconductor materials for the human semiconductor industry to add another brick.

REFERENCES

1. Rajeshwar K, De Tacconi N R, Chenthamarakshan C R. Semiconductor-Based Composite Materials: Preparation, Properties, and Performance. Chemistry of Materials, 2001, 13(9): 2765–82.
2. Hildreth O J, Lin W, Wong C P. Effect of Catalyst Shape and Etchant Composition on Etching Direction in Metal-Assisted Chemical Etching of Silicon to Fabricate 3D Nanostructures. ACS Nano, 2009, 3(12): 4033–42.
3. Cho W, Edgar T F, Lee J. Nonlinear Model Identification for Temperature Control in Single Wafer Rapid Thermal Processing. Industrial & Engineering Chemistry Research, 2008, 47(14): 4791–6.
4. Vaughn D D II, Bondi J F, Schaak R E. Colloidal Synthesis of Air-Stable Crystalline Germanium Nanoparticles with Tunable Sizes and Shapes. Chemistry of Materials, 2010, 22(22): 6103–8.
5. Rurali R. Colloquium: Structural, Electronic, and Transport Properties of Silicon Nanowires. Reviews of Modern Physics, 2010, 82(1): 427–49.
6. Wu X, Kulkarni J S, Collins G, et al. Synthesis and Electrical and Mechanical Properties of Silicon and Germanium Nanowires. Chemistry of Materials, 2008, 20(19): 5954–67.
7. Güniat L, Caroff P, Fontcuberta I, Morral A. Vapor Phase Growth of Semiconductor Nanowires: Key Developments and Open Questions. Chemical Reviews, 2019, 119(15): 8958–71.
8. Feng T, Zhou H, Cheng Z, et al. A Critical Review of Thermal Boundary Conductance across Wide and Ultrawide Bandgap Semiconductor Interfaces. ACS Applied Materials & Interfaces, 2023, 15(25): 29655–73.

9. Luo Q, Xiao K, Zhang J, et al. Direct-Current Triboelectric Nanogenerators Based on Semiconductor Structure. ACS Applied Electronic Materials, 2022, 4(9): 4212–30.

10. Barrigón E, Heurlin M, Bi Z, et al. Synthesis and Applications of III–V Nanowires. Chemical Reviews, 2019, 119(15): 9170–220.

11. Tuci G, Liu Y, Rossin A, et al. Porous Silicon Carbide (SiC): A Chance for Improving Catalysts or Just Another Active-Phase Carrier?. Chemical Reviews, 2021, 121(17): 10559–665.

12. Ashfold M N R, Goss J P, Green B L, et al. Nitrogen in Diamond. Chemical Reviews, 2020, 120(12): 5745–94.

13. Sharma S, Sunkara M K. Direct Synthesis of Gallium Oxide Tubes, Nanowires, and Nanopaintbrushes. Journal of the American Chemical Society, 2002, 124(41): 12288–93.

14. Chun H J, Choi Y S, Bae S Y, et al. Controlled Structure of Gallium Oxide Nanowires. The Journal of Physical Chemistry B, 2003, 107(34): 9042–6.

15. Ryder C R, Wood J D, Wells S A, et al. Chemically Tailoring Semiconducting Two-Dimensional Transition Metal Dichalcogenides and Black Phosphorus. ACS Nano, 2016, 10(4): 3900–17.

16. Zhao Y, Kong X, Shearer M J, et al. Chemical Etching of Screw Dislocated Transition Metal Dichalcogenides. Nano Letters, 2021, 21(18): 7815–22.

17. Bersin R L. Eliminating Solvents and Acids in Wafer Processing. in: Green Engineering, Paul T. Anastas. American Chemical Society. 2000: 29–41.

18. Lenigk R, Carles M, Ip N Y, et al. Surface Characterization of a Silicon-Chip-Based DNA Microarray. Langmuir, 2001, 17(8): 2497–501.

19. Kulkarni M S. A Review and Unifying Analysis of Defect Decoration and Surface Polishing by Chemical Etching in Silicon Processing. Industrial & Engineering Chemistry Research, 2003, 42(12): 2558–88.

20. Lee K S, Lee J, Chin I, et al. Control of Wafer Temperature Uniformity in Rapid Thermal Processing Using an Optimal Iterative Learning Control Technique. Industrial & Engineering Chemistry Research, 2001, 40(7): 1661–72.

21. Maria Angela V, Anjali A, Harshini D, et al. Organic Light-Emitting Transistors: From Understanding to Molecular Design and Architecture. ACS Applied Electronic Materials, 2021, 3(2): 550–73.

22. Arjmand T, Legallais M, Nguyen T T, et al. Functional Devices from Bottom-Up Silicon Nanowires: A Review. Nanomaterials, 2022, 12(7): 1043.

23. Sanders D P. Advances in Patterning Materials for 193 nm Immersion Lithography. Chemical Reviews, 2010, 110(1): 321–60.

24. Song S K, Kim J-S, Margavio H R M, et al. Multimaterial Self-Aligned Nanopatterning by Simultaneous Adjacent Thin Film Deposition and Etching. ACS Nano, 2021, 15(7): 12276–85.

25. Khan M B, Shakeel S, Richter K, et al. Atomic Layer Etching of Nanowires Using Conventional Reactive Ion Etching Tool. Journal of Physics: Conference Series, 2023, 2443(1): 012004.

26. Brunet M, Aureau D, Chantraine P, et al. Etching and Chemical Control of the Silicon Nitride Surface. ACS Applied Materials & Interfaces, 2017, 9(3): 3075–84.

27. Huang C, Li Y, Wang N, et al. Progress in Research into 2D Graphdiyne-Based Materials. Chemical Reviews, 2018, 118(16): 7744–803.

28. Reiss P, Carrière M, Lincheneau C, et al. Synthesis of Semiconductor Nanocrystals, Focusing on Nontoxic and Earth-Abundant Materials. Chemical Reviews, 2016, 116(18): 10731–819.

29. Srdanov V I, Blake N P, Markgraber D, et al. Alkali Metal and Semiconductor Clusters in Zeolites. in: Jansen J C, Stöcker M, Karge H G, et al. Studies in Surface Science and Catalysis. Elsevier. 1994: 115–44.

30. Duvvury C. A Guide to Short-Channel Effects in MOSFETs. IEEE Circuits and Devices Magazine, 1986, 2(6): 6–10.

31. Pichon A. Surface Effects. Nature Chemistry, 2013, 5(7): 551.

32. Kavokine N, Marbach S, Siria A, et al. Ionic Coulomb Blockade as a Fractional Wien Effect. Nature Nanotechnology, 2019, 14(6): 573–8.

33. Clarke W R, Simmons M Y, Liang C T, et al. Ballistic Transport in 1D GaAs/AlGaAs Heterostructures [M]. Reference Module in Materials Science and Materials Engineering. Elsevier. 2016.

34. Von Klitzing K, Chakraborty T, Kim P, et al. 40 Years of the Quantum Hall Effect. Nature Reviews Physics, 2020, 2(8): 397–401.

35. Kharissova O V, Kharisov B I, Oliva González C M. Carbon–Carbon Allotropic Hybrids and Composites: Synthesis, Properties, and Applications. Industrial & Engineering Chemistry Research, 2019, 58(10): 3921–48.

36. Hobbs R G, Petkov N, Holmes J D. Semiconductor Nanowire Fabrication by Bottom-Up and Top-Down Paradigms. Chemistry of Materials, 2012, 24(11): 1975–91.

37. Wu Y, Zhao X, Shang Y, et al. Application-Driven Carbon Nanotube Functional Materials. ACS Nano, 2021, 15(5): 7946–74.

38. Bao H, Wang L, Li C, et al. Structural Characterization and Identification of Graphdiyne and Graphdiyne-Based Materials. ACS Applied Materials & Interfaces, 2019, 11(3): 2717–29.

39. Zheng X, Chen S, Li J, et al. Two-Dimensional Carbon Graphdiyne: Advances in Fundamental and Application Research. ACS Nano, 2023, 17(15): 14309–46.

40. Kang J, Wei Z, Li J. Graphyne and Its Family: Recent Theoretical Advances. ACS Applied Materials & Interfaces, 2019, 11(3): 2692–706.

41. Jia Z, Li Y, Zuo Z, et al. Synthesis and Properties of 2D Carbon—Graphdiyne. Accounts of Chemical Research, 2017, 50(10): 2470–8.

42. Liu X, Cho S M, Lin S, et al. Constructing Two-Dimensional Holey Graphyne With Unusual Annulative π-Extension. Matter, 2022, 5(7): 2306–18.

43. Tian H, Wang J, Lai G, et al. Renaissance of Elemental Phosphorus Materials: Properties, Synthesis, and Applications in Sustainable Energy and Environment. Chemical Society Reviews, 2023, 52(16): 5388–484.

44. Liu Y, Cui D, Chen M, et al. Synthesis of Red and Black Phosphorus Nanomaterials. in: Fundamentals and Applications of Phosphorus Nanomaterials, Hai-Feng (Frank) Ji. American Chemical Society. 2019: 1–25.

45. Zhu Q, Wang H, Yang J, et al. Red Phosphorus: An Elementary Semiconductor for Room-Temperature NO2 Gas Sensing. ACS Sensors, 2018, 3(12): 2629–36.

46. Fung C-M, Er C-C, Tan L-L, et al. Red Phosphorus: An Up-and-Coming Photocatalyst on the Horizon for Sustainable Energy Development and Environmental Remediation. Chemical Reviews, 2022, 122(3): 3879–965.

47. Ricciardulli A G, Wang Y, Yang S, et al. Two-Dimensional Violet Phosphorus: A p-Type Semiconductor for (Opto)electronics. Journal of the American Chemical Society, 2022, 144(8): 3660–6.

48. Ning J, Zhang B, Siqin L, et al. Designing Advanced S-scheme CdS QDs/La-Bi2WO6 Photocatalysts for Efficient Degradation of RhB. Exploration, 2023, 3(5): 20230050.

49. Ji X-Y, Sun K, Liu Z-K, et al. Identification of Dynamic Active Sites Among Cu Species Derived from MOFs@CuPc for Electrocatalytic Nitrate Reduction Reaction to Ammonia. Nano-Micro Letters, 2023, 15(1): 110.

50. Ambrosi A, Chua C K, Bonanni A, et al. Electrochemistry of Graphene and Related Materials. Chemical Reviews, 2014, 114(14): 7150–88.

11 Fundamentals and Advanced Concepts of Microprocessors

Farida A. Ali and Sabita Mali

11.1 INTRODUCTION

Microprocessors have emerged as the unsung champions of modern technology in the 21st century, silently but profoundly influencing nearly every aspect of our existence. These minuscule electronic processors, often referred to as central processing units (CPUs), are at the core of a staggering variety of devices, including smartphones, laptops, industrial apparatus, and space-based equipment. Their function and significance in the modern world are nothing short of revolutionary.

A microprocessor is, at its core, a semiconductor device that conducts the fundamental computations and data processing duties that power modern technology. It executes a series of instructions, processes data, and manages the different operations necessary for a device to perform its intended function. This function spans from powering the applications on your smartphone and executing complex algorithms on your laptop to controlling industrial robotics and ensuring the proper operation of essential medical equipment. Microprocessors are, in essence, the digital minds of these devices, responsible for their functionality, performance, and efficiency.

In addition, they play a crucial role in our interconnected world by powering the communication networks and smartphones that have revolutionized the way we share information globally. The intricate interaction between microprocessors in data centers, routers, and network switches enables the Internet, which has become an indispensable aspect of our existence. The Internet of Things (IoT) is a swiftly expanding ecosystem of interconnected devices that is propelled by microprocessors. These devices rely on microprocessors to process data, make decisions, and communicate with other devices and systems. Examples include smart thermostats, wearable fitness monitors, industrial sensors, and autonomous vehicles. The result is a world that is more connected and efficient, where data-driven insights and automation improve our daily lives.

Furthermore, microprocessors are essential to the operation of advanced medical devices in the healthcare industry. In MRI machines, pacemakers, and insulin pumps, they process medical data, monitor vital signs, and provide precise control. In the automotive industry, microprocessors are used to operate engine control units (ECUs) that optimize engine performance, advanced driver-assistance systems (ADAS) that increase safety, and infotainment systems that enhance the driving experience. Even in space exploration, microprocessors enable scientific research and interplanetary missions by controlling satellites, spacecraft, and rovers.

Microprocessors' significance extends beyond their mere prevalence. They continually improve in efficacy, energy efficiency, and adaptability. Microprocessors have integrated billions of transistors into a single device as a result of Moore's law and unrelenting research and development efforts. This miniaturization enables devices to become smaller and lighter while retaining exceptional computational capability. Moreover, energy efficiency has become a top priority, resulting in the development of low-power designs, dynamic voltage and frequency scaling, and power gating techniques. As a consequence, the battery life of portable devices is increased and the energy consumption of data centers is decreased, contributing to environmental sustainability and cost savings.

The last-but-not-least aspect of microprocessors' significance is their versatility. They are programmable, which means that their functionality can be modified via software modifications. This adaptability makes them cost-effective solutions for a vast array of applications, ranging from

DOI: 10.1201/9781003450146-11

consumer electronics to industrial automation. Additionally, it fosters innovation in disciplines such as artificial intelligence, machine learning, augmented reality, and quantum computation. Microprocessors provide the processing capacity required for the development of these revolutionary technologies, reshaping industries and creating new opportunities.

Microprocessors are the driving force behind the digital revolution, enabling the development of sophisticated and intelligent electronic devices. Their architecture, functionality, and programmability make them versatile and adaptable to different applications. As technology continues to advance, microprocessors will play an increasingly vital role in shaping the future of computing and transforming various industries, from artificial intelligence and robotics to the IoT and beyond. This primer seeks to offer a fundamental grasp of microprocessors, their relevance, and their function in numerous applications [1, 2].

11.2 A SHORT HISTORY

The fascinating history of microprocessors stretches over half a century, documenting the evolution of computing technology from its rudimentary beginnings to the cutting-edge innovations as depicted in Figure 11.1 defines the modern era. The introduction of the Intel 4004 as the first commercially available microprocessor in the early 1970s marked a turning point in the history of computing. With its modest 2,300 transistors, this 4-bit processor paved the way for a new era of miniaturized and integrated CPUs. These early microprocessors, initially utilized in calculators and simple embedded systems, paved the way for a technological revolution.

As the 1970s progressed, so did the capabilities of the microprocessor. The introduction of 8-bit processors such as the Intel 8008 and 8080 made personal computing possible. These processors, though rudimentary by today's standards, propelled the first microcomputers and ignited a flame that would eventually lead to the personal computer revolution of the 1980s. The year 1978 marked

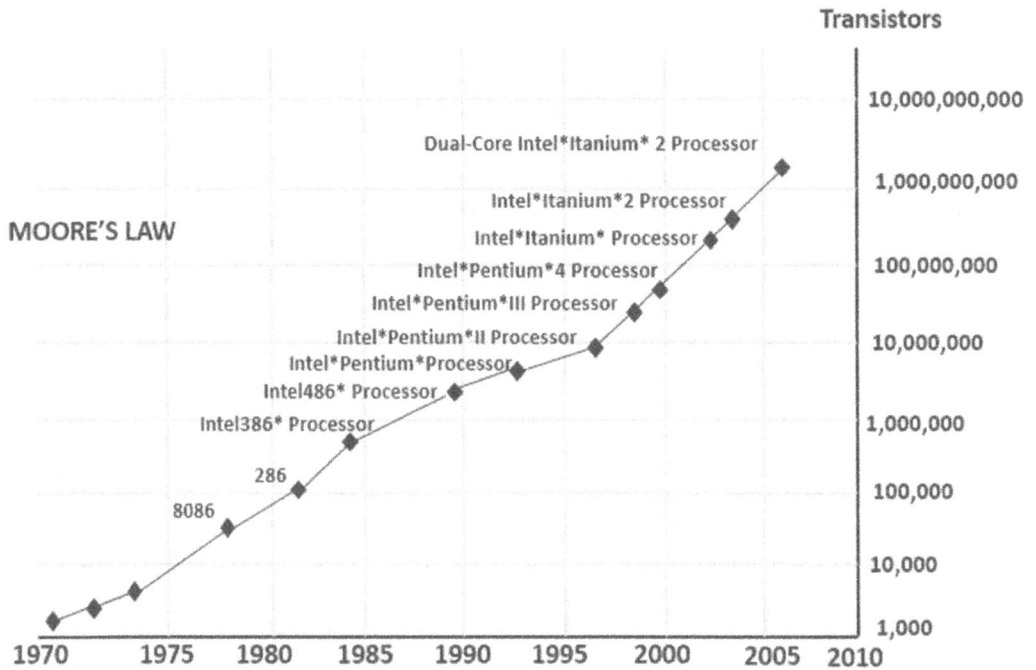

FIGURE 11.1 History and evolution of the microprocessor.

the introduction of the x86 architecture with the Intel 8086, amidst these developments. This 16-bit CPU would eventually evolve into the x86 architecture that currently underpins the vast majority of personal computers. The x86 architecture evolved from 16-bit to 32-bit to 64-bit, continually stretching the limits of computing capacity and efficacy.

In the 1980s, two distinct approaches to microprocessor design emerged: reduced instruction set computer (RISC) and complex instruction set computer (CISC). RISC processors, such as those based on the Million Instructions per Second (MIPS) architecture and Scalable Processor Architecture (SPARC), prioritized simplicity and optimized instruction execution for performance. In contrast, CISC processors, led by the x86 family, prioritized versatility and maintained software backward compatibility. This architectural schism fostered competition and innovation, thereby accelerating the evolution of the microprocessor [3, 4].

In the subsequent decade, the 1990s, energy efficiency became the focal point. Microprocessor manufacturers embarked on a search for more sustainable technologies in response to the proliferation of mobile devices and the growing demand for energy-efficient computation. To reduce energy consumption, techniques such as dynamic voltage and frequency scaling (DVFS) and power gating were implemented. In addition to extending the battery life of portable devices, these innovations also contributed to more environmentally favorable data center operations [5].

As the 21st century dawned, microprocessors initiated the on-score revolution. Faced with physical constraints on clock speed increases, designers have turned to multicore architectures. These processors integrated multiple CPU cores into a single device, thereby heralding in the era of parallel computation. The era of multi-core processors has not only improved computational capabilities but also made multitasking and parallel workload execution more efficient.

In addition, the pursuit of miniaturization and energy efficiency has persisted unabatedly. The incorporation of billions of transistors into a single microprocessor device was made possible by the nanometer scale reduction in feature sizes. This miniaturization not only facilitates the development of ever-more-powerful processors but also of compact, more energy-efficient devices [6].

The evolution of microprocessors exemplifies the unrelenting pursuit of innovation and advancement in the field of computing. From the Intel 4004's humble beginnings to today's complex multicore processors, microprocessors have continuously pushed the limits of what is possible in the technological world. Not only are they the center of modern computation, but they are also the generators of innovation in innumerable industries. Microprocessors will continue to evolve as time passes, influencing the future of technology and our digital world [7, 8].

11.3 EMERGING TRENDS AND FUTURE DIRECTIONS FOR ADVANCED PROCESSORS

Microprocessors are becoming more powerful, compact, and energy-efficient. They are present in a broad range of devices, from smartphones and tablets to servers and embedded systems, and are driving technological innovation in a variety of industries. Since their introduction in the early 1970s, microprocessors have advanced tremendously. Microprocessors have altered the world of computing, from the basic beginnings of 4-bit processors to the powerful and sophisticated designs of today. Their on-going growth has resulted in quicker processing speeds, expanded capabilities, and better integration, allowing technology to progress and influencing the digital era we live in.

The constant push for better functionality, efficiency, and adaptability is what keeps the cutting-edge processor industry moving forward. The landscape of sophisticated processors is being shaped by a number of new trends and directions [9]. This section will discuss these fascinating breakthroughs and their possible effects on the future of computing.

11.3.1   MULTI-CORE AND MANY-CORE PROCESSORS

Multiple processing cores on a single silicon chip (also called a multi-core processor) are now standard in most electronic gadgets. However, many-core CPUs are becoming more popular as they pack more processing cores onto a single silicon die. Applications like AI, data analytics, and scientific simulations are driving this change because they need more parallelism and faster performance. Extreme computing capacity is available from many-core computers, allowing for quicker and more efficient execution of complicated tasks.

11.3.2   HETEROGENEOUS COMPUTING

Convergence was the dominating trend in computer science throughout the last decade. The integration of Serial Attached SCSI (SAS) and Fiber Channel (FC) with the Ethernet fabric has reached a level of maturity that allows for convergence. The access times for cutting-edge data storage were measured in milliseconds as the instructions were executed by the CPU cores. Over the last decade, there has been a shift storage standards also, with the introduction of non-volatile memory express over fibres (NVMeoF), which has substituted SAS/FC protocols. Solid state devices (SSDs) have reduced data access times from milliseconds to microseconds or less, with little protocol cost relative to native device latencies. Moore's law and efficient algorithmic programming reduce software execution times, emphasizing the importance of data strategy. The feeds, or interconnects, are developing slowly as a core explosion occurs. The CPU-core-based homogenous computing paradigm is under pressure. Application programmers' hunt for optimal hardware is driving hybrid cloud computing worldwide. Heterogeneous computing is replacing homogeneous computing at the micro level. Heterogeneous computing optimizes speed and efficiency by combining Central Processing Units (CPUs), Graphics Processing Units (GPUs), Field Programmable Grid Arrays (FPGAs), and accelerators. Using each processor's best features boosts performance [10].

11.3.3   QUANTUM COMPUTING

The discipline of quantum computing is an advanced area of study and technological development that utilizes the principles of quantum physics to perform information processing in a fundamentally distinct manner compared to traditional computers. Fundamentally, quantum computing is based on the use of quantum bits, often referred to as qubits, which possess the ability to exist in several states concurrently due to the phenomenon known as superposition. The aforementioned characteristic allows quantum computers to execute certain categories of computations at a significantly accelerated rate compared to conventional computers. Furthermore, entanglement, a distinct quantum phenomenon, facilitates the interconnection of qubits in a manner where the state of one qubit is instantaneously linked with the state of another, irrespective of the spatial separation between them. The presence of this distinctive attribute provides an opportunity to address intricate challenges in fields such as encryption, optimization, and materials science that were previously deemed computationally insurmountable. Although quantum computing is now in its early stages and encounters several technological obstacles, its ability to potentially transform various domains, including finance and drug development, makes it an exceptionally captivating frontier in contemporary science and technology [11].

11.3.4   NEUROMORPHIC COMPUTING

Quantum computing is a cutting-edge field of research and technological advancement that uses the laws of quantum physics to process data in a fundamentally new way from classical computers. Qubits are the fundamental building blocks of quantum computing; they may exist in a superposition of states, each with its own unique properties. Due to the aforementioned quality, quantum computers can do certain types of calculations at an extremely fast pace compared to classical computers. In addition, entanglement, a unique quantum phenomenon, makes it possible for qubits to be connected in a way that links the state of one qubit instantly to the state of another, regardless of their physical distance from one another. This unique quality allows for the possibility of solving computationally intractable problems in areas like cryptography, optimization, and materials research. Quantum computing is an intriguing new frontier in modern science and technology because of its potential to revolutionize a wide range of fields, from finance to drug discovery. However, the field is still in its infancy and faces a number of technical challenges [12].

11.3.5 ENERGY-EFFICIENT DESIGNS

Minimizing power usage while maintaining performance on modern CPUs is essential. The rising need for more potent and versatile CPUs has made energy efficiency optimization an essential goal. This plan covers all the bases, including advances in semiconductor manufacturing techniques, improved architecture, and new power management techniques. Reduced transistor sizes and innovative materials, such as Fin-shaped Field Effect Transistor (FinFET) technology, are incorporated into today's CPUs to provide increased performance with less power consumption. By improving computing efficiency and optimizing the allocation of workloads, new architectural developments like out-of-order execution and multi-core designs hope to cut down on energy consumption. In addition, advanced power management features are built in to dynamically adjust CPU clock speeds and voltages in response to workload demands, maximizing performance only when it's really needed. Collectively, these activities provide a major contribution to the research and development of more power-efficient CPUs. This is crucial because it allows for longer battery life on mobile devices, lowers data center power consumption, and helps mitigate the environmental impacts of computing [13].

11.3.6 SECURITY AND PRIVACY ENHANCEMENTS

Advanced processors have improved security and privacy, easing our connected society's concerns. Trusted Execution Environments (TEEs), secure enclaves, and hardware-based cryptographic accelerators are among the hardware-level security features described. TEEs isolate processor data and code from external vulnerabilities. Intel's Software Guard Extensions (SGX) and AMD's Secure Encrypted Virtualization (SEV) allow programs to run in encrypted memory sectors, protecting against unauthorized access even by privileged applications. Current CPUs feature hardware components that support current encryption protocols, improving safe data transfer and storage. These methods protect data and combat vulnerabilities like Spectre and Meltdown. Additionally, processors use differential privacy, secure boot methods, and increased access controls. As society becomes increasingly dependent on data-driven technology, these precautions protect user data and privacy. Technology improves security and privacy, enhancing confidence in our digital world.

Without hardware, software, and academic collaboration, contemporary CPUs cannot reach their full potential. By carefully embracing these changes and overcoming their difficulties, we may create a future where powerful processors help us solve complex problems, make new discoveries, and enhance our lives and interactions. The continual reduction in nanometer measured feature sizes is another major advancement in microprocessor technology. Moore's law, which states that computer chip transistors double every two years, drives this trend. Smaller feature sizes allow more transistors on a semiconductor, increasing processing power and usefulness. The upcoming section anticipates the revolutionary effects of miniaturization on computers and society from the futuristic CPUs [14].

11.4 IMPLICATIONS OF SMALLER FEATURE SIZES AND HIGHER PERFORMANCE

Advanced processors with smaller feature sizes and higher performance affect many technical and social domains. Processor power efficiency improves with smaller feature sizes, extending mobile device battery life and reducing data center energy use. This promotes environmental sustainability.

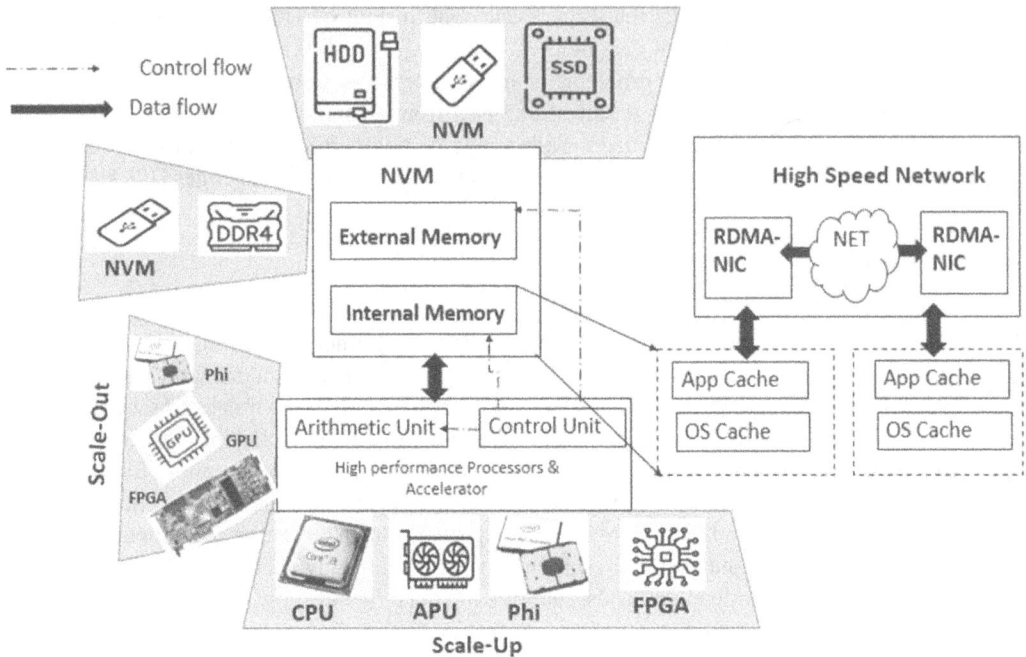

FIGURE 11.2 Integration of lower feature size and higher performance for future computing platforms.

Improved performance speeds up complex computations and data processing, advancing artificial intelligence, scientific investigation, and computer graphics.

Technology increases heat generation and requires more advanced cooling devices, which presents certain challenges. Security and privacy concerns are developing as processors get more powerful. This requires stronger encryption and cybersecurity. Since certain people or locations may not have access to the latest and most powerful processors, rapid technological advancement may increase digital access disparities. The integration of lower feature sizes and higher performance, as represented in Figure 11.2 for the future computing platforms, is important, but it also raises important issues that must be addressed to provide a fair and well-rounded technical environment [15]. Thus, the implications can be discussed in a broader sense as mentioned in the following subsections.

11.4.1 HIGHER TRANSISTOR DENSITY

By increasing transistor density, semiconductor manufacturing advancements have driven computer technology's rapid expansion. The increase in transistor density per unit space on integrated circuits has enabled powerful CPUs and microchips. Modern devices can efficiently do complex tasks like AI-based computations and HD video processing. Due to improved transistor density, electronic components are smaller, lighter, and more energy-efficient, making smartphones, tablets, and wearables possible. In emerging technologies like 5G, IoT, and autonomous vehicles, tiny, energy-efficient CPUs are required for rapid data processing and networking. The constant increase in transistor density is redefining electronics' limits, allowing breakthroughs that affect our daily lives and technological development [16].

11.4.2 LOW POWER CONSUMPTION

Power consumption is affected by semiconductor technology's smaller feature sizes and higher performance. Smaller transistors reduce switching energy, improving computer power efficiency.

Battery-operated gadgets like smartphones and laptops need to reduce power usage to improve battery life and energy efficiency. However, modern CPUs' improved performance may offset some of these benefits. Computers may perform processes quicker by employing higher clock speeds or more cores, which increases power consumption. The shrinkage of transistors may also raise concerns about leakage currents, which increase idle power usage. Workload, design efficiency, and power management affect power consumption when feature sizes are lowered and performance is improved. Designers must include advanced power management algorithms and optimizations into CPU architectures to effectively manage power usage and maximize these advancements.

11.4.3 Increased Performance

Because there are more transistors available, microprocessors can execute instructions in parallel and handle bigger data sets more effectively. This improves performance in activities like multitasking, calculation, and multimedia processing.

Increased microprocessor performance has various advantages.

11.4.3.1 Quicker Task Execution

Enhanced microprocessor performance has greatly improved digital work execution speed and efficiency. Processors have become more powerful because of smaller transistors, multi-core architectures, and improved instruction sets. Processes that were formerly slow are now fast. The improved speed of computer systems makes 3D visualizations, scientific data analysis, and complex machine learning algorithms quicker and more responsive. Faster acceleration improves financial market decision-making and autonomous car response. Additionally, it has enabled the development of cutting-edge technologies like augmented and virtual reality, which demand continuous, speedy processing to provide immersive experiences. In essence, microprocessors have improved everyday computer operations, boosted creativity, and opened new doors in our technologically evolved world.

11.4.3.2 Advanced Computing Applications

The exponential growth in microprocessor performance enabled advanced computer applications. Advanced transistor technology, parallel processing, and optimization have allowed these high-performance CPUs, offering unmatched potential. AI and machine learning have built algorithms that interpret spoken language, recognize photos, and make autonomous decisions. Science has progressed due to speedier simulations and data processing. Genetics and climate models have advanced. This enhances entertainment-sector 3D graphics, virtual reality, and HD video rendering. Healthcare, banking, and cybersecurity employ advanced computer systems to analyze enormous data sets for diagnosis, risk assessment, and threat detection. Improved microprocessor speed has sped up current programs and allowed the introduction of novel solutions that are transforming our work, daily lives, and technology interactions [17].

11.4.3.3 Clock Speed

Processor performance depends on clock speed. A CPU's cycle rate is measured in gigahertz. Faster clocks increase data processing and job efficiency. The speed at which a processor executes instructions speeds up computational tasks and improves responsiveness. This is useful for quick processing in gaming, video editing, and science simulations. However, increasing clock speeds to boost performance has limits. Clock speed increases power consumption and heat generation, requiring more complicated cooling solutions and lowering energy efficiency. Some tasks may be parallelized and use multi-core processors; therefore, higher clock rates affect application performance differently. Thus, clock speed, power consumption, and multi-core design must be balanced to optimize microprocessor performance for different processing demands [18].

11.4.3.4 Improved Instruction Set

Improved instruction sets boost microprocessor performance in several ways. An optimized instruction set may boost a processor's efficiency, speeding up and improving job performance. These improvements improve system performance, task execution, and power consumption. Specialized instructions for common tasks or emerging technologies, such as multimedia processing or encryption, improve processor performance and reduce the need for additional, less efficient instructions. This speeds up certain tasks and enhances system responsiveness. Enhanced instruction sets may also help software developers maximize processor capabilities, enabling more creative and resource-efficient applications. An enlarged instruction set is crucial to microprocessor performance. It drives advancement in consumer electronics, scientific computers, and other fields. An enlarged instruction set helps computer systems meet modern computing needs by promoting technological improvements.

11.4.3.5 Memory and Cache Performance

Memory and cache efficiency impact CPU speed. Memory retrieval efficiency affects CPU performance and responsiveness. Larger, quicker L1, L2, and L3 caches store frequently requested data and instructions closer to CPU cores in advanced CPUs. The CPU and main memory are close, minimizing data retrieval latency and accelerating task execution and multitasking. Games, video editing, and database administration need frequent data retrieval; therefore, cache efficiency is essential. More data in larger caches reduces access to slower main memory. This may remove the bottleneck. Bandwidth and latency enhancements in memory subsystems increase CPU performance. With enormous datasets and complicated applications, modern computing demands memory and cache optimization. Data management and processing by the processor are necessary for rapid and responsive job completion. Memory and cache technologies improve microprocessor performance and enable technology to fulfill digital needs.

11.4.3.6 Parallel Processing

Parallelism altered CPU performance. Many processor cores tackle subtasks simultaneously in this method. Parallel processing boosts computationally demanding program performance. Once time-consuming and computationally expensive processes can now be done quickly. Scientific computing, AI, and simulations need vast processing power for large datasets and complicated algorithms. Parallel processing boosts computer speed and job completion. Smartphones and high-end servers multitask and run resource-intensive applications on multicore CPUs. Parallel processing may challenge developers with software optimizations and parallel techniques. However, parallel processing speeds up microprocessors, advancing technology and solving intractable issues. Faster CPUs and smaller feature sizes have improved computing device development. Tiny characteristics boost transistor density, power efficiency, and utility. Work execution, sophisticated computer programs, and new technologies increase with performance. Better microprocessors will make computers more powerful, energy-efficient, and capable.

11.5 RECOGNIZING ENERGY EFFICIENT DESIGNS

Energy efficiency in microprocessors is a system's ability to calculate and operate with less power. Optimizing many components, circuits, and design methods reduces energy usage while retaining performance and functionality. Today's technology-driven world prioritizes energy efficiency in microprocessor and computer design. High demand for portable devices, longer battery life, and environmental sustainability have increased the focus on energy-efficient constructions and low-power solutions. Here are some energy-efficiency techniques and effects of energy-efficient architectures and low-power designs in microprocessors.

11.5.1 Dynamic Voltage and Frequency Scaling (DVFS)

Energy-efficient advanced processors use DVFS. This innovative technology allows processors to dynamically adjust operating voltage and clock frequency to the workload. To conserve power and heat, DVFS lowers voltage and clock frequency when workload is low. DVFS boosts voltage and frequency for peak computation. The tricky performance-power economy balance makes DVFS critical. By enhancing processor performance, DVFS saves energy, extends mobile device battery life, and reduces data center energy use. This strategy is crucial when processors speed up and power efficiency increases. It helps upgraded CPUs meet modern application performance criteria while decreasing cooling and environmental impact. DVFS helps modern processors be energy-efficient for smarter, more sustainable computing.

11.5.2 Clock Gating

Modern CPUs use clock gating for energy efficiency. Clock signal gating during data processing dormancy is this technique. Reducing unneeded clock pulses saves CPU power and heat. Modern processors with numerous transistors utilize clock gating to selectively activate CPU components to save energy. Energy efficiency and performance integrity are advantages of modern CPU clock gating. Powering down inactive semiconductor components saves energy, increases processor life, and lowers temperature. Mobile devices and data centers with massive computing arrays need this for battery life and energy savings. In conclusion, energy-efficient CPUs need clock gating. Modern computer systems depend on the approach, which balances power consumption and computational performance for high-speed processors.

11.5.3 Power Gating

Energy-efficient CPU designs need power gating. It involves gating CPU power during inactivity. Energy usage and heat production are reduced by cutting power to inactive central processing unit parts. Complex high-performance computers with high power demands benefit from power gating. Power gating enhances advanced CPU power efficiency and performance, making it crucial. Reduce power consumption in idle or inactive chip areas to save energy, processor life, and heat. Mobile devices need this feature for battery life. In data centers with enormous processor clusters, energy savings are crucial. Power gating is essential for energy-efficient CPU architecture. Advanced CPUs combine power and performance. Modern computer systems depend on them, particularly in an energy-efficient and sustainable era.

11.5.4 Pipelining and Parallelism

Energy-efficient CPUs need pipelining and parallelism. Pipelining simplifies complicated instructions into one-task stages. Multi-level pipeline instructions may be performed concurrently. This method optimizes step execution to boost CPU performance and reduce idle time and power consumption. In contrast, parallelism runs several instructions or processes. ILP and DLP enable sophisticated processors to execute several instructions or data bits concurrently. Thereby, completing the processes faster, save electricity and enhance processing efficiency. Newer CPUs save energy through pipelining and parallelism. They improve instruction execution and allow the CPU to multitask, reducing active state duration and idle power consumption. Optimizing mobile device battery life and data center energy usage, where processors work 24/7, is critical. Finally, energy-efficient processor designs in current computer systems that prioritize speed and sustainability need pipelining and parallelism technologies [19].

11.5.5 Low-Power Modes

Energy-efficient CPU designs need low-power modes. CPUs utilize less power in sleep or idle states. CPU cores and caches may be partly or totally deactivated on low power while retaining data and state. This strategy saves energy and heat, prolonging CPU life and simplifying cooling. Low-power modes are essential for performance and energy economy. Long-term connectivity and productivity need mobile device energy-saving settings. In data centers, low-power modes minimize server farm power usage while idle. This strategy is key to energy and cost savings. Energy-efficient CPU designs need low-power modes. Advanced processors can dynamically adjust workloads in these modes, boosting sustainability and resource efficiency [8]. In an era of electrical usage and environmental impact, this matters.

Energy-efficient architectures and low-power designs are critical for portable devices like smartphones, laptops, and wearable technologies. Microprocessors may improve battery life by optimizing power use, giving customers more usage time, and minimizing the need for frequent recharging [20, 21]. Besides the energy-efficient techniques, energy-efficient architectures and low-power designs also have their own implications as discussed below.

11.5.5.1 Environmental Sustainability

Modern processors with energy-efficient and low-power architectures must be assessed for environmental sustainability. These design principles enhance processor efficiency and reduce carbon emissions and environmental impact. By improving power utilization, these processors cut computing equipment energy use and greenhouse gas emissions. Given global climate change and energy reduction initiatives, this is vital. These CPUs and other components last longer and need fewer hardware repairs due to their decreased heat output and operating efficiency, thereby improving electronic equipment reliability and longevity. In data centers, where thousands of CPUs run continuously, these architectural choices save energy and reduce cooling needs, promoting sustainability. In an age of increased technological dependency given the growing requirement for complex processors, sustainability must be considered in their design. Environmentally responsible and sustainable futures depend on these design choices.

11.5.5.2 Heat Dissipation and Cooling

Heat dissipation and cooling are essential to new CPUs' energy-efficient architectures. Although intended to consume less power and heat, modern CPUs create heat. Cooling and heat dissipation affect CPU performance and longevity. Energy-efficient CPUs minimize heat through DVFS, clock gating, and power gating when processing demand is low. These CPUs use temperature sensors and throttling algorithms to optimize performance and avoid overheating. CPU cooling solutions vary per device/application. Mobile electronics may use heat sinks and thermal pads for passive cooling. Laptops and servers cool via fans or liquid. This assures optimum performance, increases processor life, and lowers heat output.

11.5.5.3 Scalability and Integration

Scalability and integration are essential for low-power CPUs. The following design approaches meet dynamic computing needs. CPU performance may be scalable to meet individual needs, saving power and resources. Multiple cores or processing units may be parallelized effortlessly. Data centers need scalable processing for workloads. However, microchip integration involves several parts and technologies. Processors with CPU cores, GPUs, memory controllers, and hardware accelerators are advanced. Because components are nearby, data transmission consumes less power. Energy-efficient and low-power CPU architectures scale and integrate well with developing technologies [18, 22].

Energy-efficient buildings and low-power technologies are needed to meet portable device demand, improve battery life, and ensure environmental sustainability. Microprocessors may optimize power use without sacrificing performance via dynamic voltage and frequency scaling, clock gating, power gating, and low-power modes. Battery life, temperature control, scalability, and computer platform compatibility increase with energy-efficient designs. Microprocessor design will emphasize energy efficiency as technology advances to promote innovation and sustainable computing.

11.6 ADOPTION OF CUTTING-EDGE TECHNOLOGY

Neuromorphic computing and quantum computing are two cutting-edge technologies that have the potential to completely transform the computer sector. While their approaches and ideas differ, both strive to solve complicated computational problems that standard computer structures are incapable of handling. Let us investigate these fascinating disciplines and comprehend their significance for the future advance microprocessors.

11.6.1 Cognitive Neuromorphic Computing

Neuromorphic computing is an interdisciplinary field that involves the development of customized computer systems by leveraging the structural and operational principles seen in the human brain [23]. The objective is to replicate the neural networks and cognitive capabilities of the human brain, enabling computers to process information in a manner like that of the human brain. The advancement of neuromorphic systems has promise for augmenting artificial intelligence, robotics, sensor technologies, and brain-machine interfaces. These systems possess the capacity to provide efficient and advanced processing capabilities, enabling the development of novel applications in autonomous systems and cognitive computing [24]. Key characteristics features of neuromorphic computing are as shown in Figure 11.3 and as discussed further.

11.6.1.1 Spiking Neural Networks

Spike-based neural networks are often used in neuromorphic computing architectures to effectively interpret input data that is represented as discrete spikes or pulses of activity. This phenomenon

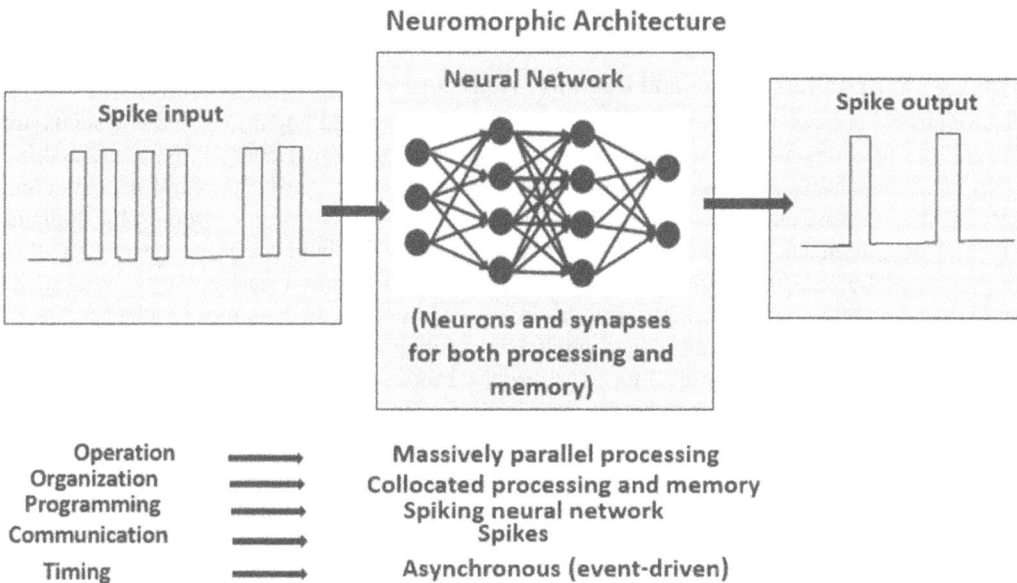

FIGURE 11.3 Fundamental features in the cognitive neuromorphic computing system.

exhibits similarities to the way in which neurons inside the brain participate in communication and coordination [1].

11.6.1.2 Parallelism and Efficiency

Parallelism and efficiency are key to neuromorphic computing. Due to the brain's outstanding parallel processing skills, these systems can manage vast amounts of data and several tasks at once. Neuromorphic hardware and algorithms strive to mimic the brain's coupled and distributed neurons. This approach allows simultaneous calculations and consistent data flow. Neuromorphic computing uses parallelism to process complex data patterns faster and save energy. These systems use less electricity than von Neumann architectures by spreading computations over several cores or neuromorphic units. The brain's ability to process information simultaneously enhances neuromorphic devices' energy efficiency. Neuromorphic computing systems excel in low-energy, real-time data processing, making them ideal for edge computing, robotics, and sensor networks. Artificial intelligence and cognitive computing differ in their ability to spot patterns, adapt to changing environments, and learn from datasets. In summary, neuromorphic computing systems that integrate parallelism and efficiency provide a groundbreaking computing paradigm that mimics the brain's cognitive capacities and improves speed, energy conservation, and flexibility. This opens the door to innovative artificial intelligence, machine learning, and other applications.

11.6.1.3 Hardware Acceleration

Hardware accelerators are crucial for neuromorphic computing. The above specialist hardware components are designed to perform artificial neural network and cognitive computing tasks more efficiently than standard CPUs. Neuromorphic computing relies on hardware accelerators to simulate neural connections and synaptic plasticity in the brain. Hardware accelerators in neuromorphic systems speed up artificial neural network training and inference. Matrix operations, convolutions, and other deep learning model computations can be performed faster than on CPUs or GPUs. Speed boosts neuromorphic system functioning and optimizes energy usage, which is vital in cognitive computing. Hardware accelerators help achieve real-time processing and reduce latency in image and audio recognition, natural language understanding, and sensor data analysis. Neuromorphic systems excel at processing complex data patterns quickly and efficiently because of their specialist components. Neuromorphic computing systems employ hardware accelerators to make use of specialized hardware for cognitive tasks for improving artificial intelligence and cognitive computing applications.

11.6.1.4 Pattern Recognition and Cognitive Activities

Neuromorphic computer systems focus on pattern recognition and cognition. These systems are inspired by the human brain and excel in processing and identifying complex patterns, making them ideal for sensory perception, learning, and decision-making. Artificial neural networks and specialized hardware emulate the brain's coupled neurons and synapses in neuromorphic computing. This emulation lets computers analyze and interpret data like humans. These systems excel in image and audio recognition, natural language processing, and sensor data analysis. AI systems excel in unstructured or chaotic input circumstances because they can dynamically adapt and learn from their environment. Neuromorphic computing's energy-efficient design matches the brain's low power consumption, making it ideal for edge devices, robots, and other energy-constrained applications. Artificial intelligence may benefit from neuromorphic computing, which allows robots to think, solve problems, and make context-aware decisions. These technologies demonstrate a major leap in intelligent computers that can interact with and understand the world like humans.

11.6.2 COMPUTING AT THE QUANTUM LEVEL

Quantum computing leverages principles from quantum physics to do computational tasks that are beyond the capability of conventional computers. In contrast to conventional computers that rely on

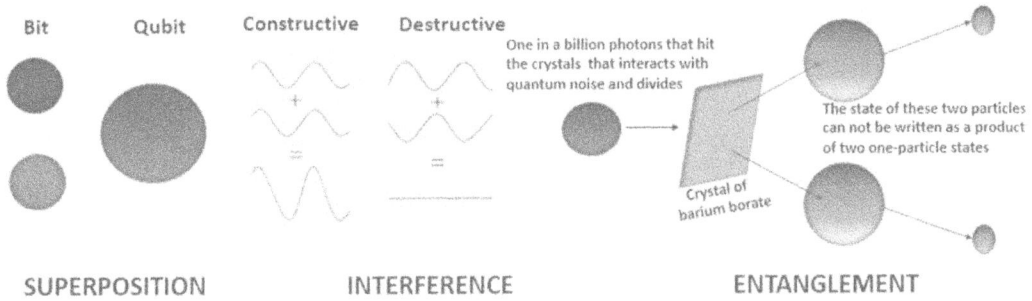

Bit   Qubit   Constructive   Destructive

One in a billion photons that hit the crystals that interacts with quantum noise and divides

The state of these two particles can not be written as a product of two one-particle states

Crystal of barium borate

SUPERPOSITION   INTERFERENCE   ENTANGLEMENT

FIGURE 11.4 Features of quantum computing.

bits to represent binary values of 0 or 1, quantum computers use qubits, which may simultaneously exist in several states due to the principles of quantum superposition and entanglement as shown in Figure 11.4. The field of quantum computing holds promise for addressing complex problems that now remain intractable for classical computing systems. Applications such as cryptography, optimization, drug development, materials research, and quantum system modeling have been identified in the field. The development of fault-tolerant quantum computers and quantum algorithms will play a pivotal role in driving further progress in the field [25]. Key characteristics of quantum computing are as follows.

11.6.2.1 Superposition and Entanglement

Qubits within the context of a quantum computer have the ability to occupy a state of superposition, whereby they may simultaneously represent several possibilities. The phenomenon of entanglement enables the coupling of qubits, hence facilitating computations that use the collective behavior of these qubits.

11.6.2.2 Quantum Gates and Quantum Circuits

Quantum computing involves the manipulation and transformation of qubits via the use of quantum gates, which have resemblance to conventional logic gates. Quantum circuits are constructed using these gates in order to execute quantum algorithms.

11.6.2.3 Quantum Parallelism and Speedup

Quantum computers use the concepts of quantum parallelism to do several computations concurrently. In some scenarios, such as the process of factoring large numbers and addressing optimization problems, quantum algorithms have the potential to provide exponential improvements in computational performance compared to traditional methods.

11.6.2.4 Error Correction and Quantum Decoherence

Quantum systems are susceptible to errors due to the influence of environmental interactions and the phenomenon of decoherence. The objective of quantum error correction methodologies is to mitigate these imperfections while maintaining the fidelity of quantum computations.

Neuromorphic computing and quantum computing are two distinct and captivating areas of computer research that have significant implications and future prospects across several disciplines. Neuromorphic computing draws inspiration from the anatomical and functional characteristics of the brain, while quantum computing leverages principles of quantum physics to address complex computational challenges. Both areas possess the capacity to revolutionize computer capabilities and pave the path for novel applications across diverse industries. Ongoing research and development in these domains will shape the trajectory of computers and provide novel approaches for addressing complex issues.

11.7   CONCLUSION

The current trajectory of sophisticated processors is driving the expansion of computing capabilities, with new trends and future directions playing a pivotal role in this progression. The sector is seeing significant developments facilitated by multi-core and many-core processors, heterogeneous computing, quantum computing, neuromorphic computing, energy-efficient designs, and increased security and privacy features. These advancements will have a transformative impact on several fields, including artificial intelligence, scientific inquiry, data analysis, and beyond.

The realization of the full potential of improved processors in the context of technological growth necessitates the imperative cooperation of hardware designers, software developers, and researchers. By responsibly adopting these developing trends and effectively tackling the associated issues, we have the potential to facilitate a future in which enhanced processors enable us to effectively handle intricate problems, uncover novel discoveries, and enrich our lifestyles and interactions with the environment. The potential of improved processors is very promising, and there is much anticipation of the revolutionary effects they will bring to the field of computing and society at large.

REFERENCES

1. N. L. Vintan, Neural branch prediction: From the first ideas, to implementations in advanced microprocessors and medical applications. In Proceedings of the Romanian Academy, Series A: Mathematics, Physics, Technical Sciences, Information Science, 20(2), 200–207, 2019.
2. A. Lorusso, & G. Domenico, IoT System for Structural Monitoring, International Conference New Technologies, Development and Application, Cham: Springer International Publishing, 2022.
3. C. Mazuré, G. K. Celler, C. Maleville, & I. Cayrefourcq, Advanced SOI Substrate Manufacturing, In 2004 International Conference on Integrated Circuit Design and Technology (IEEE Cat. No. 04EX866), pp. 105–111. IEEE, 2004.
4. J. McMahon, S. Crago, & D. Yeung, Advanced Microprocessor Architectures. In High Performance Embedded Computing Handbook (pp. 499–521). CRC Press (2018).
5. https://oercommons.org/courseware/lesson/69411/student/?section=2
6. C. Lichtenau, A. G. Ortiz, & T. Pfluger, Technological and architectural power optimizations for advance microprocessors, In International Symposium on Signals, Circuits and Systems, 2005 (ISSCS 2005), vol. 1, pp. 11–14. IEEE, 2005.
7. A. K. Ray, & K. M. Bhurchandi, Advanced Microprocessors and Peripherals: Architecture, Programming and Interfacing. McGraw-Hill/Irwin (2006).
8. B. P. Singh, & R. Singh, Advanced Microprocessors and Microcontrollers. New Age International (2008).
9. A. K. Ganguly, Architecture, Programming and Applications of Advanced Microprocessors. Alpha Science International, Ltd (2013).
10. J. L. Hennessy, & A. P. David, Computer Architecture: A Quantitative Approach. Elsevier (2011).
11. J. Singh, & M. Singh, Evolution in quantum computing, In 2016 International Conference System Modeling & Advancement in Research Trends (SMART) (pp. 267–270). IEEE, (2016, November).
12. C. Pan, C. Y. Wang, S. J. Liang, Y. Wang, T. Cao, P. Wang, & F. Miao, Reconfigurable logic and neuromorphic circuits based on electrically tunable two-dimensional homojunctions. Nature Electronics, 3(7), 383–390 (2020).
13. J. Stokes, Inside the Machine: An Illustrated Introduction to Microprocessors and Computer Architecture. No Starch Press, 2007.
14. I. A. Young, Analog mixed-signal circuits in advanced nano-scale CMOS technology for microprocessors and SoCs. In 2010 Proceedings of ESSCIRC (pp. 61–70). IEEE, (2010, September)
15. W. Pan, Z. Li, & Y. Zhang, The new hardware development trend and the challenges in data management and analysis. Data Science and Engineering, 3, 263–276 (2018).
16. M. Horstmann, M. Wiatr, A. Wei, J. Hoentschel, T. Feudel, T. Scheiper, & M. Raab, Advanced SOI CMOS transistor technology for high performance microprocessors, In 2009 10th International Conference on Ultimate Integration of Silicon (pp. 11–14). IEEE, (2009, March)
17. A. M. Veronis, Survey of Advanced Microprocessors. Springer Science & Business Media (2012).
18. A. FOG, The Microarchitecture of Intel, AMD, and VIA CPUs, 2014, available online at http://www.agner.org/optimize/microarchitecture.pdf (accessed on August 7, 2018).

19. S. Hill, The ARM10 family of advanced microprocessor cores. In Symposium on High-Performance Chips at Stanford University (HOT Chips 13), (2001, August).
20. M. T. Bohr, & Y. A. El-Mansy, Technology for advanced high-performance microprocessors. IEEE Transactions on Electron Devices, 45(3), 620–625 (1998).
21. B. W. Liu, S. M. Chen, & D. Wang, Survey on advance microprocessor architecture and its development trends. Jisuanji Yingyong Yanjiu/Application Research of Computers, 24(3), 16–20 (2007).
22. C. Jacobi, A. Saporito, M. Recktenwald, A. Tsai, U. Mayer, M. Helms, A. B. Collura, P. K. Mak, R. J. Sonnelitter, M. A. Blake, & T. C. Bronson, Design of the IBM z14 microprocessor. IBM Journal of Research and Development, 62, 2–3 (2018).
23. N. K. Upadhyay, H. Jiang, Z. Wang, S. Asapu, Q. Xia, & J. J. Yang, Emerging memory devices for neuromorphic computing. Advanced Materials Technologies, 4(4), 1800589 (2019).
24. C. D. Schuman, S. R. Kulkarni, & M. Parsa, Opportunities for neuromorphic computing algorithms and applications. Nature Computational Science, 2, 10–19 (2022).
25. https://insights.daffodilsw.com/blog/quantum-computing-applications-for-software-development

12 Fundamentals and Recent Advancements in Photodetectors Using Semiconductors

Xi Lin and Xiaoguang Luo

12.1 INTRODUCTION

Semiconductor photodetectors are optoelectronic devices based on semiconductors, which are usually used to detect light by the conversion ability (from light to electrical signal) of semiconductors. As the core components of modern optoelectronics, semiconductor photodetectors have a wide range of potential applications in military (including infrared imaging, missile guidance, navigation, etc.) and civilian fields (including environmental monitoring, optoelectronic imaging, optical communication, etc.) [1–7]. When light is impinged on the surface of the semiconductor, the photons will be absorbed by electrons or lattices through electron transition or the photo-thermoelectric effect, then forming the electron-hole pairs. With the assistance of the electric field (either a built-in or applied electric field), the electron-hole pairs will be separated as photogenerated electrons and holes and collected by the electrodes, forming the photocurrent or photovoltage. Therein, the band-to-band transition in semiconductors is widely employed to design specific photodetectors. Under this mechanism, only the photons with energy larger than the band gap of semiconductors can be absorbed, and therefore the working spectrum of photodetectors is dramatically limited by the band gap of semiconductors. Fortunately, the working spectrum can be extended beyond the band gap by the interaction among photons, electrons, and phonons, leading to a rise in temperature and a temperature-induced change in electron conductivity.

Currently, the most widely used commercial photodetectors are based on bulk Silicon (Si), Germanium (Ge), and III-V semiconductors. However, the intrinsic nature of these traditional semiconductors is also challenging some of their applications in photodetection. Due to the indirect and specific band gap, Si and Ge materials cannot effectively absorb light, and their response spectrum is out of long wavelength. Still, under the premise in high crystallinity, the size of these traditional semiconductors is too difficult to reach the nanoscale or sub-nanoscale in technique, although the miniaturization for devices is an important and promising trend for technological advancement currently and in the future. The main restraining factors are the dangling bonds, defects, impurities, unstable interface, and quantum effect [8]. So far, some low-dimensional semiconductors such as quantum dots, nanowire, and two-dimensional semiconductors are emerging in the field of photodetection [9, 10]. Based on the single low-dimensional semiconductors or the hybrid structure with traditional semiconductors, various nanoscale photodetectors have been designed for high performance and exploited for the potential applications.

In this chapter, the main performance parameters of a photodetector are first introduced, and then the main mechanisms of photodetection are reviewed and illustrated with some new study progress of photodetectors, including photovoltaic effect, photoconductive effect, photogating effect, photo-thermoelectric effect, and photo-bolometric effect.

DOI: 10.1201/9781003450146-12

12.2   PERFORMANCE PARAMETERS OF PHOTODETECTORS

There are many performance parameters for a photodetector; here, the general ones are introduced to evaluate the devices from one to another, such as the carrier mobility (μ), photocurrent (I_{ph}), responsivity (R), external quantum efficiency (EQE), response time (τ), photoconductive gain (G), signal-to-noise ratio, noise equivalent power (NEP), and specific detectivity (D^*) [11–14].

As an optoelectronic device, the carrier mobility is very important for both the fundamental analysis and the performance study of a photodetector. It is an intrinsic parameter of a semiconductor that is used to evaluate the transport speed of carriers, and a high mobility is indicative of rapid response speed and even high photoconductive gain. For a field-effect transistor configuration, the mobility in the linear region for both electrons and hole can be expressed by the formula of

$$\mu = \frac{\Delta I_{ds}}{\Delta V_g} \frac{L}{W} \frac{1}{CV_{ds}} \tag{12.1}$$

where L and W are the length and width of channel, I_{ds} and V_{ds} are source-drain current and voltage, respectively, V_g is the gate voltage, $C = \varepsilon_0 \varepsilon / d$ is dielectric layer capacitance density, ε_0 is permittivity of vacuum, and ε and d are the relative dielectric constant and thickness of the dielectric layer, respectively.

The directly measured parameter for a photodetector is photocurrent or photovoltage. Here, we focus on the former (unless otherwise specified), which is defined as the current difference with and without light illumination, expressed as

$$I_{ph} = I_{illu.} - I_{dark} \tag{12.2}$$

where the dark current I_{dark} is related to the noise of the device.

The responsivity (in the unit of A/W) is a very important parameter of a photodetector to characterize the input-output efficiency, which is defined as the ratio of photocurrent to effective light power:

$$R = \frac{I_{Ph}}{P_{in} A} \tag{12.3}$$

where P_{in} is the light power density and A is the effective active area of the photodetector. A high responsivity denotes a high convert efficiency from light to electrical signal, which can also be described by the external quantum efficiency. EQE is defined as the number ratio of collected carriers to the incident photons, i.e.,

$$EQE = \frac{n_q}{n_{ph}} = R \frac{hc}{q\lambda} \tag{12.4}$$

where h is Planck's constant, q is the element charge, and c and λ are light speed and wavelength, respectively. It can be found EQE is proportional to the responsivity. Generally different from the internal quantum efficiency, EQE can exceed 100% at the existence of photoconductive gain, with which the responsivity can be enhanced by several orders. In some literatures, EQE is regarded as the photoconductive gain [15]. However, the incident photons are not absorbed totally; thus, EQE/η_{abs} can also be adopted after considering the light absorbance η_{abs} [16].

The photoconductive gain can also be estimated by the response time and the transit time ($\tau_{tran} = L^2/\mu V_{ds}$) of photogenerated carriers, especially for the phototransistors. As the light is turned on and off, the photocurrent will rise and decay accordingly. The rise time (τ_r) and decay time (τ_d) can be extracted using two methods, including the duration time from 10% to 90% its maximum value, and fitting factor τ by the exponential term of $e^{-t/\tau}$. τ_d is generally larger than the rise time and is actually the lifetime τ_{life} of photogenerated carriers. Then the photoconductive gain can be calculated by

$$G = \frac{\tau_{\text{life}}}{\tau_{\text{tran}}} \tag{12.5}$$

To a photodetector, sensitivity is a crucial figure of merit, which concerns the detection limit at a certain level of noise. With the decrease of the light intensity, the photocurrent will be reduced accordingly, as well as the signal-to-noise ratio. The detection limit is often defined as the light intensity when the signal-to-noise ratio lowers to unity at the output bandwidth of 1 Hz, which is called noise equivalent power (in unit of W·Hz$^{-1/2}$) [4]. *NEP* can be calculated by the formula of

$$NEP = \frac{i_n}{R} \tag{12.6}$$

where the noise current i_n (in a unit of A·Hz$^{-1/2}$) can be calculated by the spectral noise density [17]

$$i_n^2(f) = 2q\langle i_{\text{dark}}\rangle + \frac{4k_B T}{R_{\text{shunt}}} + i_{1/f}^2(f) \tag{12.7}$$

and three terms denote shot noise, thermal noise, and 1/f noise, respectively. Here, k_B is Boltzmann constant, T is temperature, R_{shunt} is shunt resistance, and Δf is the electrical bandwidth (i.e., the measurement bandwidth). In experiments, the spectral noise density can be measured directly under dark conditions or calculated through Fourier transform with dark current [18]. Then one can calculate the noise current by $i_n = \sqrt{\frac{1}{\Delta f}\int_0^{\Delta f} i_n^2(f)df}$, and finally get *NEP*. The minimum *NEP* value is the sensitivity of the photodetector.

The sensitivity parameters *NEP* can be replaced by the specific detectivity (in unit of cm·Hz$^{1/2}$·W^{-1} or Jones), which is defined as the normalized signal-to-noise ratio for a photodetector with an active area of 1 cm^2 at the incident power of 1 W when the electrical bandwidth is 1 Hz, i.e.,

$$D^* = \frac{\sqrt{A}}{NEP} = R\frac{\sqrt{A}}{i_n} = R\frac{\sqrt{A\Delta f}}{\sqrt{\int_0^{\Delta f} i_n^2(f)df}} \tag{12.8}$$

This formula is a universal assessment method of specific detectivity for any kinds of photodetectors, and it can be approximated as $D^* = R\sqrt{A/i_n^2(\Delta f)}$ at a small bandwidth. While, for photodiodes, shot noise component is dominating, thus the specific detectivity can be further approximated as $D^* = R\sqrt{A/2q\langle i_{\text{dark}}\rangle}$.

12.3 WORKING MECHANISMS OF SEMICONDUCTOR PHOTODETECTORS AND THEIR PROGRESS

When light is absorbed by the semiconductor photodetectors, plenty of electron-hole pairs will be generated. The photocurrent can only be formed with the effective separation and transmission processes. There are so many device designs for high-efficiency separation and transmission based on the mechanisms of photovoltaic effect (PVE), photoconductive effect (PCE), photogating effect (PGE), photo-thermoelectric effect (PTE), and photo-bolometric effect (PBE) [9, 11, 12]. In the following, these mechanisms are illustrated one by one with the specific examples.

12.3.1 PHOTOVOLTAIC EFFECT

Almost all the self-powered or self-driven photodetectors are based on PVE, where the photocurrent is produced spontaneously under illumination even at the absence of bias voltage. The driven force for separation and transmission originates from the built-in electric field, which only exists in asymmetric electrical conditions. So many efforts have been made by researchers to realize the asymmetric electrical conditions with semiconductors, with the common strategy of fabricating photodiode with heterojunctions and homojunctions, such as PN, P⁺P, N⁺N junctions, Schottky junctions, and so on [19]. Figure 12.1 shows the PVE schematics in a PN junction photodiode. Under illumination, photons with energy greater than the band gap of the semiconductor are absorbed, leading to the generation of electron-hole pairs. The photogenerated electron-hole pairs are separated with the assistance of a built-in electric field and finally collected by electrodes to form a photocurrent [11, 20, 21]. The typical characteristic parameters of the PVE are the short-circuit current (I_{SC}) and open-circuit voltage (V_{OC}), which are also important in the field of solar cells. PVE photodetectors usually work at photovoltaic mode (no bias) with low-energy consumption, low dark current, linear response, and fast response speed. However, the input-output efficiency is relatively low because of the absence of photoconductive gain. Thus, the photodiodes are sometimes studied under photoconductive mode (reverse bias) [22] or avalanche mode (large reverse bias) [23] to gain higher responsivity.

To realize the lateral PN junction, the half of N-type InSe channel have been doped with Cu by solid-state reaction technology [24], thereby forming the InSe/CuInSe$_2$ lateral PN heterojunction, as shown in Figure 12.2a. The built-in field is produced at the junction between N-type InSe and P-type CuInSe$_2$, which facilitates the PVE of the device. Figure 12.2b displays the output curves ($I_{ds} - V_{ds}$) of the device under dark and illumination conditions. It can be observed that the dark current shows the typical rectification behavior of the PN junction photodiode, with $I_{ds} = 0$ at $V_{ds} = 0$. However, the current deviates from zero point under illumination, where the non-zero current at $V_{ds} = 0$ is the short-circuit current and the non-zero voltage at $I_{ds} = 0$ is the open-circuit voltage. The extracted I_{SC} can be regarded as the photocurrent because of the low dark current, and Figure 12.2c shows that I_{SC} increases linearly with the light power density, indicating a good linear response of the PVE device. Nevertheless, the responsivity at photovoltaic mode is relatively low (with the power conversion efficiency ~ 3.5%), which will be improved to 4.2 A/W after applying a very large bias voltage (−10 V). One main reason for this low efficiency of the lateral PN heterojunction is the small active area which can be enlarged by the vertical configuration of the heterojunction. Wu et al. [25] designed

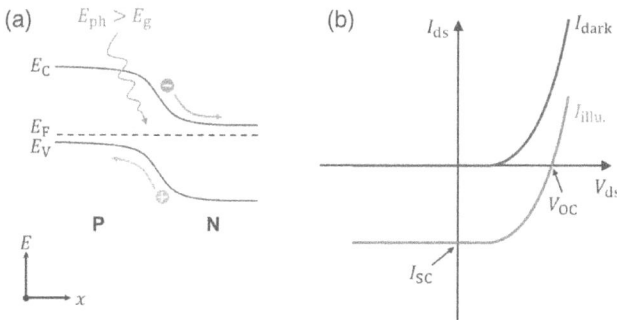

FIGURE 12.1   Photovoltaic effect schematics of a PN junction photodiode. (a) Band alignment and carrier behavior in the PN junction. (b) Output curves under dark and illumination conditions, respectively.

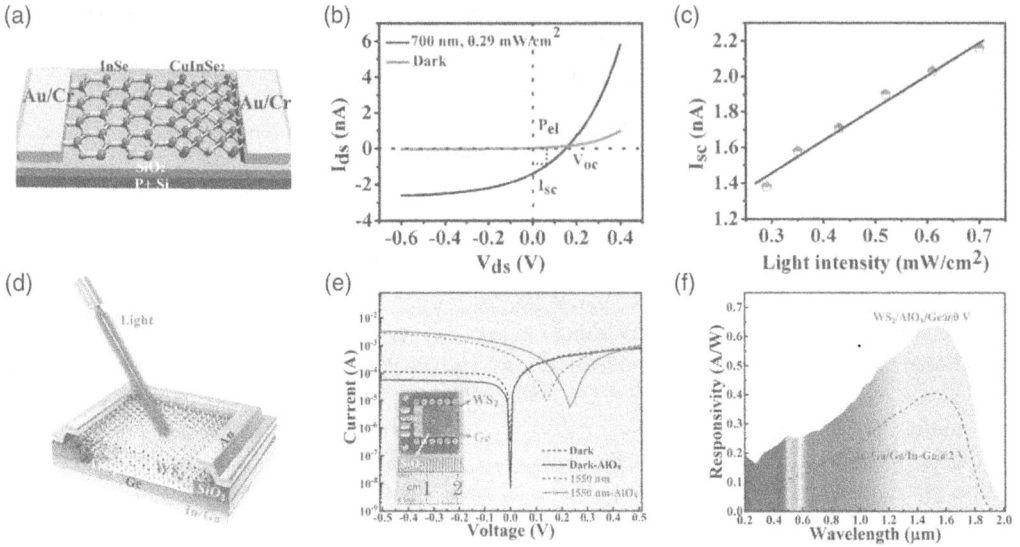

FIGURE 12.2 (a–c) Structure schematic of InSe/CuInSe2 PN junction photodiode, output curves of the photodiode under dark and illumination conditions, and short-circuit current as a function of light intensity, respectively [24]. (Copyright [2015] American Chemical Society.) (d–f) Schematic of the $WS_2/AlO_x/Ge$ NN photodiode, output curves of the photodiode under dark and illumination conditions, and ultrabroadband photoresponse characteristics, respectively [25]. (Copyright [2021] American Chemical Society.)

the photodiode with a $WS_2/AlO_x/Ge$ vertical NN heterojunction for ultrabroadband (wavelength from 200 nm to 4.6 µm) and high-detectivity (as high as 4.3×10^{11} Jones) photodetection, as shown in Figures 12.2d–12.2f. Clear PVE can be confirmed by the deviation of current from zero point ($I_{ds} = 0$ when $V_{ds} = 0$), and the interface passivation layer AlO_x is beneficial to lower dark current and enhances the photoresponse. The ultrabroadband response (Figure 12.2f) is ascribed to the narrow band gap of Ge and the tunable band gap of WS_2 (effectively narrowed by controlling the concentration of S vacancies), and the fast response speed (on 10 µs order) is due to the strong built-in field and the reduction of recombination at the WS_2/Ge interface. However, the linearity of the photoresponse in this NN photodiode is poor, where the responsivity decreases with the light intensity. Effective control for the photogenerated carriers may be a way for good linearity. Dou et al. [26] prepared a PIN photodiode based on organic-inorganic hybrid perovskite of $CH_3NH_3PbI_{3-x}Cl_x$, where the I-type perovskite is sandwiched between a P-type hole-transporting material and an N-type electron-transporting material. In addition to the high $EQE \sim 70\%$ and detectivity $\sim 10^{14}$ Jones, the device also achieves a linear dynamic range of over 100 dB based on the formula of

$$LDR = 20\log \frac{P_u}{P_l} \tag{12.9}$$

where the light power densities of P_u and P_l are the upper bound and lower bound of the linear region, respectively.

12.3.2 Photoconductive Effect

Different from PVE, an external bias voltage is required when the photodetection is carried out with PCE-based photodetectors, otherwise no photocurrent can be detected. According to the existence of photoconductive gain or not, the PCE is commonly divided into two types, with the former and the latter being recalled as PGE and PCE here, respectively. Figures 12.3a and 12.3b illustrate the

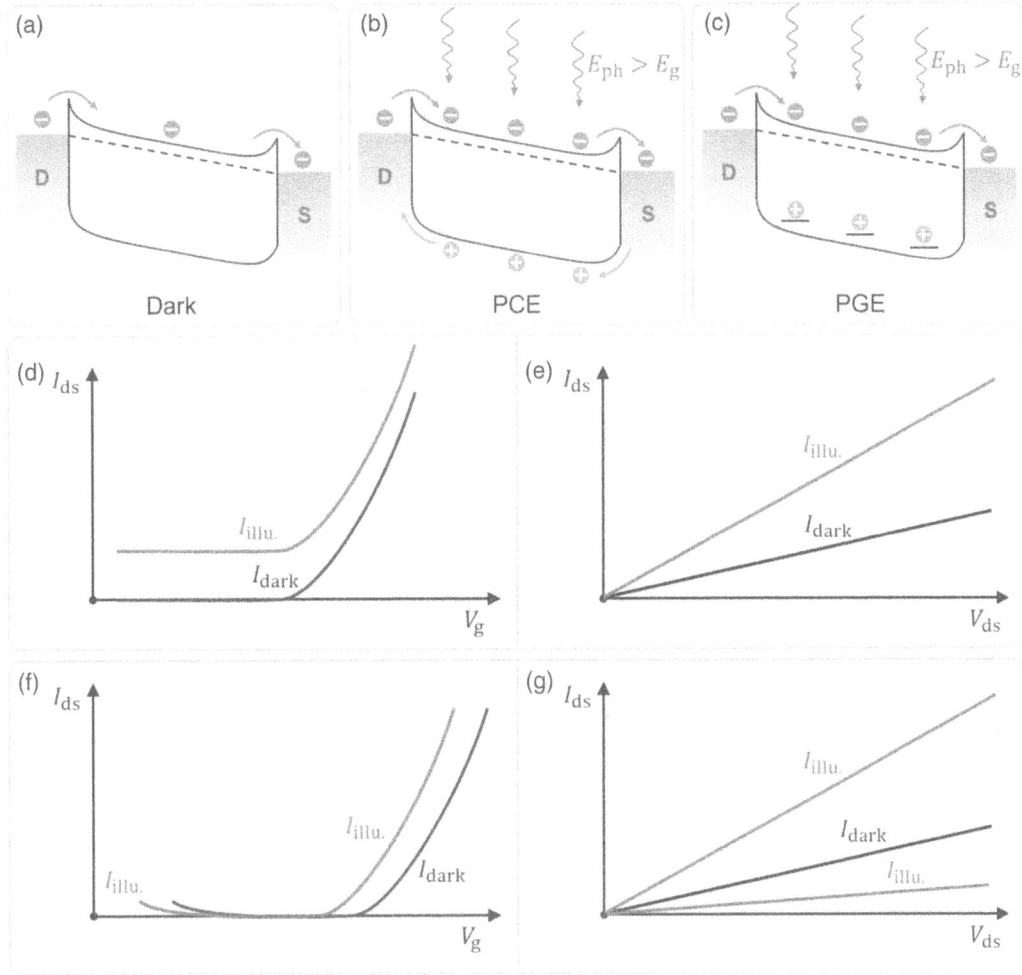

FIGURE 12.3 Schematic of the photoconductive effect and photogating effect. (a) Band diagram and carrier behavior of a phototransistor with external bias under dark condition. (b) Carrier behavior at photoconductive mode under illumination. (c) Carrier behavior at photogating mode under illumination. (d–e) Transfer curves and output curves of the photoconductive mechanism under dark and illumination conditions. (f–g) Transfer curves and output curves of the photogating mechanism under dark and illumination conditions.

carrier behavior of PCE in an N-type phototransistor. Under dark conditions, the current caused by the external bias voltage is attributed to the transport of electrons. When the device is illuminated, the electron-hole pairs are generated and then separated with the assistance of the applied electric field. As a result, the carrier concentration increases accordingly, leading to the photocurrent [11, 20, 21]. In this mechanism, as the incident light power increases, the accumulation of photogenerated carriers leads to a decrease in resistance of the channel material and an increase in conductivity. Figures 12.3d and 12.3e display the typical transfer and output curves of the N-type PCE phototransistor. The transfer curve under illumination is lifted as compared to that under dark condition, while all output curves pass through the zero point.

Without the photoconductive gain, the *EQE* of phototransistors based on PCE cannot exceed 100%, which limits the responsivity. However, linear response and fast response speed can be achieved because of the weak interaction in those carriers, defects, impurities, and other forms of charge. Figure 12.4 shows an example of an In_2Se_3 phototransistor for realizing the photodetection based on PCE [27]. For this N-type phototransistor, the carriers transport can be turned from

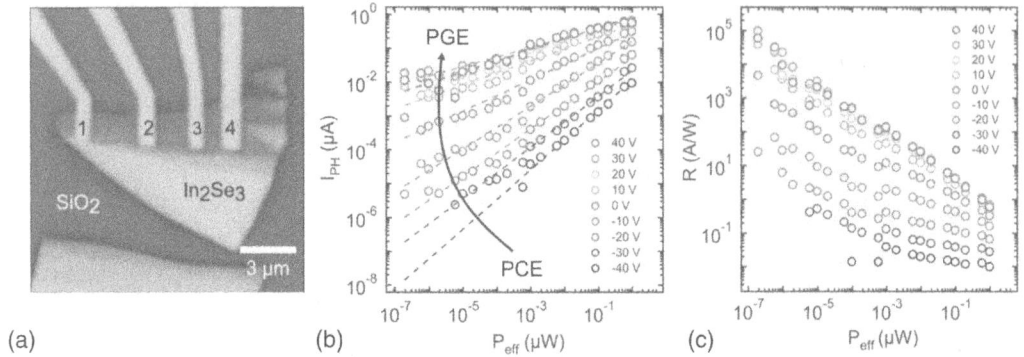

FIGURE 12.4 Realizing the photodetection based on the photoconductive effect. (a) Optical image of an In$_2$Se$_3$ phototransistor. (b) Photocurrent and (c) responsivity versus laser power for different gate voltages under 640 nm illumination [27]. (Copyright [2015] American Chemical Society.)

"OFF" to "ON" when the gate voltage changes from −40 to 40 V, and the working mechanism is tuned from PCE to PGE accordingly. At the intrinsic PGE mechanism, the responsivity for 640 nm illumination is only about 0.02 A/W as compared to the high responsivity ~10^5 A/W under PCE. However, the response speed ~30 ms for PGE is much faster than that ~9 s for PCE.

12.3.3 PHOTOGATING EFFECT

PGE refers to a particular case of PCE in semiconductor phototransistors, with the typical feature of huge photoconductive gain. The key to achieving the PGE is the trapping of minority carriers. Under illumination, photogenerated minority carriers (electrons for P-type channel and holes for N-type channel) can be captured by defect states inside or other materials outside the semiconductor channel. These captured carriers then act as localized gates, allowing for the control of channel carriers through the gating effect. Figure 12.3c shows the representative PGE in an N-type phototransistor through defect trapping, where the photogenerated holes are trapped by defects and form the gating effect on the channel. Therefore, the lifetime of trapped holes has a significant impact on the photoconductive gain (equal to the ratio of lifetime of trapped holes to transit time of transmitted electrons). It is known that the majority carriers in a semiconductor have a higher mobility, corresponding to a shorter transit time. The photogenerated majority carriers can drift across the channel more quickly than the minority carriers. In order to maintain the charge neutrality, more majority carriers are supplied from the opposite electrode during the life of holes. Therefore, prolonging the lifetime of minority carriers and improving the mobility of majority carriers are effective strategies for getting high photoconductive gain. However, long lifetime means slow response speed. Thus, a trade-off must be made between gain and response speed to achieve reasonable performance of a phototransistor. The transfer and output curves of an N-type dominating phototransistor (light P-type) are shown in Figures 12.3f and 12.3g. The transfer curve under illumination is the shift of transfer curve at dark toward negative direction for hole trapping and positive direction for electron trapping. For the case of hole trapping, the positive photocurrent is produced at N-branch and negative photocurrent can be found at P-branch.

It is worth noting that PGE is particularly prominent in nanoscale materials with a large surface-to-volume ratio, such as colloidal quantum dots, nanowires, and two-dimensional layered materials. By employing the defects of the channel materials or the nanomaterials with strong quantum confinement effect, minority carriers can be trapped inside or outside the channel, respectively. Lopez-Sanchez et al. [28] developed a monolayer MoS$_2$ phototransistor and achieved the high responsivity of ~880 A/W at 561 nm illumination when the channel was at "OFF" state. Such high responsivity implies the high photoconductive gain, which can be attributed to the hole trapping-induced PGE in the MoS_2 channel. The PGE can also be confirmed by the long photocurrent decay time of ~9 s. The responsivity exhibits an exponential decay with respect to the light intensity, which is the typical nonlinear phenomenon of PGE because of the saturated filling of deep defects [29]. It should be noted that 880 A/W is not the limit of this phototransistor. Higher responsivity can be extracted at "ON" state for the channel. However, large dark current and high noise level are also produced.

As aforementioned, both large mobility of majority carriers and long lifetime of minority carriers are the key to achieve high photoconductive gain. Thus, the assembly of graphene and quantum dots is a promising strategy for high-performance phototransistors. Konstantatos et al. [30] have developed a hybrid phototransistor with graphene PbS colloidal quantum dots by using spin casting. Once illuminated, the transfer curves turn to the positive direction of the gate voltage, identifying the electron trapping in the quantum dots. Combined with the ultra large hole mobility in graphene and the electron trapping in the quantum dots, a high photoconductive gain of ~10^8 is realized under 532 nm illumination, resulting in the high responsivity ~10^7 A/W and the high specific detectivity ~7×10^{13} Jones. However, the long response time (on the order of several seconds) challenges high-speed applications of this hybrid phototransistor, unless reset with the pulsed gate bias. Other hybrid photodetectors such as MoS_2/ZnO phototransistors [31] and WSe_2/InSe phototransistors [32] also show great enhancement on the responsivity through PGE. Under the premise in high photoconductive gain, the lifetime under PGE can be shortened with some methods, such as the clear interface [33] and the strain manipulation [18]. Liu et al. [33] reported an organic-inorganic hybrid phototransistor with the epitaxial ultrathin organic crystals (C_8-BTBT) on Graphene, as shown in Figures 12.5a–12.5c, which exhibited a high gain of >10^9 when C_8-BTBT was thicker than 5 layers. Hole trapping in C_8-BTBT film can be identified by the transfer curves change of the phototransistor, i.e., the shift in the negative direction after illuminated. The clean C_8-BTBT/Graphene interface allows a fast response speed, with the response time spanning from 25 ms to 830 ms for the C_8-BTBT layer from 1 to 6. Even for the 1-layer C_8-BTBT, the responsivity is still as large as 1.57×10^4 A/W.

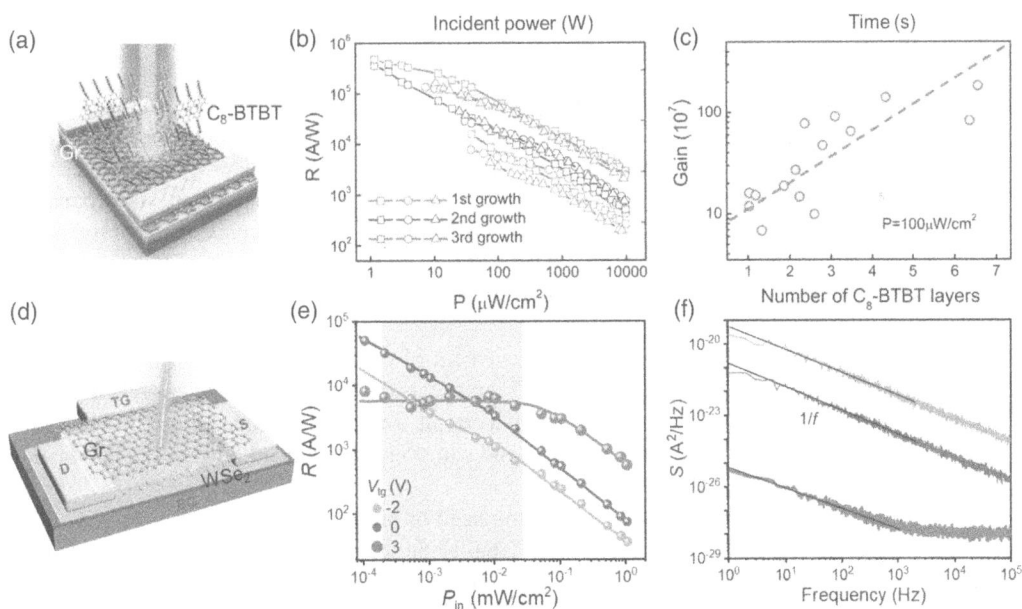

FIGURE 12.5   (a–c) Schematic of a C_8-BTBT/graphene hybrid phototransistor, responsivities for different grown times, and photoconductive gains versus C_8-BTBT layers under 355 nm illumination [33]. (Copyright [2016], Wiley-VCH.) (d–f) Schematic of a dual-gate WSe_2 phototransistor, responsivities at different top gate voltage when the Bottom gate voltage is −70 V, and the corresponding spectral noise density, where the light wavelength is 532 nm [16]. (Copyright [2021] Wiley-VCH.)

Strain distribution in the channel plane can also shorten the response time. After studying the phototransistor with grown monolayer WS_2 on Si_3N_4 substrate, Liu et al. [18] found that the responsivity reached up to 1.58×10^5 A/W, while the response time was just on the order of 40 ms. This fast response speed is probably attributed to the partially suspended characteristics of the WS_2 channel on the porous Si_3N_4 surface (revealed by atomic force microscopy). The strain distribution in the partially suspended WS_2 induces an in-plane built-in electric field and less carrier mass, benefiting the separation of electron-hole pairs and the carrier mobility.

The nonlinearity of PGE impedes the applications of the phototransistor in some fields, such as quantitative light detection, high-resolution imaging, machine vision technology, and so on. However, realizing the robust linearity with high gain is still a challenge. Several attempts have been reported for linearity with the vertically heterojunctioned phototransistors, and all of them require rigorous gating conditions. Ho et al. [34] fabricated the Ta_2O_5/graphene phototransistor and realized the linear response at weak illumination (wavelength from 200 nm and 10.6 μm) when $V_g = 7$ V, with $LDR \sim 80$ dB at 940 nm. Surprisingly, the responsivity retains a high value of $\sim 10^5$ A/W. The reason to the linear response is the light-tunable built-in field at the Ta_2O_5/graphene heterostructure, which influences the lifetime of trapped minority carriers. At the specific gate voltage, the photoconductive gain holds a constant value in a range of light intensity, resulting in the linear response. However, the dark current is on the order of 100 μA, resulting in a high noise level and low detectivity. To lower the noise level, the configuration of vertical dual-gate WSe_2 phototransistors has been studied [16], as shown in Figures 12.5d–12.5f. With the help of the vertical dual-gating, electrons and holes accumulate to the two interfaces of the WSe_2 channel and form the vertical homojunction. The separation of electrons and holes further weakens the interaction between carriers, resulting in a low noise level (Figure 12.5f).

Large dark current is an intrinsic nature of phototransistors because of the applied bias voltage, which, however, goes against the sensitivity. Therefore, achieving both high gain and high sensitivity is a promising research direction for phototransistors in the future.

12.3.4 PHOTO-THERMOELECTRIC EFFECT

PTE refers to the uneven heating of the channel material, resulting in a temperature gradient ΔT across the channel. For symmetric photodetectors, ΔT is often generated by localized illumination with a focused laser spot with dimensions much smaller than those of devices [35, 36]. For asymmetric devices, global illumination can also produce ΔT due to the uneven light absorbance or thermal conductivity [20]. According to the Seebeck effect, a voltage difference

$$\Delta V_{PTE} = S\Delta T \tag{12.10}$$

is generated across the semiconductor by the movement of charge carriers (from the hot to cold terminal due to the temperature gradient), where S represents the Seebeck coefficient. The PTE current will be formed once the device is connected to a circuit. To a PTE photodetector, the generation of ΔT is of great importance, and the response band can even be extended beyond the bandgap of the semiconductor channel.

Similar to the PVE, a PTE photodetector has no need of bias voltage, and PTE voltage is often focused for the performance evaluation. Figures 12.6a–12.6c show the example of a symmetric PTE phototransistor based on NbS_3 [37]. The temperature of one Au/NbS_3 contact will rise after illuminated by a focused laser spot. Then the charge carriers flow from one contact to the other through the NbS_3 channel, leading to the difference in Fermi levels of two terminals and the PTE voltage (or current) as well. The PTE voltage and PTE current can be extracted from the output curves under dark and illumination, i.e., the deviation from zero point, as depicted in Figure 12.6b. A strong PTE response is produced due to the short thermal decay length and low thermal loss in the device, showcasing a considerable performance from ultraviolet (375 nm) to terahertz (118.8 μm), as shown

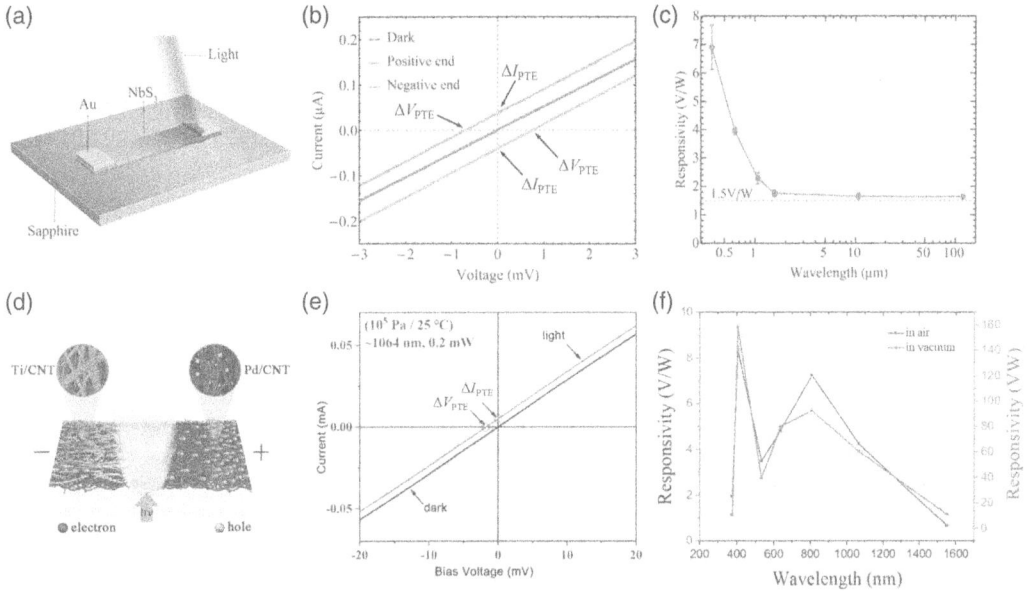

FIGURE 12.6 (a–c) Schematic of NbS$_3$-based PTE photodetector, output characteristics of the device under dark and localized 532 nm illumination, and responsivity versus light wavelength [37]. (Copyright [2020] American Chemical Society.) (d–f) Schematic of the Ti-CNT-Pd PTE photodetector, output characteristics of the device in the air under dark and global illumination, and responsivity versus light wavelength in air and in vacuum [38]. (Copyright [2022] American Chemical Society.)

in Figure 12.6c. The responsivities for all examined wavelengths are larger than 1.5 V/W, and the response time is less than 10 ms.

The localized illumination is not controllable, especially for the micro- or even nano-devices. Therefore, so many asymmetric PTE photodetectors have been designed for photodetection under global illumination. Liu et al. [38] utilized suspended carbon nanotubes (CNTs) to achieve a broadband (from ultraviolet [375 nm] to terahertz [118.8 μm]) PTE photodiode with faster response speed (several milliseconds), low noise (the smallest *NEP* is ~0.05 nW·Hz$^{-1/2}$), and high responsivity (from 0.35 to 158 V/W), as shown in the Figures 12.6d–12.6f. Generally, high electrical conductivity and low thermal conductivity benefit the PTE. In this device, the thermal conductivity of CNTs is effectively reduced by depositing Ti and Pd on terminals. After that, N-type and P-type doping are formed at the respective terminals due to different work functions of Ti, Pd, and CNTs. In addition, the introduction of metals also leads to thermal localization, increasing ΔT between the hot terminal and cold terminal, enhancing the PTE response of the device. Owing to the heat exchange between the device and the surrounding environment, the responsivities in air are at least one order smaller than those in vacuum (Figure 12.6f). Photodetectors based on the PTE exhibit so many advantages, such as wide spectral response beyond semiconductor band gap, operating at zero bias voltage, detection at room temperature, and so on. Hence, the PTE photodetectors are expected to be applied in infrared and even terahertz band photodetection.

12.3.5 PHOTO-BOLOMETRIC EFFECT

PBE primarily occurs in thermosensitive semiconductors under global illumination, where the temperature rise occurs in the semiconductors and results in the resistance change of the semiconductor [11]. Generally, either positive or negative photocurrent can be generated under PBE. The resistance variation induced by the incident light is governed by two ways, including changes in carrier mobility and changes in the number of carriers due to the temperature variations. The magnitude of the PBE is jointly determined by the degree of conductivity variation with temperature in the semi-conductor and how much the temperature increases by the illumination [39]. The photodetection based on PBE requires the bias voltage, which is similar to PCE and PGE and different from PTE. The sign of the photoresponse is also very different between PTE and PBE. The former is related to the difference in Seeback coefficients between the components of the junction and the type of charge carriers in the semiconductor material, while the latter is related to the change in the mate-rial conductivity with temperature. Nevertheless, both these two kinds of thermoelectric effects can be widely applied in the broadband detection beyond the semiconductor bandgap, because the photoresponse is based on the temperature-induced carrier generation (namely hot holes or hot elec-trons), rather than photoexcited carrier generation.

Yin et al. [40] reported an ultrabroadband phototransistor up to 10.6 μm with 2D Fe_3O_4 nanosheets. The lift of transfer curves under illumination implies the existence of PCE or PBE, and the dominating PBE is verified with the similar lift of transfer curves with the increase of the substrate temperature. The PBE phototransistor displays a high-performance detecting capabil-ity. When illuminated by 10.6 μm laser, the photoresponsivity, external quantum efficiency, and detectivity reach 561.2 A/W, $6.6 \times 10^3\%$ and 7.42×10^8 Jones, respectively. However, the response time (on the order of second) is relatively long, which may be ascribed to the slow behavior of the phonons. Through the study of black phosphorus phototransistors, Miao et al. [41] found that the carrier mobility and carrier concentration under PBE contributed jointly to the conductivity. No photodetector can operate just based on PBE, and the photoresponse is often from complementary mixing mechanisms. For example, in the investigation of the Bi_2O_2Se phototransistor, Yang et al. [42] clarified three kinds of contribution to the photocurrent (including PBE, PCE, and PTE) by local heating method [43, 44], and concluded that the PBE plays a dominant role in the entire pho-togeneration process of Bi_2O_2Se.

12.4 CONCLUSIONS

In summary, we introduced the main mechanisms of photodetection based on three-dimensional and low-dimensional semiconductors, including PVE, PCE, PGE, PTE, and PBE. In addition, we also clarified the common performance parameters used to evaluate the photodetectors. The car-rier generation, electron-hole pair separation, and carrier transmission are the main processes of a photodetector, which definitively influence the detection performance. Combined with emerg-ing achievements in the field of photodetection, the mechanisms with corresponding performance parameters were discussed in detail, as well as the advantages and disadvantages of those mecha-nisms. For instance, PVE and PTE are nice choices for low-power and fast-speed detection due to their ability to operate without bias voltage. However, the input-output efficiency is very low under the absence of photoconductive gain. PGE is good for enhancing efficiency based on the huge gain, while its large dark current results in high noise level and low sensitivity. PCE and PBE are always contributed together to the photoresponse, and PTE and PBE can be employed to realize the broad-band photodetection beyond the semiconductor bandgap. With the development of materials science and semiconductor physics, we believe that more surprising achievements about photodetection will be made in the future, which may further promote the advances of electronics and optoelectronics.

REFERENCES

1. S. Chen, C. Teng, M. Zhang, Y. Li, D. Xie, G. Shi, A flexible UV-vis-NIR photodetector based on a perovskite/conjugated-polymer composite, Advanced Materials, 28 (2016) 5969–5974.
2. H. Zhao, Y. Zhang, T. Li, Q. Li, Y. Yu, Z. Chen, Y. Li, J. Yao, Self-driven visible-near infrared photo-detector with vertical $CsPbBr_3$/PbS quantum dots heterojunction structure, Nanotechnology, 31 (2020) 035202.

3. Z. Huang, W. Zhou, J. Tong, J. Huang, C. Ouyang, Y. Qu, J. Wu, Y. Gao, J. Chu, Extreme sensitivity of room-temperature photoelectric effect for terahertz detection, Advanced Materials, 28 (2016) 112–117.

4. F.H.L. Koppens, T. Mueller, P. Avouris, A.C. Ferrari, M.S. Vitiello, M. Polini, Photodetectors based on graphene, other two-dimensional materials and hybrid systems, Nature Nanotechnology, 9 (2014) 780–793.

5. G.A. Rance, D.H. Marsh, R.J. Nicholas, A.N. Khlobystov, UV–vis absorption spectroscopy of carbon nanotubes: Relationship between the π-electron plasmon and nanotube diameter, Chemical Physics Letters, 493 (2010) 19–23.

6. C. Chen, P. Zhou, N. Wang, Y. Ma, H. San, UV-assisted photochemical synthesis of reduced graphene Oxide/ZnO nanowires composite for photoresponse enhancement in UV photodetectors, Nanomaterials, 8 (2018) 26.

7. G. Li, L. Liu, G. Wu, W. Chen, S. Qin, Y. Wang, T. Zhang, Self-powered UV-near infrared photodetector based on reduced graphene Oxide/n-Si vertical heterojunction, Small, 12 (2016) 5019–5026.

8. M.-Y. Li, S.-K. Su, H.S.P. Wong, L.-J. Li, How 2D semiconductors could extend Moore's law, Nature, 567 (2019) 169–170.

9. F. Wang, Y. Zhang, Y. Gao, P. Luo, J. Su, W. Han, K. Liu, H. Li, T. Zhai, 2D metal chalcogenides for IR photodetection, Small, 15 (2019) 1901347.

10. S.-H. Jo, H.-Y. Park, D.-H. Kang, J. Shim, J. Jeon, S. Choi, M. Kim, Y. Park, J. Lee, Y.J. Song, S. Lee, J.-H. Park, Broad detection range rhenium diselenide photodetector enhanced by (3-aminopropyl) triethoxysilane and triphenylphosphine treatment, Advanced Materials, 28 (2016) 6711–6718.

11. M. Long, P. Wang, H. Fang, W. Hu, Progress, challenges, and opportunities for 2D material based photodetectors, Advanced Functional Materials, 29 (2018) 1803807.

12. J. Wang, J. Han, X. Chen, X. Wang, Design strategies for two-dimensional material photodetectors to enhance device performance, InfoMat, 1 (2019) 33–53.

13. Z. Cheng, C.-S. Pang, P. Wang, S.T. Le, Y. Wu, D. Shahrjerdi, I. Radu, M.C. Lemme, L.-M. Peng, X. Duan, Z. Chen, J. Appenzeller, S.J. Koester, E. Pop, A.D. Franklin, C.A. Richter, How to report and benchmark emerging field-effect transistors, Nature Electronics, 5 (2022) 416–423.

14. S. Kaushik, R. Singh, 2D layered materials for ultraviolet photodetection: A review, Advanced Optical Materials, 9 (2021) 2002214.

15. G. Konstantatos, E.H. Sargent, Nanostructured materials for photon detection, Nature Nanotechnology, 5 (2010) 391–400.

16. J. Xu, J. Luo, S. Hu, X. Zhang, D. Mei, F. Liu, N. Han, D. Liu, X. Gan, Y. Cheng, W. Huang, Tunable linearity of high-performance vertical dual-gate vdW phototransistors, Advanced Materials, 33 (2021) 2008080.

17. Y. Fang, A. Armin, P. Meredith, J. Huang, Accurate characterization of next-generation thin-film photodetectors, Nature Photonics, 13 (2019) 1–4.

18. F. Liu, J. Xu, Y. Yan, J. Shi, S. Ahmad, X. Gan, Y. Cheng, X. Luo, Highly sensitive phototransistors based on partially suspended monolayer WS2, ACS Photonics, 10 (2023) 1126–1135.

19. G. Konstantatos, E.H. Sargent, Erratum: Nanostructured materials for photon detection, Nature Nanotechnology, 5 (2010) 885–885.

20. M. Buscema, J.O. Island, D.J. Groenendijk, S.I. Blanter, G.A. Steele, H.S.J. van der Zant, A. Castellanos-Gomez, Photocurrent generation with two-dimensional van der Waals semiconductors, Chemical Society Reviews, 44 (2015) 3691–3718.

21. J. Jiang, Y. Wen, H. Wang, L. Yin, R. Cheng, C. Liu, L. Feng, J. He, Recent advances in 2D materials for photodetectors, Advanced Electronic Materials, 7 (2021) 2001125.

22. S. Hu, Q. Zhang, X. Luo, X. Zhang, T. Wang, Y. Cheng, W. Jie, J. Zhao, T. Mei, X. Gan, Au–InSe van der Waals Schottky junctions with ultralow reverse current and high photosensitivity, Nanoscale, 12 (2020) 4094–4100.

23. A. Gao, J. Lai, Y. Wang, Z. Zhu, J. Zeng, G. Yu, N. Wang, W. Chen, T. Cao, W. Hu, D. Sun, X. Chen, F. Miao, Y. Shi, X. Wang, Observation of ballistic avalanche phenomena in nanoscale vertical InSe/BP heterostructures, Nature Nanotechnology, 14 (2019) 217–222.

24. W. Feng, W. Zheng, X. Chen, G. Liu, W. Cao, P. Hu, Solid-state reaction synthesis of a InSe/CuInSe$_2$ lateral p–n heterojunction and application in high performance optoelectronic devices, Chemistry of Materials, 27 (2015) 983–989.

25. D. Wu, J. Guo, C. Wang, X. Ren, Y. Chen, P. Lin, L. Zeng, Z. Shi, X.J. Li, C.-X. Shan, J. Jie, Ultrabroadband and high-detectivity photodetector based on WS$_2$/Ge heterojunction through defect engineering and interface passivation, ACS Nano, 15 (2021) 10119–10129.

26. L. Dou, Y. Yang, J. You, Z. Hong, W.-H. Chang, G. Li, Y. Yang, Solution-processed hybrid perovskite photodetectors with high detectivity, Nature Communications, 5 (2014) 5404.
27. J.O. Island, S.I. Blanter, M. Buscema, H.S.J. van der Zant, A. Castellanos-Gomez, Gate controlled photocurrent generation mechanisms in high-gain In$_2$Se$_3$ phototransistors, Nano Letters, 15 (2015) 7853–7858.
28. O. Lopez-Sanchez, D. Lembke, M. Kayci, A. Radenovic, A. Kis, Ultrasensitive photodetectors based on monolayer MoS$_2$, Nature Nanotechnology, 8 (2013) 497–501.
29. J. Jiang, C. Ling, T. Xu, W. Wang, X. Niu, A. Zafar, Z. Yan, X. Wang, Y. You, L. Sun, J. Lu, J. Wang, Z. Ni, Defect engineering for modulating the trap States in 2D photoconductors, Advanced Materials, 30 (2018) 1804332.
30. G. Konstantatos, M. Badioli, L. Gaudreau, J. Osmond, M. Bernechea, F.P.G. de Arquer, F. Gatti, F.H.L. Koppens, Hybrid graphene–quantum dot phototransistors with ultrahigh gain, Nature Nanotechnology, 7 (2012) 363–368.
31. X.-L. Zhang, J. Li, B. Leng, L. Yang, Y.-D. Song, S.-Y. Feng, L.-Z. Feng, Z.-T. Liu, Z.-W. Fu, X. Jiang, B.-D. Liu, High-performance ultraviolet-visible photodetector with high sensitivity and fast response speed based on MoS$_2$-on-ZnO photogating heterojunction, Tungsten, 5 (2022) 91–99.
32. T. Lei, H. Tu, H. Lv, H. Ma, J. Wang, R. Hu, Q. Wang, L. Zhang, B. Fang, Z. Liu, W. Shi, Z. Zeng, Ambipolar photoresponsivity in an ultrasensitive photodetector based on a WSe$_2$/InSe heterostructure by a photogating effect, ACS Applied Materials & Interfaces, 13 (2021) 50213–50219.
33. X. Liu, X. Luo, H. Nan, H. Guo, P. Wang, L. Zhang, M. Zhou, Z. Yang, Y. Shi, W. Hu, Z. Ni, T. Qiu, Z. Yu, J.-B. Xu, X. Wang, Epitaxial ultrathin organic crystals on graphene for high-efficiency phototransistors, Advanced Materials, 28 (2016) 5200–5205.
34. V.X. Ho, Y. Wang, M.P. Cooney, N.Q. Vinh, Graphene-Ta$_2$O$_5$ heterostructure enabled high performance, deep-ultraviolet to mid-infrared photodetection, Nanoscale, 13 (2021) 10526–10535.
35. J. Park, Y.H. Ahn, C. Ruiz-Vargas, Imaging of photocurrent generation and collection in single-layer graphene, Nano Letters, 9 (2009) 1742–1746.
36. M. Buscema, M. Barkelid, V. Zwiller, H.S.J. van der Zant, G.A. Steele, A. Castellanos-Gomez, Large and tunable photothermoelectric effect in single-layer MoS$_2$, Nano Letters, 13 (2013) 358–363.
37. W. Wu, Y. Wang, Y. Niu, P. Wang, M. Chen, J. Sun, N. Wang, D. Wu, Z. Zhao, Thermal localization enhanced fast photothermoelectric response in a quasi-one-dimensional flexible NbS$_3$ photodetector, ACS Applied Materials & Interfaces, 12 (2020) 14165–14173.
38. Y. Liu, Q. Hu, Y. Cao, P. Wang, J. Wei, W. Wu, J. Wang, F. Huang, J.-L. Sun, High-performance ultrabroadband photodetector based on photothermoelectric effect, ACS Applied Materials & Interfaces, 14 (2022) 29077–29086.
39. C. Xie, C. Mak, X. Tao, F. Yan, Photodetectors based on two-dimensional layered materials beyond graphene, Advanced Functional Materials, 27 (2016) 1603886.
40. C. Yin, C. Gong, J. Chu, X. Wang, C. Yan, S. Qian, Y. Wang, G. Rao, H. Wang, Y. Liu, X. Wang, J. Wang, W. Hu, C. Li, J. Xiong, Ultrabroadband photodetectors up to 10.6 μm based on 2D Fe$_3$O$_4$ nanosheets, Advanced Materials, 32 (2020) 2002237.
41. J. Miao, B. Song, Q. Li, L. Cai, S. Zhang, W. Hu, L. Dong, C. Wang, Photothermal effect induced negative photoconductivity and high responsivity in flexible black phosphorus transistors, ACS Nano, 11 (2017) 6048–6056.
42. H. Yang, C. Tan, C. Deng, R. Zhang, X. Zheng, X. Zhang, Y. Hu, X. Guo, G. Wang, T. Jiang, Y. Zhang, G. Peng, H. Peng, X. Zhang, S. Qin, Bolometric Effect in Bi$_2$O$_2$Se Photodetectors, Small, 15 (2019) 1904482.
43. M.-J. Lee, J.-H. Ahn, J.H. Sung, H. Heo, S.G. Jeon, W. Lee, J.Y. Song, K.-H. Hong, B. Choi, S.-H. Lee, M.-H. Jo, Thermoelectric materials by using two-dimensional materials with negative correlation between electrical and thermal conductivity, Nature Communications, 7 (2016) 12011.
44. H. Yuan, X. Wang, B. Lian, H. Zhang, X. Fang, B. Shen, G. Xu, Y. Xu, S.-C. Zhang, H.Y. Hwang, Y. Cui, Generation and electric control of spin–valley-coupled circular photogalvanic current in WSe$_2$, Nature Nanotechnology, 9 (2014) 851–857.

13 Semiconductor Photoelectrochemistry

Mohsen Lashgari

13.1 INTRODUCTION

Before delving into semiconductor photoelectrochemistry, it is essential to remember that any phenomenon involving electron transfer can be viewed as an electrochemical process. Concerning the electron transport phenomena, we can exemplify two well-known electrochemical reactions: proton reduction at the cathode and hydroxide oxidation at the anode of an electrochemical cell. In the first reaction (Equation 13.1), protons gain electrons and evolve into hydrogen gas, while in the second, hydroxide anions lose their electrons and are transformed into water and oxygen molecules (Equation 13.2).

$$2H^+ + 2e^- \rightarrow H_2 \uparrow \tag{13.1}$$

$$2OH^- \rightarrow H_2O + 2e^- + 1/2O_2 \uparrow \tag{13.2}$$

Electrochemical processes are a type of reduction-oxidation (redox) reaction. They can occur within an electrochemical cell, where the reduction reaction happens at the cathode – the electrode connected to the negative pole – while the oxidation reaction simultaneously takes place at the opposite electrode (called anode), which is connected to the positive pole of a power supply. Of course, spontaneous redox phenomena can occur without the need for an external power supply (a dc rectifier or a battery) and the use of an electrode. These redox reactions can happen through the direct transfer of electrons between reducing and oxidizing species. Such electron transferring reactions can be referred to as electroless phenomena [1]. As an example of electroless reactions, we can refer to the combustion process [2], in which a fuel (e.g., H_2) is oxidized and the resulting electrons are transferred to the oxidant (O_2). The outcome of this process is energy release and product formation (water):

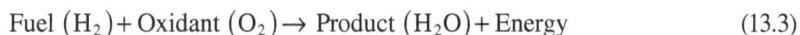

$$\text{Fuel } (H_2) + \text{Oxidant } (O_2) \rightarrow \text{Product } (H_2O) + \text{Energy} \tag{13.3}$$

If this spontaneous (thermodynamically favored) redox process occurs on two separate electrodes inside an electrochemical reactor (fuel cell), it results in the fuel oxidation and generation of electrons (negative charges) at the anode. Simultaneously, the oxidant undergoes reduction at the cathode, creating a positive polarity on its terminal. By generation of a potential difference and flow of electrons between anode and cathode, the chemical energy stored in the fuel is released as electricity.

During redox reactions that occur inside an electrochemical reactor/cell, a potential gradient exists between the anodic and cathodic terminals of the reactor. In the case of non-spontaneous processes (such as electrolysis and conversion of water into oxygen and hydrogen), the mentioned potential difference is supplied by connecting the reactor terminals to an external potential source (e.g., a dc rectifier). For spontaneous redox processes (such as those reactions occurring in a fuel cell or battery), by contrast, there is no need to employ a power supply and spend electricity. The potential gradient (E) is naturally generated between the electrodes' terminals, as follows:

$$E = \frac{\Delta_r G}{-nF} \tag{13.4}$$

DOI: 10.1201/9781003450146-13

where n is the number of electrons being transferred during the redox process, F is the Faraday constant (charge of one mole electrons; 96485 C), and $\Delta_r G$ is the Gibbs free energy change accompanying the redox reaction.

By connecting the negative pole of a DC power source to the cathode and the positive pole to the anode of an electrochemical cell, electrons start to accumulate on the cathode surface. In contrast, they become depleted on the anode surface, causing reduction and oxidation processes to occur simultaneously at the cathode and anode surfaces, respectively. Furthermore, it is worth highlighting that the matter of electron depletion and anodic reactions in electrochemical systems can be reconsidered by introducing the concept of the hole to these systems. Unlike an electron, which is a physical reality and a fundamental particle with the smallest unit of charge, a hole is merely an abstract concept with the opposite charge, being described as the lack of electrons in a system ($e^- + h^+ = 0$). Based on this concept, an anode can be redefined as an electrode with positive polarity that accommodates holes on its surface. By transferring holes to a species located at the electrode/solution interface, oxidation takes place. Briefly, we can say by applying a potential gradient, a charge separation occurs between the electrodes. Electrons are accommodated on the cathode and holes on the anode, leading to reduction and oxidation reactions on the mentioned electrodes, respectively.

To make a charge separation and generate an e/h pair in an electrochemical system, instead of consuming electricity and applying a potential difference between the anode and cathode, the charge (e/h) separation/generation can be photonically attained by utilizing semiconductor materials. In this environmentally friendly approach, photons provide the necessary energy for redox reactions, inducing an electronic excitation and generating holes in the ground and electrons in the excited level. As a result, redox processes occur on the semiconductor surface without the need for an external power source (dc rectifier). This green strategy, which relies on the use of semiconductor materials and sunlight, is extremely important from technological, environmental, and academic perspectives. The field that scientifically studies this approach is known as semiconductor photoelectrochemistry.

13.2 BAND STRUCTURE AND ENERGY DIAGRAM OF e/h (REDUCING/OXIDIZING AGENT) IN SEMICONDUCTING MATERIALS: A PHYSICOCHEMICAL PERSPECTIVE AND CONDUCTIVITY TYPE

Similar to molecules and atoms, whose electronic energy levels are quantized, a semiconducting material can be thought of as a hyper-molecule composed of a large number of atoms (on the order of Avogadro's number). Its energy levels are so closely spaced that we observe them in a continuous form, rather than as discrete levels. These continuous energy levels are referred to as bands, and the electronic structure of semiconductor materials is known as the band structure (Figure 13.1a). As we know from general chemistry and molecular orbital (MO) theory [3, 4], the highest electron-occupied MO energy level is called HOMO and the lowest unoccupied one is named LUMO. Regarding the energy levels of semiconducting materials, we should state that the band consisting of the HOMO and energy levels below it (naturally occupied by electrons) is referred to as a valence band (VB). The band consisting of the LUMO and higher energy levels (which are normally empty of electrons) is named a conduction band (CB). Therefore, the maximum of valence bands and minimum of conduction bands, denoted by VBM and CBM, correspond to the HOMO and LUMO energy levels, respectively. Due to the absence of a quantum state, the gap between the valence and conduction bands is called a forbidden region, implying that electrons cannot exist within this energy range. The presence of this gap (distance between VBM and CBM) is of great significance in the realm of semiconductors and their applications in science and technology. This energy gap ($E_g = E_{CBM} - E_{VBM}$) serves as the origin of electric potential difference, being created under stimulated conditions within semiconducting materials. When photons with appropriate energy ($h\nu \geq E_g$)

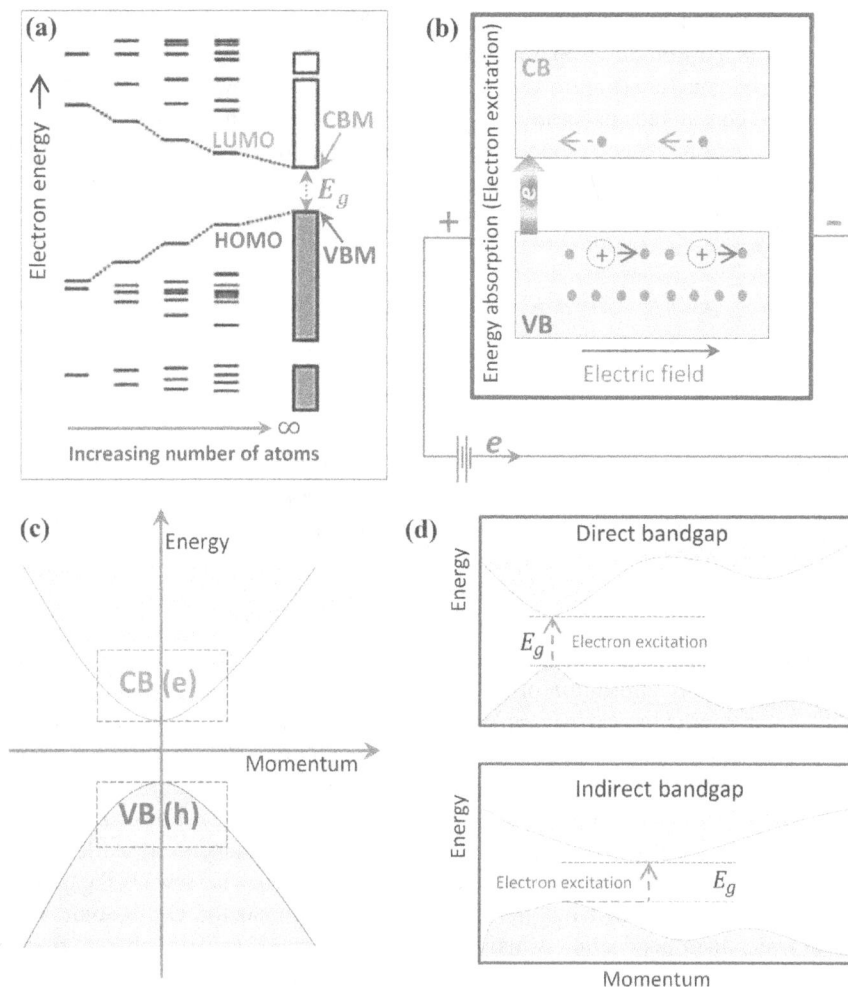

FIGURE 13.1 Semiconductors' energy quantization (a), conduction (b), energy diagram (c), and bandgap determination (d).

strike a semiconductor surface, they excite electrons, creating holes (oxidizing agent) in VB and electrons (reducing agent) in the CB of the material. Under an external electric field, the organized motion of electrons in the CB and holes in the VB leads to the flow of electricity through the semiconductor (Figure 13.1b).

The irradiation of photons causes an increase in the number of electrons in the CB and holes in the VB, thereby enhancing the conductivity of the semiconducting material. This property can be used as a simple method to determine whether a material is a semiconductor. Regarding the band structure of materials, which is schematically drawn and routinely utilized in the context of semiconductor photoelectrochemistry, the vertical axis represents energy (in eV), while the horizontal axis (often not depicted) is wave vector/momentum ($\vec{p} = \hbar \vec{k}$). Assuming that the behavior of an electron in a semiconductor can be described using a simple model of a free particle (no interaction with the surrounding environment), its energy will be different depending on what momentum (wave vector) it has (Equation 13.5).

$$E = \frac{p^2}{2m} = \frac{\hbar^2}{2m} k^2 \tag{13.5}$$

where m is the mass of electron and $\hbar$ is reduced Planck's constant $\left(\frac{h}{2\pi}\right)$. In this model, an upward parabolic trend with a minimum is obtained between the energy of the electron and the magnitude of the wave vector/momentum (Figure 13.1c). In the case of a hole, however, since it is an imaginary particle, its mass is considered apparently to be negative. Therefore, the coefficient of k^2 in Equation 13.5 gets negative, causing its corresponding parabola to point downward and exhibit a maximum. Furthermore, once the band structure of a semiconducting material becomes known, the effective mass (m^*) of its particles (electron/hole) can be defined and calculated using this formula:

$$\frac{\partial^2 E}{\partial k^2} = \frac{\hbar^2}{m} \rightarrow \boxed{m^* := \frac{\hbar^2}{\dfrac{\partial^2 E}{\partial k^2}}} \tag{13.6}$$

Due to the interaction of the particles (e/h) with the surrounding environment, the band structure diagram in reality is more complicated than what is obtained from the simple free particle model [5, 6]. This complexity of the band structure is schematically drawn in Figure 13.1d. This figure clearly demonstrates that in real systems, the position (momentum) of the CB curve minimum can differ from that of the VB curve maximum. In such cases, the electronic excitation is accompanied by a change in the wave momentum. This phenomenon is referred to as an indirect transition. In the case of semiconductors with a direct band gap, however, the electronic transition is not accompanied by a change in the momentum of the wave vector. Figure 13.1d also shows that to find the energy gap of a semiconductor with a complex band structure, two parallel lines are normally drawn from the maximum point of the VB and the minimum point of the CB. The distance between these lines represents the band gap.

Regarding the types of conductivity in semiconducting materials, it is worth mentioning that materials whose conductivity is mainly based on holes (positive charges) are called p-type semiconductors. Examples from this category of materials include CuO, a narrow band gap ($E_g \sim 1.7eV$) semiconductor, and NiO, a wide band gap ($E_g \sim 3.1eV$) semiconductor. On the other hand, semiconductors in which their conduction is mainly based on electrons (negative charge) are referred to as n-type semiconductors. Examples of n-type semiconductors include hematite (α-Fe_2O_3, a narrow band gap semiconductor with $E_g \sim 1.7eV$), TiO_2, and ZnO (a wide band gap semiconductor with $E_g \sim 3.2eV$) [7, 8].

The reason why the compounds like TiO_2 are often n-type semiconductors and why compounds like CuO are often p-type semiconductors can be explained in terms of lattice defects and the type of vacancies present in these materials [9, 10]. The vacancy in TiO_2 is typically of the oxygen type (V_O), while in CuO, it is of the metal type (V_M). Since there is naturally a negative charge (O^{2-}) at the location of oxygen atoms in the semiconductor lattice, the creation of an oxygen vacancy in TiO_2 causes the electrons of the oxide anion to be localized in the vacancy. This phenomenon results in the formation of new quantum states (energy levels) in the forbidden region, near the conduction band minimum (CBM). These vacancy-related quantum states, which are occupied by the excess electrons of the oxide anions that leave the lattice as oxygen molecules, serve as electron-donating centers that can easily transfer their electrons to the semiconductor's CB [2]. These vacancy-induced electrons are mobile in the CB and are responsible for the n-type conductivity of TiO_2. In the case of CuO, unlike TiO_2, the vacancy is of a metal type. Furthermore, since there is naturally a positive charge in the place of the metal cation, when the metal atom leaves the lattice, the positive charge (hole) is left in the vacancy of this metal deficient semiconductor. This results in the creation of new quantum states (energy levels) in the gap region, near the valence band maximum (VBM). These energy levels (related to metal vacancies) act as electron acceptors and, by gaining electrons from the VB, holes appear in the semiconductor's VB [2]. This causes the semiconductor to exhibit a p-type nature.

Based on the explanations provided above, we can conclude that the synthesis method of a material determines its semiconducting property. When the synthesis route results in a metal vacancy (deficiency), the semiconductor will be of the p-type, whereas if it creates an anion vacancy, the resulting semiconductor will be of the n-type. For instance, consider pyrite (FeS_2): It exhibits n-type behavior when synthesized in bulk form but switches to p-type when electrosynthesized as a film [11, 12].

13.3 SEMICONDUCTORS IN ELECTROCHEMICAL AND PHOTOREDOX SYSTEMS

In electrochemical setups, the difference between a conductor and a semiconductor is that, depending on the polarity of applied potential, the conductive (metal) electrode can act as either a cathode (negative polarity) or an as anode (positive polarity). In contrast, semiconductor electrodes behave differently. In n-type semiconductors under dark conditions, because the majority charge is electron, these electrodes can act as a cathode when connected to the negative pole of a power supply. Conversely, in p-type semiconductors, since the majority charge is a hole, these electrodes can serve as an anode in the absence of light by connecting them to the positive pole of the power supply. Regardless of the type of semiconductor, in the application of semiconducting materials in photoelectrochemical (PEC) systems, as mentioned earlier, electronic excitation occurs under illuminated conditions. This generates holes (oxidizing agent) in the valence band (VB) and electrons (reducing agent) in the conduction band (CB) of the semiconducting material. Therefore, like a traditional electrolyzer, redox reactions take place in a PEC reactor, composed of semiconducting particles and photons as energy source; see Figures 13.2a and 13.2b.

In another common application of semiconductors in PEC systems, instead of using dispersed semiconductor particles in the electrolyte solution, thin-film semiconductors are fabricated and used

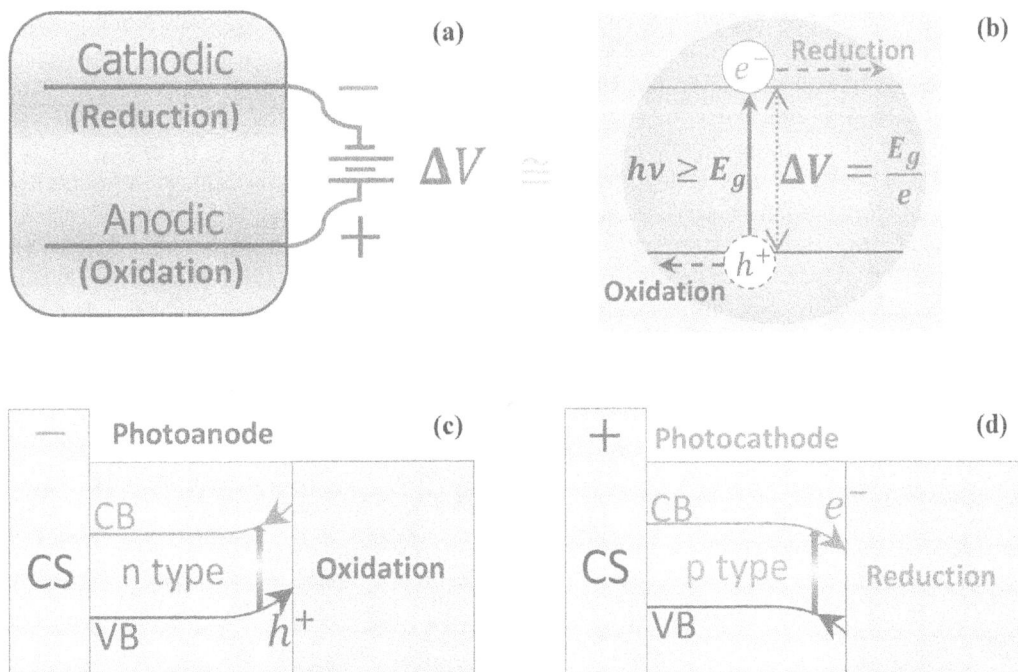

FIGURE 13.2 Electrolysis through traditional (a) and modern semiconductor-based light-induced (b) approaches. (c) Photoanode and (d) photocathode.

as photoelectrodes. Although in classical applications of electrochemistry, p-type semiconductor electrodes act as anodes and n-type as cathodes, the behavior of these electrodes changes under illuminated conditions. A p-type electrode serves as a photocathode and conducts a reduction process, whereas an n-type electrode acts as a photoanode and carries out an oxidation reaction. This different behavior of a semiconductor electrode under dark and light conditions can be explained in terms of band bending [13]. As shown in Figure 13.2c, for the n-type semiconductor electrode, the band bending is upward at the semiconductor/solution interface. When photons strike and generate e/h in the CB and VB, the electrons are pushed toward the conductor support (electrode terminal), while the holes are consumed in an oxidation process on the electrode surface. In the case of the p-type electrode (Figure 13.2d), however, the band bending and hence the redox process are reversed. Reduction reaction (electron consumption) occurs on the electrode surface, and the photogenerated holes accumulate in the photoelectrode, making its polarity positive.

In the absence of light, when a negative polarity is applied to an n-type electrode, electrons (the majority charge) gain the necessary energy to overcome the energy barrier of the interface region. This causes electrons to be transported from the electrode and consumed in a reduction (cathodic) process on the electrode surface. On the other hand, when a positive polarity is applied to a p-type electrode, species at the interface region are oxidized and the resulting electrons are neutralized by holes (the majority charge carriers) that reach the surface of the electrode.

13.4 PN JUNCTION AND ITS IMPORTANCE IN PHOTOELECTROCHEMICAL PHENOMENA AND CHARGE SEPARATION

Compared to powdery systems (so-called photocatalytics; Figure 13.2b), photoelectrodics (Figures 13.2c, 13.2d; routinely referred to as PEC or SPEC) exhibit superior charge separation due to band bending at the electrode/solution interface. As a result, in the latter system, the photogenerated charges (e/h) can be more effectively utilized in redox processes. In relation to photocatalytic (powdery) systems, it is worth mentioning that their performance can be enhanced by applying the same concept of band bending and charge separation improvement through the synthesis of p-n junction composites [7, 14]. A pn junction photocatalyst is a composite of p-type and n-type semiconductors, joined together at atomistic scale. Under this condition, there is an exchange of electrons (and holes) between two semiconducting components. Because of higher electronic chemical potential $(\mu_e^n > \mu_e^p)$ [15, 16], electrons migrate from n to p, in the direction of decreasing electronic pressure (concentration) in the n component, where the majority charges are electrons. Due to the migration/transport of electrons (and holes), the junction of pn semiconductors becomes polarized, and an electric field forms at the interface region (see Figure 13.3). The transfer of e/h continues until the forces of chemical potential $(\vec{\nabla}\mu_e$; also known as thermodynamic force [17]) and electric field $(e\vec{E})$ are balanced. Upon this circumstance, the Fermi levels of the p and n semiconductors equalize, and band alignment occurs. Under illuminated conditions, the electric field of the pn interface exerts

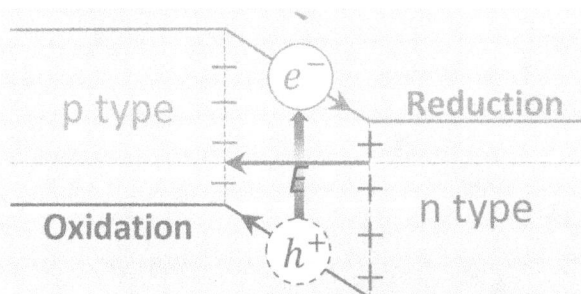

FIGURE 13.3   Electron/hole separation and photoredox phenomena in a pn-junction system.

an electrostatic force on the photogenerated electrons and holes, pushing them in opposite directions. By improving their separation, the photogenerated charges can be more effectively utilized as reducing/oxidizing agents in redox processes.

13.5   PHOTOVOLTAGE AND PHOTOCURRENT RESPONSE, AND SEMICONDUCTOR TYPE RECOGNITION

When a semiconductor electrode in an electrochemical cell is exposed to light, a potential difference called photovoltage is created. This light-induced voltage is attributed to the existence of a band gap and the generation of electron-hole pairs in the CB and VB of the semiconducting material. In a PEC system, the reducing and oxidizing potency of the photogenerated electrons and holes depends on the semiconductor's CBM and VBM energy levels, respectively. So, the photogenerated electrons with higher CBM (more negative potential value) would have more energy and thus exhibit a greater reducing ability. On the other hand, a lower VBM (more positive potential value) would result in a higher oxidizing power for its photogenerated holes. Therefore, from an energy perspective, to determine whether a semiconductor is suitable for a specific redox process, it is important to consider not just the band gap, but also its CBM and VBM. For example, for a semiconductor with the ability of water splitting, its band gap should be greater than 1.23 eV (E(in volt) $\leq \frac{\Delta G^\circ}{-nF}$):

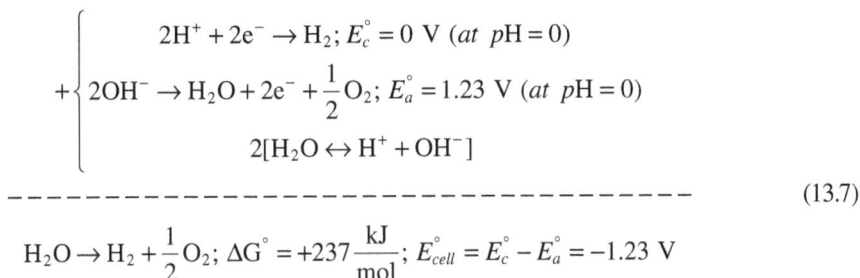

$$+\begin{cases} 2H^+ + 2e^- \rightarrow H_2; E_c^\circ = 0 \text{ V } (at \ pH = 0) \\ 2OH^- \rightarrow H_2O + 2e^- + \frac{1}{2}O_2; E_a^\circ = 1.23 \text{ V } (at \ pH = 0) \\ 2[H_2O \leftrightarrow H^+ + OH^-] \end{cases}$$

$$- \quad (13.7)$$

$$H_2O \rightarrow H_2 + \frac{1}{2}O_2; \Delta G^\circ = +237 \frac{kJ}{mol}; E_{cell}^\circ = E_c^\circ - E_a^\circ = -1.23 \text{ V}$$

Furthermore, the semiconductor will be able to reduce protons and evolve hydrogen gas if its CBM potential becomes negative (less than zero, the redox potential of hydrogen under standard condition). This semiconductor can also perform the oxidation of hydroxide anions and produce oxygen if the potential corresponding to the semiconductor's VBM becomes greater than +1.23 V (see Equation 13.7). Besides photovoltage, which is a thermodynamic quantity and influenced by the semiconductor band gap, the kinetics of a photo-redox process can be determined by measuring photocurrent [8]. A larger photocurrent indicates a faster light-induced redox process in a PEC reactor.

A straightforward method to identify the type of a semiconductor electrode is to measure its open circuit potential (OCP) in the absence and presence of light. For p-type electrodes, since the band bending is downward, upon light irradiation, the photogenerated electrons are consumed in a reduction process and holes are transferred to the electrode terminal, resulting in the OCP shift toward positive values. For n-type electrodes, however, OCP changes toward negative values. This is because for n-type electrodes, the band bending is upward, holes are consumed in an oxidation (anodic) process but electrons are accumulated at the electrode terminal (see Figure 13.4).

Besides photovoltage (Figures 13.4a and 13.4b), photocurrent (Figures 13.4c and 13.4d) can also be used to determine the type of semiconductor electrode. In the case of n-type electrodes (h^+ is the minority charge), the anodic current is low in the absence of light. However, the current increases significantly upon exposure to photons by generating additional holes (oxidizing agents) on the electrode surface. Similarly, in the case of p-type electrodes (e^- is the minority charge), the irradiation of light generates extra electrons. These electrons act as a reducing agent and consequently boost the cathodic current. Therefore, by shining light on the surface of the semiconductor electrode, if the

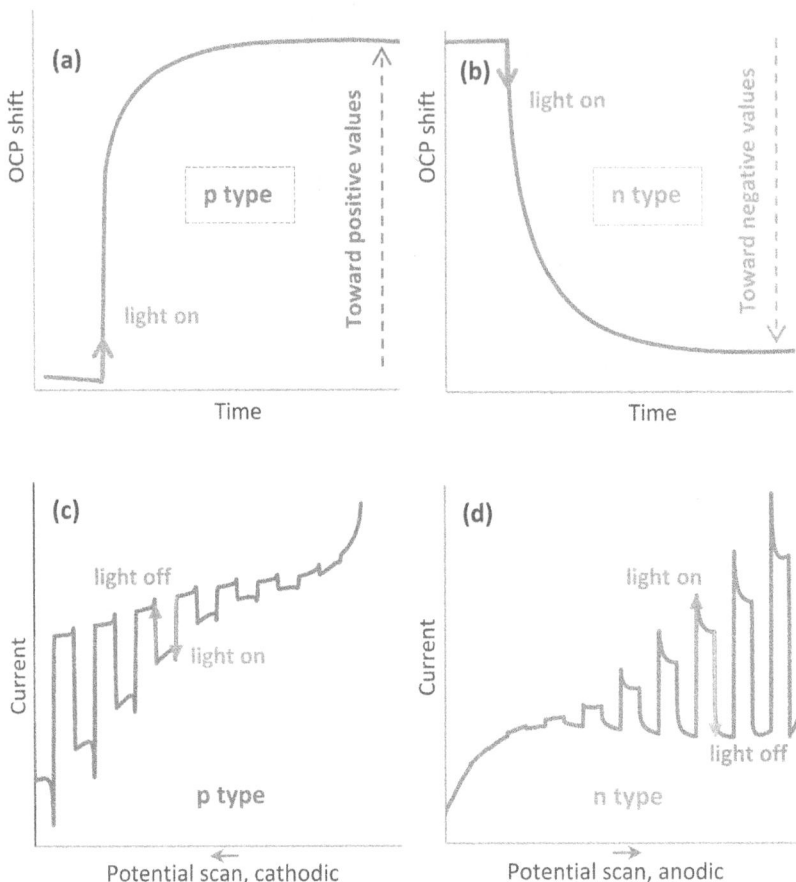

FIGURE 13.4 Light-induced potential and current for p-type (a, c) and n-type (b, d) semiconductors.

anodic current is strengthened, the semiconductor is n-type, and if the cathodic current is increased, the semiconductor is p-type.

13.6 IDENTIFYING THE TYPE OF SEMICONDUCTOR BY THE MOTT-SCHOTTKY METHOD

Before explaining the Mott-Schottky (MS) method, it is necessary to note that due to the different electronic chemical potential across the semiconductor/solution interface ($\mu_e^{SC} \neq \mu_e^{Sol}$), electron migration occurs in the interface region. This causes the energy of electrons on the semiconductor surface to become different from its value in the semiconductor bulk. The result of this electron migration process is band bending and a difference in the electric potential between the surface of the semiconductor and its bulk ($\Delta\phi_{sc} \neq 0$). Furthermore, it is worth mentioning that by applying negative and positive potentials to the semiconductor electrode, the energy of electrons within the semiconductor can either increase or decrease. This, in turn, can intensify, weaken, or even make the band bending disappear. A potential where the energy of electron on the surface of the semiconductor becomes equal to its energy in the bulk ($\Delta\phi_{sc} = 0$), is called flat band potential (E_{fb}).

In the previous section, we used photovoltage and photocurrent to determine the semiconductor type. Now, we are going to introduce the MS approach, which does not require light. This method is based on measuring the system capacitance (C_{obs}) at different potential biases. Before delving into the details, it is necessary to note that one of the major differences between semiconductor and metal electrodes is that in the case of metals, the movement of electric charge is confined to the metal surface. By conducting a faradaic process (redox reaction) and consuming charge on the metal surface, electrons (or holes) are immediately replenished. As a result, no polarization (charge separation) occurs within the metal electrode. In the case of semiconductor electrodes, however, due to their low electrical conductivity, the consumed charges (e/h) are not immediately restored. This leads to charge diffusion and polarization issues within the semiconductor material near the electrode/electrolyte interface. In other words, in electrochemical systems based on semiconductor electrodes, the charge (e/h) density on the semiconductor surface differs from its value in the semiconductor bulk. This difference leads to the formation of a region called space charge (SC), near the surface of the electrode. In the SC region, unlike the semiconductor bulk, the net charge density (ρ) and hence the electric potential (ψ; $\rho = -\kappa\varepsilon_\circ \nabla^2\psi$) is not zero, but it changes exponentially with distance [1]:

$$\psi(x) = \psi_s.e^{-\frac{x}{L_D}}; \; L_D = \left(\frac{\kappa\varepsilon_\circ k_B T}{2n^\circ e_\circ^2}\right)^{0.5} \quad (13.8)$$

where ψ_s is the electric potential at the semiconductor surface, L_D Debye length, n° charge concentration (number per unit volume), κ dielectric constant of the semiconductor, $e_\circ$ electron charge, $\varepsilon_\circ$ vacuum permittivity, k_B Boltzmann's constant, and T temperature in Kelvin. It should also be noted that by having L_D, the thickness of the space charge (d_{SC}) can be calculated under different potentials (being applied to the semiconductor electrode) using this formula [18]:

$$d_{sc} = 2L_D\left(\frac{e_\circ\Delta\phi_{sc}}{k_B T} - 1\right)^{0.5}; \; \Delta\phi_{sc} = E - E_{fb} \quad (13.9)$$

Equations (13.8) and (13.9) indicate that in the case of metals, which can be considered as semiconductors with a high electron concentration (n°), the thicknesses of L_D and d_{sc} are both small. Consequently, the space-charge region disappears and electrons accumulate on the metal surface. These equations also reveal that by applying different electrical potentials to a semiconductor electrode, the values of d_{sc}, $\psi(x)$, and hence the SC capacitance (C_{sc}) are changed. Furthermore, it is crucial to note that the interface between a semiconductor electrode and an electrolyte comprises two capacitors in series: one for the space charge (C_{sc}) and another for the Helmholtz layer (C_H). Given that the thickness of the space-charge capacitor is significantly larger than that of the Helmholtz layer (approximately 100 times), the charge distribution in the space-charge capacitor spans a greater distance. As a result, C_{sc} has a smaller value $\left(C_\downarrow = \kappa\varepsilon_\circ \frac{A}{d_\uparrow}\right)$ and can be overlooked in comparison to C_H ($C_{SC} \ll C_H$). However, when considering the inverse of the capacitance, the relationship is reversed ($C_H^{-1} \ll C_{sc}^{-1}$). Therefore, the observed equivalent capacitance (C_{obs}) equals the space-charge capacitance (C_{sc}), which can be used in practical studies.

$$C_{obs}^{-1} = C_H^{-1} + C_{sc}^{-1} \xrightarrow{C_H^{-1} \ll C_{sc}^{-1}} \boxed{C_{obs} \approx C_{sc}} \quad (13.10)$$

For a semiconductor electrode, by measuring the system capacitance (per unit area of the electrode surface) at different applied potentials and plotting the data as the inverse square of C_{sc} versus potential, a linear dependency is expected according to the MS formula (Equation 13.11) [18, 19]:

$$\boxed{C_{sc}^{-2}} = \frac{2k_B T}{n^\circ\kappa\varepsilon_\circ e_\circ} \cdot \left(\boxed{E} - E_{fb} - \frac{k_B T}{e_\circ}\right) \quad (13.11)$$

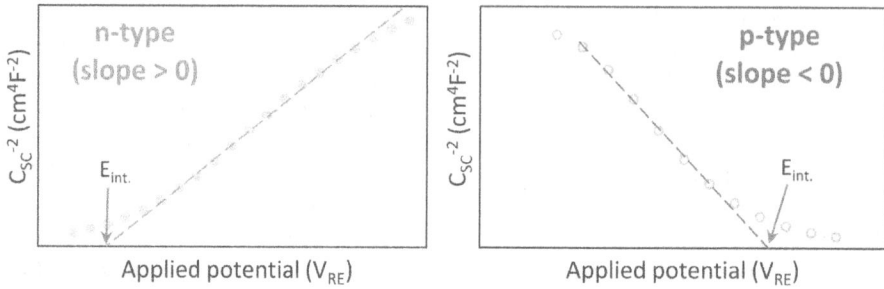

FIGURE 13.5   Mott-Schottky and semiconductor electrodes.

In the MS analysis, the observation of lines with positive and negative slopes (see Figure 13.5) indicates that the semiconductor is of n and p type, respectively [2]. The concentration of electron donor (N_d) and acceptor (N_a) can also be determined for the mentioned semiconductors from the slope of MS lines and dielectric constant data of the semiconducting material (see Equation 13.12).

$$n^\circ = \frac{1}{Slope} \times \frac{2k_BT}{\kappa\varepsilon_\circ e_\circ} = \begin{cases} N_d, \text{ for } n \text{ type} \\ -N_a, \text{ for } p \text{ type} \end{cases} \quad (13.12)$$

Furthermore, from the intersection of the MS line with the potential axis, E_{fb} (the flat band potential) is determined:

$$E_{fb} = E_{int.} - \frac{k_BT}{e_\circ} \quad (13.13)$$

where, $E_{int.}$ corresponds to the potential at which the MS line intersects the x-axis (see Figure 13.5) and, $\frac{k_BT}{e_\circ}$ equals 25.7 mV at 298 K.

13.7 APPLYING FARADAIC AND ELECTROSTATIC BIASES TO A SEMICONDUCTOR PHOTOELECTRODE: IMPACTS AND ELECTRICITY CONSUMPTION

As mentioned earlier, by applying an electric potential to a semiconductor electrode, the energy of electrons can be altered and by displacing the semiconductor's band structure, the band bending is changed. For example, when the negative pole of a power supply is connected to a semiconductor electrode and electrons are transferred to the semiconductor, its Fermi level increases. This shifts the band structure toward higher energy values, which increases the reducing ability of photoproduced electrons and decreases the oxidizing ability of holes. Electron injection into the semiconductor or electron removal from the semiconductor can be done in two ways: electrostatic and faradaic [20]. In the first method (electrostatic bias), the system behaves like a capacitor. Depending on the type of induced charges, whether they are electrons or holes, the semiconductor electrode will acquire either negative or positive charges. To achieve the desired charge, the appropriate pole of a power supply is connected to the semiconductor electrode, while the other pole is connected to earth (ground). In the second method (faradaic bias), in addition to the presence of the semiconductor photoelectrode (SPE), it is necessary to have an auxiliary electrode in the cell (reactor). This electrode is electrically connected to the SPE, thereby completing the system's electrical circuit. Moreover, in the latter method, the potential bias is applied to the photoelectrode with respect to a

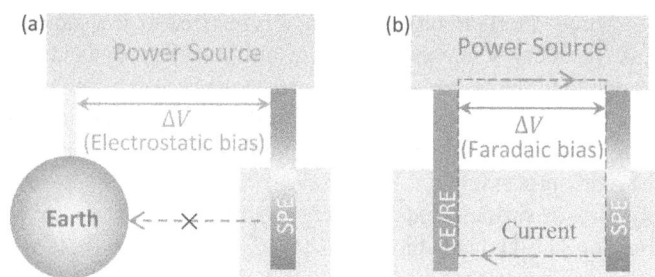

FIGURE 13.6 Applying electrostatic (a) and faradaic (b) biases to a semiconductor photoelectrode (SPE).

reference or auxiliary (counter) electrode. By passing current through the PEC reactor, electricity is consumed and the electrical work/energy is calculated using this formula:

$$W = \int_0^\tau I \Delta V (dt)$$ (13.14)

where ΔV is a potential difference between working (photoelectrode) and counter electrodes, I is the electrical current passing through the PEC reactor, and τ is the reactor operation period.

In the electrostatic bias method, unlike the faradaic, the system is single-electrode in nature and the electrical circuit is incomplete. As a result, the current passing through the system and hence the electricity consumption becomes almost zero (Figure 13.6). By contrast, in the case of faradaic bias, the completeness of the circuit allows for the passage of electricity. This leads to the consumption of electrical energy as an additional, non-photonic source for the generation of e/h pairs, i.e., reducing/oxidizing agents on the electrodes' surface (cf. Figure 13.2a and 13.2b).

13.8 TYPES OF SEMICONDUCTOR PHOTOELECTROCHEMICAL SYSTEMS AND THEIR APPLICATIONS

As deduced from Section 13.3 (Figure 13.2), semiconductor photoelectrochemical systems (SPECs) can be classified into two main categories. The first category, popularly referred to as photocatalyst or photocatalysis, consists of powdery semiconducting particles that are dispersed/suspended in an electrolyte medium. The second category, called a single-electrode photoreactor, is made up of a semiconductor thin film that is supported on a conductive substrate and immersed in the electrolyte medium. In both systems, due to direct contact of the anodic and cathodic zones of the semiconductor (either as separated particles or as a unified thin film), the light-induced charges (electricity), being transferred across the zones, cannot be externally utilized. In the case of a thin-film semiconductor, however, if the conductive support (CS) is not in direct contact with the electrolyte solution and its connection is done through an external circuit via an auxiliary electrode, then the anodic and cathodic processes occur upon two separate electrodes. Hence, the flow of electricity between them can also be exploited. The latter system, which is usually made of two or three electrodes (the third is a reference electrode), is routinely referred to as a PEC reactor. If the semiconductor electrode is of the p-type, as mentioned earlier, it is termed a photocathode and a reductive reaction is performed on its surface. Conversely, an n-type electrode is called a photoanode, and an oxidative reaction occurs upon it. Furthermore, for each of these electrodes, their complementary reactions are simultaneously carried out on the system's counter electrode. In a PEC reactor, the semiconductor and counter electrodes may be connected directly by a wire (short circuit), or they may be connected indirectly by inserting a power supply and applying a potential bias between the electrodes (Figure 13.6b). If no bias is needed to carry out a desired redox process and the reactor can operate independently under a wired (short circuit) condition, it is referred to as a standalone PEC. Notably, photocatalyst systems (powdery semiconductors) are also considered standalone, as no external electrical biases are applied to the system. Furthermore, in the case of a single-electrode PEC, even though an electrostatic bias is applied to the photoelectrode (Figure 13.6a), it can still be considered standalone because its circuit is incomplete and no significant electricity (non-photonic energy) is consumed during the redox process.

SPECs are widely used in various fields of energy and fuel production, as well as in pollutant removal and its conversion to value-added materials [8, 12, 21, 22]. In all these applications, as mentioned before, the basic and important point is that when photons strike the semiconductor surface, reducing (electron) and oxidizing (hole) agents are created in the CB and VB of the semiconducting material, respectively; see Equation 13.15.

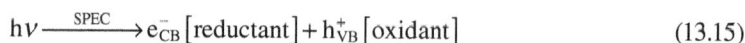

$$h\nu \xrightarrow{\text{SPEC}} e_{CB}^{-}[\text{reductant}] + h_{VB}^{+}[\text{oxidant}] \qquad (13.15)$$

These photogenerated charges (electrons and holes) may directly participate in a desired redox process or be transformed into H atoms and hydroxyl radicals in aqueous media (Equation 13.16), which can further serve as powerful reducing and oxidizing chemicals, respectively [23].

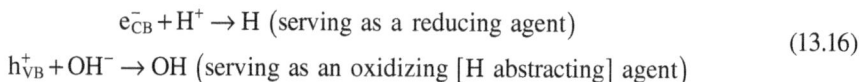

$$e_{CB}^{-} + H^{+} \rightarrow H \text{ (serving as a reducing agent)}$$
$$h_{VB}^{+} + OH^{-} \rightarrow OH \text{ (serving as an oxidizing [H abstracting] agent)} \qquad (13.16)$$

The transiently generated H atoms are indeed reactive chemical reducers. They can be effectively employed in the atomistic hydrogenation of environmentally important molecules such as CO_2 and N_2. This process allows for their conversion to value-added products such as methanol (CH_3OH), C/H/O based chemicals/fuels, ammonia (NH_3), and hydrazine (N_2H_4) [2, 23, 24]. These H atoms can also be used in the hydrodesulfurization of S-containing fuels [25, 26]. If no additional molecules (e.g., CO_2, N_2) are present in the reaction medium, the reactive H atoms recombine and evolve as H_2 gas [27, 28]:

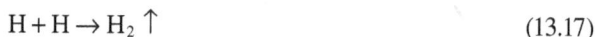

$$H + H \rightarrow H_2 \uparrow \qquad (13.17)$$

Regarding the transiently generated OH radicals, they are potent oxidants capable of H abstraction. These radicals can effectively destroy and mineralize water pollutants, including dye molecules, antibiotics, and bacteria [20, 29, 30]. This capability of OH radicals can also be utilized for the oxidative desulfurization of organosulfur compounds (e.g., dibenzothiophene) mixed with hydrocarbon fuel, converting them into polar sulfones. The oxidation products are subsequently separated from the reaction medium through a solvent extraction method [31]. In the absence of additional molecules in the reaction medium, hydroxyl radicals are recombined, and oxygen is evolved [11]:

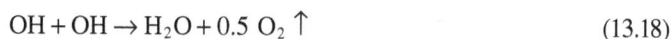

$$OH + OH \rightarrow H_2O + 0.5\ O_2 \uparrow \qquad (13.18)$$

SPEC systems can also be employed to resolve the H_2S issue in sour oil/gas industries. They can convert this corrosive hazardous gas into hydrogen clean fuel and a value-added semiconductor or sulfur in an economic and safe manner [13, 32, 33]. In addition, these systems can be applied in the development of green technologies and conversion of atmospheric pollutants such as SO_x to sulfuric acid inside a standalone PEC reactor [21].

At the end, it should be highlighted that besides photosynthesis and production of H-based solar fuels/chemicals, semiconductor photoelectrochemistry has also been employed in electricity generation and fabrication of PEC solar cells and photovoltaics. Depending on the type of semiconductor and band structure engineering, it is possible to make efficient photocathodes/photoanodes and ultimately fabricate high-performance solar cells [34–37].

REFERENCES

1 J. O'M. Bockris, A. K. N. Reddy, *Modern Electrochemistry 2B: Electrodics in Chemistry, Engineering, Biology and Environmental Science*. Kluwer Academic Publishers, New York, 2004 (Ch. 10).

2. M. Lashgari, "Fundamental aspects of CO_2 transformation into C/H/O based fuels/chemicals" in *Nanomaterials for CO_2 Capture, Storage, Conversion and Utilization*. (eds. P. N. Tri, H. Wu, T. A. Nguyen, S. Barnabé, P. Bénard). Elsevier, Amsterdam, 2021 (pp. 283–305)

3. M. S. Silberberg, *Principles of General Chemistry*. McGraw-Hill Higher Education, Boston, 2007 (Ch. 11).

4. I. N. Levine, *Quantum Chemistry*. Pearson, Boston, 2013 (Ch. 17).

5. R. Dovesi, B. Civalleri, C. Roetti, V. R. Saunders, R. Orlando, Ab initio quantum simulation in solid state chemistry, *Rev. Comput. Chem.* 21 (2005) 1–125.

6. P. Kratzer, J. Neugebauer, The basics of electronic structure theory for periodic systems, *Front. Chem.* 7 (2019) 106.

7. M. Lashgari, P. Elyas-Haghighi, M. Takeguchi, A highly efficient pn junction nanocomposite solar-energy-material [nano-photovoltaic] for direct conversion of water molecules to hydrogen solar fuel, *Sol. Energy Mater. Sol. Cells* 165 (2017) 9–16.

8. M. Lashgari, S. Soodi, CO_2 conversion into methanol under ambient conditions using efficient nano-composite photocatalyst/solar-energy materials in aqueous medium, *RSC Adv.* 10 (2020) 15072–15078.

9. V. C. Anitha, A. N. Banerjee, S. W. Joo, Recent developments in TiO_2 as n- and p-type transparent semiconductors: Synthesis, modification, properties, and energy-related applications, *J. Mater. Sci.* 50 (2015) 7495–7536.

10. A. Živković, N. H. de Leeuw, Exploring the formation of intrinsic p-type and n-type defects in CuO, *Phys. Rev. Mater.* 4 (2020) 074606.

11. M. Lashgari, P. Zeinalkhani, Ammonia photosynthesis under ambient conditions using an efficient nanostructured FeS_2/CNT solar-energy-material with water feedstock and nitrogen gas, *Nano Energy* 48 (2018) 361–368.

12. M. Lashgari, P. Zeinalkhani, Electrostatic promotion of the catalyst activity for ammonia photosynthesis upon a robust affordable nanostructured uni-electrodic photodevice/reactor, *Catal. Sci. Technol.* 10 (2020) 7998–8004.

13. Z. Zhang, J. T. Yates Jr, Band bending in semiconductors: Chemical and physical consequences at surfaces and interfaces, *Chem. Rev.* 112 (2012) 5520–5551.

14. M. Lashgari, M. Ghanimati, An excellent heterojunction nanocomposite solar-energy material for photocatalytic transformation of hydrogen sulfide pollutant to hydrogen fuel and elemental sulfur: A mechanistic insight, *J. Colloid Interface Sci.* 555 (2019) 187–194.

15. S. W. Boettcher, S. Z. Oener, M. C. Lonergan, Y. Surendranath, S. Ardo, C. Brozek, P. A. Kempler, Potentially confusing: Potentials in electrochemistry, *ACS Energy Lett.* 6 (2020) 261–266.

16. R. G. Pearson, The electronic chemical potential and chemical hardness, *J. Mol. Struct. (THEOCHEM)* 255 (1992) 261–270.

17. P. W. Atkins, J. de Paula, *Atkins' Physical Chemistry*. Oxford University Press, Oxford, 2006 (Ch. 21).

18. R. Memming, *Semiconductor Electrochemistry*. Wiley-VCH, Weinheim, 2015 (Ch. 5).

19. K. Gelderman, L. Lee, S. W. Donne, Flat-band potential of a semiconductor: Using the Mott–Schottky equation, *J. Chem. Educ.* 84 (2007) 685–688.

20. M. Lashgari, S. Naseri-Moghanlou, T. Khanahmadlou, R. Hempelmann, Electrostatic boosting of ionic dye pollutant removal from aquatic environment using a single electrode photoreactor, *NPJ Clean Water* 6 (2023) 10.

21. M. Lashgari, S. Afshari, M. Ghanimati, J. Seo, SO_2 pollutant conversion to sulfuric acid inside a standalone photoelectrochemical reactor: A novel, green, and safe strategy for H_2SO_4 photosynthesis, *J. Ind. Eng. Chem.* 121 (2023) 529–535.

22. M. Lashgari, M. Ghanimati, Pollutant photo-conversion strategy to produce hydrogen green fuel and valuable sulfur element using H_2S feed and nanostructured alloy photocatalysts: Ni-dopant effect, energy diagram and photo-electrochemical characterization, *Chem. Eng. Res. Des.* 162 (2020) 85–93.

23. M. Lashgari, P. Zeinalkhani, Photocatalytic N_2 conversion to ammonia using efficient nanostructured solar-energy-materials in aqueous media: A novel hydrogenation strategy and basic understanding of the phenomenon, *Appl. Catal. A Gen.* 529 (2017) 91–97.

24. M. Lashgari, S. Soodi, P. Zeinalkhani, Photocatalytic back-conversion of CO_2 into oxygenate fuels using an efficient ZnO/CuO/carbon nanotube solar-energy-material: Artificial photosynthesis, *J. CO2 Util.* 18 (2017) 89–97.

25. H. Tominaga, M. Nagai, Mechanism of thiophene hydrodesulfurization on clean/sulfided β-Mo_2C (001) based on density functional theory—cis- and trans-2-butene formation at the initial stage, *Appl. Catal. A Gen.* 343 (2008) 95–103.

26. P. Zheng, A. Duan, K. Chi, L. Zhao, C. Zhang, C. Xu, Z. Zhao, W. Song, X. Wang, J. Fan, Influence of sulfur vacancy on thiophene hydrodesulfurization mechanism at different MoS_2 edges: A DFT study, *Chem. Eng. Sci.* 164 (2017) 292–306.

27. M. Lashgari, D. Matloubi, Atomistic understanding of hydrogen loading phenomenon into palladium cathode: A simple nanocluster approach and electrochemical evidence, *J. Chem. Sci.* 127 (2015) 575–581.

28. M. Lashgari, M. Ghanimati, Efficient mesoporous/nanostructured Ag-doped alloy semiconductor for solar hydrogen generation, *J. Photonics Energy* 4 (2014) 044099.

29. S. Gligorovski, R. Strekowski, S. Barbati, D. Vione, Environmental implications of hydroxyl radicals (•OH), *Chem. Rev.* 115 (2015) 13051–13092.

30. C. P. Huang, C. Dong, Z. Tang, Advanced chemical oxidation: Its present role and potential future in hazardous waste treatment, *Waste Manage* 13 (1993) 361–377.

31. I. Shafiq, S. Shafique, P. Akhter, G. Abbas, A. Qurashi, M. Hussain, Efficient catalyst development for deep aerobic photocatalytic oxidative desulfurization: Recent advances, Confines, and outlooks, *Catal. Rev.* 64 (2022) 789–834.

32. M. Lashgari, M. Ghanimati, Photocatalytic degradation of H_2S aqueous media using sulfide nanostructured solid-solution solar-energy-materials to produce hydrogen fuel, *J. Hazard. Mater.* 345 (2018) 10–17.

33. M. Lashgari, M. Sabeti-Khabbazmoayed, M. Konsolakis, A cost-effective H_2S pollutant electro-transformation to hydrogen clean fuel and value-added semiconducting materials: A green alternative to Claus process, *J. Ind. Eng. Chem.* 122 (2023) 326–333.

34. R. N. Pandey, K. C. Babu, O. N. Srivastava, High conversion efficiency photoelectrochemical solar cells, *Prog. Surf. Sci.* 52 (1996) 125–192.

35. P. V. Kamat, K. Tvrdy, D. R. Baker, E. J. Radich, Beyond photovoltaics: Semiconductor nanoarchitectures for liquid-junction solar cells, *Chem. Rev.* 110 (2010) 6664–6688.

36. M. Lashgari, N. Shafizadeh, P. Zeinalkhani, A nanocomposite p-type semiconductor film for possible application in solar cells: Photo-electrochemical studies, *Sol. Energy Mater. Sol. Cells* 137 (2015) 274–279.

37. S. D. Tilley, Recent advances and emerging trends in photo-electrochemical solar energy conversion, *Adv. Energy Mater.* 9 (2019) 1802877.

14 Semiconductors for Transparent Conductors

*Jnanraj Borah, Dipak Barman,
and Bimal K. Sarma*

14.1 INTRODUCTION

The inherently high electrical conductivity is generally exhibited by metals, which is credited to the presence of a sea of free electrons with optimum mobility. High carrier density in metals also enables high plasma frequency limiting the transparency to the visible light [1, 2]. Transparent conductors find widespread applications in a diverse array of optoelectronic devices. Transparent conductors are equally vital in display technologies, serving as transparent electrodes in organic light-emitting diode (OLED) displays, liquid crystal displays (LCDs), and electroluminescent displays, ensuring that these screens produce clear and vibrant images. Some of the important applications are presented in Figure 14.1a. These electrodes collect the electricity generated by the solar cells while permitting sunlight to transmit and reach the active layer for energy conversion, making solar energy capture efficient. Moreover, in the pursuit of energy-efficient buildings, transparent conductors are deployed in smart windows and electrochromic devices. These applications utilize the material's ability to control light transmission and regulate indoor temperatures. All these applications require components possessing coexistence of high electrical conductivity and visible light transparency [3–5]. Conceptually, wide band gap semiconductors are transparent in the region of interest and hence projected as possible candidates provided their electrical conductivity can be enhanced, which otherwise falls below the metallic regime [6]. This type of conductor is realized by methods such as doping or incorporating some carrier injecting system like metal nanoparticles, thin metallic layer (~5–10 nm), mesh made of metallic nanowires, etc. in the semiconductors to supply extra free carriers. A typical transparent conductor possesses a resistivity of ~10^{-3} $\Omega\cdot$cm or lower and a transparency of 80% or higher in the visible region [7]. A transparent conductor allows light with frequencies higher than the plasma frequency to pass through, rendering them transparent at higher frequencies (Figure 14.1b) [8]. Figure 14.1c displays the spectral behavior of wide band gap degenerate semiconductors in the ultraviolet (UV), visible, and near infrared (NIR). It illustrates key characteristics of the material: fundamental absorption in the UV, high visible light transparency, and free carrier absorption in the NIR.

Recently, there was a surge in research and development related to transparent conductors, marked by a substantial increase in the research publications and patents and a broader range of materials and methodologies being explored. Metal oxide semiconductors are regarded as front runners in the race of components like TCO [9]. Currently, Sn-doped In_2O_3, commonly known as indium tin oxide (ITO), is commercially used, but it suffers from major drawbacks of scarcity, toxicity, brittleness, and inferior mechanical flexibility, and efforts are being made to find alternatives to ITO [10]. Graphene and Ag nanowires are promising transparent conductor candidates due to their favorable electrical conductivity, optical transparency, and flexibility. Ag nanowires, forming efficient networks, provide flexibility, making them suitable for emerging technologies like flexible displays and solar panels [11, 12]. Conductive polymers like poly(3,4-ethylenedioxythiophene)-polystyrene sulfonate (PEDOT:PSS) deserve mention here, combining transparency and electrical conductivity for use in organic electronics [13].

DOI: 10.1201/9781003450146-14

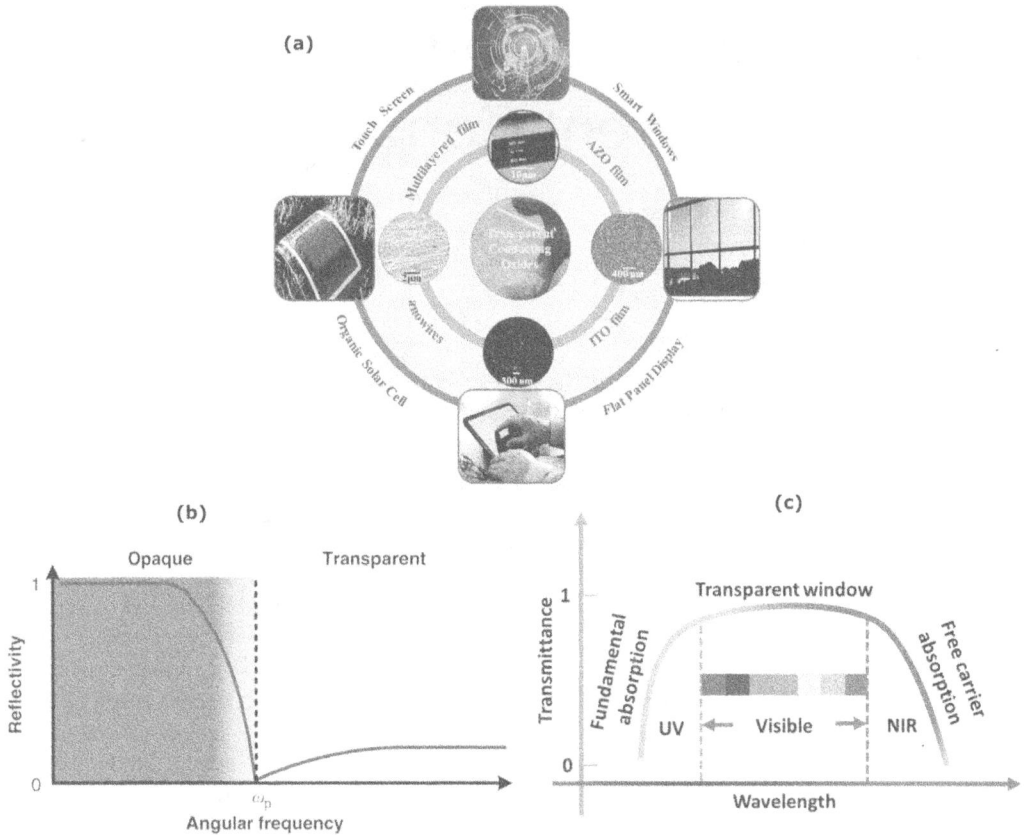

FIGURE 14.1   (a) Different applications of transparent conductors. Figure 14.1a adapted with permission from [5]. Copyright the Authors, some rights reserved; exclusive licensee [RSC publishing]. Distributed under a Creative Commons Attribution License 3.0 (CC BY). (b) Reflectivity as a function of frequency of electro-magnetic waves. Figure 14.2b adapted with permission from [8]. Copyright the Authors, some rights reserved; exclusive licensee [Springer Nature]. Distributed under a Creative Commons Attribution License 4.0 (CC BY). (c) Spectral response of a wide band gap semiconductor-based TCO.

The challenges like fragility, oxidation susceptibility, high sheet resistance, and complex syn-thesis accompany these materials. Al-doped ZnO (AZO) stands out for its economic, eco-friendly, and electro-optical performances, making it an appealing alternative to ITO in transparent conduc-tors. Various deposition techniques ensure precise deposition of transparent conductors on diverse surfaces [14]. This review delves into semiconductor characteristics tailored for transparent conduc-tors, covering electronic structures, synthesis methods, microstructure, electro-optical behavior, aging stability, and mechanical flexibility of ITO and AZO. The chapter concludes with emerging applications of TCOs.

14.2   ELECTRONIC STRUCTURE OF A TCO

Here, In_2O_3 and ZnO are mainly considered, although there exists a multitude of TCO materials. In_2O_3 adopts the bixbyite structure, characterized by a close-packed arrangement of O and In atoms situated at both sixfold and fourfold interstitial positions. The overall symmetry of this structure is cubic, albeit with a notably large unit cell comprising 40 atoms [15]. The band structure for In_2O_3 is shown in Figure 14.2a. The first accurate calculations for In_2O_3 were conducted by Shigesato et al. followed by subsequent work by Mi et al., and Mryasov and Freeman [16–18]. In_2O_3 possesses

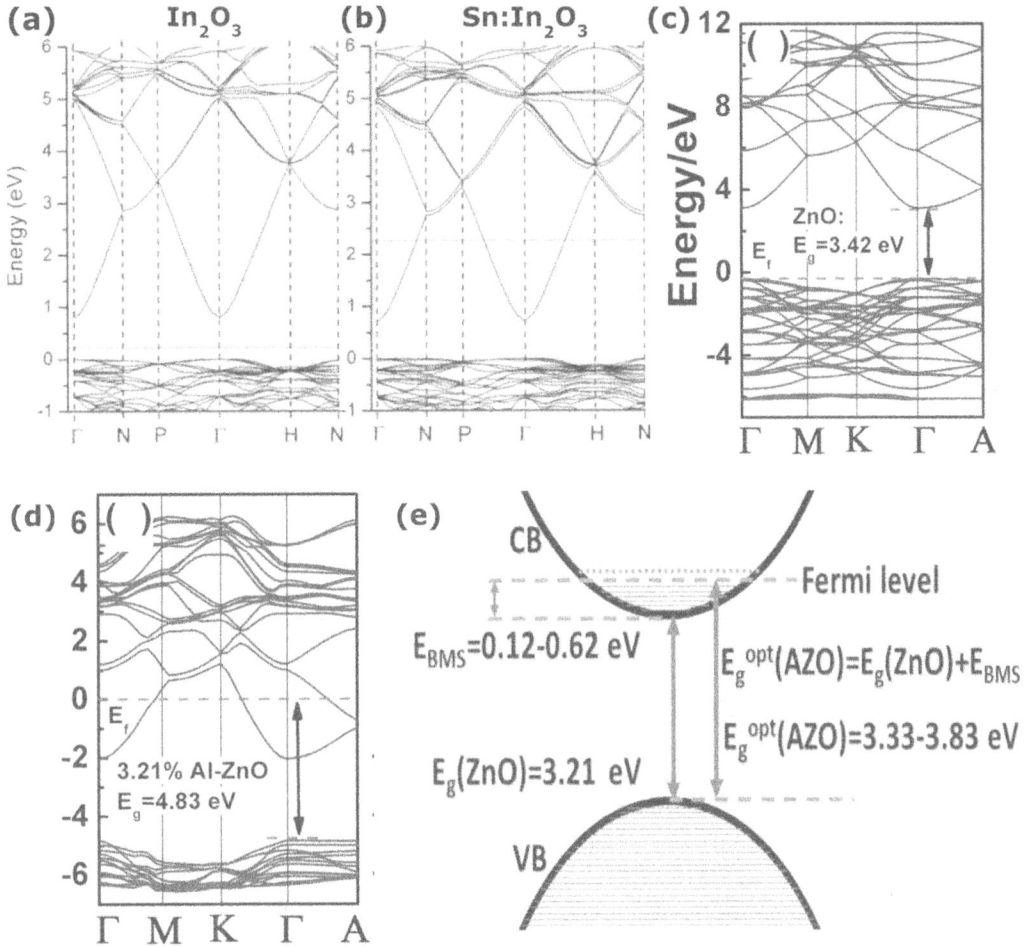

FIGURE 14.2 Band structures of (a) In_2O_3, (b) Sn: In_2O_3. (Adapted with permission [15], Copyright [2009], American Physical Society.) (c) ZnO and (d) Al:ZnO (AZO). (Adapted with permission from [16]. Copyright the Authors, some rights reserved; exclusive licensee [Springer Nature]. Distributed under a Creative Commons Attribution License 4.0 [CC BY].) (e) Schematic representation of the increased optical band gap of AZO owing to the Burstein-Moss effect. (Adapted with permission [23], Copyright [2021], Elsevier.)

transparency to visible light but exhibits robust optical absorption in the ultraviolet region, characterized as a direct and wide band gap of 3.7 eV. Interestingly, this specific value, although erroneously, has been extensively disseminated as the primary band gap of In_2O_3. Furthermore, there is a notably weaker absorption at lower energy levels, which effectively obstructs the blue portion of the visible light spectrum when dealing with thick films or bulk substrates. This lower energy absorption has historically been interpreted as an indirect band gap measuring 2.6 eV, with the valence band maximum (VBM) positioned away from the Γ-point. Recent experiments employing X-ray photoelectron spectroscopy (XPS) challenged the previously widely accepted band gap value (derived from optical absorption in thin films), revising it to a range of 2.7 to 2.9 eV [15]. Angle-resolved photoelectron spectroscopy (ARPES) has ultimately substantiated a band gap of approximately 2.7 eV, precisely located at the Γ-point of the Brillouin zone. In line with this, recent ab initio calculations questioned the notion of a distinct indirect band gap. The observations indicate the potential existence of either a direct band gap or an indirect band gap, whereby the VBM is slightly raised by a mere 50 meV in comparison to its position at the Γ-point. The discord between optical absorption data and band structure data has been resolved by identifying a minimal dipole matrix toward the parity-restricted direct optical transition between the conduction band minimum (CBM) and the VBM at the Γ-point. The first observable strong optical transition occurs between the CBM and valence bands, positioned approximately 0.8 eV below the VBM. Whether the faint absorption initiation at 2.6 eV arises from the parity-prohibited direct band gap or represents a slightly indirect band gap remains a subject of ongoing and contentious debate [17, 18].

ZnO, another significant TCO material, historically used in ancient China for brass production, exhibits three structures: rock salt, zinc blende, and hexagonal wurtzite.

The hexagonal wurtzite structure is more prevalent due to its stability at ambient conditions [19]. A polar hexagonal axis is designated as the c-axis, parallel to the [001] direction. The O atoms are situated inside a tetrahedral group of four Zn atoms, while the Zn atoms are packed in close proximity to one another in a nearly hexagonal pattern. ZnO structure with space group P63mc have several planes in which Zn^{2+} and O^{2-} ions are tetrahedrally coordinated and alternately placed along the c-axis [19, 20]. ZnO has a direct band gap of 3.37eV, and its electronic structure is explored using the density functional theory (DFT) and DFT+U (Figure 14.2c). The band gap of ZnO exceeds 3.0 eV, depending on the method used. Experimental results align with calculated band structures, confirming ZnO as a direct band gap semiconductor [21, 22].

Doping involves adding impurities to semiconductors to alter their physical and chemical attributes. Sn doping in In_2O_3 increases free carrier density, boosting electrical characteristics, conductivity, and introducing plasmonic behavior [15]. Figure 14.2b illustrates the band gap alteration compared to In_2O_3. Undoped ZnO is highly transparent but lacks sufficient conductivity for next-gen electronics. The n-type doping with Group III elements increases the electron density while preserving visible light transparency. Ga or Al addition to ZnO significantly alters its electrical properties. The substitutional incorporation of Al in the ZnO lattice facilitates to tune the electro-optical properties required for different applications. Due to the trivalent nature of Al, doping adds additional electrons, ultimately raising the free carrier density (Figure 14.2d). The level of Al doping is adjusted as per the requirement of the end application. The band structure is highly dependent on the Al concentration and synthesis methods. The Fermi level of intrinsic ZnO is positioned deep below the conduction band. However, due to the introduction of Al as a dopant, the Fermi level moves upward and rises above the CBM upon heavy doping, and AZO becomes a degenerate semiconductor behaving like metals [23]. The upward shift of the Fermi level is known as the Burstein-Moss shift (BMS), as shown in Figure 14.2e. The free carrier density in AZO can reach as high as 10^{20} cm^{-3} [24, 25].

14.3 PREPARATIONS OF TCO

The methods used to grow TCO have a significant influence on their electro-optical properties. When considering two TCOs with the same chemical composition, their electrical and optical properties can vary depending on the deposition techniques employed. Magnetron sputtering yields smoother, more uniform films compared to alternative methods. There exist several deposition methods for the development of transparent conductors, and a brief elucidation of a few of these techniques is presented below.

14.3.1 SPRAY PYROLYSIS

The spray pyrolysis represents a commonly employed and cost-effective approach for creating thin films. TCO films are fabricated using the spray pyrolysis method. In this procedure, a thin film is crafted by spraying a solution onto a heated surface, where the elements within the solution undergo reactions to produce the desired thin solid film. This method proves highly efficient in the production of transparent conducting films. For details, readers may check the excellent review articles, and one such review is included in ref. [26].

14.3.2 Chemical Vapor Deposition

Chemical vapor deposition (CVD) is another prevalent method for depositing solid thin films. High-performance thin films in a vacuum environment can be prepared using CVD. Materials generated via the CVD method possess inherent high purity, as this method falls within the category of vapor transfer processes. Here, gaseous precursors decompose at the heated substrate, resulting in the deposition of the film. A detailed presentation of CVD in the context of TCO deposition is provided in ref. [27].

14.3.3 Pulsed Laser Deposition

Pulsed laser deposition (PLD) stands out as an effective approach for depositing TCO thin film coatings. However, this technique is less widely used as a commercial technique due to limited uniformity. Here, photon energy in the form of a laser characterized by pulse duration and frequency is absorbed by the target material, which gets ablated and deposited on a substrate. Interested readers can see reference [28] for details.

14.3.4 Sputtering

Sputtering is a widely utilized method for the deposition of thin films onto substrates. This technique relies on ion bombardment of a source material known as the target. The impact of ion bombardment leads to the liberation of target atoms and nucleation on a substrate. A practical method for growing thin films by sputtering involves the use of a magnetron source for confinement of plasma near the target. The bombarding ions can be energized by providing bias in various ways, spanning from direct current (DC) for conductive targets to pulsed DC and radio frequency (RF) for nonconductive targets. Sputtering offers advantages over methods like PLD and molecular beam epitaxy, especially in the manufacturing of large-area films for industrial applications. Synthesis pathways to high-quality transparent and conductive oxides through sputtering have been achieved, contributing to the scientific advancement as presented in refs. [10, 29, 30].

14.4 TCO AND MULTILAYER TRANSPARENT ELECTRODE

A variety of TCO materials are realized, with certain ones achieving widespread commercial success, while others remain in the research phase. However, as of now, transparent conductors with p-type conductivity having practical applications are not realized. NiO and Cu_2O are some very commonly speculated semiconductors expected to show p-type conductivity [31, 32]. The main challenge in achieving p-type conductivity is getting a favorable conduction path in the hole-dominated TCOs, which are formed by filled valence bands. The presence of almost no overlapping oxygen ions in the valence bands of oxides, ascribed to the fact that most of the filled 2p orbitals contain nonbonding states, limits the formation of the conduction paths for the carriers in usual conditions [32]. Hence, till now, the focus has been on developing n-type transparent conductors. The current scenario of In-based and ZnO-based TCOs is presented in the following sub-sections including mechanically flexible multilayer transparent electrodes.

14.4.1 Indium-based TCOs

A crucial element in all flat-panel displays, ranging from basic liquid-crystal displays found in calculators to expensive liquid-crystal color screens, is the electrode responsible for governing the alignment of liquid-crystal molecules. These electrodes, in turn, determine whether light gets easy access or is blocked, essentially controlling the on-off state of the display. It is essential for at least one of these electrodes to be transparent. This transparency is imperative as either the surrounding ambient light or transmitted light, such as backlighting, needs to traverse the device and arrive at the viewer's sight. The favored choice for this transparent conductor has been ITO. Nonetheless, alternative transparent conductors like SnO_2 are also available. ITO is predominantly preferred within this category of materials because it boasts an achievable level of high transparency for visible light while simultaneously exhibiting low electrical resistivity. As a result, it finds applications not only in flat-panel displays but also in energy-efficient window coatings. Through appropriate processing methods, the free carrier density in ITO can be increased to approximately $1–2 \times 10^{21}$ cm^{-3} by heavily doping it with Sn at levels ranging from 5 to 10 at.%. These free carriers originate from two distinct sources within the material: substitutional Sn ions and oxygen vacancies. When it comes to producing top-quality thin films of ITO, the versatile methods are magnetron sputtering and activated electron beam (EB) evaporation. Films prepared by these techniques typically exhibit a resistivity in the range of $1.5–2.0 \times 10^{-4}$ $\Omega \cdot cm$ [33, 34].

The current change on the road to enhance the performance electrodes in flat-panel devices has given rise to novel display technologies characterized by significantly thinner electrodes. Consequently, there is an increased demand for further refinement in the properties of ITO and the conditions under which it is processed. These cutting-edge and sophisticated technologies necessitate even lower film resistivities, aiming for levels as low as 1×10^{-4} $\Omega \cdot cm$. Achieving this requires the use of lower substrate temperatures during the deposition process. The need for reduced deposition temperatures is primarily driven by the emergence of newer device designs, such as flexible displays, which require the deposition of ITO films onto polymer substrates incapable of withstanding vacuum processing temperatures exceeding 100°C. Meeting this requirement calls for the development of fresh approaches to enhance traditional deposition techniques. The DC magnetron sputtering through amplifying the magnetic field intensity at the cathode or introducing the RF discharge can be vital in attaining deposition on polymeric substrates. The use of tungsten electron emitters or arc plasma generators may further enhance EB evaporation, leading to plasma-assisted EB evaporation with a high density, resulting in high-density plasma-assisted EB evaporation [33]. Amorphous ITO films can only be reliably deposited at high total gas pressures, typically in the range of 3–5 Pa, while maintaining a substrate temperature around room temperature. Presently, amorphous indium zinc oxide (IZO) films with a fully amorphous structure can be produced through DC magnetron sputtering using an oxide target comprising 89.3% In_2O_3 and 10.7% ZnO by weight [18]. This process offers high reproducibility across a wide range of deposition conditions, including total gas pressure and substrate temperatures of up to 300°C. As a result, IZO is emerging as a promising candidate for transparent electrodes, particularly in applications such as thin-film transistor LCDs or OLEDs [33, 35].

14.4.2 ZnO-based TCOs

Although ITO has enjoyed immense success and widespread use as a TCO material thanks to its outstanding electro-optical characteristics, the escalating demand for ITO as a TCO thin film has driven up its price. Simultaneously, the availability of indium, a key component of ITO, is diminishing. Additionally, concerns about the toxicity associated with indium pose significant challenges related to ITO, especially for large-scale uses. ZnO, characterized by its cost-effectiveness and eco-friendliness as a wide band gap semiconductor (> 3.0 eV), emerges as a promising alternative to ITO. ZnO, in its pristine, stoichiometric state, boasts impressive transparency to visible light but grapples with electrical resistivity. An alternative strategy involves harnessing the intrinsic defects in ZnO, which induce n-type conductivity through oxygen vacancies. However, the conductivity stemming from these native vacancies falls short of the requisites for a TCO [29, 30]. ZnO is composed of elements that are abundantly available, with Zn and O making up 132 ppm and 49.4%, respectively, of the Earth's crust. This abundance is crucial for large-scale applications of ZnO, such as its use as transparent electrodes in thin-film solar cells or flat-panel displays. It also presents an advantage over ITO as In has a much lower abundance in the Earth's crust, at only 0.1 ppm [36]. Therefore, obtaining the required low resistivity, often referred to as degenerate doping, to make ZnO films suitable for use as transparent electrodes can be achieved through two separate approaches [37]:

i. Generating intrinsic donors via lattice imperfections, such as oxygen vacancies or the presence of zinc atoms occupying interstitial lattice positions.
ii. Introducing extrinsic dopants involves incorporating foreign elements, such as metals with an oxidation state of three substituting for metal lattice positions.

However, it is worth noting that films doped intrinsically are not well-suited for practical applications. Firstly, their resistivity typically falls within the range of 10^{-2} to 10^{-3} $\Omega\cdot$cm. Additionally, these films exhibit poor stability, especially at elevated temperatures. This instability is attributed to the reoxidation of the films, which are oxygen-deficient or enriched in zinc, leading to a significant increase in resistivity [38].

It is essential to heavily dope ZnO with Group III elements like Al and Ga, and studies indicate that AZO films offer economic and environmental viabilities, while achieving high conductivity through a significant increase in the concentration of free charge carriers. Interestingly, the visible light transmittance of AZO films remains nearly unchanged compared to pure ZnO films. Optimal doping concentration maintains transparency without introducing additional states that would absorb visible light in this degenerate semiconductor [23, 39]. A study, conducted by the authors revealed high-quality indium-free TCO films primarily composed of ZnO with Al as the sole dopant [29]. AZO transparent conductors with appealing electro-optical properties were developed using AZO sputtering targets, incorporating dopant concentrations corresponding to 0.5–5.0 wt.% Al_2O_3, which involves subjecting the surface of the AZO target to high-energy Ar ions produced through a low-pressure gas discharge utilizing a magnetron cathode (Figure 14.3a). Microstructural and morphological properties unveil highly crystalline films with dense surfaces (Figures 14.3b and 14.3c). Optimal electro-optical characteristics include a sheet resistance of 2.3 Ω/sq, visible light transmittance exceeding 90%, and a figure of merit of 75.9 $m\Omega^{-1}$ for AZO thin films. The resistivity of AZO films depends on dopant concentrations, with the lowest value measured at 1.66×10^{-4} $\Omega\cdot$cm. Comparing the figure of merit of AZO films with that of ITO suggests potential applications for AZO as a TCO. The high transparency of AZO films in the near-UV spectral region is limited by strong fundamental absorption. In another study, AZO thin films were investigated as potential alternatives to ITO for transparent electrodes in optoelectronic devices [30]. The deposition of AZO thin films was carried out using pulsed DC magnetron sputtering, with variations in sputtering power and pulse frequencies. These AZO thin films exhibit exceptional transparency in the visible light range but limited transparency in the NIR region due to free carrier absorption (Figure 14.3g).

The sheet resistance falls within the range of 9–45 Ω/sq, which is acceptable for TCOs to qualify for their use in optoelectronics. The outstanding figure of merit and stability exhibited by AZO thin films make them well-suited for serving as transparent electrodes in photovoltaic applications. It was reported [31] that beyond 1.0 wt.% of Al_2O_3 as the dopant material, a notable shift of the onset of band-to-band absorption occurs, primarily due to the BMS effect (Figures 14.3h and 14.3i). Bulk plasmon frequencies fall in the NIR, affecting the transmittance of AZO films for light frequencies below that of visible light. This significant free carrier absorption suggests the existence of a conduction electron gas upon doping, making AZO a promising material for infrared plasmonic applications. Flexible electronics require a transparent conductor, which should maintain suitable electro-optical properties while being flexible. Typically, dielectric/metal/dielectric multilayers display transparency, conductivity, and stability. Kim and co-workers have created a transparent conducting electrode using ZnO/Ag nanowire (NW)/ZnO multilayers with a sheet resistance of 8.0 Ω/sq demonstrating exceptional thermal stability and flexibility (Figures 14.4a–14.4c) [39]. A highly flexible transparent electrode composed of AZO/Ag nanoparticles (NP)/AZO with very low resistivity (3.3×10^{-5} $\Omega\cdot$cm) was developed in one go for three-step deposition by magnetron sputtering without exposure to ambient oxidizing atmosphere [40]. These films exhibit excellent

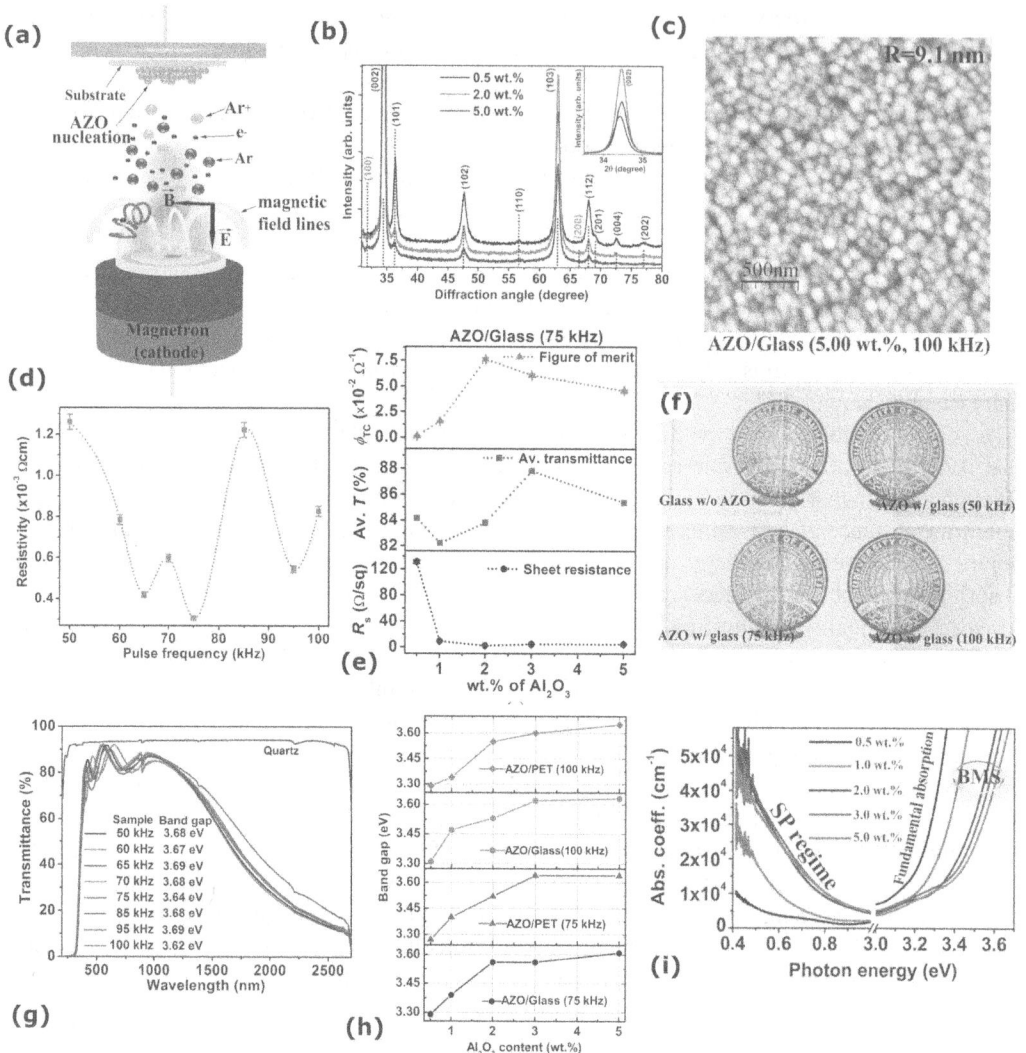

FIGURE 14.3 (a) A schematic of the AZO target surface bombarded by high-energy Ar ions generated through low-pressure gas discharge utilizing a magnetron cathode. (b) XRD patterns of sputtered AZO thin films with varying dopant content. (c) Two-dimensional atomic force microscopy image of an AZO film. (d)–(e) Resistivity, sheet resistance, average transmittance, and figure of merit of AZO TCO films prepared under different conditions. (f) Digital photographs of the glass substrate (without AZO) and AZO coated glass. (g) UV-Vis-NIR transmission spectra of AZO deposited on quartz. (h) Dependence of optical band gaps of AZO on the dopant content. (i) Absorption coefficients of AZO films showing absorption in the UV and NIR. (Adapted with permission [29], [30], Copyright [2019, 2020], Elsevier.)

morphology and bending stability, with no visible microcracks even after few hundreds of bending cycles (Figures 14.4d–14.4m).

A significant concern with AZO is its limited electrical stability in an ambient oxidizing environment. During the deposition, unwanted defects and impurities may be introduced, potentially altering the electro-optical properties of AZO films. The density of defects and surface imperfections in AZO films are influenced by the growth methods and, more importantly, by the deposition and post-deposition conditions. The authors conducted an extensive study to create aging stable AZO transparent conductors by pulsed DC magnetron sputtering [23]. The achievement of a high figure

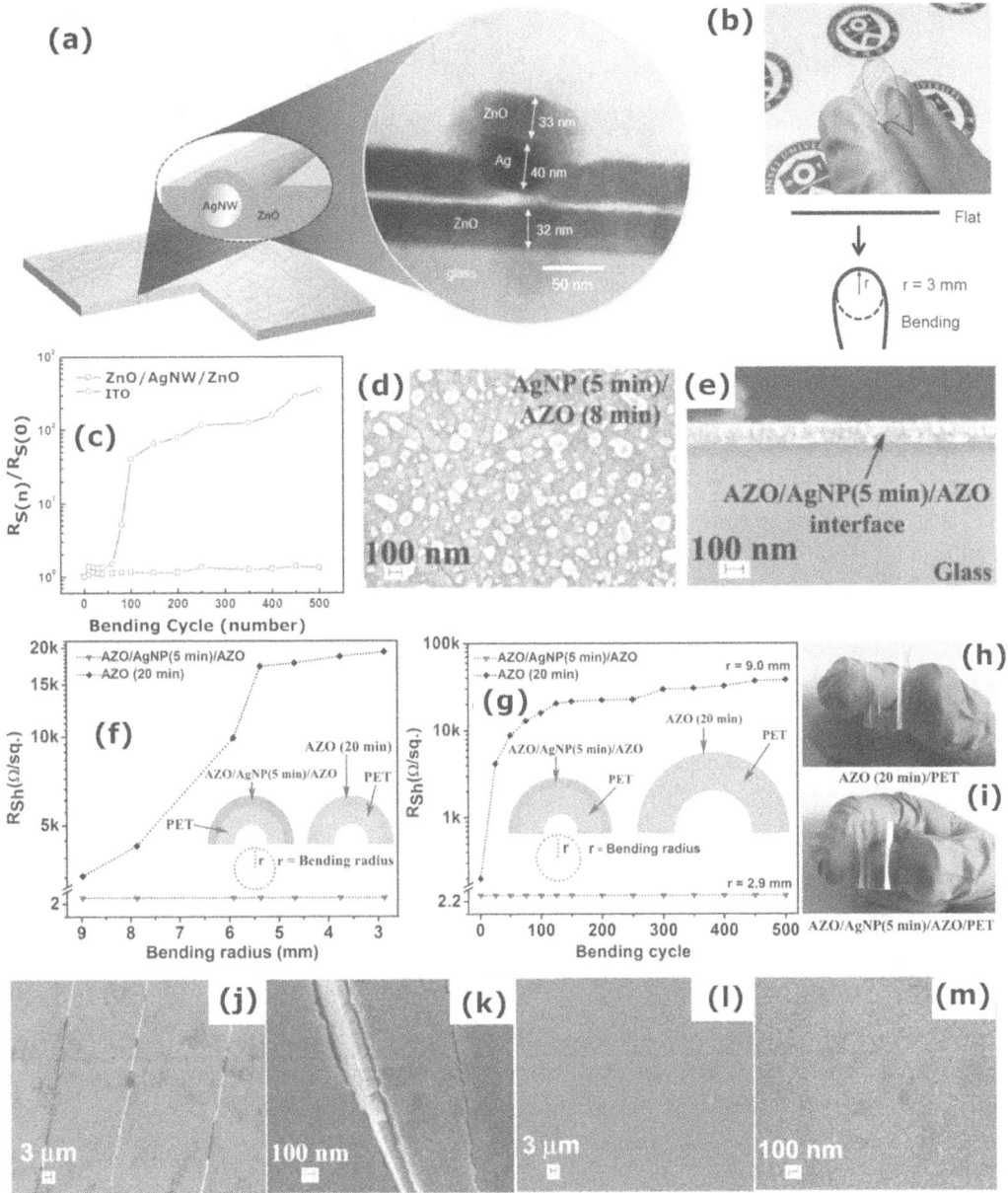

FIGURE 14.4 (a) Schematic of ZnO/AgNW/ZnO multilayer transparent conductor. (b) Digital photograph and (c) mechanical bending result of ZnO/AgNW/ZnO multilayer transparent conductor. (Adapted with permission [39], Copyright [2013], American Chemical Society.) (d, e) Surface morphology and cross-sectional view of AZO/AgNP/AZO multilayer composite structure. (f)-(i) mechanical bending test results and digital photographs of AZO and AZO/AgNP/AZO multilayer film. Surface morphology of (j, k) single layer AZO, (l, m) AZO/AgNP/ AZO multilayer composite structure after bending. (Adapted with permission [40], Copyright [2020], Elsevier.)

of merit, typically on the order of 10^{-2} Ω^{-1}, which results from the favorable balance between low electrical resistivity and high optical transparency, is significantly influenced by both the sputtering process conditions and the levels of doping content.

The electrical stability of the TCO films is enhanced by their highly crystalline microstructure, characterized by a uniform morphology and highly regular columnar growth (Figures 14.5a, 14.4b).

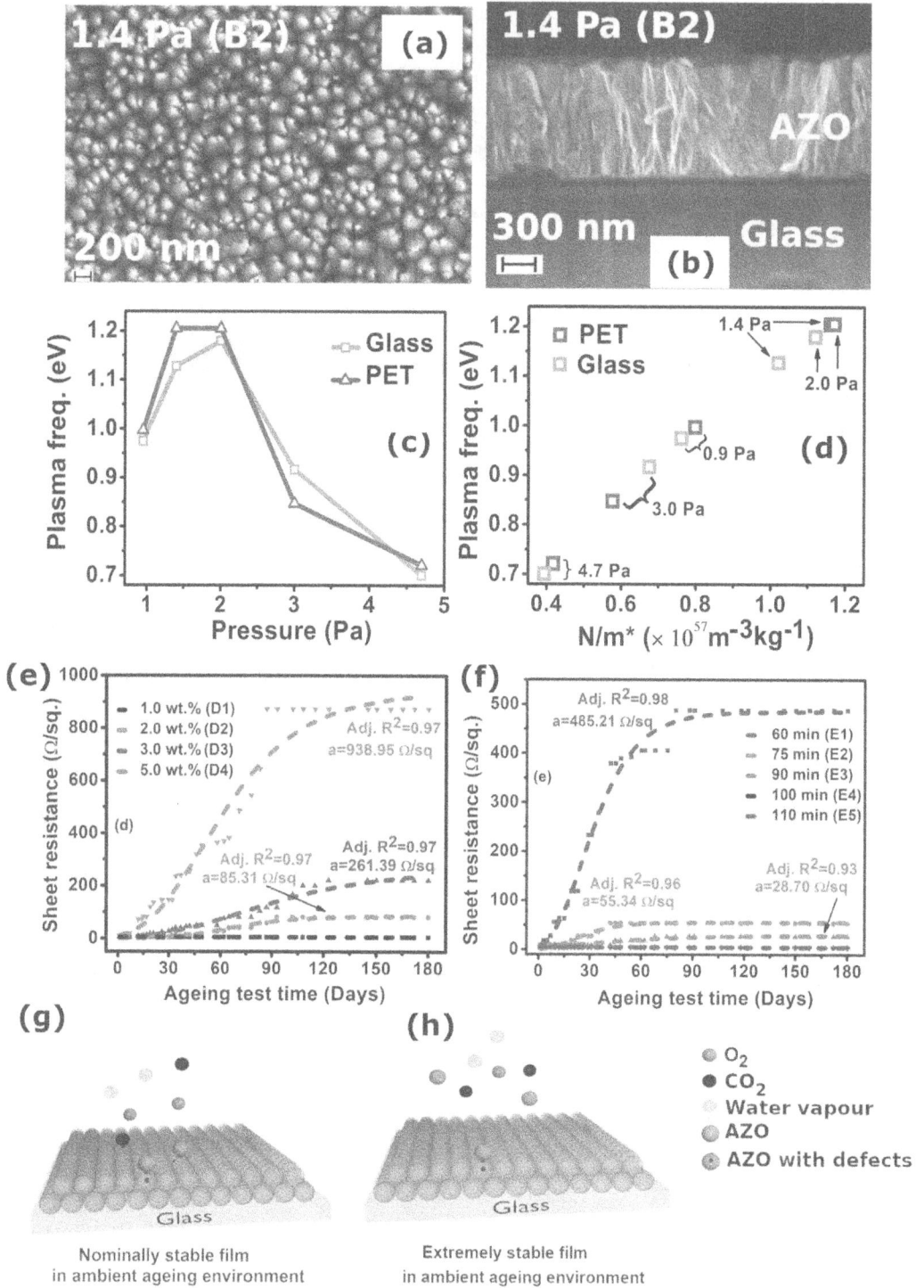

FIGURE 14.5 (a)–(b) Morphology and cross-sectional view of AZO TCO films grown on glass. (c) Plot of plasma frequency against working pressure. (d) Variations of plasma frequency with the carrier concentration to the effective mass ratio in AZO films. (e, f) Results of an aging stability study of sheet resistance of AZO coated glass in an ambient oxidizing environment. (g, h) Schematic representation of adsorption centers and its linkage with aging stability of AZO. (Adapted with permission [23], Copyright [2021], Elsevier.)

An insight into the plasma frequency is presented in Figures 14.5c and 14.5d, which suggests free carrier absorption in the NIR, and it is tunable by controlling the effective carrier concentration by the deposition conditions, like sputtering pressure. The influence of interface chemistry on achieving aging stable AZO films is critically examined and supported by the adsorption kinetics of cationic and anionic dyes at the AZO surface. The results shed light on the intricate connection between the excellent stability of AZO films, the specific process conditions employed, and the dopant material content in the AZO sputtering targets. The findings affirm that the sheet resistances of particular AZO films remain unchanged, consistently measuring at 4.5 Ω/sq, even after an aging period extending over 180 days of aging (Figures 14.5e–14.5h). Surface roughness, defects, and impurities serve as potential adsorption sites for molecules like O_2, CO_2, and water vapor. This, in turn, leads to a gradual rise in the sheet resistance of AZO films over time.

14.5 APPLICATIONS OF TCOs

14.5.1 Transparent Heaters

TCOs can be used to convert electric energy to thermal energy where the temperature is controlled by the voltage applied to the material. These electrical heaters find applications in many important areas like temperature control in industrial processes, heating of microchannel chips, and defogging surfaces. TCO coatings need to have either very low resistance (~ 1.0 Ω/sq) or a very high voltage power source. TCO heaters are used as defrosters in aircraft windshields, as they are advantageous over traditional hot air blowers for faster defrosting and uniform heating over larger areas [41]. Figures 14.6a–14.6c nicely illustrate the use of ITO as a transparent heater.

14.5.2 Solar Cells

TCOs play a pivotal role in the domain of solar cell technology, exerting a profound impact on the efficiency and functionality of photovoltaic systems. In particular, ITO has emerged as a prominent choice for this function due to its exceptional transparent plus conductive properties. A representative use of transparent electrodes in solar cells is presented in Figures 14.6d–14.6h, and ZnO/AgNW/ZnO emerges as a suitable front electrode in the tandem structure [39]. Moreover, TCOs can be intricately engineered with nanostructures or textured surfaces to optimize light entrapment within the solar cell. This technique is employed to enhance light collection ability of a solar cell, which improves the power conversion efficiency. TCOs are indispensable in the next generation flexible and thin-film solar cells, enabling the creation of lightweight and bendable solar panels that seamlessly integrate into curved and flexible substrates. Additionally, TCOs are progressively penetrating emerging solar cell technologies like perovskite solar cells and organic photovoltaics.

14.5.3 Photodetectors

Photodetectors, designed to convert incident light into electrical signals, benefit extensively from TCO integration. TCOs predominantly serve as transparent conductive electrodes in photodetectors, functioning as essential front and back contacts that permit light penetration while efficiently collecting and transporting electrical charges generated during the photon absorption process. Materials like ITO and AZO are preferred due to their exceptional blend of electrical and optical attributes (Figures 14.6i–14.6k). Furthermore, engineered TCO surfaces with textures and nanostructures enhance light absorption, mitigating reflection and scattering, thereby improving photodetector efficiency, particularly in low-light conditions. Flexible photodetectors benefit from TCOs, a boon for wearable technology and curved displays. A transparent electrode is a vital component that augments the performance, sensitivity, and adaptability of photodetectors across diverse applications [42].

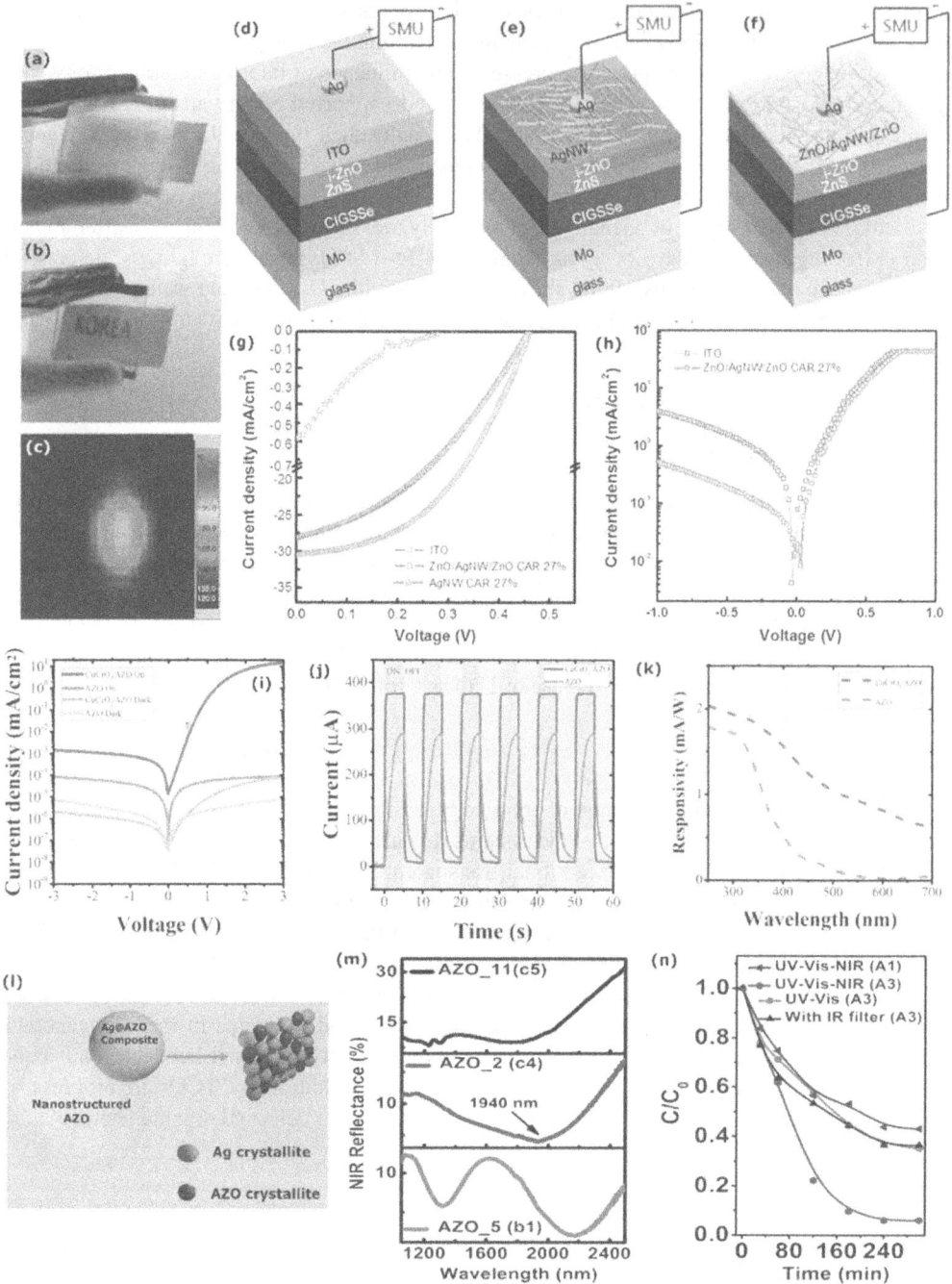

FIGURE 14.6 An ITO-based transparent heater lying on a block of dry ice (a) before and (b) after applying voltage and (c) its thermal image after voltage applied. (Adapted with permission from [41], Copyright [2010], Elsevier.) (d)–(h) Schematic structures and current density vs. voltage characteristics of solar cells with transparent electrodes. (Adapted with permission from [39], Copyright [2013], American Chemical Society.) (i)–(k) J-V characteristics, current-time measurement, and responsivity of $CuCrO_2$ photovoltaic and photoconductive photodetectors using AZO electrodes. (Adapted with permission from [42]. Copyright the Authors, some rights reserved; exclusive licensee [Springer Nature]. Distributed under a Creative Commons Attribution License 4.0 [CC BY].) (l)–(n) Schematic of possible arrangement of AZO and Ag crystallites, reflectance spectra in NIR, and photocatalysis results of AZO and AZO-Ag@AZO nanocomposite-based broadband photocatalyst. (Adapted with permission [44], Copyright [2023], Elsevier.)

14.5.4 SEMICONDUCTOR PLASMONICS

Transparent conductors have fascinating applications in the field of semiconductor plasmonics. TCOs assume a vital role as integral components, facilitating the manipulation of surface plasmon polaritons (SPPs) and contributing significantly to the advancement of sophisticated optoelectronic devices.

TCO nanocrystals can be employed to engineer plasmonic nanostructure, which enables the localization of optical fields within subwavelength regions, thereby improving the efficiency of various processes such as light absorption, emission, and nonlinear optics. These advancements hold a great promise for the development of exceptionally efficient photodetectors, photocatalysts, and light-emitting devices [43]. For instance, AZO-Ag@AZO broadband photocatalysts are highly sensitive to NIR light as depicted in Figures 14.6l–14.6n. A series of photocatalysis experiments reveal the role of NIR photons exciting NIR plasmonics for the efficient degradation of rhodamine 6G dye [44].

14.6 CONCLUSION

Semiconductors for transparent conductors represent a dynamic and evolving field with far-reaching implications for various technological applications. This chapter has explored the fundamental principles, material properties, and fabrication techniques that underpin the development of TCOs and their integration into optoelectronic devices. The indispensable use of transparent conductors in solar cells to photodetectors, displays, and beyond have reshaped the landscape of transparent electronics [45]. The pursuit of novel materials and innovative fabrication methods continues to drive progress in this field, promising even higher performance, lower costs, and sustainability. Future endeavors will undoubtedly focus on addressing the challenges of reducing material costs, enhancing electrical and optical performance, and broadening the range of applications especially in the field of miniaturized devices and sensors.

REFERENCES

1. J. Gao, K. Kempa, M. Giersig, E.M. Akinoglu, B. Han, R. Li, Physics of transparent conductors. Adv. Phys. 65 (2016) 553–617.
2. R.A. Afre, N. Sharma, M. Sharon, M. Sharon, Transparent conducting oxide films for various applications: A review. Rev. Adv. Mater. Sci. 53 (2018) 79–89.
3. H. Liu, V. Avrutin, N. Izyumskaya, U. Özgür, H. Morkoç, Transparent conducting oxides for electrode applications in light emitting and absorbing devices. Superlattices Microstruct. 48 (2010) 458–484.
4. S.C. Dixon, D.O. Scanlon, C.J. Carmalt, I.P. Parkin, n-Type doped transparent conducting binary oxides: An overview. J. Mater. Chem. C 4 (2016) 6946–6961.
5. G.K. Dalapati, H. Sharma, A. Guchhait, N. Chakrabarty, P. Bamola, Q. Liu, M. Sharma, Tin oxide for optoelectronic, photovoltaic and energy storage devices: A review. J. Mater. Chem. A 9 (2021) 16621–16684.
6. S. Lany, A. Zunger, Dopability, intrinsic conductivity, and nonstoichiometry of transparent conducting oxides. Phys. Rev. Lett. 98 (2007) 045501.
7. D.S. Ginley, C. Bright, Transparent conducting oxides. MRS Bulletin 25 (2000) 15–18.
8. T. Kim, G. Kim, H. Kim, H.J. Yoon, T. Kim, Y. Jun, W. Shim, Megahertz-wave-transmitting conducting polymer electrode for device-to-device integration. Nat. Commun. 10 (2019) 653.
9. H. Hosono, K. Ueda, Transparent Conductive Oxides (p. 1). Springer Handbook of Electronic and Photonic Materials (2017).
10. T. Minami, Present status of transparent conducting oxide thin-film development for indium-tin-oxide (ITO) substitutes. Thin Solid Films 516 (2008) 5822–5828.
11. N.N. Rosli, M.A. Ibrahim, N.A. Ludin, M.A.M. Teridi, K. Sopian, A review of graphene based transparent conducting films for use in solar photovoltaic applications. Renew. Sust. Energ. Rev. 99 (2019) 83–99.
12. Z. Chen, X. Su, H. Luo, A. Ade, H. Zhu, Y. Zhang, L. Yu, Transparent conductive film of silver nanowires employed for SERS detection of mercury ions. Mater. Chem. Phys. 309 (2023) 128335.

13. M.B. Lee, C.T. Lee, W.W.F. Chong, S.M. Sanip, A quick and facile solution-processed method for PEDOT: PSS transparent conductive thin film. IIUM Eng. J. 24 (2023) 170–182.
14. K. McLellan, Y. Yoon, S.N. Leung, S.H. Ko, Recent progress in transparent conductors based on nanomaterials: Advancements and challenges. Adv. Mater. Technol. 5 (2020) 1900939.
15. J. Rosén, O. Warschkow, Electronic structure of amorphous indium oxide transparent conductors. Phys. Rev. B 80 (2009) 115215.
16. S.M. Sun, W.J. Liu, Y.F. Xiao, Y.W. Huan, H. Liu, S.J. Ding, D.W. Zhang, Investigation of energy band at atomic-layer-deposited ZnO/β-Ga2O3 (201) heterojunctions. Nanoscale Res. Lett. 13 (2018) 1–6.
17. H. Odaka, S. Iwata, N. Taga, S. Ohnishi, Y. Kaneta, Y. Shigesato, Study on electronic structure and optoelectronic properties of indium oxide by first-principles calculations. Japanese J. Appl. Phys. 36 (1997) 5551.
18. J. Robertson, B. Falabretti, Electronic structure of transparent conducting oxides. In Handbook of Transparent Conductors (pp. 27–50). Boston, MA: Springer US (2010).
19. Ü. Özgür, Y.I. Alivov, C. Liu, A. Teke, M.A. Reshchikov, S. Doğan, A.H. Morkoç, A comprehensive review of ZnO materials and devices. J. Appl. Phys. 98 (2005) 041301 (1-103).
20. Ü. Özgür, D. Hofstetter, H. Morkoc, ZnO devices and applications: A review of current status and future prospects. Proc. IEEE 98 (2010) 1255–1268.
21. P. Sikam, P. Moontragoon, Z. Ikonic, T. Kaewmaraya, P. Thongbai, The study of structural, morphological and optical properties of (Al, Ga)-doped ZnO: DFT and experimental approaches. Appl. Surf. Sci. 480 (2019) 621–635.
22. K. Qi, X. Xing, A. Zada, M. Li, Q. Wang, S.Y. Liu, G. Wang, Transition metal doped ZnO nanoparticles with enhanced photocatalytic and antibacterial performances: Experimental and DFT studies. Ceram. Int. 46 (2020) 1494–1502.
23. J. Borah, B.K. Sarma, Design strategy and interface chemistry of ageing stable AZO films as high quality transparent conducting oxide. J. Colloid Interface Sci. 582 (2021) 1041–1057.
24. T. Jan, S. Azmat, A.U. Rahman, S.Z. Ilyas, A. Mehmood, Experimental and DFT study of Al doped ZnO nanoparticles with enhanced antibacterial activity. Ceram. Int. 48 (2022) 20838–20847.
25. L.C. Damonte, G.N. Darriba, M. Rentería, Structural and electronic properties of Al-doped ZnO semiconductor nanopowders: Interplay between XRD and PALS experiments and first-principles/DFT modeling. J. Alloys Compd. 735 (2018) 2471–2478.
26. P.S. Patil, Versatility of chemical spray pyrolysis technique. Mater. Chem. Phys. 59 (1999) 185–198.
27. T. Shirahata, T. Kawaharamura, S. Fujita, H. Orita, Transparent conductive zinc-oxide-based films grown at low temperature by mist chemical vapor deposition. Thin Solid Films 597 (2015) 30–38.
28. K.B. Masood, P. Kumar, M.A. Malik, J. Singh, A comprehensive tutorial on the pulsed laser deposition technique and developments in the fabrication of low dimensional systems and nanostructures. Emergent Mater. 4 (2021) 737–754.
29. B.K. Sarma, P. Rajkumar, Al-doped ZnO transparent conducting oxide with appealing electro-optical properties–Realization of indium free transparent conductors from sputtering targets with varying dopant concentrations. Mater. Today Commun. 23 (2020) 100870.
30. B. Sarma, D. Barman, B.K. Sarma, AZO (Al: ZnO) thin films with high figure of merit as stable indium free transparent conducting oxide. Appl. Surf. Sci. 479 (2019) 786–795.
31. K.H. Zhang, K. Xi, M.G. Blamire, R.G. Egdell, P-type transparent conducting oxides. J. Phys. Condens. Matter. 28 (2016) 383002.
32. M. Johnson-Groh, Comprehensive review of p-and n-type transparent conducting materials. Scilight 2021 (2021) 341105.
33. Y. Shigesato, In based TCOs. In: Ginley, D. (ed) Handbook of Transparent Conductors. Boston, MA: Springer (2011).
34. I. Hamberg, C.G. Granqvist, Evaporated Sn-doped In2O3 films: Basic optical properties and applications to energy-efficient windows. J. Appl. Phys. 60 (1986) R123–R160.
35. T. Sasabayashi, N. Ito, E. Nishimura, M. Kon, P.K. Song, K. Utsumi, Y. Shigesato, Comparative study on structure and internal stress in tin-doped indium oxide and indium-zinc oxide films deposited by rf magnetron sputtering. Thin Solid Films 445 (2003) 219–223.
36. K. Ellmer, Transparent conductive zinc oxide and its derivatives. In: Handbook of Transparent Conductors (pp. 193–263) Springer US (2011).
37. K. Ellmer, A. Bikowski, Intrinsic and extrinsic doping of ZnO and ZnO alloys. J. Phys. D: Appl. Phys. 49 (2016) 413002.
38. M.D. McCluskey, S.J. Jokela, Defects in ZnO J. Appl. Phys. 106 (2019) 071101 (1-13).

39. A. Kim, Y. Won, K. Woo, C.H. Kim, J. Moon, Highly transparent low resistance ZnO/Ag nanowire/ZnO composite electrode for thin film solar cells. ACS Nano. 7 (2013) 1081–1091.

40. D. Barman, B.K. Sarma, Thin and flexible transparent conductors with superior bendability having Al-doped ZnO layers with embedded Ag nanoparticles prepared by magnetron sputtering. Vacuum 177 (2020) 109367.

41. K. Im, K. Cho, J. Kim, S. Kim, Transparent heaters based on solution-processed indium tin oxide nanoparticles. Thin Solid Films 518 (2010) 3960–3963.

42. M. Ahmadi, M. Abrari, M. Ghanaatshoar, An all-sputtered photovoltaic ultraviolet photodetector based on co-doped $CuCrO_2$ and Al-doped ZnO heterojunction. Sci. Rep. 11 (2021) 18694.

43. A. Calzolari, A. Ruini, A. Catellani, Transparent conductive oxides as near-IR plasmonic materials: The case of Al-doped ZnO derivatives. ACS Photonics 1 (2014) 703–709.

44. J. Borah, B.K. Sarma, Realization of Al: ZnO (AZO)-Ag nanocomposite as a novel photocatalyst capable of broadband photon harnessing driven by near-infrared plasmonics. J. Alloys Compd. 956 (2023) 170312.

45. X. Zhao, H. Shen, Y. Zhang, X. Li, X. Zhao, M. Tai, H. Lin, Aluminum-doped zinc oxide as highly stable electron collection layer for perovskite solar cells. ACS Appl. Mater. Interfaces 8 (2016) 7826–7833.

15 Semiconductor-based Photodiodes

Adem Kocyigit

15.1 INTRODUCTION TO PHOTODIODES

Light is an electromagnetic wave that propagates linearly in a medium and can give information about any condition or any object around. The detection and manipulation of light is important in optics and electronics. Harnessing the power of light is not only essential for human vision but also important for various applications [1]. Light sensors are employed to detect the light. One of the most fundamental elements to detect light is semiconductor photodiodes. A photodiode is a specialized diode designed to convert incoming photons of light into an electrical current. This unique ability has made photodiodes indispensable in a wide range of applications as eyes of electronic systems for capturing and processing optical information [2]. Photodiodes are important for scientists and engineers in the field of optics, photonics, electronics, and telecommunications. Photodiodes are crucial for various today's applications such as high-speed data transmission systems, sensitive light sensors for scientific instruments, or digital cameras [3]. Although photodiodes were invented in the middle of the 20th century, studies on photodiodes have gained great interest due to advancements in semiconductors and nanotechnology [4, 5]. Nowadays, increasing their performance (responsivity, specific detectivity, quantum efficiency, and noise equivalent power) and enhancing their usability in smart technologies such as mobile phones, smart homes, and unmanned vehicles are very popular [6].

In this chapter, the principles, structures, and applications of photodiodes will be explained in detail. The chapter also explores the underlying physics behind photodetection, discusses materials that are used in photodiodes, and explains their types, key characteristics and parameters, and emerging trends in photodiode technology.

15.2 PRINCIPLES OF PHOTODETECTION AND WORKING PRINCIPLES OF PHOTODIODES

A fundamental understanding of the photodetection principles for photodiode operation will be provided in this section. Thus, the working principle of the photodiode can be understood. Photodetection is the sensing of light by capturing photons with an active material and generating measurable electrical signals. The transformation from photons to electrical signals is fundamental to various technologies and applications ranging from simple light sensors to advanced imaging systems [7].

The operation of photodiodes relies on the fundamental principles of semiconductor physics. It is important to know some of the terms, such as the physics of semiconductors, P-N junctions, biasing, and photoelectric effect. The photoelectric effect is very important at the heart of photodetection. When photons with sufficient energies strike the surface of a semiconductor, it causes electron-hole pairs according to this effect. These pairs provide the flow of electrical current in photodiodes under biasing. Photodiodes are typically constructed on semiconductor materials like silicon or gallium arsenide and their electrical conductivity is increased by doping. They have a band gap energy level higher than metal and lower than insulators. Photons with greater energies than this band gap can release electrons to higher energy levels, allowing them to contribute to the electrical current.

DOI: 10.1201/9781003450146-15

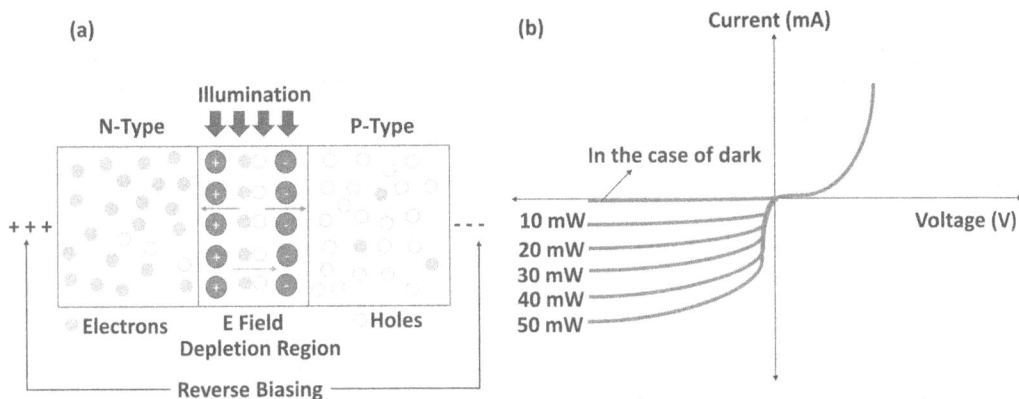

FIGURE 15.1 (a) Working principle of the P-N photodiode. (b) I-V characteristics of a photodiode for increasing light power intensity.

A P-N junction is fundamental to understanding the operation of photodiodes. P- and N-type are doped semiconductors with other atoms such as boron (B) for P-type conductivity or phosphorus (P) for N-type conductivity as an example of silicon. Thus, N- and P-regions have an excess of negatively charged electrons and positively charged holes, respectively. In the case of unbiased conditions, there is a depletion region in the interface of the junction. The depletion region does not have mobile charges except for ionized atoms, and an internal electric field from the N-region to the P-region is formed. When photons are absorbed in the depletion region of this junction, electron-hole pairs are created. An internal electric field causes these charge carriers to separate and generate a photocurrent [8]. Typically, a reverse external biasing is applied on the photodiode to widen the depletion region and increase the chances of photon absorption. This bias provides a low-dark-current mode to decrease noise or separate better photoinduced charges by illumination. Nowadays, there is a popularity of self-powered photodiodes without biasing due to easy operation in the literature [9]. Working principles of the P-N photodiodes have been represented in Figure 15.1a. Figure 15.1b shows the current-voltage characteristics (I-V) of a photodiode. In dark condition, the device works as a diode that conducts current at forward biases and blocks current at reverse biases. However, the current increases by the intensity of light power by illumination at reverse biases. The working principle of photodiodes was explained by the example of P-N photodiodes here. Other photodiodes have almost the same working principle as P-N photodiodes.

In summary, the working of a photodiode starts with photon absorption, it forms electron-hole pairs, and photogenerated charges cause an electrical current. Understanding these fundamental principles is essential for utilizing and designing photodiodes effectively in a wide area of applications [10].

15.3 PHOTODIODE MATERIALS, STRUCTURES, FABRICATION, AND TESTING

15.3.1 MATERIALS IN PHOTODIODES

Photodiodes are used for converting light into electrical signals. To carry out this conversion effectively, the choice of semiconductor is important for photodiode performance. Common semiconductors employed for the fabrication of photodiodes include silicon, germanium (Ge), gallium arsenide (GaAs), indium gallium arsenide (InGaAs), etc. They can be separated from each other depending on some of the behaviors such as costs, abundance in the world, speeds, dark current, and spectral responses. The other active materials with semiconductors are 2D materials such as transition metal dichalcogenides (TMDs), graphene metal oxides, etc. [11]. Germanium photodiodes are sensitive to longer wavelengths, particularly in the infrared (IR) region. They have high absorption coefficients but often require cooling due to increasing dark current [12]. Silicon photodiodes are widely used for their excellent responsivity in the visible and near-IR spectrum, and they have low dark current and good quantum efficiency [13]. InGaAs photodiodes are optimized for IR applications and exhibit excellent sensitivity in the 1.0 to 2.6 µm range. GaAs photodiodes are suitable for high-speed applications in the near IR. They offer fast response times but may have higher dark currents and high costs [14].

Two-dimensional (2D) materials have gained significant attention in photodiodes due to their unique optical, electrical, and mechanical properties. These materials, which consist of a single or few layers of atoms or molecules, have demonstrated promising characteristics for various photodetection applications [15]. Some 2D materials used in photodiodes include graphene, TMDs, and black phosphorus (BP). A graphene is a single layer of carbon atoms arranged in a hexagonal lattice. Graphene-based photodiodes are highly sensitive to light due to the direct bandgap in graphene. Graphene photodiodes have fast response times and wide spectral response and can be used in high-speed optical communications [16]. TMDs, such as molybdenum disulfide (MoS_2) and tungsten diselenide (WSe_2), are layered materials with a direct band gap. TMD-based photodiodes can be designed to have high sensitivity in the visible and near-IR range. They offer good photodetection performance and are suitable for applications like image sensors and wearable devices [17]. BP is a 2D layered structure and has an adjustable band gap that can be tuned by changing the number of layers. BP-based photodiodes can cover a broad range of wavelengths from visible to IR. They exhibit high responsivity and tunable photoresponse, making them suitable for multispectral imaging and optoelectronic applications [18]. 2D materials can be integrated with conventional semiconductor materials to create hybrid photodiodes. Combining different 2D materials as heterostructures can lead to enhanced photodetector performance. For example, stacking graphene and TMDs can create a Schottky junction photodiode with improved responsivity and tunable spectral sensitivity [19]. Since 2D materials are thin and flexible, they are suitable for flexible and wearable photodetectors. These devices can be incorporated into clothing or wearable devices for various applications, including health monitoring and augmented reality [20].

15.3.2 Photodiode Structures

Photodiodes have various structures according to specific applications. Figures 15.2a–15.2e display photodiodes structures of PN, PIN, avalanche, Schottky photodiodes, and metal-semiconductor-metal (MSM) Schottky configuration, respectively. PN photodiodes are the most common photodiode structures, consisting of a P-N junction (Figure 15.2a). Their simplicity makes them suitable for a wide range of applications. PIN photodiodes have a P-N structure with an intrinsic region between the P- and N-type layers (Figure 15.2b). PIN photodiodes offer high-speed performance and reduced capacitance. Avalanche photodiodes (APDs) are specialized photodiodes that use avalanche multiplication to amplify the photocurrent (Figure 15.2c). They are ideal for low-light applications but require higher bias voltages. A Schottky photodiode uses a metal-semiconductor junction (Schottky barrier) instead of a P-N junction (Figure 15.2d). It is known for its fast response time and low capacitance, making it suitable for high-frequency applications. Schottky photodiodes are often used in photodetectors for radio frequency (RF) and microwave applications. The MSM configuration of the Schottky photodiode contains two metal electrodes (usually made of materials like gold or aluminum) on top of a semiconductor (Figure 15.2e). The metal-semiconductor interface forms Schottky barriers, which are responsible for the photoresponse of the device. These are some of the common photodiode structures, and they have specific applications based on sensitivity, speed, and spectral range. The choice of photodiode structure depends on the requirements of the optical or optoelectronic system. Figure 15.2f exhibits an electrical circuit diagram of the photodiodes. When

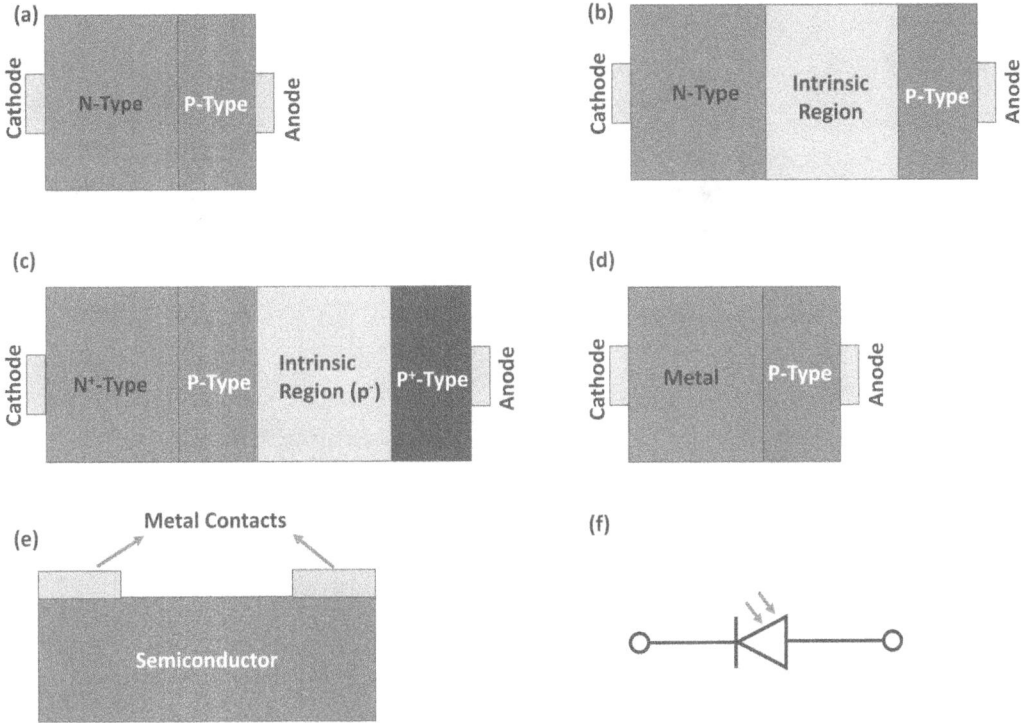

FIGURE 15.2 General structure geometries of photodiodes (a) PN, (b) PIN, (c) avalanche, (d) Schottky photodiodes, (e) MSM configuration of Schottky photodiodes, and (f) electrical circuit diagram of a photodiode.

the light strikes the photodiode, electrical current flows through it. This structural circuit diagram represents this condition [21].

15.3.3 FABRICATION AND TESTING OF PHOTODIODES

15.3.3.1 Fabrication of the Photodiodes

Photodiode fabrication requires precise semiconductor manufacturing techniques such as photolithography, etching, and ion implementation. The semiconductor material is doped with specific impurities to create the P- and N- regions for altering its electrical properties by ion implementation, diffusion, and epitaxial methods. Metal contacts are deposited on the photodiodes to provide electrical connections. A protective layer such as SiO_2 is applied to prevent surface contamination and enhance the reliability of the device [22].

Photodiode fabrication processes start with selecting a suitable semiconductor substrate such as silicon (Si), indium gallium arsenide (InGaAs), gallium arsenide (GaAs), etc. depending on the performance requirements. The cleaning procedure is applied to the substrate to remove contaminations. This procedure involves cleaning alcohol and deionized water by an ultrasonic cleaner and dumping diluted acidic solvents such as H_2O_2 or HF in a short time [23]. Another method to clean substrate is Radio Corporation of America (RCA), which included removal organic contaminations, oxide layers and ionic contaminations [24]. Epitaxial growth provides the depositing of a thin layer of material on top of the substrate. This growth is usually achieved by various techniques such as metal-organic chemical vapor deposition (MOCVD), liquid phase epitaxy (LPE), and molecular beam epitaxy (MBE). The epitaxial layer is formed for composing specific electrical and optical properties [25]. Other techniques such as spin coating, atomic layer deposition, spray coating, physical vapor deposition, etc. can also be used to grow thin film layers for photodiode applications.

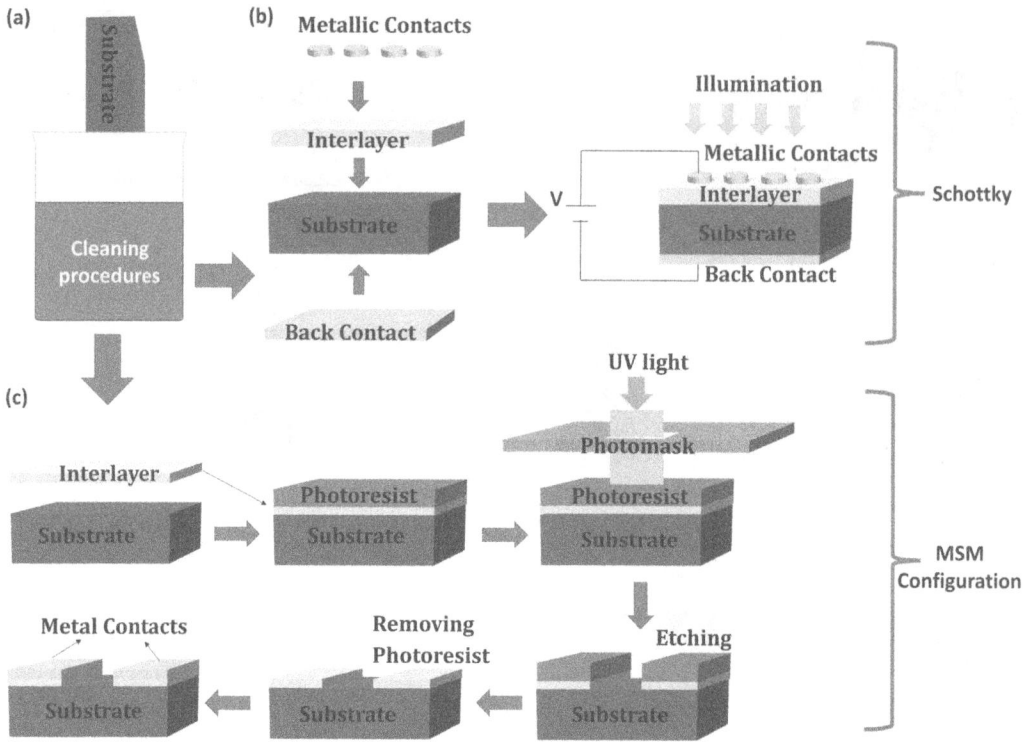

FIGURE 15.3 (a) Cleaning procedure of substrates, (b) fabrication procedure of Schottky photodiodes, and (c) MSM configuration.

In the case of P-N photodiodes, ion implantation or diffusion techniques are used to obtain P- or N-type regions. Photolithography is used to restrict or determine structural regions of the photodiode. A photoresist is coated on the substrate of surface and exposed to ultraviolet (UV) light through a photomask to create a pattern. This pattern defines the active region of the photodiode. Plasma or chemical etching methods are used to remove the related layer not protected by the photoresist. Metal contacts are deposited onto the semiconductor surface to create electrical connections to compose anode and cathode. The fabricated photodiode is typically packaged with a transparent window (for incident light) to protect it from environmental conditions and external contaminants. Figures 15.3a–15.3c display the cleaning and fabrication processes of Schottky photodiodes and MSM configuration. While Figure 15.3a represents the cleaning of the substrate, Figure 15.3b shows the fabrication procedure of the Schottky photodiodes, and Figure 15.3c displays the formation of MSM configuration. While the Schottky photodiode has back and metallic contact growth with an interlayer thin film, the MSM configuration needs photomasking with photoresist, UV exposure, etching of interlayer, and deposition of metallic contacts. Sometimes, photodiodes may be integrated with other electronic components or systems to form more complex devices like optoelectronic integrated circuits (OEICs) or photodetector arrays, and the fabrication method can be changed according to complex devices.

15.3.3.2 Testing and Characterization of the Photodiodes
The specific details of the photodiode fabrication process can vary depending on the desired application, the semiconductor material, and advancements in semiconductor manufacturing technology. Photodiodes undergo rigorous testing to ensure their performance meets specifications in the performance parameters such as responsivity, detectivity, dark current, and quantum efficiency after

fabrication. Characterization equipment such as a source meter for I-V characteristics is used to measure their performance under various operating conditions such as dark or changing illuminated power [26]. Dark current is used to test photodiodes and is critical for low-light applications. It is typically measured at different bias voltages and temperatures. Noise measurements, including shot noise and thermal noise, help assess signal quality with dark current noise. Responsivity is another parameter for assessing the ability of a photodiode for conversion of light into electrical current and is given in the A/W unit. Frequency response and bandwidth testing are important for high-speed applications. The linearity of a photodiode reveals how responsivity changes depending on optical power. Spectral response helps to determine the sensitivity of a photodiode for various wavelengths. Photodiodes used in space or nuclear environments are tested for resistance to ionizing radiation [27]. Photodiodes can be fabricated for specific applications with high reliability and performance depending on the above parameters. The key parameters for testing the performance of a photodiode are discussed in the next section.

15.4   KEY PARAMETERS OF PHOTODIODES

To comprehend the performance of a photodiode, it is essential to determine some of the key parameters such as photosensitive area and material, responsivity, spectral response, dark current, quantum efficiency (QE), noise equivalent power (NEP), bandwidth, reverse bias voltage, capacitance, rise and fall times, and linearity to obtain high-performance photodiodes [28].

A photosensitive area or active area is a portion of the photodiode that is sensitive to light and generates photocurrent when the photodiode is illuminated. It is usually a thin, circular, or rectangular region on the surface of the photodiode. The size of this area affects the sensitivity of photodiodes, and larger areas generally capture more photons. The photosensitive area is generally protected by encapsulation materials to shield it from environmental factors. Proper alignment between light and photosensitive provides capturing the maximum amount of light for high efficiency. In the case of imaging, the size and arrangement of the photosensitive areas on a photodiode array are very important for spatial resolution. Smaller and densely packed photosensitive areas can capture finer details in an image [29]. Another important thing is photosensitive materials for light detection. Various materials such as quantum dots, 2D materials, etc. increase photodiode performance. These materials have been explained in the previous sections.

The dark current (I_{dark}) of a photodiode indicates a flowing electrical current over the photodiode in the dark. Particularly in low-light-level applications, dark current is very important to detect low light. Thermally generated electron-hole pairs cause the dark current within the photodiode because electrons can gain enough energy to jump from the valence band to the conduction band toward higher temperatures. Increasing temperatures typically lead to higher dark currents which have an exponential relationship with temperature. Low dark current is important in applications where the background noise must be minimized. For example, dark current can limit the ability to detect faint objects in low-light imaging or astronomical observations [30].

The photocurrent refers to the electrical current generated when photons strike the photosensitive area of a photodiode and cause electron-hole pairs in the interface. These pairs contribute to the flow of photocurrent. Its magnitude is directly related to the number of generated electron-hole pairs. Therefore, the intensity of the incident light directly influences the photocurrent. The photocurrent is a crucial parameter for light detection in some applications such as spectroscopy, optical communication, imaging, and photometry. The photocurrent formula is given by the following expression:

$$I_{ph} = I_{light} - I_{dark} \qquad (15.1)$$

Photosensitivity (K) implies the ability of a photodiode to respond and interact with light. Photosensitivity (K) represents the signal (photocurrent) to noise (dark current) ratio of the photodiode.

The level of photosensitivity is an essential factor in the performance and design of many optical and optoelectronic devices. Photosensitivity (K) can vary with the wavelength of incident light, and it is essential in various applications, including photography, astronomy, remote sensing, communication systems, medical imaging, and renewable energy (solar panels) [31]. The formula of photosensitivity is given below.

$$K = \frac{I_{ph}}{I_{dark}} \tag{15.2}$$

Responsivity is another key parameter of a photodiode that quantifies its sensitivity to light. It represents a relationship between the incident optical power and the resulting photocurrent. Responsivity indicates how efficiently a photodiode converts light into a photocurrent. It can be expressed as

$$R = \frac{I_{ph}}{PA} \tag{15.3}$$

where P is the power intensity of incident light, and A is the active area. It is important to note that the responsivity of a photodiode can vary with the wavelength of the incident light. Responsivity is a critical parameter when choosing a photodiode for various applications such as light detection, optical communication, spectroscopy, and photometry.

Specific detectivity, denoted as D^*, is a figure of merit used to evaluate the performance of a photodiode, especially in low-light conditions. It quantifies how effectively a photodetector can distinguish a weak optical signal from background noise. Specific detectivity considers various parameters such as responsivity, noise characteristics, and area of a photodiode. The formula for specific detectivity is as follows:

$$D^* = R\sqrt{\frac{A}{2qI_{dark}}} \tag{15.4}$$

where q is the charge of an electron. The unit of the D^* is Jones (cm.$\sqrt{\text{Hz}}$/W). The specific detectivity provides a valuable metric for comparing the performance of different photodiodes in low-light conditions. Photodiodes with higher specific detectivity values are often preferred for extremely weak signal applications such as astronomical observations, remote sensing, and scientific instrumentation [32].

NEP is a measure of the smallest distinguishable optical power level that can be detected by photodiodes in the presence of noise. NEP considers both the responsivity and noise characteristics of the photodiode. It is important in high-sensitivity applications such as scientific experiments, spectroscopy, and optical communication, provides a measure of the noise performance, and is typically expressed in units of watts per square root of hertz (W/$\sqrt{\text{Hz}}$). The NEP is calculated as follows:

$$NEP = \frac{\sqrt{A\Delta f}}{D^*} \tag{15.5}$$

where Δf is bandwidth. It represents the range of frequencies or modulation speeds. Δf is typically specified in hertz (Hz) or megahertz (MHz). To achieve the best performance in the case of NEP, it needs to consider factors like temperature control and signal processing to minimize noise and maximize sensitivity.

QE is another crucial parameter for photodiodes and shows how many incident photons are converted into photocurrent. QE is expressed as a percentage and represents the ratio of the number of electron-hole pairs generated by incoming photons to the total number of incident photons. The QE formula is given by:

$$QE = \frac{N_e}{N_v} \tag{15.6}$$

where N_e and N_v are the number of electrons produced and the number of photons absorbed, respectively. It can be calculated as a percentage multiplied by 100. The maximum possible QE is 100%, which would mean that every incident photon generates an electron-hole pair. However, achieving 100% QE in practice is challenging because of surface reflections, material properties, and design limitations.

Rise and fall times describe the time it takes for the photocurrent to rise from a low level to a specified percentage (generally 90%) of its final value (rise time) or fall from a high level to the same specified percentage (fall time). The rise and fall times of a photodiode are important parameters that exhibit the response speed to incident light intensity. They are particularly relevant in applications where the photodiode needs to capture rapidly changing optical signals. However, shorter rise and fall times of a photodiode may lead to a decrease in the responsivity [33].

Linearity is another important parameter for photodiodes and shows how accurately its output current or voltage response corresponds to changes in incident light intensity. The relationship between the light input and electrical output should be proportional and predictable in a linear photodiode. The linear dynamic range (LDR) of a photodiode indicates that the response of the photodiode is linear in the range of incident light intensities. The relationship between the incident light power and the photocurrent is approximately linear in this range. However, beyond this range, the response may become nonlinear due to saturation effects [34].

Other key parameters are reverse bias, capacitance, and packaging. Photodiodes are generally operated in reverse bias to increase their speed and reduce their capacitance. The junction capacitance of the photodiode affects its response time and bandwidth. Lower capacitance is desirable for high-speed applications. The package of the photodiode can also be important, especially for applications where protection from environmental factors or mechanical robustness is required. These are some of the key parameters required for obtaining high-performance photodiodes.

15.5 TYPES OF PHOTODIODES

Photodiodes are fabricated in various types and designed to cater to specific applications and operational requirements, depending on sensitivity, speed, wavelength range, etc. Thus, photodiode types can change according to their application areas or specific needs. In this section, common photodiode types of PN, PIN, avalanche, and Schottky photodiodes are explained in the case of their advantages and disadvantages.

A PN photodiode is the most basic type of photodiode and works according to obtaining electrons when exposed to photons with sufficient energy. The structure, fabrication, and working principle of PN photodiodes have been discussed above sections. PN photodiodes have several advantages, which make them widely used in optoelectronic applications. They can be fabricated easily and have high responsivity due to high sensitivity in reverse biases. PN photodiodes also have fast response times, typically in the nanosecond-to-microsecond range depending on the specific design and application. This makes them suitable for high-speed applications due to capturing rapid changes in light intensity. They typically exhibit low noise characteristics. Depending on the semiconductor material used, PN photodiodes can be designed to be sensitive to a wide range of wavelengths changing from UV to IR. They generally exhibit good linearity over a wide range of light intensity. PN photodiodes can have a long operational lifespan, making them reliable for long-term applications. PN photodiodes are often cost-effective compared to some other types of photodiodes such as APDs or photomultiplier tubes (PMTs). They can be easily integrated into electronic circuits and systems [35].

A PIN photodiode works with the same fundamental principles as standard PN photodiodes but with the addition of a lightly doped intrinsic (I) layer placed between the N- and P-type layers. This layer serves as a wide, low-doped region that enhances the performance of the photodiode by behaving as an insulator layer for electric field strength. The intrinsic layer reduces the capacitance of the diode, enabling faster response times compared to standard PN photodiodes. PIN photodiodes are known for their low capacitance, low reverse bias voltage, high quantum efficiency, larger bandwidth, and high-speed response. They have a wide linear range, allowing them to accurately measure a broad range of light intensities without saturating, and are often used in high-frequency applications like optical communication and spectroscopy. However, PIN photodiodes are less sensitive, have slower recovery time and high power loss, and their sensitivity is affected so much with varying temperatures [36].

An APD is a specialized photodiode designed to provide internal signal amplification through avalanche breakdown. When a high reverse bias voltage is applied, the electrons and holes generated by incident light collide with other atoms, creating additional electron-hole pairs. This avalanche effect leads to higher sensitivity and lower noise, making APDs suitable for low-light-level detection, such as in optical communication systems and photon-counting applications. They can detect very weak optical signals, making them suitable for applications involving low-light conditions, such as long-range optical communication and light detection and ranging (LIDAR). While APDs offer excellent sensitivity, they have a limited linear range compared to standard photodiodes. In high light power conditions, they can saturate due to the avalanche multiplication effect. APD performance can be affected by temperature variations, and careful temperature control may be necessary in some applications [37].

Schottky photodiodes form a metal-semiconductor junction instead of a P-N junction. They are also known as Schottky barrier photodiodes, which use a metal-semiconductor junction to convert incident light into photocurrent. Schottky photodiodes operate on the principle of the Schottky diodes but work at reverse biases. They offer very fast response times, typically in the picosecond to nanosecond range. This rapid response makes them suitable for high-speed photodetection applications, such as laser pulse measurements and time-resolved spectroscopy. Due to the absence of depletion region seen in PN photodiodes, Schottky photodiodes have lower capacitance. This low capacitance allows them to operate at high frequencies and respond quickly to changes in incident light intensity. Although Schottky photodiodes are good for their speed and sensitivity, they may not be as widely used in low-light-level applications compared to APDs or PIN photodiodes due to higher dark current and light reflection from metal layer [38].

Sometimes, photodiodes can be classified in the case of UV, IR, and visible ranges. The categorization of photodiodes depending on spectral response can be thought of for various applications. Furthermore, fabricated array and multi-junction photodiodes can also be included as types of photodiodes [39, 40].

15.6 APPLICATIONS OF PHOTODIODES

Photodiodes find widespread usage in various applications due to converting photons into photocurrent with high sensitivity and speed. The general application areas are optical communications, sensing applications, monitoring, and imaging. Figure 15.4 shows general applications of photodiodes. Photodiodes are integral components in optical communication systems, including fiber-optic networks. Photodiodes are extensively used in long-distance and high-speed fiber-optic communication because they serve as the core component in optical receivers. Optical receivers are an important part of optical communication systems, including fiber-optic communication networks [41]. Photodiodes are also used for visible light communication (VLC) receivers. Photodiodes for VLC applications are critical for achieving high-speed, secure, and energy-efficient data communication using visible light. Careful consideration of the photodiode characteristics and performance parameters is essential to ensure the success of VLC systems in various real-world applications such as indoor positioning, underwater communication, and secure data transmission in sensitive environments [42].

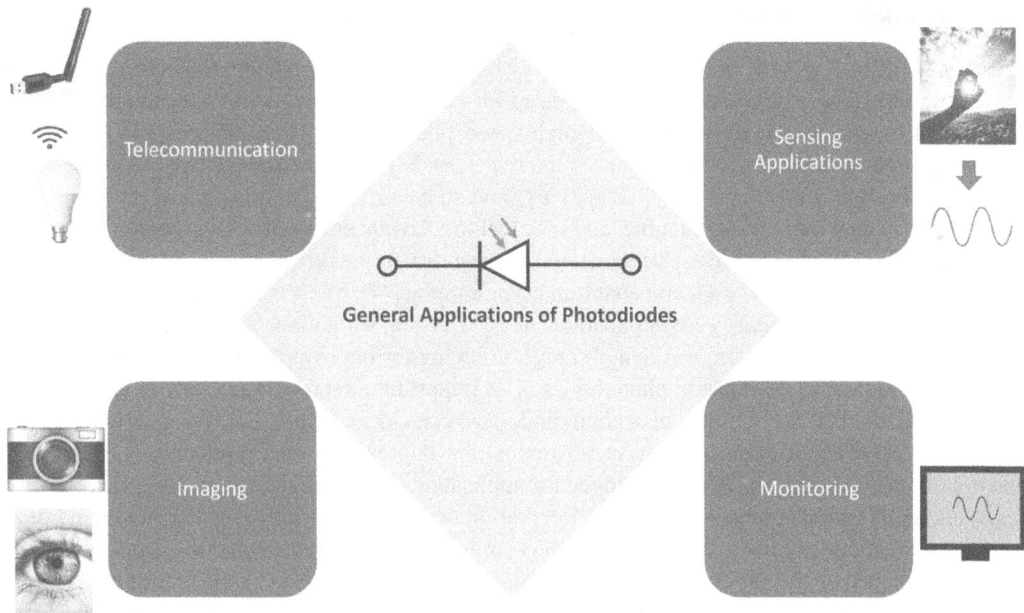

FIGURE 15.4 General applications of photodiodes.

Photodiodes are commonly used as the core sensing elements in light sensors or photodetector sensors. These sensors are designed to detect and measure light levels. Photodiodes are used in light sensors for various applications, such as automatic brightness control in displays, ambient light sensing or proximity sensors in smartphones, and motion detection in security systems. The other applications of light sensors are healthcare and medical devices, gesture recognition, automotive lighting applications, flame detection, solar light tracking, daylight harvesting, etc. [43].

Photodiodes are used in various imaging applications, both as primary imaging sensors and as components within imaging systems. The ability to convert light into electrical current makes them valuable in capturing visual information. Photodiodes are at the heart of image sensors in digital cameras, including charge-coupled device (CCD) and complementary metal-oxide-semiconductor (CMOS) sensors. These sensors capture light to form digital images with various resolutions and qualities. In medical applications, photodiodes are used in imaging systems like X-ray detectors and endoscopes to capture images of the human body for diagnostic and surgical purposes. Photodiodes are used in astronomical cameras and space applications to capture images of celestial objects. These are just a few examples of the diverse imaging applications of photodiodes. The choice of photodiode and imaging system design depends on factors such as the desired spectral range, sensitivity, resolution, speed, and environmental conditions specific to each application [3].

Photodiodes are used in various monitoring applications to detect some signals by measuring light levels, changes in illumination time, or detectivity of specific wavelengths. Monitoring applications typically involve real-time or periodic measurement and data recording for analysis, control, or safety purposes. These applications can be widened from environmental and automation monitoring in the industry for waste and pollution. Photodiodes are applied to monitor changes in natural light levels for weather forecasting, climate research, and environmental studies. Furthermore, they are employed in air quality sensors to measure pollutant levels, such as particulate matter and gas concentrations in the case of environmental monitoring. They are utilized in pulse oximeters to monitor blood oxygen levels and heart rate by measuring light absorption through body tissues in medical applications [44]. These diverse applications highlight the significance and versatility of photodiodes in modern technology across various industries.

15.7 EMERGING TRENDS IN PHOTODIODE TECHNOLOGY

Photodiode technology continues to develop, driven by the demand for higher performance, improved efficiency, and various new applications. The need for faster data transmission rates in optical communication has led to the development of high-speed photodiodes. These devices can operate at extremely high frequencies, enabling faster internet speeds [45]. Quantum dot photodiodes offer tunable absorption wavelengths, high sensitivity, and reduced noise, making them promising for applications in quantum optics, imaging, and sensing [46]. Advancements in single-photon avalanche diodes (SPADs) and other single-photon detectors have driven breakthroughs in quantum sensing, quantum information processing, and quantum cryptography [47]. Short-wave infrared (SWIR) photodiodes are gaining popularity for applications in night vision, surveillance, and imaging. They can detect light in the 1.0 to 2.6 μm wavelength range, allowing for improved visibility in low-light conditions [48]. Reducing the noise in photodiodes is an important attention, especially in applications requiring high sensitivity. Lower-noise photodiodes are critical for improving signal-to-noise ratios in low-light-level applications such as spectroscopy and fluorescence measurements [49]. Flexible and organic photodiodes are being developed for applications like wearable technology, flexible displays, and conformable image sensors. The integration of photodiode arrays on-chip for signal processing and readout circuits is becoming more common. These integrated solutions simplify system design and reduce noise, making them ideal for portable and miniaturized applications [44].

Photodiodes are increasingly being customized for specific applications. This includes tailoring the spectral response, sensitivity, and packaging to meet the exact requirements of industries such as aerospace, medical, and automotive. In space and nuclear applications, radiation-hardened photodiodes are essential. These devices are designed to withstand the harsh radiation environments encountered in space missions and nuclear facilities. Photodiodes are being used in energy harvesting applications, where they convert ambient light into electrical energy to power low-power devices, sensors, and IoT nodes. Photodiode-based sensors are increasingly being integrated with artificial intelligence and machine learning algorithms for advanced data processing, pattern recognition, and autonomous decision-making in various applications, including robotics and autonomous vehicles [4]. These emerging trends in photodiode technology reflect the ongoing efforts to improve performance, expand the range of applications, and meet the evolving needs of industries ranging from telecommunications and healthcare to aerospace and beyond. As technology continues to advance, photodiodes will play an increasingly critical role in enabling innovative solutions.

REFERENCES

1. N.R. Council, Harnessing Light, National Academies Press, 1998. https://doi.org/10.17226/5954.
2. J. Liu, S. Cristoloveanu, J. Wan, A review on the recent progress of silicon-on-insulator-based photodetectors, Phys. Status Solidi A. 218 (2021) 2000751. https://doi.org/10.1002/pssa.202000751.
3. T. Shan, X. Hou, X. Yin, X. Guo, Organic photodiodes: Device engineering and applications, Front. Optoelectron. 2022 15:1. 15 (2022) 1–33. https://doi.org/10.1007/S12200-022-00049-W.
4. R. Douhan, K. Lozovoy, A. Kokhanenko, H. Deeb, V. Dirko, K. Khomyakova, Recent advances in Si-compatible nanostructured photodetectors, Technologies (Basel). 11 (2023) 17. https://doi.org/10.3390/technologies11010017.
5. R.F. Pires, V.D.B. Bonifácio, Photodiodes : Principles and recent advances, J. Mat. NanoSci. 6 (2019) 38–46.
6. W. Yang, J. Chen, Y. Zhang, Y. Zhang, J.H. He, X. Fang, Silicon-compatible photodetectors: Trends to monolithically integrate photosensors with chip technology, Adv. Funct. Mater. 29 (2019) 1808182. https://doi.org/10.1002/adfm.201808182.
7. A. Ferron, O. Boulade, O. Gravrand, G. Destefanis, C. Cervera, J.P. Zanatta, N. Baier, Discussion about photodiode architectures for space applications, in: Proceedings Volume 10563, International Conference on Space Optics — ICSO 2014, SPIE, 2017: p. 206. https://doi.org/10.1117/12.2304266.
8. S.M. Sze, Physics of Semiconductor Devices, 2. edition, Wiley, New York, 1981.

9. H. Lin, A. Jiang, S. Xing, L. Li, W. Cheng, J. Li, W. Miao, X. Zhou, L. Tian, Advances in self-powered ultraviolet photodetectors based on P-N heterojunction low-dimensional nanostructures, Nanomaterials. 12 (2022). https://doi.org/10.3390/nano12060910.

10. D.A. Neamen, Semiconductor Physics and Devices : Basic Principles, McGraw-Hill, New York, 2012.

11. P. Martyniuk, P. Wang, A. Rogalski, Y. Gu, R. Jiang, F. Wang, W. Hu, Infrared avalanche photodiodes from bulk to 2D materials, Light Sci. Appl. 12 (2023) 1–26. https://doi.org/10.1038/s41377-023-01259-3.

12. Q. Durlin, A. Aliane, L. André, H. Kaya, M. Le Cocq, V. Goudon, C. Vialle, M. Veillerot, J.M. Hartmann, Fabrication and characterisation of the PiN Ge photodiode with poly-crystalline Si:P as n-type region, Opto-Electronics Rev. 31 (2023) 1–7. https://doi.org/10.24425/opelre.2023.144550.

13. T.H. Kim, J.-K. Lee, J. Park, I.-S. Kang, G. Sim, Broadband responsivity enhancement of Si photodiodes by a plasmonic antireflection bilayer, Optics Express. 29(17), (2021) 26634–26644. https://doi.org/10.1364/OE.432689.

14. X. Chen, Y. Gu, Y. Zhang, X. Chen, Y. Gu, Y. Zhang, Epitaxy and Device Properties of InGaAs Photodetectors with Relatively High Lattice Mismatch, Epitaxy. (2017). https://doi.org/10.5772/INTECHOPEN.70259.

15. M. Malik, M.A. Iqbal, J.R. Choi, P.V. Pham, 2D materials for efficient photodetection: Overview, mechanisms, performance and UV-IR range applications, Front. Chem. 10 (2022) 905404. https://doi.org/10.3389/FCHEM.2022.905404/BIBTEX.

16. F. Luo, M. Zhu, Y. tan, H. Sun, W. Luo, G. Peng, Z. Zhu, X.-A. Zhang, S. Qin, High responsivity graphene photodetectors from visible to near-infrared by photogating effect, AIP Adv. 8 (2018) 115106. https://doi.org/10.1063/1.5054760.

17. F. Özel, E. Arkan, H. Coskun, İ. Deveci, M. Yıldırım, M. Yıldırım, İ Orak, M.O. Erdal, A. Sarılmaz, T.T. Ersöz, A. Koçyiğit, A. Karabulut, A. Özen, A. Aljabour, M. Kus, M. Ersöz, Refractory-metal-based chalcogenides for energy, Adv. Funct. Mater. 32 (2022). https://doi.org/10.1002/adfm.202207705.

18. X. Chen, X. Lu, B. Deng, O. Sinai, Y. Shao, C. Li, S. Yuan, V. Tran, K. Watanabe, T. Taniguchi, D. Naveh, L. Yang, F. Xia, Widely tunable black phosphorus mid-infrared photodetector, Nat. Commun. 8 (2017) 1672. https://doi.org/10.1038/s41467-017-01978-3.

19. D.H. Shin, S.H. Choi, Graphene-based semiconductor heterostructures for photodetectors, Micromachines (Basel). 9 (2018) 350. https://doi.org/10.3390/mi9070350.

20. J. Yao, G. Yang, Flexible and high-performance All-2D photodetector for wearable devices, Small. 14 (2018). https://doi.org/10.1002/smll.201704524.

21. S. Donati, Photodetectors: Devices, Circuits and Applications, Wiley, 2020. https://doi.org/10.1002/9781119769958.

22. R. Peng, S. Jiao, D. Jiang, H. Li, L. Zhao, Dark current mechanisms and spectral response of SiO2-passivated photodiodes based on InAs/GaSb superlattice, Thin Solid Films. 629 (2017) 55–59. https://doi.org/10.1016/J.TSF.2017.03.045.

23. A. Kocyigit, D.E. Yıldız, A.A. Hussaini, D.A. Kose, M. Yıldırım, Cu and Mn centered nicotinamide/nicotinic acid complexes for interlayer of Schottky photodiode, Curr. Appl. Phys. 45 (2023) 53–63. https://doi.org/10.1016/j.cap.2022.11.001.

24. E. Brachmann, M. Seifert, S. Oswald, S. Menzel, T. Gemming, Evaluation of surface cleaning procedures for CTGS substrates for SAW technology with XPS, Materials. 10 (2017) 1373. https://doi.org/10.3390/ma10121373.

25. H. Kum, D. Lee, W. Kong, H. Kim, Y. Park, Y. Kim, Y. Baek, S.-H. Bae, K. Lee, J. Kim, Epitaxial growth and layer-transfer techniques for heterogeneous integration of materials for electronic and photonic devices, Nat. Electron. 2 (2019) 439–450. https://doi.org/10.1038/s41928-019-0314-2.

26. Z. Bielecki, K. Achtenberg, M. Kopytko, J. Mikołajczyk, J. Wojtas, A. Rogalski, Review of photodetectors characterization methods, Bull. Pol. Acad. Sci. Tech. Sci. 70 (2022). https://doi.org/10.24425/BPASTS.2022.140534.

27. I.S. Amiri, F.M.A.M. Houssien, A.N.Z. Rashed, A.E.-N.A. Mohammed, Temperature effects on characteristics and performance of near-infrared wide bandwidth for different avalanche photodiodes structures, Results Phys. 14 (2019) 102399. https://doi.org/10.1016/j.rinp.2019.102399.

28. A. Chetia, J. Bera, A. Betal, S. Sahu, A brief review on photodetector performance based on zero dimensional and two dimensional materials and their hybrid structures, Mater Today Commun. 30 (2022) 103224. https://doi.org/10.1016/J.MTCOMM.2022.103224.

29. A. Rogalski, Z. Bielecki, J. Mikołajczyk, J. Wojtas, Ultraviolet photodetectors: From photocathodes to low-dimensional solids, Sensors. 23 (2023) 4452. https://doi.org/10.3390/s23094452.

30. J. Kublitski, A. Hofacker, B.K. Boroujeni, J. Benduhn, V.C. Nikolis, C. Kaiser, D. Spoltore, H. Kleemann, A. Fischer, F. Ellinger, K. Vandewal, K. Leo, Reverse dark current in organic photodetectors and the major role of traps as source of noise, Nat. Commun. 12 (2021) 551. https://doi.org/10.1038/s41467-020-20856-z.

31. W. Ran, L. Wang, S. Zhao, D. Wang, R. Yin, Z. Lou, G. Shen, An integrated flexible all-nanowire infrared sensing system with record photosensitivity, Adv. Mater. 32 (2020). https://doi.org/10.1002/adma.201908419.

32. S. Veeralingam, L. Durai, P. Yadav, S. Badhulika, Record-high responsivity and detectivity of a flexible deep-ultraviolet photodetector based on solid state-assisted synthesized hBN nanosheets, ACS Appl. Electron. Mater. 3 (2021) 1162–1169. https://doi.org/10.1021/acsaelm.0c01021.

33. H.T.D.S. Madusanka, H.M.A.M.C. Herath, C.A.N. Fernando, High photoresponse performance of self-powered n-Cu2O/p-CuI heterojunction based UV-visible photodetector, Sens. Actuators A Phys. 296 (2019) 61–69. https://doi.org/10.1016/j.sna.2019.07.008.

34. R. Nie, X. Deng, L. Feng, G. Hu, Y. Wang, G. Yu, J. Xu, Highly sensitive and broadband organic photodetectors with fast speed gain and large linear dynamic range at low forward bias, Small. 13 (2017) 1603260. https://doi.org/10.1002/smll.201603260.

35. H. Lin, A. Jiang, S. Xing, L. Li, W. Cheng, J. Li, W. Miao, X. Zhou, L. Tian, Advances in self-powered ultraviolet photodetectors based on P-N heterojunction low-dimensional nanostructures, Nanomaterials. 12 (2022) 910. https://doi.org/10.3390/nano12060910.

36. A. Mousazadeh, M. Kafaee, M. Ashraf, Ranking of commercial photodiodes in radiation detection using multiple-attribute decision making approach, Nucl. Instrum. Methods Phys. Res. A. 987 (2021) 164839. https://doi.org/10.1016/j.nima.2020.164839.

37. A.H. Jones, S.D. March, S.R. Bank, J.C. Campbell, Low-noise high-temperature AlInAsSb/GaSb avalanche photodiodes for 2-μm applications, Nat. Photonics. 14 (2020) 559–563. https://doi.org/10.1038/s41566-020-0637-6.

38. N. Biyikli, I. Kimukin, B. Butun, O. Aytür, E. Ozbay, ITO-schottky photodiodes for high-performance detection in the UV-IR spectrum, IEEE J. Sel. Top. Quantum Electron. 10 (2004) 759–765. https://doi.org/10.1109/JSTQE.2004.833977.

39. J. Liu, P. Liu, T. Shi, M. Ke, K. Xiong, Y. Liu, L. Chen, L. Zhang, X. Liang, H. Li, S. Lu, X. Lan, G. Niu, J. Zhang, P. Fei, L. Gao, J. Tang, Flexible and broadband colloidal quantum dots photodiode array for pixel-level X-ray to near-infrared image fusion, Nat. Commun. 14 (2023) 1–9. https://doi.org/10.1038/s41467-023-40620-3.

40. S. Schidl, M. Hofbauer, K. Schneider-Hornstein, H. Zimmermann, Advances in triple junction photo diodes, in: 2014 Microelectronic Systems Symposium, MESS 2014 - Conference Proceedings, Institute of Electrical and Electronics Engineers Inc., 2015. https://doi.org/10.1109/MESS.2014.7010251.

41. O.F. Chukwujekwu, O.I. Nkemdilim, Technical report: Comparative analysis of photodetectors for appropriate usage in optical communication applications, Int. J. Trend Sci. Res. Dev. 5 (2021) 569–582. https://doi.org/10.5281/ZENODO.5518079.

42. C. He, C. Chen, A review of advanced transceiver technologies in visible light communications, Photonics. 10 (2023) 648. https://doi.org/10.3390/photonics10060648.

43. J.M. López-Higuera, Sensing using light: A key area of sensors, Sensors. 21 (2021) 6562. https://doi.org/10.3390/s21196562.

44. S. Cai, X. Xu, W. Yang, J. Chen, X. Fang, Materials and designs for wearable photodetectors, Adv. Mater. 31 (2019). https://doi.org/10.1002/adma.201808138.

45. S. Lischke, A. Peczek, J.S. Morgan, K. Sun, D. Steckler, Y. Yamamoto, F. Korndörfer, C. Mai, S. Marschmeyer, M. Fraschke, A. Krüger, A. Beling, L. Zimmermann, Ultra-fast germanium photodiode with 3-dB bandwidth of 265 GHz, Nat. Photonics. 15 (2021) 925–931. https://doi.org/10.1038/s41566-021-00893-w.

46. R. Guo, M. Zhang, J. Ding, A. Liu, F. Huang, M. Sheng, Advances in colloidal quantum dot-based photodetectors, J. Mater. Chem. C Mater. 10 (2022) 7404–7422. https://doi.org/10.1039/d2tc00219a.

47. F. Ceccarelli, G. Acconcia, A. Gulinatti, M. Ghioni, I. Rech, R. Osellame, Recent advances and future perspectives of single-photon avalanche diodes for quantum photonics applications, Adv. Quantum Technol. 4 (2021). https://doi.org/10.1002/qute.202000102.

48. F. Cao, L. Liu, L. Li, Short-wave infrared photodetector, Mater. Today. 62 (2023) 327–349. https://doi.org/10.1016/j.mattod.2022.11.003.

49. K. Zhu, N. Solmeyer, D.S. Weiss, A low noise, nonmagnetic fluorescence detector for precision measurements, Rev. Sci. Instrum. 83 (2012). https://doi.org/10.1063/1.4765745.

16 Semiconductors for Solar Cells

Santosh V. Patil and Kshitij Bhargava

16.1 INTRODUCTION

The sun is an abundant source of renewable energy in terms of thermal and electromagnetic radiation and is also pollution-free, noiseless, and freely available on the earth. Renewable energy resources can be used for the compensation of diminutive non-renewable energy derived from natural resources such as natural gas, crude oil, coal, etc. Such an abundant and clean energy source should be fully utilized for the development of humanity. The sunlight can be directly harvested into electrical energy using semiconductor-based miniaturized solar cells. In 1839 the French physicist Alexandre Edmund Becquerel experimentally demonstrated for the first time the photovoltaic (PV) effect in liquid electrolytes [1], and therefore solar cell has been well-known as a PV cell. Thus, by receiving light or photons from sunlight, solar cells produce electrical direct current (DC) and voltage. The phenomena of electrical power generation continue till light is shined on the front glass of the solar cell and it stops under dark conditions. Later in 1954, C. S. Fuller et al. reported the first practical 6% efficient Si-based solar cell fabricated by Bell Laboratory, US. In 1959, Hoffman Electronics from the United States reported a conversion efficiency of 10% for Si solar cells. Alferov et. al. reported the development of the first heterojunction GaAs solar cells in 1970 [2]. In the previous five decades, the first generation of silicon-based thick-film solar cells dominated the PV market. However, the limited power conversion efficiency and expensive film growth techniques have attracted great research interests worldwide. In 1980, IEC developed the first thin-film solar cell using Cu_2S and CdZnS with more than 10% efficiency [3]. However, the problems related to encapsulation and stability of these devices have acquired greater attention toward the second-generation technology that comprised hydrogenated amorphous silicon (a-Si H), cadmium telluride (CdTe), and chalcopyrite copper indium di-selenide ($CuInSe_2$), which often require chemical vapor deposition for thin-film growth at high energy. Meanwhile, in 1985, a crucial breakthrough was achieved as power conversion coefficient efficiency (PCE) for Si solar cells exceeded 20% under standard sunlight conditions. In 1998, National Renewable Energy Laboratory (NREL) reported 19% efficient Cu(In, Ga)S e_2 solar cells [4]. Third-generation PV technology includes polymer or organic solar cells (OSC) and dye-sensitized solar cells (DSSCs) as solution-processed solar cells. Lastly, the fourth generation includes the organic-inorganic-metal-halide perovskite solar cells, which have become more popular than any other generation solar cell in a very short period as they offer high PCE at low manufacturing cost. Perovskite is a quaternary compound semiconductor material incorporated in perovskite solar cells with a low density of bulk defects, low interfacial defects, excellent absorption coefficient, and large carrier diffusion length [5]. Semiconductor-based solar cells offer the advantage of operating at ambient temperatures with no fire hazards or safety issues. No air and noise pollution, no radioactive waste generation, and the ability to be installed at load centers with ease are the advantages of PV systems. They have a lifetime of more than 20 to 25 years [5], and the energy source is abundant and free of cost. However, PV systems face challenges such as high installation costs, power reliability that depends on the reliability of other peripherals, and low power conversion efficiency. Additionally, stability, reliability, and toxicity are significant concerns with thin-film solar cells.

DOI: 10.1201/9781003450146-16

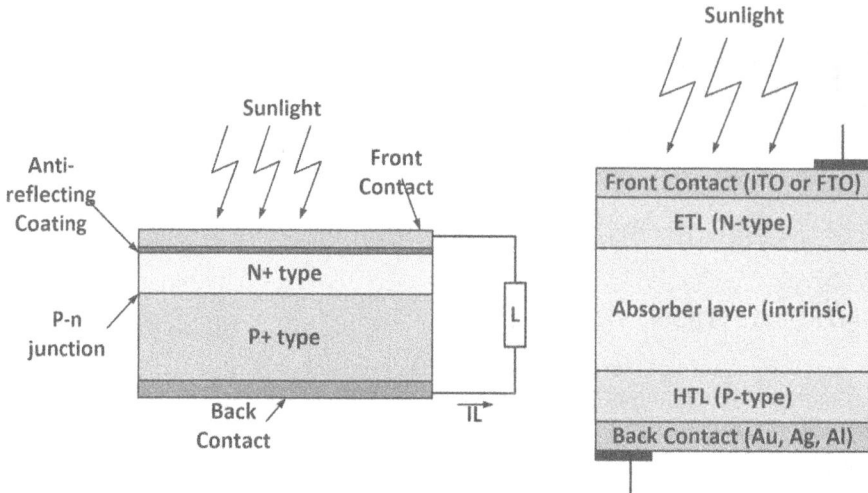

FIGURE 16.1 Structural representation of thick film and thin film semiconductor-based solar cells with p-n type and conventional n-i-p type layers, respectively.

16.2 OPERATING PRINCIPLE OF SOLAR PHOTOVOLTAIC CELLS

The typical solar irradiance spectrum received on the earth's surface consists of ultraviolet, visible, and infrared regions measured under typical test conditions i.e., AM 1.5G, total irradiance $E_{STC} = 1000$ W/m^2, which is more than useful to extract maximum power output of a PV module [6]. The PV conversion in the solar cell takes place mainly in two steps (1) creation of excitons and (2) separation of charged free carriers. The device shows poor power conversion efficiency when any one of the processes gets impeded. Therefore, a proper selection of absorber and transport layer materials in solar cells becomes an important as well as a challenging task. Figure 16.1 shows the band gap alignment of the heterojunction architecture of a solar cell used to produce the depletion region. Doping concentrations of p- and n-type layers determine the width of the depletion region. Low doping concentration causes a wide depletion region as compared to heavy concentration. For thin-film solar cells to perform optimally, the depletion region typically has a width of 2 μm. Therefore, to control the doping, heterostructures facilitate the absorption of different regions of the solar spectrum due to the stacking of materials of distinct bandgap values. For this, a heterostructure consists of the inverted planar (p-i-n) and planar type (n-i-p) structured solar cells with an intrinsic "i" layer that is sandwiched between the electron transport and hole transport layers. Therefore, by controlling the doping concentration, a strong diffusion length and electric field can be developed across the heterojunction.

16.2.1 CARRIER GENERATION POST PHOTON ABSORPTION

The electron-hole pairs will be generated as the light incident on a solar cell and absorption of photons inside the absorber material. However, photons are being absorbed only when photon energy ($E_{ph} = h\nu$) would be larger than or equal to the bandgap energy Eg of the absorber layer where "h" represents Planck's constant and "ν" is the frequency of light. The band gap is the difference between energy levels, viz. conduction band (E_C) and valence band (E_V) [7]. The following three cases are possible depending upon the photon energy and band gap energy of the absorber layer.

    i. If $E_{ph} = E_g$, the absorption of photons will take place and electron-hole pairs will be generated without any loss of photon energy.

ii. If $E_{ph} > E_g$, the photons will be generated when photon energy is equal to band gap energy, and the remaining excessive energy will be lost in terms of heat (thermalization loss).

iii. If $E_{ph} < E_g$, the photon energy is insufficient to enable the transition of electrons from valence to conduction band, resulting in the majority of photons being either reflected or lost.

16.2.2 Separation of Charge Carriers

Post exciton generation, the electron-hole pairs get separated into free carriers which get pushed toward the respective electrodes due to the presence of built-in voltage (V_{bi}) across the depletion region and are always smaller than (E_g/q). Figure 16.2 demonstrates the flow of charge carriers at the time of charge separation in single-junction silicon thick- and thin-film solar cells. Typically, under appropriate doping conditions, silicon solar cells exhibit a built-in voltage of around 1 V. Figure 16.3 shows the variations of space charge density represented by N_D^+ and N_A^- and electric field strength to position (x) associated with homojunction solar cells and heterojunction-based solar cells. The charge separation process is more efficient in thin-film solar cells as the electron transport layers or buffer layers perform a significant role in charge transportation. The large value of material band gap energy "E_g" produces a strong electric field and built-in voltage. Also, the charge transportation becomes very smooth when a strong electric field is established within the space charge region as the proper bandgap alignment is maintained.

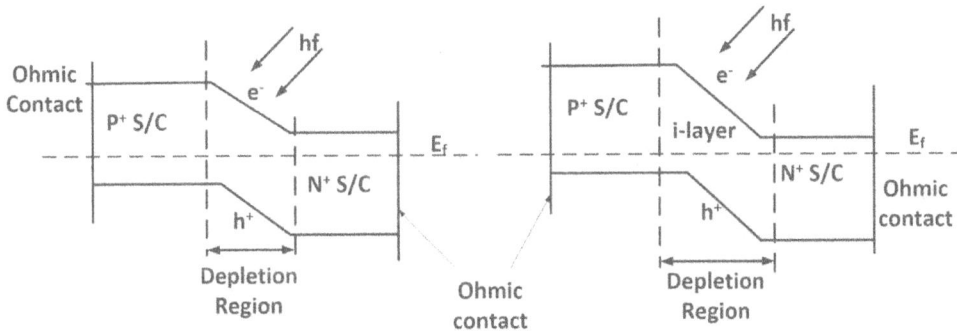

FIGURE 16.2   Energy band diagram for p-n junction or homojunction and p-i-n heterojunction-based solar cell.

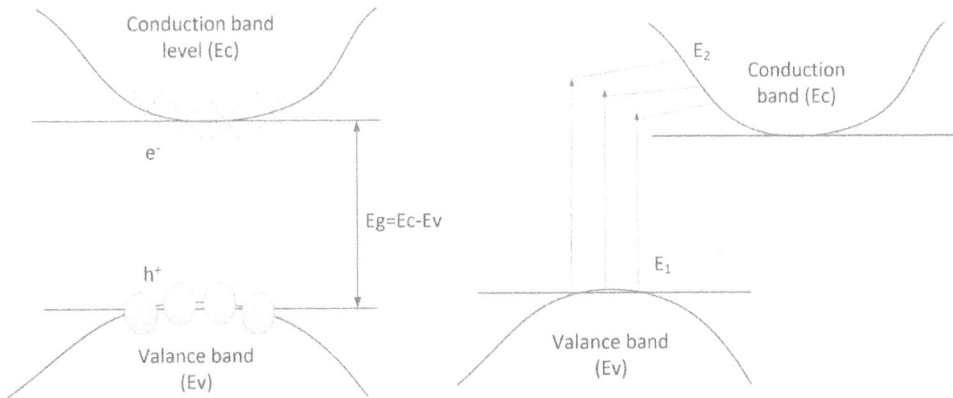

FIGURE 16.3   Band energy ϵ versus crystal momentum (p) diagram at T > 0 K for direct bandgap and indirect bandgap semiconductor.

16.3 FUNDAMENTALS OF SEMICONDUCTORS USED IN SOLAR CELLS

The physical properties of the semiconductor materials used in the solar cells have been described here. Some of the semiconductor materials like Si, CdTe, GaAs, GaInP, and Cu(InGa)Se$_2$ are commonly used in the fabrication of solar cells based on their high absorption capability and affordable cost. Some important terms related to the performance of solar cells are explained as follows.

16.3.1 CRYSTAL STRUCTURE

The crystallinity of an electronic-grade semiconductor is distinguished by the periodic arrangement of atoms in an array. The periodicity of atoms governs the electronic properties of semiconductor thin films. Moreover, silicon is an element that lies in Group 14, possessing four electrons in the valance band that can be shared with neighboring Si atoms. In this way, the covalent bonds form inside the silicon crystal. In silicon crystals, the atoms are positioned in a diamond lattice with tetrahedral bonding. The angle made by the bonds with each Si atom is 109.5°. Two face-centered cubic (FCC) unit cells can describe this arrangement. Besides this, the second FCC cell is relocated at a quarter of the distance along the diagonal of the first FCC cell along the body. This displacement along the body diagonal results in a one-of-a-kind configuration within the crystal lattice. To completely grasp the concept, it's useful to look at the use of Miller indices and the definition of the lattice constant. Miller indices are a set of three metrics (hkl) that describe the orientation of a crystallographic plane within a crystal lattice. They are used to uniquely identify and characterize distinct planes in a crystal structure. In the context of FCC unit cells, they assist us in locating certain planes and orientations inside the cubic lattice [8]. The stacking of these unit cells is responsible for the entire component lattice structure. Similarly, in the case of zinc blende lattice, different semiconductors from columns III-V and II-VI are incorporated such as CdTe (II-VI compound) and GaAs (III-V compound). For example, in CdTe, one face-centered cubic (FCC) unit cell is occupied by cadmium atoms, while the rest are filled entirely with tellurium atoms. As a result, CdTe has an average valency of four for each compound, thus forming four covalent bonds with neighboring atoms.

16.3.2 ENERGY BAND ALIGNMENT

Generally, the physics of charge transport in solar cells is well understood by studying the electronic properties of crystalline semiconductor layers. Considering the semiconductor crystal as a 3D box, each crystal has a multifaceted interior structure that accommodates an electron as a particle. There exists a potential field surrounding an atom's nucleus. The time-independent Schrodinger equation was used to determine the electron wave function, ψ as,

$$\nabla^2 \psi + \frac{2m}{h^2}\left[E - U(\vec{r})\right]\psi = 0 \qquad (16.1)$$

where m denotes electronic mass, E indicates the energy of an electron, h symbolizes Planck's constant, and U(r) represents the potential energy of the semiconductor film that exhibits periodicity. The electron wave function, ψ, establishes the dynamics of an electron. The band structure of the semiconductor can be defined with the solution of the above equation. Using this band structure, the approximation for the motion of an electron in a crystal and an electron moving in the free space can be estimated. When the effective mass, m^*, is substituted for the mass, m, of an electron, Newton's law of motion dictates that the speed remains constant. The formula relates the applied force (F) and the acceleration (a) of an electron that yields F = m × a. The band structure can be realized by a plot between allowed electron energies and the crystal momentum, p = hk, where the wave function solution is correlated by the wave vector designated by k. Here, Figure 16.3 shows the band gap energy diagram that represents the valence band and conduction energy bands of a direct band gap semiconductor material in which electrons are fully occupied in the energy bands below the valence band, while electrons are empty (o) in those above the conduction band. The effective mass, m^* can be depicted by

$$m^* \equiv \left[\frac{d^2E}{dp^2} \right]^{-1} = \left[\frac{1}{h^2} \frac{d^2E}{dk^2} \right]^{-1} \tag{16.2}$$

The effective mass varies within each band and in particular, the effective mass is negative near the peak point of the valence band. Thermal energy causes valence band electrons (*) to jump to the conduction band, creating empty states (o) at the top of the valence band. An empty state can be described as a positively charged carrier of a current, also known as a hole. These holes have a positive effective mass and classically positively charged particles. As a result of their parabolic shape, the electron and hole effective masses m_n^*, m_p^* remain constant at the bottommost of the conduction band and the topmost of the valence band.

In direct band gap semiconductors, the conduction band minima and valence band maxima occur simultaneously at the same crystal momentum, whereas if the minimum energy level of the conduction band and maximum energy level of the valence band occur at different values of the crystal momentum, then the semiconductor is referred to as an indirect band gap semiconductor. In a semiconductor material, the concept of indirect band gap means a lot for the absorption of light.

16.3.3 LIGHT ABSORPTION

The absorption of sunlight is essential for the operation of solar cells as they generate electron-hole pairs. A fundamental absorption occurs when an electron moves from the valence band to the conduction band, resulting in a hole left behind. The momentum of all charge carriers and the photon energy should be conserved during the process of absorption. It has been evident that the range of crystal momentum $p = h/l$ is comparatively higher than the photon momentum, $p\lambda = h/\lambda$ during the photon absorption process and ensures the concept of energy conservation. The absorption coefficient α in terms of the photon energy, $h\upsilon$, is proportional to the transition of an electron from the initial state E_1 to the final state E_2, the probability, P_{12}, the density of electrons in the initial state $g_v(E1)$, and the density of available final states, and is then summed over all possible transitions between states where $E_2 - E_1 = h\upsilon$.

$$\alpha(h\upsilon) \propto \sum P_{12} g_v(E_1) g_c(E_2) \tag{16.3}$$

Assume that initially, the valance band is filled with electrons and the conduction band is filled with holes. After the photon absorption, electrons are excited to the conduction band and leave behind the positively charged holes, which results in the generation of electron-hole pairs. The photon absorption process is shown in Figure 16.3 for an indirect bandgap semiconductor, such as GaInP, GaAs, Cu(In, Ga)Se$_2$, and CdTe. There must be a conservation of energy and crystal momentum during all transitions in the crystal. As a result of E_1 energy and p_1 crystal momentum in the valence band, each initial electronic state can be associated with energy E_2 and crystal momentum p_2 in the conduction band of the final state. Due to the conservation of electron momentum, the net momentum of the crystal in the initial state and the final state remains the same, $p_1 \approx p_2 = p$.

As a result of energy conservation, the photon energy is determined by the following equation:

$$h\upsilon = E_2 - E_1 \tag{16.4}$$

By assuming the parabolic bands,

$$Ev - E_1 = \frac{p^2}{2m_p^*} \tag{16.5}$$

$$E_2 - E_c = \frac{p^2}{2m_n^*} \tag{16.6}$$

Combining Equations (16.4), (16.5) and (16.6) produces

$$hv - E_G = \frac{p^2}{2}\left(\frac{1}{m_n^*} + \frac{1}{m_p^*}\right) \tag{16.7}$$

An absorption coefficient characterizes direct transitions as

$$\alpha(hv) \approx A^*(hv - E_G)^{1/2} \tag{16.8}$$

where A^* is a constant. A quantum selection rule in some semiconductor materials does not permit transitions at p = 0, but permits them if p = 0 in the case

$$\alpha(hv) \approx \frac{B^*}{hv}(hv - E_G)^{3/2} \tag{16.9}$$

where B^* is a constant.

Figure 16.3 shows the E vs. p graph for the indirect bandgap semiconductors such as Si and Ge, where the maxima of the valence band occur at a different minimum of the crystal momentum (p) of the conduction band. There is an extra particle that retains the electron momentum in the photon absorption process. The low energy particles having high momentum cause lattice vibrations in the semiconductor. Keep in consideration that phonon absorption or phonon emission constitutes light absorption. During the phonon absorption, it is associated with an absorption coefficient, which is defined by

$$\alpha_a(hv) = \frac{A(hv - E_G + E_{ph})^2}{e^{E_{Ph}/kT} - 1} \tag{16.10}$$

$$\alpha_e(hv) = \frac{A(hv - E_G - E_{ph})^2}{1 - e^{E_{Ph}/kT}} \tag{16.11}$$

When a phonon is emitted the absorption coefficient is

$$\alpha(hv) = \alpha_a(hv) + \alpha_e(hv) \tag{16.12}$$

Therefore, it is concluded that in the case of direct band gap semiconductor materials, the absorption process is quite convenient as the maxima of valance band momentum are the same as that of the minima of crystal momentum of the conduction band. Consequently, absorption happens easily with less energy required for the electron transition from E_1 to E_2. Conversely, in the case of indirect semiconductor materials, photon absorption is less as the maxima of crystal momentum does not interact with the minima of crystal momentum of the conduction band [9].

16.3.4 CHARGE RECOMBINATION

Recombination is a process in which electrons fall back from the conduction band to the valence band when a material is taken out of the thermal equilibrium (occurs due to illumination or injecting the current into the semiconductor material). There are three important recombination mechanisms observed in semiconductor-based solar cells, viz. single-level trap recombination, radiative recombination, and Auger recombination.

16.3.4.1 Single-level Trap Recombination (R_{STL})

Single-level trap recombination is also known as trap-assisted recombination or Shockley-Read-Hall (SRH) recombination. In this case, a charge carrier encounters an occupied trap state of immobile charge of the opposite sign [1]. The recombination centers produced by foreign atoms are also known as traps. Such type of recombination occurs due to the presence of a material defect or lattice structure defect or the presence of foreign atoms, which produces an additional defect energy level within the forbidden zone, as shown in Figure 16.4. The general expression of the SRH recombination can be given by [10]

$$R_{SLT} = \frac{pn - n_i^2}{\tau_{SLT},n\left(p + n_i e^{(E_i - E_T)/kT}\right) + \tau_{SLT},p\left(n + n_i e^{(E_i - E_T)/kT}\right)} \quad (16.13)$$

and

$$\tau = \frac{1}{\sigma \times N_t \times v_{th}}$$

where N_t is trap density; D is the diffusion coefficient, E_t refers to defect energy level, and σ and v_{th} denote the capture cross-section and thermal velocity of electrons and holes, respectively. Additionally, the relationship between the carrier lifetime (τ) and diffusion length is $L = \sqrt{D \times \tau}$. From Equation (16.13), it is obvious that the SRH recombination is dependent upon the density

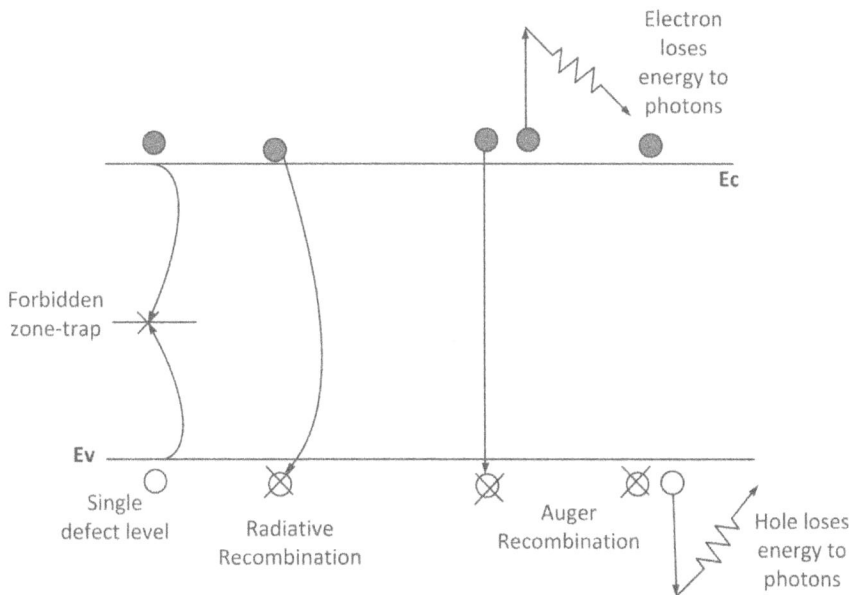

FIGURE 16.4 Illustration of different types of recombination processes inside a semiconductor film.

of trap states, the density of trapped electrons and holes, and the lifetime of trap states (τ_e and τ_h). The rate of recombination is inversely proportional to the defect density or trap density. The rate of recombination increases with an increased value of trap density in semiconductor materials.

16.3.4.2 Radiative Recombination

In this mechanism, rather than being trapped at some intermediate trap, electrons present in the conduction band fall directly back into the valence band (Figure 16.4), and due to this reason, it is referred to as the band-to-band recombination. Meanwhile, the radiative recombination results in the emission of photons yielding applications such as light-emitting diodes and laser devices. The general equation for the net radiative recombination process is given by,

$$R_\lambda = B\left(pn - n_i^2 \right) \tag{16.14}$$

16.3.4.3 Auger Recombination

In Auger recombination, the excess energy is given to other electrons or holes, and again these carriers get relaxed by losing their energy in the form of heat and photons (as illustrated in Figure 16.4). According to calculations involving Auger processes for recombination rates, as described by

$$R_{Auger} = \left(\wedge_n n + \wedge_p p \right)\left(pn - n_i^2 \right) \tag{16.15}$$

Thus, each of these recombination processes is responsible for the reduction in generation or carrier transport in solar cells. Thus, the total recombination rate is

$$R = \left[\sum_{trapsi} R_{SLT,i} \right] + R_\lambda + R_{Auger} \tag{16.16}$$

16.3.5 Equations Related to Solar Cells

In numerical modeling, equivalent mathematical equations like Poisson's and continuity equations for holes and electrons are solved iteratively to evaluate the performance of the solar cell. Poisson's equation is mathematically expressed by [11],

$$\frac{\partial}{\partial x}\left(\varepsilon(x)\frac{\partial \psi}{\partial x} \right) = -\frac{q}{\varepsilon_0}\left[-n + p - N_D^- + \frac{1}{q}\rho def\left(n, p \right) \right] \tag{16.17}$$

where ψ denotes electrostatic potential in volts (V), q the electronic charge in coulombs (Q), p and n refer to hole and electron concentrations (cm^{-3}), N_A and N_D are acceptors and donor doping concentrations in cm^{-3}, $\rho def(n, p)$ denote the defect distribution in cm^{-3}, and ε denotes effective dielectric permittivity. Equation (16.18) and (16.19) represent the carrier continuity equation iteratively for the two types of charge carriers, holes and electrons [11],

$$-\frac{\partial Jn}{\partial x} + G - Un\left(n, p \right) = \frac{\partial n}{\partial t} \tag{16.18}$$

$$-\frac{\partial Jp}{\partial x} + G - Un\left(n, p \right) = \frac{\partial p}{\partial t} \tag{16.19}$$

where, J_n and J_p denote the current density of the electron and hole in mA/cm^2, respectively; U_n refers to the total rate of recombination, which is expressed in cm^{-3}s^{-1}, and G is the rate of generation which is expressed in cm^{-3} s^{-1}. The carrier density gradient produces the diffusion current flowing through the semiconductor. Equations (16.20) and (16.21) show that electric field (E) yields drift current density attributed to the drift velocity of the charge carriers, viz. electrons and holes,

$$J_{nd} = qD_n \frac{\partial n}{\partial x} + qn\mu_n E \tag{16.20}$$

$$J_{pd} = -qD_p \frac{\partial p}{\partial x} + qp\mu_p E \tag{16.21}$$

J_{nd} and J_{pd} denote the current density due to the majority and minority electrons and holes in the n and p-type regions, respectively; D_n and D_p denote the diffusion constant of the electron and hole, and μ_n and μ_p refer to electron and hole mobility. Equation (16.22) shows that the net current density is equal to the addition of J_{nd}, and J_{pd}, concerning the diffusion length limits (X_n and X_p) in the n and p regions, respectively.

$$J = J_{nd}\left(X_n\right) + J_{pd}\left(X_p\right) \tag{16.22}$$

16.4   TFSC MATERIALS REVIEW

A thin-film solar cell (TFSC) is comprised of many layers of various semiconductor materials whose thickness varies from 50 nm to 3000 nm [12]. For a thin-film solar cell to operate, there are several components necessary, including a substrate, a transparent conducting oxide layer, a window layer (p- or n-type), a layer of active intrinsic material or absorber (i-type), and metal connections. Each semiconductor layer possesses different chemical and physical properties. Thus, the overall performance of a solar cell is influenced by the semiconductor properties, and it becomes essential to understand the typical behavior of these layers while designing a cell. The involved layers in a solar cell possess different diffusion coefficients, lattice constant, electron affinity or work function, crystal structure, chemical affinity and mobility, thermal expansion coefficient, mechanical adhesion, carrier mobility, microstructure, etc. Moreover, each layer produces bulk and interfacial defects, light reflection, transmission, and scattering, which also prove to be critical in the context of PV performance of the cells. Figure 16.5. shows the different layers inside a thin-film solar cell structure.

16.4.1   SUBSTRATE

There are two types of structures generally meant for fabricating thin-film solar cells, viz. superstrate structure and substrate structure. In a superstrate structure, a transparent conducting oxide is used as the substrate and contact. In substrate configuration, a metal coating is used on a glass substrate where metal coating acts as the contact. A substrate is an inert, passive layer with mechanical

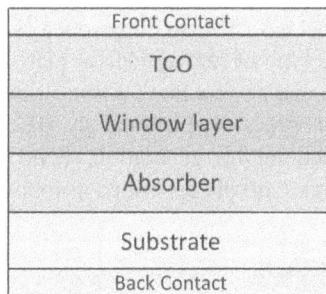

FIGURE 16.5   Schematic diagram of a cross-section view of a thin-film solar cell.

stability and is insensitive to thermal expansion. The substrate can be electrically conducting for front or rear contacts. Polymer materials are preferred for the fabrication of flexible substrates which can be used in flexible solar cells. A substrate is chosen such that it should be sustained during the high-temperature deposition process up to 600°C. The commonly utilized substrates include stainless steel, molybdenum (Mo), nickel (Ni), copper (Cu), and others.

16.4.2 TRANSPARENT CONDUCTING OXIDE (TCO)

According to the solar spectrum, transparent conducting oxide (TCO) is generally regarded as one of the most effective n-type semiconductors due to its high conductivity and transparency. TCO has a low sheet resistance and allows the complete transmission of most of the photons to the intrinsic absorber layer. It is important to note that the conductivity of TCO depends on the carrier concentration and carrier mobility. The transparency of the TCO layer is reduced if the charge carrier concentration is increased beyond a certain level. This means that, without affecting the carrier concentration, the carrier mobility can be adjusted to control the conductivity and transparency of the TCO. Thus, to increase the quality of the TCO layer, the only good solution is to increase its crystalline carrier mobility. Apart from these, the mechanical, thermal, and chemical stability and passivity of TCO are some other important considerations in the context of TCO. The typical examples of transparent conducting oxide materials are indium oxide (In_2O_3), zinc oxide (ZnO), tin oxide (SnO_3), etc.

16.4.3 WINDOW LAYER

Most of the incident radiation passes through the window layer, which is generally made up of a heavily doped n-type semiconductor. No charge carriers are generated inside the window layer while it admits the maximum amount of light to an absorber layer. With the high value of the band gap energy of the window and possibly keeping the low value of thickness, low series resistance has been ensured. Commonly used window layer materials are CdS, CdZnS, a-SiC: H, etc.

16.4.4 ABSORBER

An absorber layer is an intrinsic layer that creates a strong electric field or built-in potential in the depletion region. Absorber materials should have a good absorption coefficient, i.e., absorber material should absorb the photons corresponding to the wide range of solar spectrum in UV, visible, and IR regions, large carrier diffusion length, and high carrier mobility. The bandgap energy of the absorber should be such that it absorbs the maximum photon energy corresponding to the wide range of the solar spectrum. A typical value of the absorber bandgap energy varies from 1.1 eV to 4 eV. The popular absorber layer materials are $Cu(Ga, In)Se_2$, CdTe, a-Si: H, $CH_3NH_3PbI_3$, etc.

16.5 CLASSIFICATION OF THE SOLAR CELL TECHNOLOGIES

The classification of technologies that have been in the practice of solar cells is shown in Figure 16.6, which has been made based on the type of semiconductor material and processing methods used. The highly efficient, cost-effective, and environment-friendly solar cells can be fabricated through the proper selection of materials and processing technology. Based on it, there are four generations of PV technological evolutions called the first-generation, second-generation, third-generation, and fourth-generation solar cells, which are briefly described as follows.

16.5.1 FIRST-GENERATION SOLAR CELLS

The first generation of solar cells was manufactured with crystalline silicon, which had ruled the PV market for the previous 50 years. The silicon solar cell processing cost is too expensive for an

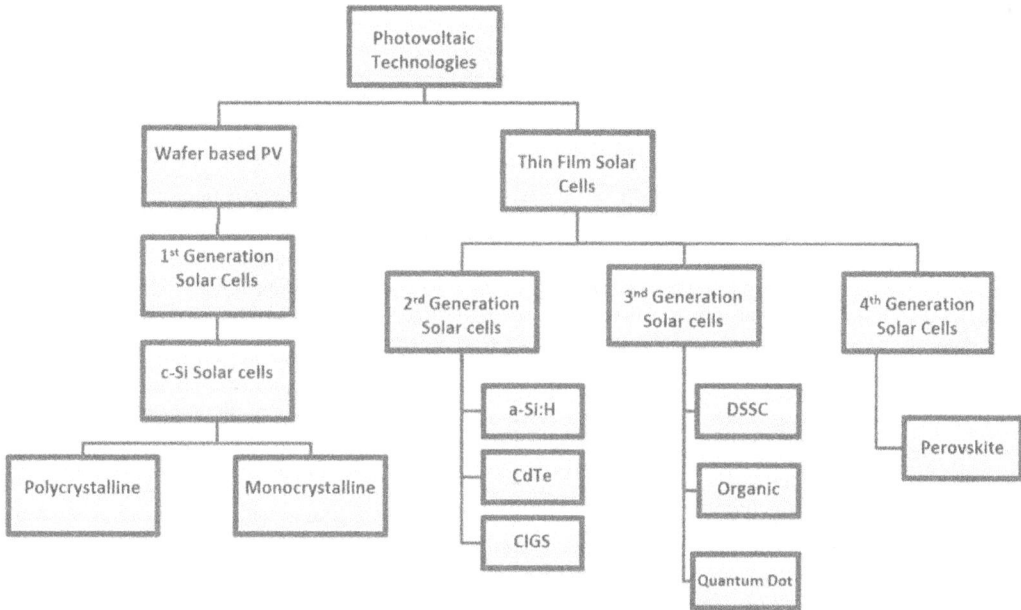

FIGURE 16.6 Classification of the solar cell technologies.

efficiency of less than 24%. Again, there are two subtypes of silicon solar cells, viz. monocrystalline or single-crystal and polycrystalline. Monocrystalline types are costly since they utilize the silicon wafer that has a regular lattice structure, continuous, unbroken, homogeneous, and pure form than polysilicon. These single crystals are grown using the Czochralski technique, and their power conversion efficiency varies between 17% to 24%. Polycrystalline solar cells or panels are composed of several small silicon crystals. Polysilicon is made from metallurgical grade silicon using a high-temperature chemical purification process. It has less purity as compared to single-crystal silicon. Hence, due to this the grain boundaries and defect concentration are very high. Therefore, their power conversion efficiency is also limited from 13% to 17%. Si wafer releases poisonous gases during the processing of the single crystal, so it is environmentally unfriendly [13].

16.5.2 SECOND GENERATION OF SOLAR CELLS

The CdTe, CIGS-based, and a-Si: H, thin-film solar cells are regarded as the second generation of solar cells. In 1980, the first Cu_2S/CdS-based thin-film solar cell was developed to yield >10% (USA) efficiency value. In 1986, the first a-Si G4000-based commercial thin-film power module was developed by Arco Solar (USA). CIGS solar cells are efficient inorganic thin-film solar cells. A CIGS cell is manufactured by depositing copper, indium, gallium, and sulfide on a soda-lime glass or polymer substrate. $CuInGaSe_2$ solar cells have a wide absorption coefficient in the range of 10^5 cm^{-1} with a bandgap value of 1.5 eV. In 2014, the National Renewable Energy Laboratory (NERL) recorded the highest efficiency of 21.4%. The a-Si: H solar cell has an amorphous absorber semiconductor thin-film material for a solar cell that provides an efficiency lower than that of a c-Si cell. However, it has a low manufacturing cost due to the roll-to-roll deposition technique, and the a-Si: H solar cells have been fabricated on the lime soda glass as well as on a plastic substrate [14]. Till 2019, the efficiency of a-Si: H cells reached up to 14.2%. CdTe, CIGS, and a-Si: H-based solar cells often require chemical vapor deposition methods during film growth that consumes high energy.

16.5.3   THIRD GENERATION OF SOLAR CELLS

The third generation of solar cells remains to be popularly discovered as solution-processable thin-film PV cells, which are fabricated using organic/polymer materials at a drastically reduced cost. These include quantum dot solar cells (QDSCs), organic or polymer PV cells (OPVs), and organic dye-sensitized solar cells (DSSCs). In DSSCs, organic dyes are used as photosensitizers or absorber materials. The charges are produced similarly to the process of photosynthesis in plants, although at high temperature electrolyte gets expanded and vaporizes. Also, at low temperature, the electrolyte freezes. Hence, the efficiency of DSSC is very low and also presents stability issues. In QDSCs, PbS2-based quantum dots are used as absorber material. Quantum dots have tunable band gap energy throughout an extensive range of energy levels, which is regulated by the size of the Q-dot. Till 2019, the maximum efficiency of QDSCs achieved exceeded 16.5%. However, the QDSCs suffer from weak absorption ability at room temperature. Organic solar cells use organic or polymer materials for thin-film growth. OPVs are also used for fabricating flexible solar cells. Generally, solution-based deposition methods are used, which have very low cost and stability. OPVs are under the research phase due to the limited range of power conversion efficiency varying from 1% to 13% achieved experimentally. The QDSCs, DSSCs, and OPVs offer a low power conversion efficiency, PCE < 20% due to the low light absorption and charge collection [14].

16.5.4   FOURTH GENERATION OF SOLAR CELLS

The fourth generation of solar cells consists of halide-based perovskite solar cells which are specially designed to improve the power conversion efficiency and intended to reduce the cost of the solution processing methods. The chemical formula commonly used for metal halide perovskite materials is ABX_3, where A represents a monovalent or organic cation such as $CH_3NH_3^+$ or $CH(NH_2)^+$ and B represents a divalent metal cation (such as Pb^{2+} or Sn^{2+}). X represents a halide anion (Cl^-, Br^-, I^-) [15]. In 2009, Kojima and Miyasaka et al testified the primary perovskite solar cells were made using iodide or bromide-based electrolytes with the sensitizers being $CH_3NH_3PbBr_3$ and $CH_3NH_3PbI_3$. They tailored an improved PCE of 3.81% for the $CH_3NH_3PbBr_3$ and the $CH_3NH_3PbI_3$-based perovskite devices at 3.13% [16]. Organometal perovskite solar cells became very popular in a short period because of their high spectral absorption, large diffusion length, and low trap density, which results in a high percentage of power conversion efficiency. In 2019, Shuyan Shao, et al. reported the highest power conversion efficiency of 25.2% [15].

16.6   PERFORMANCE PARAMETERS OF THE SOLAR CELLS

To determine the performance parameters of solar cells, the J-V characteristic is generally utilized. The importance and equations of performance metrics are explained as follows.

16.6.1   SHORT CIRCUIT CURRENT (I_{SC})

Short circuit current (I_{SC}) can be defined as the maximum value of the current that flows in its external part of the circuit when metal contact, viz. anode and cathode of a solar cell has been short-circuited, i.e., V = 0. Short-circuit current depends upon the following factors,

1. Photon flux incidence (solar radiation): The standard solar radiation on the earth is 1000 W/m^2 at AM1.5G, i.e., $I_{SC} = I_{Ph}$ which is proportional to the solar radiation intensity.
2. When sunlight falls on the surface area of the solar cell, it causes the rise of I_{sc}. As the surface area of the solar cell increases, more electron-hole pairs will be generated when the solar cell is exposed to light.

Ideally, the maximum value of the short-circuit current flow through a crystalline silicon solar cell is 46 mA/m². However, the experimentally obtained value of I_{SC} in a solar cell is 35 mA/m² [17]

16.6.2 OPEN CIRCUIT VOLTAGE

Open circuit voltage (V_{OC}) is the voltage that is attained at the solar cell terminal when the current is zero through the external load circuit.

$$V_{OC} = \frac{nK_BT}{q}\ln\left(\frac{I_{Ph}}{I_0}+1\right) \tag{16.23}$$

i.e.,

$$V_{OC} = \frac{nK_BT}{q}\ln\left(\frac{I_{Ph}}{I_0}\right) \text{ When } I_{Ph} \gg I_0 \tag{16.24}$$

Generally, V_{OC} is dependent upon I_{Ph} and I_0, where I_0 is a dark current reverse saturation current and I_{ph} is a photocurrent. I_0 is dependent upon the rate of recombination in the device and the temperature. As the temperature of the device increases, I_0 increases and V_{OC} will be decreased. In addition, the temperature of the cells is inversely proportional to the V_{OC}. Due to the charge carriers' rapid movement, the open-circuit voltage rises as the temperature drops.

16.6.3 FILL FACTOR

The fill factor (FF) is the ratio of maximum power generated by the solar cell to the product of V_{OC} and I_{SC}.

$$FF = \frac{P_{MPP}}{V_{OC}*I_{SC}} = \frac{V_{MPP}*I_{MPP}}{V_{OC}*I_{SC}} \tag{16.25}$$

Operating a solar cell at its maximum power point (MPP) can optimize its performance. The FF can frequently be reduced due to unwanted resistive losses. Consider that the solar cell behaves like an ideal diode, with the FF being a function of the open-circuit voltages V_{OC} [6].

$$FF = \frac{V_{OC} - In(V_{OC}+0.72)}{V_{OC}+1} \tag{16.26}$$

Here, V_{OC} is the normalized voltage (approximately V_{OC} >10) for an ideal value of the FF. FF for the GaAs solar cell is 0.89 and for Si, FF is 0.85.

16.6.4 POWER CONVERSION EFFICIENCY (η)

The ratio of the maximum generated power to the optical incident power (Pi) is known as the power conversion efficiency (PCE). Corresponding to the AM1.5G spectrum, 1000 w/m² is the solar cell irradiance on the solar surface, i.e., P_{in} = 1000 w/m².

$$\eta = \frac{P_{max}}{P_{in}} = \frac{I_{MPP}*V_{MPP}}{P_{in}} = \frac{I_{SC}*V_{OC}*FF}{P_{in}} \tag{16.27}$$

Here, P_{in} = (Solar constant) × (Area of the solar cell's surface [W/m²]).

16.7   FUTURE OF EMERGING SOLAR TECHNOLOGIES

Most of the sunlight falling on the earth's surface gets diffused or dispersed in the atmosphere. Due to this fact, the complete utilization of the solar cell performance is not being accomplished. However, we can significantly improve solar cell efficiency at a lower cost. The maximum amount of power that may be obtained through a single junction PV cell is 33%, following the Shockley-Queisser limit [18]. NREL, a part of the US Department of Energy (DOE), has improved solar cell efficiency through meticulous material design in the cell stack. Their study in Cell Reports Physical Science details the growth of a gallium arsenide (GaAs) heterojunction solar cell using dynamic hydride vapor phase epitaxy (D-HVPE), setting a new record with a certified efficiency of 27% for a single-junction GaAs cell produced with this method [19].However, the capability of solar cells can be enhanced through two approaches, viz. using the multi-junction structure and structure with different band gap semiconductor materials (tandem solar cells). In tandem solar cells, each layer of the cell is band gap graded to absorb the wide range of solar energy spectrum. A tandem solar cell involving GaInP/GaAs/Ge layers has been designed with 32% conversion efficiency [20], which is very close to the Shockley-Quissier efficiency limit.

16.8   CONCLUSION

Nowadays, more communities are using thin-film solar cells than their conventional thick-film silicon-based counterparts. However, thin-film solar cells have lower fabrication cost as it involves a lesser amount of material consumption and economical solution processing fabrication methods. Since 2009, the PV market share of thin-film solar cells has risen from 17% to 38%. Currently, organometal halide solar cells are becoming more popular and on the verge of commercialization. Researchers have designed perovskite solar cells intending to maximize efficiency at a much-reduced cost; however, the main problem with perovskite solar cell technology is their long-term stability. Apart from the rigid PV technology, flexible solar cells are also fast catching up due to their ability to be folded or rolled. Therefore, a lot of research activities are going on toward the improvement of substrate material quality by improving their durability, mechanical breakdown strength, and high-temperature processing ability. The flexible solar panels offer novel applications such as wall shades, window shades, wallpapers, curved roofs, etc.

REFERENCES

1. S. Hegedus, and A. Luque, *Handbook of Photovoltaic Science and Engineering*, vol. 129. WILEY Online Library (2003).
2. Z. I. Alferov, "The double heterostructure: Concept and its applications in physics, electronics, and technology," *Int. J. Mod. Phys. B*, vol. 16, (2002) pp. 647–675.
3. D. Frey, Vanilla vehicles, *Mach. Des.*, vol. 74, no. 21, (2002) p. 18.
4. K. Ramanathan, J. Keane, and R. Noufi, "Properties of high-efficiency CIGS thin-film solar cells," Paper presented at *The IEEE Photovoltaic Specialists Conference*, February, 2005, pp. 195–198.
5. S. A. Olaleru, J. K. Kirui, D. Wamwangi, K. T. Roro, and B. Mwakikunga, "Perovskite solar cells: The new epoch in photovoltaics," *Sol. Energy*, vol. 196, (2020) pp. 295–309.
6. K. Mertens, *Photovoltaics Fundamentals, Technology and Practice*. John Wiley & Sons (2018).
7. A. Shah, *Thin-Film Silicon Solar Cells*. CRC Press (2010).
8. I. M. Dharmadasa, *Advances in Thin-Film Solar Cells*. CRC Press (2012).
9. A. Luque, and S. Hegedus, *Photovoltaic Science Handbook of Photovoltaic Science*. WILEY Online Library (2011).
10. S. Abdelaziz, A. Zekry, A. Shaker, and M. Abouelatta, "Investigating the performance of formamidinium tin-based perovskite solar cell by SCAPS device simulation," *Opt. Mater*, vol. 101, (2020), pp. 109738(1-8).
11. S. Bhattarai, and T. D. Das, "Optimization of carrier transport materials for the performance enhancement of the MAGeI$_3$ based perovskite solar cell optimization of carrier transport materials for the performance enhancement of the MAGeI 3 based perovskite solar cell," *Sol. Energy*, vol. 217, (2021) pp. 200–207.

12. K. L. Chopra, P. D. Paulson, and V. Dutta, "Thin-film solar cells: An overview," *Prog. Photovoltaics Res. Appl.*, vol. 12, (2004) pp. 69–92.

13. P. Tonui, S. O. Oseni, G. Sharma, Q. Yan, and G. Tessema Mola, "Perovskites photovoltaic solar cells: An overview of current status," *Renew. Sustain. Energy Rev.*, vol. 91, (2018) pp. 1025–1044.

14. G. Han, S. Zhang, P. P. Boix, L. H. Wong, L. Sun, and S. Y. Lien, "Towards high-efficiency thin film solar cells," *Prog. Mater. Sci.*, vol. 87, (2017) pp. 246–291.

15. S. Shao, and M. A. Loi, "The role of the interfaces in perovskite solar cells," *Adv. Mater. Interfaces*, vol. 7, (2020), pp. 1901469 (1-31).

16. A. Kojima, K. Teshima, Y. Shirai, and T. Miyasaka, "Organometal halide perovskites as visible-light sensitizers for photovoltaic," *Journal of the American Chemical Society* vol. 131 (2009) pp. 6050–6051,

17. E. S. Han, A. Goleman, D. Boyatzis, and R. McKee, "Basic photovoltaic principles and methods," *J. Chem. Inf. Model*, vol. 53, (1982) pp. 1689–1699.

18. W. E. I. Sha, X. Ren, L. Chen, and W. C. H. Choy, "The efficiency limit of CH3NH3PbI$_3$ perovskite solar cells," *Appl. Phys. Lett.*, vol. 106, (2015), pp. 221104(1-5).

19. Kevin L. Schulte et. al., "Modeling and design of III-V heterojunction solar cells for enhanced performance," *Cell Reports Physical Science*, vol. 4, Issue 9, (2023), 101541.

20. H. Cotal *et al.*, "III-V multijunction solar cells for concentrating photovoltaics," *Energy Environ. Sci.*, vol. 2, (2009) pp. 174–192.

17 Semiconductor Lasers

Shaoteng Wu and Haizhong Weng

17.1 INTRODUCTION

17.1.1 THE BIRTH OF LASERS

The origin of laser theory can be traced back to 1916 when Einstein proposed, based on quantum theory, that when matter interacts with the electromagnetic field of light, there is not only stimulated absorption but also stimulated radiation. In 1917, he cleverly employed the concept of an equilibrium state to elucidate the phenomenon of spontaneous emission. Einstein's theory also gave a necessary condition for the realization of lasers: particle population inversion. This condition implies that there must be surplus atoms at high energy levels compared to those at low energy levels, as only these excess atoms at high energy levels can prompt the dominance of stimulated emission over stimulated absorption. In a state of thermal equilibrium, particle population inversion cannot occur naturally. Hence, an external energy source, often referred to as the "pump," is required to continually excite atoms to higher energy levels, hereby fulfilling the prerequisite condition for laser operation.

In the early 1950s, there arose a need to transition from radio technology to shorter wavelengths, prompting scientists to explore the practical implementation of lasers. For example, Charles H. Townes from the United States, and Nikolai G. Basov and Aleksander M. Prokhorov from the Soviet Union proposed leveraging Einstein's theory to amplify electromagnetic waves through stimulated radiation from atoms and molecules [1]. Later, Townes et al. introduced the concept of substituting the traditional closed resonator with an open resonator, while Nicolaas Bloembergen advocated for using a three-level atom or molecule system pumped by light to achieve particle population inversion [2]. The culmination of these efforts occurred in 1960 when the world witnessed the creation of the first ruby laser [3]. Numerous experimental results have demonstrated that lasers have excellent properties that ordinary light does not have, including excellent brightness, high monochromaticity, high directivity, and strong coherence [4].

17.1.2 SEMICONDUCTOR LASERS

Following the invention of the laser, researchers promptly initiated studies into semiconductor lasers due to the small size, high efficiency, and long lifespan. Nonetheless, early homogeneous junction semiconductor lasers, as shown in Figure 17.1a, commonly referred to as laser diodes, faced a limitation in that they could only operate in a pulsed mode at low temperatures. Overcoming this limitation was a critical focus of research and development in this area. Zhores Alferov and Rudolf Kazarinov pioneered the development of the first single-heterojunction laser [5, 6], as shown in Figure 17.1b. Then H. Kressel and Z. I. Alferov [7] proposed the structure of a double-heterojunction semiconductor laser, achieving continuous operation at room temperature with the structure depicted in Figure 17.1c. As the research and development of heterojunction lasers progressed, there was a strategy to use ultra-thin semiconductor layers in the laser's source region to induce quantum effects. Bolstered by advanced thin-film growth technologies such as molecular beam epitaxy (MBE) and metalorganic chemical vapor deposition (MOCVD) [8], this concept led to the production of quantum well (QW) lasers, as illustrated in Figure 17.1d. QW lasers are distinguished by the formation of potential wells in the source region that are smaller than the de Broglie wavelength of electrons in the material. This results in the continuous band splitting into sub-energy levels, thereby enhancing the effective carrier density.

DOI: 10.1201/9781003450146-17

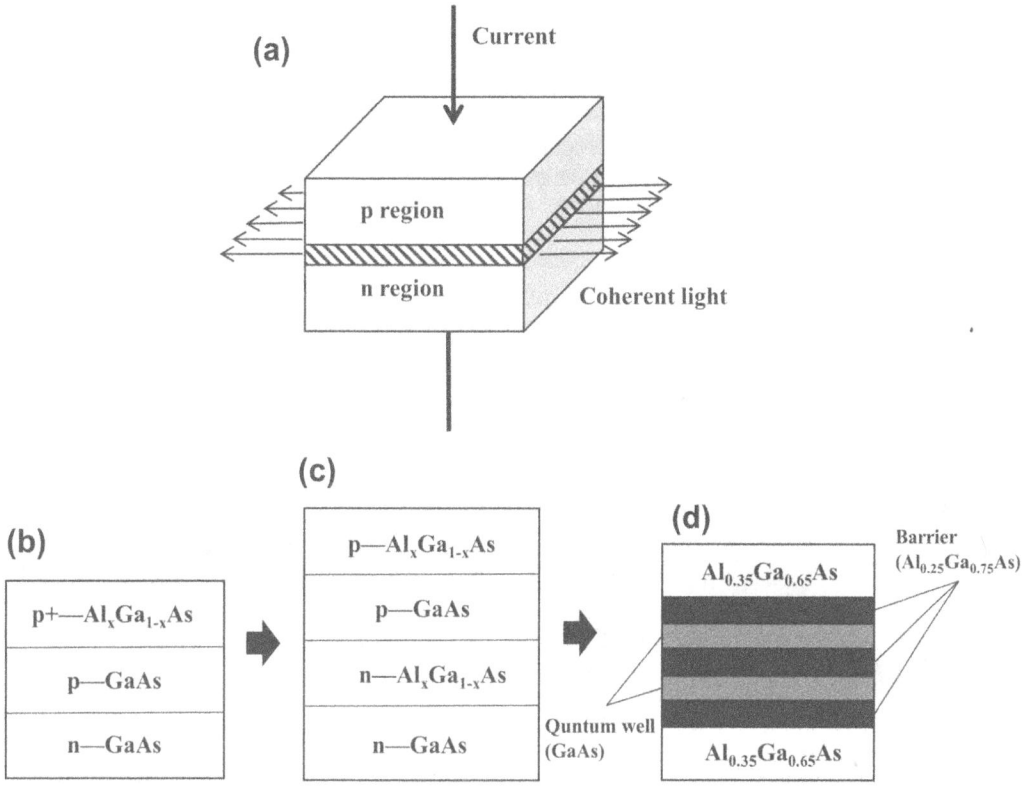

FIGURE 17.1 (a) Schematic illustration of laser diode structure; (b) Single-heterojunction laser; (c) Double-heterojunction laser; (d) Multi-quantum well (MQW) laser.

The application of semiconductor lasers in various aspects has spurred the development of a range of new semiconductor lasers, which can be broadly categorized based on their structural characteristics. These categories include the vertical-cavity surface-emitting laser (VCSEL), micro-cavity semiconductor lasers, quantum cascade lasers (QCLs), integrated semiconductor lasers, and so on. Below we will provide detailed descriptions of these lasers.

17.2 VCSEL

Traditional semiconductor edge-emitting lasers (EELs) are a type of laser diode that emits coherent light from the edge of a thin, planar semiconductor wafer, as illustrated in Figure 17.1a. These lasers typically comprise a semiconductor chip with multiple layers, including an active region where light generation occurs and two reflective facets on opposite sides of the chip. In contrast, the VCSEL has emerged as a semiconductor laser design with distinct advantages, which was proposed by Professor Kenichi Iga at Tokyo Institute of Technology in 1977 [9]. The output direction of VCSEL is perpendicular to the substrate surface, as illustrated in Figure 17.2a. VCSEL has a wide range of applications, including optical fiber communication, laser printing, sensing and pumping light sources, etc [10, 11].

As shown in Figure 17.2a, the VCSEL structure comprises p-type and n-type distributed Bragg reflectors (DBRs), a multi-quantum well (MQW) active region, an oxidation limiting layer, the substrate, and electrode contacts. The DBR in the VCSEL is constructed using a periodic layer structure consisting of alternating high and low refractive index layers, each having a thickness

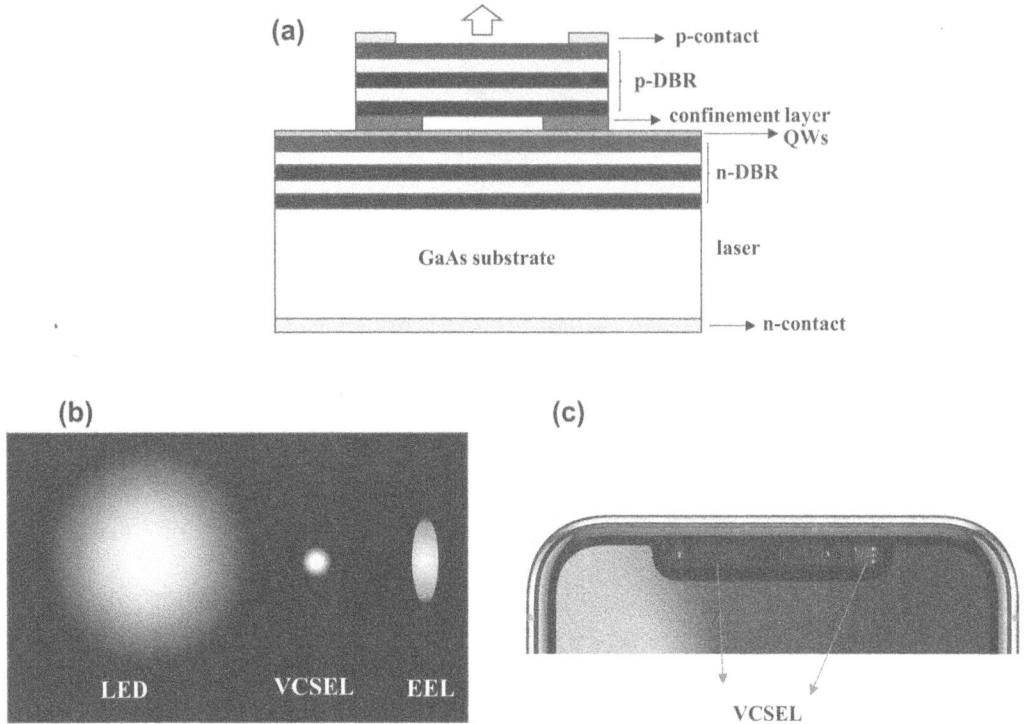

FIGURE 17.2   (a) Structure diagram of the VCSEL; (b) Comparison beween the beam shapes of different light sources; (c) Face recognition for iPhones.

of 1/4 optical wavelength. Due to the half-wave loss from the optically sparse medium to the optically dense medium, the reflected light at the two interfaces can form interference-enhanced reflection, and the reflectivity can exceed 99% [11]. The reflectivity and reflection bandwidth are related to the ratio of the two refractive indices. A greater difference in refractive indices leads to higher reflectivity and a wider bandwidth. The number of layers in the DBRs also influences reflectivity, with a higher number of layers resulting in greater reflectivity. Taking VCSELs based on GaAs as an example, AlGaAs with different doping concentrations and different Al components are commonly employed in DBR manufacturing to maximize the contrast in refractive indices. This tailored construction enhances the performance of the VCSEL by optimizing its reflective properties.

In addition, the active region consisting of MQW serves for both constraining carriers and acting as an optical cavity. The active region is placed in the middle of p- and n-DBR, forming a Fabry–Pérot (FP) cavity where the thickness is an integer multiple of half the wavelength, meeting the standing wave condition. The oxidation limiting layer plays a crucial role in controlling the single transverse mode of VCSEL [12]. Since the number of transverse modes is directly related to the transverse size, a primary method to achieve single-mode operation is filtering out the high-order modes by reducing the oxidation layer aperture. Reports suggest that achieving single-mode operation can be realized by limiting the size of the oxidation layer aperture to 3 μm [13]. However, precise control of the aperture size poses challenges, making this approach less common. To address this challenge, various alternative methods have been developed, including conical oxidation aperture VCSEL, ion implantation VCSEL, metal aperture VCSEL, Zn diffusion VCSEL, photonic crystal VCSEL, surface relief VCSEL, etc.

In addition to the convenient realization of two-dimensional arrays, VCSEL also boasts many other advantages. These include a circular beam profile (Figure 17.2b), facilitating easy coupling with optical fibers. VCSELs also exhibit small threshold currents, high power conversion efficiency, and low power consumption. Achieving single longitudinal mode operation is straightforward due to the short cavity length and large frequency spacing of longitudinal modes. Moreover, VCSELs offer a wide operating temperature range, are suitable for LED processes, facilitate commercial production, and demonstrate high reliability such as high power conversion efficiency, low power consumption, wide operating temperature range, and high reliability [14]. It has been applied in the face recognition module of the mobile phone, as depicted in Figure 17.2c. In summary, VCSELs have many advantages and applications, making them of significant value for both research and practical use.

17.3   SEMICONDUCTOR WGM MICROCAVITY LASER

Semiconductor whispering gallery mode (WGM) microcavity lasers offer promising prospects for on-chip interconnection and sensing, thanks to their advantages of a compact footprint, low threshold, and in-plane light emission. The WGM phenomenon, originally described by Lord Rayleigh to explain the sound wave behavior of sound observed in St. Paul's Cathedral [15], signifies the confinement and propagation of waves around a concave surface. Research in WGM optical microcavities occupies a prominent position in both fundamental physics investigations and the development of practical optoelectronic devices, including filters, modulators, and lasers [16]. Within an optical microresonator, the light field is strongly confined in an WGM pattern through total internal reflection at the high refractive index microcavity interface [17]. Figures 17.3a and 17.3b illustrate the ray-optics depiction of light propagation in a microring resonator and the electric field distribution of a fundamental mode near the surface and along the equatorial plane of the microring. Figures 17.3c and 17.3d present scanning electron microscopy (SEM) images of a 10-μm-radius ring resonator, viewed from the top and side angles. In such a microring resonator, the resonant WGM wavelength approximately counts as:

$$2\pi R \cdot n = m \cdot \lambda \tag{17.1}$$

where n is the refractive index, m is the integer number of wavelengths that fit around the perimeter of the resonator $2\pi R$, λ is wavelength.

We will commence by introducing the low-threshold characteristics. Following that, we will describe various semiconductor materials for the emission of diverse wavelengths in WGM microlasers. Subsequently, we will introduce the methods employed for light output coupling and achieving directional emission.

Within a WGM cavity, the spontaneous emission characteristics of excited atoms undergo significant changes. This transformation primarily arises from the presence of the optical microcavity, which modifies the spatial and spectral distribution of optical modes. When the wavelength of spontaneous emission from excited atoms aligns with the resonance wavelength of the optical cavity, spontaneous emission becomes intensified. This phenomenon was initially described by Purcell in 1946, and the degree of enhancement is quantified by the Purcell factor [21].

$$P = \frac{3}{4\pi^2}\left(\frac{\lambda}{n}\right)^3 \frac{Q}{V} \tag{17.2}$$

where the Q is the quality factor and V is the mode volume of the cavity modes. λ is the wavelength and n is the refractive index. In cases where Q/V of an optical microcavity is extremely high, it can profoundly influence the characteristics of spontaneous emission within the cavity. Consequently, there is a strong expectation to enhance the spontaneous emission factor and significantly lower the

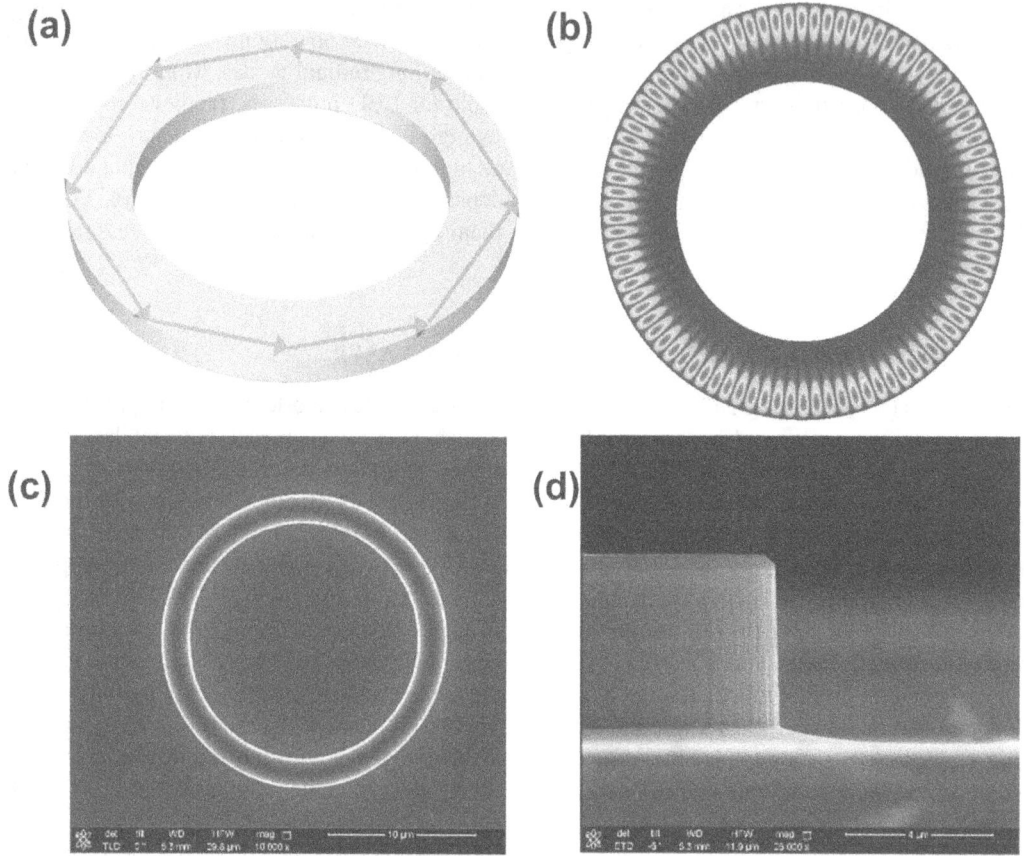

FIGURE 17.3 The light field propagation and distribution in the view of (a) geometrical optics and (b) wave optics; (c) and (d) are the top-view and side-view SEM images of a microring resonator.

laser's threshold with the WGM microcavity. These microcavities can assume various shapes, including disks, spheres, cylindricals, toroids and polygons [17]. To enable lasing within a WGM cavity, it is crucial to integrate an active layer into the cavity's structure to provide the optical gain. Our primary focus will be on the III-V semiconductors with direct band gap, making them ideal candidates for gain materials. Additionally, the refractive index of a resonator significantly differs from that of the surrounding material. In III-V semiconductor lasers, vertical optical confinement is typically achieved by exploiting the disparities in refractive indices among epitaxial layers in a sandwich-like structure. In the lateral direction, the light can be effectively limited by the low-index material such as silicon dioxide (SiO_2) or air.

The earliest demonstration of semiconductor WGM lasers can be traced back to 1992 when Bell Labs first reported an optically pumped laser utilizing a 5-μm-diameter InP/InGaAsP microdisk, as depicted in Figure 17.4a. Benefiting from the ample gain supplied by the InGaAs QW under cooling conditions, they achieved single-mode lasing at 1.3 and 1.5 μm with pump thresholds of less than 100 μW [18]. In that same year, they also reported electrically injected microdisk lasers operating with sub-milliampere threshold currents at room temperature [22]. Due to the distribution of WGMs near the boundaries of the microcavity, any surface roughness that occurs through fabrication can lead to strong light scattering. There is a pressing need for smooth cavity sidewalls to minimize non-radiative recombination at the surface and enhance the Q factors. Using SiO_2 as a hard mask and employing an optimized dry etching recipe with Cl_2-based gas can ensure the

(a) (b) (c)

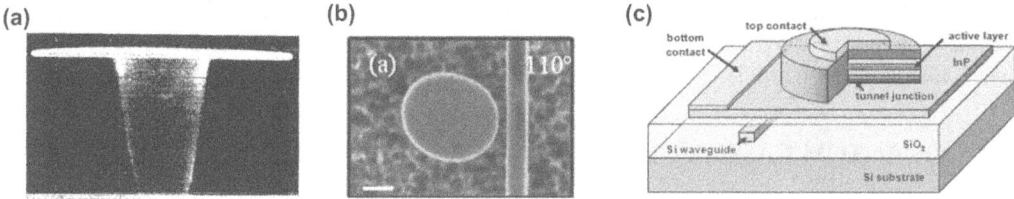

FIGURE 17.4   (a) SEM image depicting the first optically pumped microdisk laser as reported by Bell Labs. (Adapted with permission [18], Copyright [1992], AIP Publishing). (b) SEM image showcasing lateral coupling with a deformed microdisk cavity. (Adapted with permission [19], Copyright [2012], AIP Publishing). (c) Schematic illustration of a heterogeneous microdisk laser with SOI platform, facilitating vertical evanescent coupling from the laser to the silicon (Si) waveguide. (Adapted with permission [20], Copyright [2007], Optica Publishing Group).

creation of smooth etched sidewalls. In 1998, T. Baba et al. achieved the first room-temperature continuous-wave (CW) laser operation in a 3-μm-diameter InGaAsP/InP microdisk, which was created using reactive-ion etching [23]. Through the reduction of cavity sidewall roughness and the achievement of a Q factor of ~3,300, they obtained a threshold current of 150 μA. Subsequently, they demonstrated CW lasing at 1563 nm, further reducing the threshold current to just 40 μA [24]. The notable improvement was achieved by employing inductively coupled plasma (ICP) etching to form a steep microcavity profile and optimizing cavity dimensions to align the lasing mode with the peak of the gain spectrum.

By carefully manipulating alloy compositions, the emission wavelength of semiconductor WGM microlasers can be flexibly adjusted to cover broad laser spectra, spanning from the visible light to the mid-infrared region. The first WGM lasers designed for the mid-infrared range (3–4 μm) were achieved in 2005 using sub-millimeter InAsSb microrings at temperatures between 70–120 K [25]. With the optimized fabrication processes, the CW room-temperature lasing at 428 nm was successfully accomplished in a GaN microdisk, with a threshold pump power of 300 W/cm^2 [26].

Achieving high emission efficiency is crucial for microcavity lasers. However, the high-Q factor of the WGMs and circular symmetry pose challenges in achieving the directional light output. Previous efforts involved deformed cavities or introducing various scatterers to break this symmetry. Another widely employed approach leverages evanescent wave coupling between WGM output and optical waveguides, offering seamless integration with various optoelectronic devices. Depending on the specific coupling methods employed, this approach can be categorized into two main types: vertical coupling and horizontal coupling. In 2012, Cao's research group at Yale University employed electron beam exposure technology to achieve lateral coupling output between the deformed InAs quantum dot (QD) microdisk laser and the waveguide, as depicted in Figure 17.4b [19]. As shown in Figure 17.4c, researchers from Ghent University achieved directional light emission by vertically coupling evanescent waves into a waveguide on the Si substrate [20]. They first bonded the III-V material to the silicon-on-insulator (SOI) and then etched a thin InP-based microdisk laser. Recently, microdisk/microring lasers with direct-connect waveguides have also been proposed and demonstrated. This scheme effectively enhanced the directional emission efficiency while maintaining the high-Q coupled modes [27]. Additionally, square microcavities with direct-connect waveguide at one corner can support high-Q WGMs and excellent coupling efficiency, enabling high-power dual-wavelength lasing for sub-THz wave generation [28].

Recently, there have been significant advancements in the development of high-speed directly modulated WGM microlasers, including their heterogeneous integration with silicon platforms. These achievements hold great promise for enhancing their utility as on-chip laser sources in photonic integrated circuits [29]. Microcavity lasers have also found diverse applications in fields such as sensing, microwave photonics, all-optical flip-flop memory, random number generation, cavity-quantum electrodynamics, and nonlinear dynamics. Given the maturity of III-V planar photonics fabrication techniques, future research efforts should prioritize further enhancing the performance and yield of semiconductor WGM microlasers. This includes improving factors like internal efficiency, fiber collection efficiency, response speed, and energy efficiency under direct modulation.

17.4 QUANTUM CASCADE LASER

The invention of the quantum cascade laser (QCL, also known as a unipolar laser) is an important milestone in the history of semiconductor laser development. It was first born at Bell Labs in 1994 and is capable of emitting light in the mid-infrared, far-infrared, and terahertz spectral range [31]. This capability addresses a significant need for light sources in these spectral regions, where natural materials are often lacking. In recent years, with the continuous advancement of technology, the applications of QCL are becoming increasingly diverse.

The working principle of QCLs is fundamentally different from traditional semiconductor lasers. For traditional semiconductor lasers, their luminescence relies on the recombination of electrons in the conduction band and holes in the valence band. Each electron can only generate one photon, and the wavelength of the light is related to the band gap of the material. For QCLs, their luminescence relies on the transitions of electrons between the subbands in the conduction band in semiconductor quantum wells, as shown in Figure 17.5a. Each electron can generate multiple photons, and the wavelength of light can be controlled regardless of the material. Due to different structures, QCLs can be divided into three kinds: FP QCL, distributed feedback (DFB) QCL, and external cavity QCL. The basic working principles of QCLs with different structures are the same.

QCLs are made up of a series of semiconductor layers arranged in a periodic fashion. Each period typically contains an injector layer and an active layer, as illustrated in Figure 17.5b. The active layer has a structure of coupled three quantum wells, which can form three energy levels [32]. Among them, the energy difference between the upper energy level and the middle energy level is larger, and that between the lower energy level and the middle energy level is smaller. When the semiconductor thin layers have the appropriate thickness, particle population inversion can be achieved by exploiting the quantum confinement effect within the semiconductor heterojunction

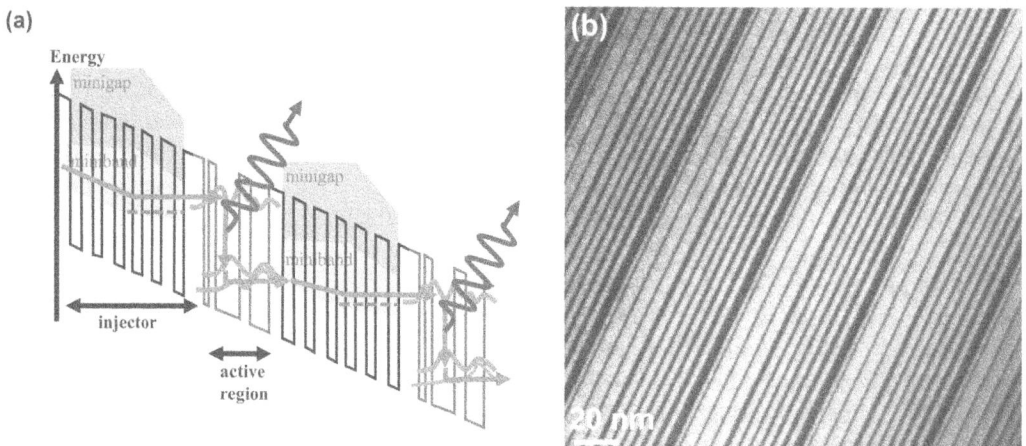

FIGURE 17.5 (a) Schematic of the conduction band structure for quantum cascade laser (QCL); (b) The dark-field transmission electron microscopy (TEM) image of the one cascade of InGaAs/AlInAs QCL heterostructure. (Figure 17.5b adapted with permission from [30]. Copyright A.V. Babichev, A.G. Gladyshev, E.S. Kolodeznyi, A.S. Kurochkin, G.S. Sokolovskii, V.E. Bougrov, L.Y. Karachinsky, I.I. Novikov, V.V. Dudelev, V.N. Nevedomsky, S.O. Slipchenko, A.V. Lutetskiy, A.N. Sofronov, D.A. Firsov, L.E. Vorobjev, N.A. Pikhtin, and A.Y. Egorov, some rights reserved; exclusive license IOP Publishing. Distributed under a Creative Commons Attribution License 3.0 [CC BY] https://creativecommons.org/licenses/by/3.0/.)

thin layer. Then, the electron transitions from the upper energy level to the middle energy level through spontaneous or stimulated emission, generating a photon and emitting light of the corresponding wavelength. After that, the electron on the middle energy level releases a phonon and relaxes to the lower energy level through resonance transport. Since the energy level difference in the active layer is related to the thickness of the semiconductor thin layers, the thickness can be adjusted for tuneable and highly specific wavelength emissions.

Compared with traditional semiconductor lasers, the advantage of QCLs lies in their cascade process. The application of an external electric field leads to the formation of minibands. When the electron descends to the lower energy level within the active layer, aided by the mentioned phonons, it has the capability to tunnel to the upper energy level of the active layer in the subsequent period through the miniband in the injector. This tunneling phenomenon aligns with the quantum tunneling effect. After that, the electron transitions again and emits photons with the same energy. The described process iterates continuously within both the injector and active region of each period. In this way, the utilization efficiency of electrons is greatly increased, and the number of photons generated by one electron is equal to the number of stages of the QCL. Since the number of periods can reach tens or even hundreds, the optical gain of QCLs can be very high with the requirement of higher external voltages.

Currently, QCLs utilizing various material systems can be manufactured through MBE and MOCVD. Since the short-wavelength limit of QCL emission is connected with the depth of the quantum well, different types of materials can be selected for various wavelength spectral ranges. The earliest QCLs in history were made of InP-based GaInAs/AlInAs materials and had very good performance in the mid-infrared spectral range [31, 33]. In addition, QCLs made of GaAs-based GaAs/AlGaAs have been proven to have excellent performance in the terahertz spectral range [34], and material systems such as InAs/AlSb/GaSb can provide shorter wavelength lasers in the mid-infrared spectral range [35].

Due to its unique design principles, QCLs offer several distinct advantages. Firstly, QCLs cover wide spectra, from mid-infrared to terahertz; secondly, QCLs have excellent wavelength tunability; thirdly, QCLs can provide high output power; and finally yet importantly, QCLs are able to work at room temperature. These advantages enable the widespread application of QCLs in various fields, such as gas detection, air pollutant monitoring [36], security, and automobile industry.

17.5   INTEGRATED SEMICONDUCTOR LASER

The Si photonics platform, represented by SOI technology, is playing an increasingly central role in data communication and high-performance computing, driven by its cost-effectiveness and compatibility with complementary metal-oxide-semiconductor (CMOS) technology. Mature SOI devices encompass a wide range of passive components like arrayed waveguide grating, optical couplers, and microring filters. Additionally, active elements like photodetectors have been successfully developed through doping. However, silicon does have a notable limitation due to its poor luminescence efficiency, stemming from its indirect band gap. To overcome this challenge and advance integrated photonic technology, there is a pressing need to incorporate laser sources onto Si platforms. Recognizing that III-V materials excel as gain mediums for laser development, integrating III-V semiconductor lasers onto group IV materials such as Si or germanium (Ge) emerges as an effective solution. These integrated III-V/Si lasers offer several crucial advantages, including narrow-linewidth lasing and seamless integration with other photonic devices manufactured with SOI, silicon nitride (Si_3N_4), and lithium niobate ($LiNbO_3$) technologies. Such capabilities are indispensable in various applications like sensing and the development of large-scale photonic circuits.

As III-V semiconductors are not compatible with CMOS technology, several methods have been proposed to enable the integration of III-V gain materials on a Si substrate. These methods include heterogeneous integration, monolithic integration, hybrid integration, and transfer printing. Among these techniques, hybrid integration, represented by flip-chip bonding, butt coupling, and photonics wire bonding, can be regarded as a micro-packing technique that eliminates the need for optical components such as lenses. This not only reduces the device's footprint but also simplifies the manufacturing process. However, it is important to note that achieving precise alignment, with an error of less than 1 μm, between the two platforms demands considerable effort, and realizing large-scale integration remains a formidable challenge. In this section, we will introduce two widely adopted integration techniques and their recent advancements in III-V/Si lasers, with a particular focus on the communication wavelength regimes.

17.5.1 Epitaxial Growth

Integrated laser through direct epitaxial growth of III-V materials on the Si substrate is the optimal solution due to the benefits of monolithic integration, low cost, and scalability. Epitaxy allows the selective growth of wafer-scale III-V material on Si without the complexities associated with alignment in hybrid integration approaches. However, there are certain challenges in achieving high growth quality in this process. Firstly, the mismatch in lattice constant and polarity between the two platforms can introduce various defects, including threading dislocations and antiphase boundaries. Additionally, the differing thermal expansion coefficients can lead to strain relaxation and thermal cracking during the cooling-down progress from the high growth temperature. These issues can significantly degrade the internal efficiency and lifetime of the laser, impacting both yield and cost [37]. To address these challenges, researchers have proposed innovative strategies. One approach for reducing the dislocation defect density from 10^{10} cm^{-2} to lower than 10^5 cm^{-2} involves the insertion of III-V QDs into the active region as defect filter layers. Another technique entails using Si substrates oriented from (100) to (110) or (111) with specific angles, which can mitigate antiphase boundaries problems. However, this introduces a new challenge of incompatibility with current CMOS technology. To reconcile this, many researchers have turned to the utilization of GaAs or GaP intermediate buffer layers grown on patterned Si (100) substrates. This approach ensures compatibility with CMOS fabrication processes while minimizing threading dislocation-related challenges. To control the stain relaxation, selective-area epitaxy was proposed to restrict the epitaxial growth in a predefined region. These combined endeavours have led to significant breakthroughs in heterogeneous epitaxial integrated lasers in recent years. So far, the performance of QD lasers grown directly on silicon substrates has seen remarkable enhancements.

Building upon extensive prior research, in 2016, Liu's research group at University College London achieved a significant milestone by successfully demonstrating a CW InAs/GaAs QD laser operating at 1.3 μm wavelength on an offcut Si substrate [38]. The integration of III-V materials onto Si was accomplished using MBE. This process involved the incorporation of a 6-nm-thick AlAs nucleation layer, a 1-μm-thick GaAs buffer layer, InGaAs/GaAs dislocation filter layers (DFLs), and five layers of InAs/GaAs dot-in-well active regions. The nucleation layer fulfills an important function in reducing the density of threading dislocations and defects. Notably, this laser exhibited an impressive 105 mW output power, a low threshold current density of 62.5 A/cm^2, and a remarkable operational lifetime exceeding 100,000 hours. These outstanding performance metrics can be attributed to the low dislocation density achieved with the DFL and the buffer layer.

Then the focus shifted from FP lasers to other variants with single-frequency lasing, expanding from the O-band to the C-band. In 2017, the researchers from Hong Kong University of Science and Technology (HKUST) and the University of California, Santa Barbara (UCSB), developed a WGM microlaser on industry-standard (001) substrate [39]. This achievement involved the growth of an InAs/InGaAs quantum dot-in-a-well active region on a Si substrate with nanometer-thick V-shaped grooves, as depicted in Figures 17.6a and 17.6b. These grooves accommodate a low lattice mismatch between GaAs and silicon. The laser demonstrated continuous-wave operation at an elevated temperature of 100°C, with an impressively low threshold of 0.6 mA for a 5-μm-radius microdisk laser. In a more recent development, Lau's research team at HKUST achieved a notable milestone. They utilized a modified InAs/InGaAs/InAlGaAs dash-in-well structure grown through MOCVD.

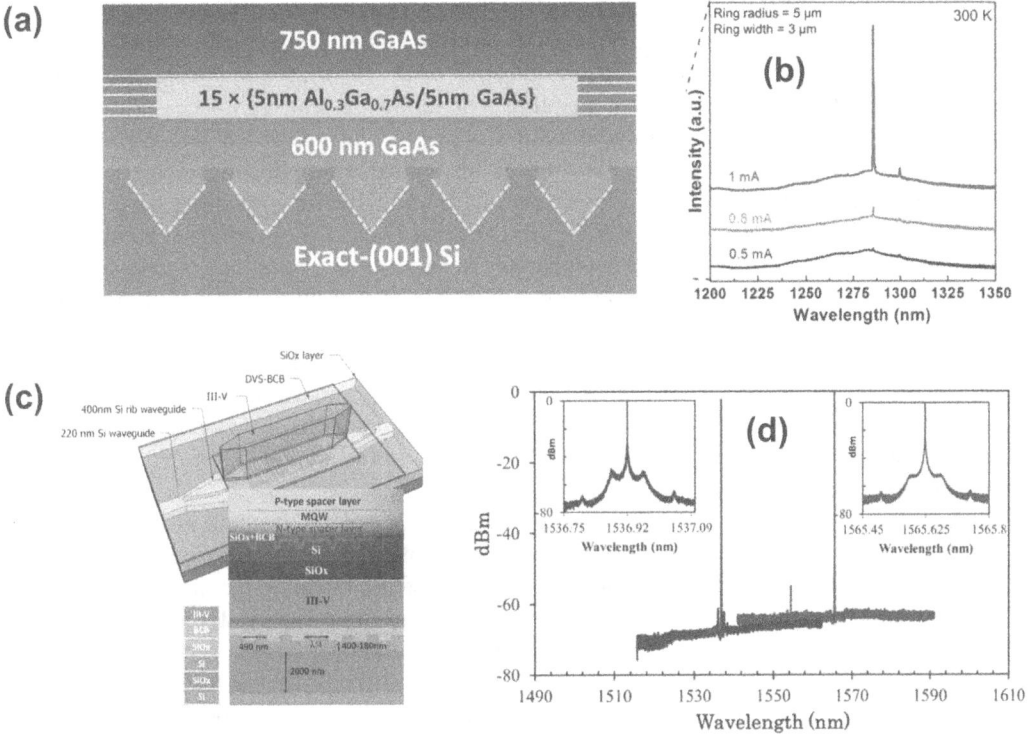

FIGURE 17.6   (a) and (b) 1.3 μm sub-mA threshold QD microring laser directly grown on (001) Si with the schematic of the GaAs-on-Si substrate and the lasing spectrum. (Adapted with permission [39], Copyright [2017], Optica Publishing Group.) (c) and (d) Hybrid DFB laser fabricated with BCB bonding technique, with the structure diagram and lasing spectrum. (Adapted with permission [40], Copyright [2013], Optica Publishing Group.)

This achievement marked the demonstration of the first electrically pumped continuous-wave laser directly grown on an on-axis (001) Si substrate, operating at 1.55 μm wavelength [41]. During that same year, Wan and colleagues at UCSB effectively presented a DFB QD laser, which was also grown on a GaAs/Si substrate. This laser featured an external modulation speed of 30 GHz, indicating significant potential for on-chip interconnect applications [42].

While the most advanced III-V/Si QD lasers grown on Si substrates have successfully met the criteria for extended operational lifetimes and high-temperature performance, it is essential to acknowledge that a persistent challenge lies in the relatively low optical coupling efficiency from III-V lasers to Si waveguides. This problem is primarily associated with thick buffer layers and V-grooved structures, and resolving it is crucial for practical applications.

17.5.2   Wafer Bonding

The bonding technique refers to the interfacial connection between different photonics platforms, facilitating enhanced integration capabilities for specific applications. In contrast to flip-chip bonding, which integrates prefabricated lasers onto Si devices, the wafer bonding does not necessitate accurate alignment and enables waver-scale processing. This direct connection enables effective evanescent wave coupling between the laser sources and Si waveguides. In the following discussion, we will explore two frequently employed bonding techniques: adhesive bonding and direct bonding.

Adhesive bonding utilizes the polymers like divinylsiloxane-bis-benzocyclobutene (DVS-BCB), and photoresist Su-8. BCB bonding offers benefits in terms of bonding strength, tolerance of roughness, and wafer surface cleanliness. In 2006, researchers from Ghent University pioneered BCB bonding between the InP/InGaAsP layer and an SOI waveguide circuit. Tapers at both layers were designed to modify the hybrid active mode, allowing it to couple into a passive Si waveguide mode through evanescent coupling. The following year, they successfully demonstrated a microdisk laser that integrated into an Si waveguide using this technique [20]. Taking advantage of III-V for its gain properties and Si for optical waveguides to create a feedback reflector, a series of innovative hybrid lasers with DFB and DBR functionalities have been successfully developed. In 2013, as shown in Figure 17.6c and 17.6d, S. Keyvaninia et al. reported an integrated DFB laser with performance comparable to native lasers, boasting a single-side output power of 14 mW and a side-mode-suppression-ratio (SMSR) exceeding 50 dB [40]. It should be mentioned that the grating was located on the Si waveguide underneath, with the III-V waveguide on top providing optical gain only. In 2016, S. Sui et al. demonstrated unidirectional emission from a BCB-bonded InP/SOI laser with a deformed WGM ring microcavity [43]. However, it is essential to keep the bonding layer as thin as possible to address challenges related to low coupling efficiency from III-V to Si and poor heat dissipation in the bonding layer owing to the low thermal conductivity of BCB.

To date, direct bonding has emerged as the most frequently employed method, whereby two material surfaces are bonded together through the robust intermolecular forces. A series of surface treatments like polishing and plasma cleaning are perquisites before initiating the bonding process. In 2006, John Bowers's group at UCSB successfully demonstrated an electrically pumped AlGaAs/SOI laser by transferring the III-V material to Si waveguides through oxygen plasma-assisted direct bonding [44]. This achievement indicates an important step toward the integration of large-scale integration of lasers on Si. Similarly, recognizing the low-loss and cost-effective nature of SOI and Si_3N_4 circuits, researchers have pioneered the development of novel lasers by designing external cavities within these passive platforms, with III-V materials providing gain. For instance, the UCSB researchers from the same group presented a hybrid widely tunable laser with a narrow linewidth of 100 kHz [45]. This was achieved by integrating a III-V semiconductor optical amplifier (SOA) on to a long-cavity loop-mirrors structure formed in SOI. The SOI external cavity not only selected the oscillation mode but also reduced the laser linewidth by increasing the photon lifetime within the whole laser cavity. More recently, they also demonstrated a heterogeneous III-V/Si/Si_3N_4 laser with a linewidth at the 1-Hz level, employing the ultra-low loss Si_3N_4 extended DBR [46]. These lasers, with ultra-narrow linewidth and ultra-low relative intensity noise, are crucial for the applications scenarios like coherent communication and sensing. Additionally, the CMOS-ready Si_3N_4 waveguide platform hints at scalable production capabilities with high yields, leveraging foundry-based technologies. Recent results displayed large-scale integrated Kerr frequency comb sources, accomplished by researchers from UCSB and Swiss Federal Institute of Technology Lausanne. This achievement involved bonding InP-based QW lasers into ultra-high-Q Si_3N_4 microring resonators [47].

The heterogeneously integrated lasers with the wafer bonding technique have been massively manufactured by Intel for high-speed transceivers. Notably, in 2022 they made a significant announcement regarding the development of an eight-wavelength integrated laser sources array with consistent wavelength separation using this technique, meeting a key requirement for optical compute interconnects and dense wavelength division multiplexing communication.

REFERENCES

1. J.P. Gordon, H.J. Zeiger, and C.H. Townes, *The maser—New type of microwave amplifier, frequency standard, and spectrometer*, Phys. Rev. 99(1955) 1264–1274
2. A.L. Schawlow, and C.H. Townes, *Infrared and optical masers*, Phys. Rev. 112(1958) 1940–1949
3. T.H. Maiman, *Stimulated optical radiation in ruby*, Nature 187(1960) 493–494
4. J. Hecht, *Short history of laser development*, Opt. Eng. 49(2010) 091002
5. H. Kroemer, *A proposed class of hetero-junction injection lasers*, Proc. IEEE 51(1963) 1782–1783

6. Z.I. Alferov, V. Andreev, V. Korol'kov, D. Trat'yakov, and V. Tuchkevich, *High-voltage p-n junctions in Ga x Al 1– x As crystals*, Fiz. Tekh. Poluprovodn. 1(1967) 0015–3222

7. Z.I. Alferov, V. Andreev, D. Garbuzov, Y.V. Zhilyaev, E. Morozov, E. Portnoi, and V. Trofim, *Effect of heterostructure parameters on the laser threshold current and the realization of continuous generation at room temperature*, Sov. Phys. Semicond. 4(1970) 1573–1575

8. P.M. Petroff, and S.P. DenBaars, *MBE and MOCVD growth and properties of self-assembling quantum dot arrays in III-V semiconductor structures*, Superlattices Microstruct. 15(1994) 15

9. F. Koyama, *Recent advances of VCSEL photonics*, J. Light. Technol. 24(2006) 4502–4513

10. R. Michalzik, *VCSEL Fundamentals*, in *VCSELs: Fundamentals, Technology and Applications of Vertical-Cavity Surface-Emitting Lasers*, R. Michalzik, Editor. 2013, Berlin, Heidelberg, Springer Berlin Heidelberg

11. R. Michalzik, and K.J. Ebeling, *Operating Principles of VCSELs*, in *Vertical-Cavity Surface-Emitting Laser Devices*, H.E. Li and K. Iga, Editors. 2003, Berlin, Heidelberg, Springer Berlin Heidelberg

12. B. Weigl, M. Grabherr, C. Jung, R. Jager, G. Reiner, R. Michalzik, D. Sowada, and K.J. Ebeling, *High-performance oxide-confined GaAs VCSELs*, IEEE J. Sel. Top. Quantum Electron. 3(1997) 409–415

13. C. Jung, R. Jäger, M. Grabherr, P. Schnitzer, R. Michalzik, B. Weigl, S. Müller, and K.J. Ebeling, *4.8 mW singlemode oxide confined top-surface emitting vertical-cavity laser diodes*, Electron. Lett. 33(1997) 1790–1791

14. J. Pozo, and E. Beletkaia, *VCSEL technology in the data communication industry*, PhotonicsViews. 16(2019) 21–23

15. L. Rayleigh, *CXII. The problem of the whispering gallery*, Lond. Edinb. Dublin Philos. Mag. J. Sci. 20(1910) 1001–1004

16. K.J. Vahala, *Optical microcavities*, Nature 424(2003) 839–846

17. L. He, ŞK. Özdemir, and L. Yang, *Whispering gallery microcavity lasers*, Laser Photonics Rev. 7(2013) 60–82

18. S. McCall, A. Levi, R. Slusher, S. Pearton, and R. Logan, *Whispering-gallery mode microdisk lasers*, Appl. Phys. Lett. 60(1992) 289–291

19. B. Redding, L. Ge, G.S. Solomon, and H. Cao, *Directional waveguide coupling from a wavelength-scale deformed microdisk laser*, Appl. Phys. Lett. 100(2012) 061125

20. J. Van Campenhout, P. Rojo-Romeo, P. Regreny, C. Seassal, D. Van Thourhout, S. Verstuyft, L. Di Cioccio, J.M. Fedeli, C. Lagahe, and R. Baets, *Electrically pumped InP-based microdisk lasers integrated with a nanophotonic silicon-on-insulator waveguide circuit*, Opt. Express. 15(2007) 6744–6749

21. E.M. Purcell, *Spontaneous Emission Probabilities at Radio Frequencies*, in Confined Electrons and Photons: New Physics and Applications, E. Burstein and C. Weisbuch, Editors. 1995, Boston, MA, Springer US

22. A. Levi, R. Slusher, S. McCall, T. Tanbun-Ek, D. Coblentz, and S. Pearton, *Room temperature operation of microdisc lasers with submilliamp threshold current*, Electron. Lett. 11(1992) 1010–1012

23. M. Fujita, K. Inoshita, and T. Baba, *Room temperature continuous wave lasing characteristics of GaInAsP/InP microdisk injection laser*, Electron. Lett. 34(1998) 278–279

24. M. Fujita, R. Ushigome, and T. Baba, *Continuous wave lasing in GaInAsP microdisk injection laser with threshold current of 40µA*, Electron. Lett. 36(2000) 1

25. V.V. Sherstnev, A.M. Monakhov, A.P. Astakhova, A.Y. Kislyakova, Y.P. Yakovlev, N.S. Averkiev, A. Krier, and G. Hill, *Semiconductor WGM lasers for the mid-IR spectral range*, Semiconductors. 39(2005) 1087–1092

26. A.C. Tamboli, E.D. Haberer, R. Sharma, K.H. Lee, S. Nakamura, and E.L. Hu, *Room-temperature continuous-wave lasing in GaN/InGaN microdisks*, Nat. Photonics. 1(2007) 61–64

27. Y.-D. Yang, S.-J. Wang, and Y.-Z. Huang, *Investigation of mode coupling in a microdisk resonator for realizing directional emission*, Opt. Express. 17(2009) 23010–23015

28. H.-Z. Weng, O. Wada, J.-Y. Han, J.-L. Xiao, Y.-D. Yang, Y.-Z. Huang, J. Li, B. Xiong, C.-Z. Sun, and Y. Luo, *Sub-THz wave generation based on a dual wavelength microsquare laser*, Electron. Lett. 53(2017) 939–941

29. X.M. Lv, Y.Z. Huang, L.X. Zou, H. Long, and Y. Du, *Optimization of direct modulation rate for circular microlasers by adjusting mode Q factor*, Laser Photonics Rev. 7(2013) 818–829

30. A.V. Babichev, A.G. Gladyshev, E.S. Kolodeznyi, A.S. Kurochkin, G.S. Sokolovskii, V.E. Bougrov, L.Y. Karachinsky, I.I. Novikov, V.V. Dudelev, V.N. Nevedomsky, S.O. Slipchenko, A.V. Lutetskiy, A.N. Sofronov, D.A. Firsov, L.E. Vorobjev, N.A. Pikhtin, and A.Y. Egorov, *Growth and optical characterization of 7.5 µm quantum-cascade laser heterostructures grown by MBE*, J. Phys. Conf. Ser. 1124(2018) 041029

31. J. Faist, F. Capasso, D.L. Sivco, C. Sirtori, A.L. Hutchinson, and A.Y. Cho, *Quantum Cascade laser*, Science. 264(1994) 553–556

32. J. Faist, F. Capasso, C. Sirtori, D.L. Sivco, A.L. Hutchinson, and A.Y. Cho, *Vertical transition quantum cascade laser with Bragg confined excited state*, Appl. Phys. Lett. 66(1995) 538–540

33. M. Razeghi, *High-performance InP-based mid-IR quantum Cascade lasers*, IEEE J. Sel. Top. Quantum Electron. 15(2009) 941–951

34. B.S. Williams, *Terahertz quantum-cascade lasers*, Nat. Photonics. 1(2007) 517–525

35. L.J. Olafsen, E.H. Aifer, I. Vurgaftman, W.W. Bewley, C.L. Felix, J.R. Meyer, D. Zhang, C.-H. Lin, and S.S. Pei, *Near-room-temperature mid-infrared interband cascade laser*, Appl. Phys. Lett. 72(1998) 2370–2372

36. E.L. Normand, and I. Howieson, *Quantum-cascade lasers enable gas-sensing technology*, Laser Focus World. 43(2007) 90–92

37. S. Pan, V. Cao, M. Liao, Y. Lu, Z. Liu, M. Tang, S. Chen, A. Seeds, and H. Liu, *Recent progress in epitaxial growth of III–V quantum-dot lasers on silicon substrate*, J. Semicond. 40(2019) 101302

38. S. Chen, W. Li, J. Wu, Q. Jiang, M. Tang, S. Shutts, S.N. Elliott, A. Sobiesierski, A.J. Seeds, I. Ross, P.M. Smowton, and H. Liu, *Electrically pumped continuous-wave III–V quantum dot lasers on silicon*, Nat. Photonics. 10(2016) 307–311

39. Y. Wan, J. Norman, Q. Li, M.J. Kennedy, D. Liang, C. Zhang, D. Huang, Z. Zhang, A.Y. Liu, A. Torres, D. Jung, A.C. Gossard, E.L. Hu, K.M. Lau, and J.E. Bowers, *1.3 μm submilliamp threshold quantum dot micro-lasers on Si*, Optica. 4(2017) 940–944

40. S. Keyvaninia, S. Verstuyft, L. Van Landschoot, F. Lelarge, G.H. Duan, S. Messaoudene, J.M. Fedeli, T. De Vries, B. Smalbrugge, E.J. Geluk, J. Bolk, M. Smit, G. Morthier, D. Van Thourhout, and G. Roelkens, *Heterogeneously integrated III-V/silicon distributed feedback lasers*, Opt. Lett. 38(2013) 5434–5437

41. Y. Xue, W. Luo, S. Zhu, L. Lin, B. Shi, and K.M. Lau, *1.55 μm electrically pumped continuous wave lasing of quantum dash lasers grown on silicon*, Opt. Express. 28(2020) 18172–18179

42. Y. Wan, J.C. Norman, Y. Tong, M.J. Kennedy, W. He, J. Selvidge, C. Shang, M. Dumont, A. Malik, H.K. Tsang, A.C. Gossard, and J.E. Bowers, *1.3 μm quantum dot-distributed feedback lasers directly grown on (001) Si*, Laser Photonics Rev. 14(2020) 2000037

43. S.-S. Sui, Y.-Z. Huang, M.-Y. Tang, H.-Z. Weng, Y.-D. Yang, J.-L. Xiao, and Y. Du, *Locally deformed-ring hybrid microlasers exhibiting stable unidirectional emission from a Si waveguide*, Opt. Lett. 41(2016) 3928–3931

44. A.W. Fang, H. Park, O. Cohen, R. Jones, M.J. Paniccia, and J.E. Bowers, *Electrically pumped hybrid AlGaInAs-silicon evanescent laser*, Opt. Express. 14(2006) 9203–9210

45. T. Komljenovic, S. Srinivasan, E. Norberg, M. Davenport, G. Fish, and J.E. Bowers, *Widely tunable narrow-linewidth monolithically integrated external-cavity semiconductor lasers*, IEEE J. Sel. Top. Quantum Electron. 21(2015) 214–222

46. J. Guo, C.A. McLemore, C. Xiang, D. Lee, L. Wu, W. Jin, M. Kelleher, N. Jin, D. Mason, L. Chang, A. Feshali, M. Paniccia, P.T. Rakich, K.J. Vahala, S.A. Diddams, F. Quinlan, and J.E. Bowers, *Chip-based laser with 1-hertz integrated linewidth*, Sci. Adv. 8(2022) eabp9006

47. C. Xiang, J. Liu, J. Guo, L. Chang, R.N. Wang, W. Weng, J. Peters, W. Xie, Z. Zhang, J. Riemensberger, J. Selvidge, T.J. Kippenberg, and J.E. Bowers, *Laser soliton microcombs heterogeneously integrated on silicon*, Science. 373(2021) 99–103

18 Light-Emitting Diode

Asmita Poddar, Madhab Roy,
and Sanjib Bhattacharya

18.1 BACKGROUND AND MOTIVATION

Light is the foundation of each and every form of life. Thus, continuous efforts haves been made to use light as an efficient and user-friendly source from the very beginning of human civilization. It is customary that scientists and engineers have to work together for the growth of research work on light sources in a more efficient way. To continue this journey, light-emitting diodes (LEDs) were introduced in the early 1950s. Development of BLUE LED in the year 1990 has been marked as commencement of a new era in the field of illumination. After that, the popularity of LEDs in the fields of medical science [1], engineering, etc., increased like no other.

LED-based lighting sources possess high luminous efficacy, which is the key reason for their popularity in the modern age. Nowadays white LEDs have surpassed the efficacy of standard incandescent lighting systems and thus have gradually replaced conventional lighting systems. For example, it can readily be observed that a 5-watt LED owns a luminous efficacy of 18–22 lumens per watt (1 m/W), whereas a conventional incandescent light bulb of 60–100 watts emits around 15l m/W, and standard fluorescent lights emit up to 100 lm/W.

The invention of a novel blue LED in September 2003, which delivered 24 mW at 20 mA, has been marked as a significant occurrence in the sector of illumination. It produced commercially packaged brightest white light during that period, which delivered 65 lm/W at 20 mA and started a new era in the commercial sector. That LED provided more than four times efficiency than any standard incandescent.

Primarily LEDs were used as indicator lamps for electronic devices as a replacement for small incandescent bulbs. Further advancements in research and development have permitted LEDs to be used in different sectors of illumination, which includes aviation lighting, automotive headlamps, advertising, general lighting, traffic signals, camera flashes, and even LED wallpaper.

18.2 HISTORY AND INVENTION OF LED

The phenomenon of electroluminescence was first discovered by the British scientist H.J. Round of Marconi Labs in 1907 while working with a crystal of silicon carbide and a cat's- whisker detector [2, 3].

Decades later, a Soviet inventor, Oleg Losev, have stated foundation of the first LED [4] in 1927 but no such practical implementation was discovered for several decades due to the very inefficient light-producing properties of silicon carbide, the semiconductor Losev used [5].

Kurt Lehovec, Carl Accardo, and Edward Jamgochian presented the first application in an instrument in 1951, utilizing a device using SiC crystals with a current source of a battery or a pulse generator and with a comparison to a different, pure crystal in 1953 [6, 7].

Indium phosphide (InP), silicon-germanium (SiGe), gallium antimonide (GaSb), and gadolinium (GaAs) alloys were used in a simple diode structure to generate infrared emission in 1955 [8], according to a remarkable discovery made by Rubin Braunstein of the Radio Corporation of America.

James R. Biard and Gary Pittman observed near-infrared (900 nm) light emission from a tunnel diode they had built on a GaAs substrate in September 1961 while working at Texas Instruments (TI) in Dallas, Texas [9].

DOI: 10.1201/9781003450146-18

By October 1961, they had successfully shown signal coupling and efficient light emission between a GaAs p-n junction light emitter and a semiconductor photodetector [10]. Based on these discoveries, Biard and Pittman submitted a patent application titled "Semiconductor Radiant Diode" on August 8, 1962. It specified a zinc-diffused p-n junction LED with a spaced cathode contact to enable effective infrared light emission under forward bias. The SNX-100, the first commercial LED device introduced by TI in October 1962, used a pure GaAs crystal to emit light with an 890 nm wavelength [9].

18.3   FUNCTION OF LED

When current passes through an LED, a semiconductor device, light is released. In essence, LEDs are PN junctions that only permit current to flow in one direction. While the N side has excess electrons on it, the P side has excess positive charge in the form of "holes," which show that there aren't any. When the PN junction is subjected to a forward voltage, holes travel toward the N area and electrons flow from the N area toward the P area. As seen in Figure 18.1, when the electrons and holes recombine close to the junction, some energy is released in the form of light. This is the basic purpose of an LED. The energy needed for electrons to bridge the semiconductor's band gap determines the hue of the light, which corresponds to the energy of the photons [11].

18.4   FACTORS OF CONSIDERATION FOR USAGE

18.4.1   POWER SOURCES

A minor change in voltage can result in a big change in current since the current in an LED or other diode increases exponentially with the voltage that is applied. To avoid damage, the LED's current flow must be controlled by an external circuit, such as a constant current source. LED fixtures must have a power converter or at the very least a current-limiting resistor, because the majority of conventional power supplies are (almost) constant-voltage sources. Small batteries' intrinsic resistance may be adequate in some applications to maintain current below the LED rating.

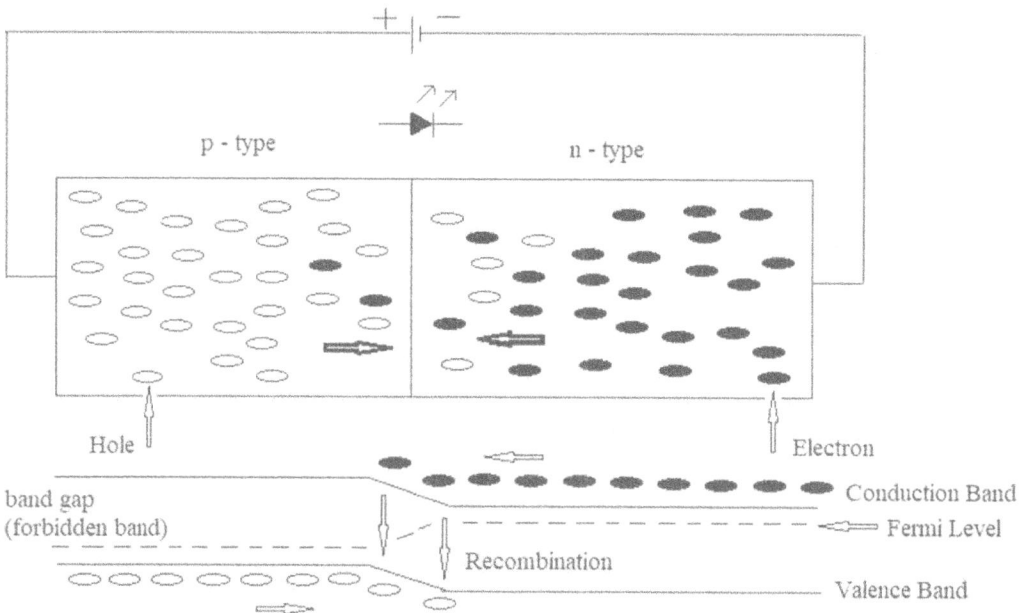

FIGURE 18.1   Schematics of a light emitting diode (LED).

18.4.2 Electrical Polarity

An LED will only illuminate when voltage is applied in the diode's forward direction, unlike a conventional incandescent lamp. In the event that voltage is supplied in the opposite direction, neither current nor light are produced. The LED will be harmed if the reverse voltage is higher than the breakdown value, which is normally around five volts. The reverse-conducting LED is a beneficial noise diode if the reverse current is suitably constrained to prevent harm.

18.4.3 Safety and Health

Specifications for eye safety state that certain cool-white and blue LEDs may exceed the safe thresholds for the so-called blue-light hazard [12]. According to one study, there is no risk in routine use at household illuminance, and caution is only necessary in specific occupational contexts or for select groups.

18.5 TYPES

LEDs are made in different packages for different applications.

18.5.1 Miniature

These single-die LED indicators are typically available in both through-hole and surface mount packages and range in size from 2 to 8 mm. The typical current ratings are between 1 mA and 20 mA or more. Ordinary tiny LEDs with a series resistor for the direct connection to a 5 V or 12 V supply are known as 5 V and 12 V LEDs.

18.5.2 High-power

Unlike ordinary LEDs, which can only be driven at currents of tens of mA, high-power LEDs (HP-LEDs) or high-output LEDs (HO-LEDs) [13] can be powered at currents of hundreds of mA to more than an ampere. Studies have shown that some of them have a 1,000+ lumen output.

Up to 300 W/cm² LED power densities have been attained. The HP-LEDs must be installed atop a heat sink, which allows for heat dissipation because overheating is harmful. An HP-LED fails in a matter of seconds if the heat it produces is not dissipated. In flashlights, one HP-LED can frequently take the place of an incandescent bulb. It can also be arranged in an array to create a potent LED lamp.

18.5.3 AC-driven

LEDs developed by some specific semiconductors can operate on AC power without a DC converter. For each half-cycle, part of the LED emits light and part is dark, and this is reversed during the next half-cycle. The efficiency of this type of HP-LED is typically 40 lm/W. A large number of LED elements in series may be able to operate directly from the line voltage. They are being driven from AC power with a simple controlling circuit. The low-power dissipation of these LEDs affords them more flexibility than the original AC LED design.

18.6 ADVANTAGES

18.6.1 Efficiency

When compared to incandescent light bulbs, LEDs produce more lumens per watt [12]. In contrast to fluorescent light bulbs or tubes, the effectiveness of LED lighting fixtures is unaffected by shape and size.

18.6.2   Color

Unlike conventional lighting techniques, LEDs may emit light of the desired color without the need for any color filters. This can result in cheaper initial expenses and is more effective.

18.6.3   Size

LEDs may be easily mounted to printed circuit boards and can be made to be very small, less than 2 mm² [12].

18.6.4   Switch on Time

A common red indicator LED reaches full brightness in less than a microsecond [12], and LEDs used in communications devices can respond much more swiftly.

18.6.5   Cycling

In contrast to incandescent and fluorescent lamps, which break down more quickly when cycled frequently, and high-intensity discharge (HID) lamps, which take a while to restart, LEDs are suited for purposes subject to frequent on-off cycling.

18.6.6   Dimming

Either reducing the forward current or pulse-width modulating are extremely simple ways to dim LEDs [14]. When seen on camera or by certain individuals, LED lights, especially car headlights, appear to flicker or flash because of this pulse-width modulation. This kind of stroboscopic effect exists.

18.6.7   Cool Light

Compared with the majority of light sources, LEDs emit very little amount of heat in the form of infrared (IR), which might harm delicate items or fabrics. Wasted energy is released through the LED's base as heat.

18.6.8   Slow Failure

Unlike incandescent bulbs, which typically fail suddenly, LEDs typically degrade by gradually dimming over time [12].

18.6.9   Lifetime

LEDs have a potential for a lengthy useful life. One study puts the useful life at 35,000 to 50,000 hours, while the period to complete failure may be either shorter or longer [12]. According to the conditions of usage, fluorescent tubes are normally rated for 10,000 to 25,000 hours of use, while incandescent light bulbs are rated for 1,000 to 2,000 hours. The payback period for an LED device is mostly influenced by decreased maintenance expenses from this increased lifetime, not energy savings, according to a number of design of experiments (DOE) demonstrations.

18.6.10   Shock Resistance

Unlike delicate fluorescent and incandescent bulbs, LEDs can withstand external trauma, since they are solid-state components.

18.6.11 Focus

The LED's sturdy packaging can be made for focusing its light. To gather light and direct it in a useful direction from incandescent and fluorescent sources, an external reflector is frequently needed. Total internal reflection (TIR) lenses are frequently utilized to achieve the same result for larger LED packages. When a lot of light is required, a lot of light sources are typically used, which sometimes may create a problem for focusing them on the same object.

18.7 DISADVANTAGES

18.7.1 High Initial Price

On a capital cost basis, LEDs are now more expensive per lumen than the majority of conventional lighting solutions.

18.7.2 Temperature Dependence

The operational environment's ambient temperatures, or "thermal management" features, have a significant impact on LED performance. When an LED is overdriven in a hot environment, the LED package may become overheated, which could eventually cause the device to malfunction. To maintain a long life, an appropriate heat sink is required. This is crucial for applications in the automotive, medical, and military industries where equipment must function in a variety of temperatures and have low failure rates.

18.7.3 Voltage Sensitivity

LEDs require a voltage supply with voltage above the threshold and current below the rating. With even a little variation in applied voltage, current and lifespan alter significantly. Therefore, they need a source with controlled current.

18.7.4 Area Light Source

Single LEDs provide a lamebrain dispersion of light instead of a spherical light distribution that comes from a point source of light. Consequently, it is challenging to apply LEDs to applications that require a spherical light field; nevertheless, diverse light fields can be adjusted by the use of various optics or "lenses." Divergence below a few degrees cannot be produced using LEDs [15]. In comparison, lasers may produce beams that diverge by little more than 0.2 degrees.

18.7.5 Electrical Polarity

Unlike incandescent light bulbs, which illuminate regardless of the electrical polarity, LEDs will only light with correct electrical polarity [15]. To automatically match source polarity to LED devices, rectifiers can be used.

18.7.6 Blue Hazard

There is a concern that blue LEDs and cool-white LEDs are now capable of exceeding safe limits of the so-called blue-light hazard as defined in eye safety specifications.

18.7.7 EFFICIENCY DROOP

As the electric current rises, LED efficiency declines. Higher currents also result in more heating, which reduces the lifetime of LEDs [16]. In high power applications, these effects place practical restrictions on the current through an LED.

18.7.8 LIGHT POLLUTION

Given that scotopic vision is more sensitive to the blue and green colors, white LEDs used for outdoor illumination produce far more sky glow than sources like high-pressure sodium vapor lamps.

18.7.9 IMPACT ON WILDLIFE

Since LEDs are so much more enticing to insects than sodium-vapor lights, there has been some speculation regarding the likelihood of food chain disruption [17].

18.8 APPLICATIONS OF LED

18.8.1 INDICATORS AND SIGNS

LEDs are used as status indicators and displays on a range of equipment and installations because of their small size, low maintenance requirements, and low energy consumption. Stadium displays, dynamic ornamental displays, and dynamic message signs on motorways are among the applications for large-area LED displays. As destination displays for trains, buses, trams, and ferries, thin, lightweight message displays are utilized in airports and train stations. For traffic lights and signals, exit signs, emergency vehicle lighting, ships' navigation lights, and LED-based Christmas lights, one-color light is ideal. LEDs have been utilized in automobile brake lights and turn signals because of their extended lifespan, quick switching times, and visibility in broad daylight due to their high brightness and focus.

18.8.2 LIGHTING

The ability to employ LEDs for lighting and illumination has been made possible by the development of high-efficiency and high-power LEDs. The US Department of Energy established the L Prize competition in 2008 to promote the use of LED lamps and other high-efficiency lighting. On August 3, 2011, the Philips Lighting North America LED bulb won the first competition. This was the culmination of 18 months of rigorous field, laboratory, and product testing.

Automotive lighting is utilized in the headlights of automobiles, motorcycles, and bicycles due to its mechanical durability and long lifespan. In parking garages and on poles, LED streetlights are used. Torraca, Italy was the first community to switch to LED street lighting [18] in 2007.

In mining operations, LEDs are employed as cap lamps to illuminate the workers. There has been research done to enhance LEDs for mining, to lessen glare and increase illumination, lowering the danger of injury to the miners [12].

LEDs are increasingly being used in educational and medical settings [19], for example, to improve mood. Even research on using LEDs to improve astronauts' health has been funded by NASA [12].

18.8.3 DATA COMMUNICATION AND OTHER SIGNALING

Both digital and analog signals can be transmitted using light. White LED illumination, for instance, can be utilized in systems that help individuals navigate through enclosed places while looking for important rooms or objects [20].

In many theaters and other similar settings, assistive listening devices transmit sound to listeners' receivers using arrays of IR LEDs. The transmission of data via many varieties of fiberoptic cable is accomplished using light-emitting diodes and semiconductor lasers [21]. LEDs can cycle on and off millions of times per second, allowing for very high data bandwidth.

Visible light communication (VLC) [22] has been suggested as a substitute for the increasingly constrained radio bandwidth. Because of this, data transmission is possible without using radio frequencies by operating in the visible region of the electromagnetic spectrum.

The indoor positioning system (IPS), a GPS analog designed to function in enclosed locations where the satellite signals necessary for the GPS to function are difficult to reach, is one promising application of VLC [22].

VLC can also be used to communicate among devices in a smart office or home. There may be interference when using standard radio waves for connectivity as the number of Internet of Things (IoT)-capable devices rises. Data and commands for such devices can be transmitted by lightbulbs having VLC capabilities.

18.8.4 MACHINE VISION SYSTEMS

Bright, uniform lighting is frequently needed for machine vision systems so that processing of interesting features is made simpler. It's common to use LEDs.

Machine vision applications most frequently include barcode scanners, and many of these scanners employ red LEDs rather than lasers. LEDs serve as the light source for the tiny camera inside optical computer mice.

For machine vision, LEDs are advantageous since they offer a portable, dependable source of light. The brightness and beam form of LED lamps can be adjusted to meet the needs of the vision system. LED lamps can be turned on and off as needed.

18.8.5 BIOLOGICAL DETECTION

The US Army Research Laboratory (ARL)'s discovery of radiative recombination in aluminum gallium nitride (AlGaN) alloys inspired the development of ultraviolet (UV) LEDs for inclusion in light-induced fluorescence sensors used for biological agent identification [23].

One of the most reliable methods for the quick, real-time detection of biological aerosols is UV-induced fluorescence. Under a UV light beam, biological aerosols glow and scatter light. The applied wavelength and the biochemical fluorophores present in the biological agent affect the observed fluorescence. A quick, precise, effective, and logistically feasible method of finding biological agents is UV-induced fluorescence. This is due to the fact that UV fluorescence is reagent less, meaning it doesn't need any additional chemicals to initiate a reaction, doesn't utilize any consumables, and doesn't result in any chemical by-products.

18.8.6 OTHER APPLICATIONS

Since LEDs' light can be rapidly manipulated, they are widely employed in optical fiber and free space optics communications. This applies to remote controls for devices like televisions, which frequently include infrared LEDs. An LED and a photodiode or phototransistor are used in opto-isolators to create a signal route with electrical isolation between two circuits. This is particularly helpful in medical equipment where the signals from a low-voltage sensor circuit in contact with a living organism must be electrically isolated from any potential electrical failure in a recording or monitoring device operating at potentially hazardous voltages. These sensor circuits are typically battery-powered.

Light is frequently used as the signal source in sensor systems. Due to the specifications of the sensors, LEDs are frequently the best choice for a light source. IR LEDs are used in the sensor bar of the

Nintendo Wii. They are used by pulse oximeters to gauge oxygen saturation. Instead of the standard cold-cathode fluorescent lamp as the light source, some flatbed scanners use arrays of RGB LEDs [24]. The scanner can calibrate itself for a more accurate color balance without the need for warm-up, thanks to the independent control of the three lighted colors. Furthermore, since only one color of light is ever used to illuminate the page being scanned at once, its sensors just need to be monochromatic.

LEDs can be utilized for both photo emission and detection because they can also be employed as photodiodes. For instance, a touchscreen that measures reflected light from a finger or stylus may take advantage of this [25].

Light sensitivity dependence is seen in a wide range of substances and biological systems. LEDs are used in grow lamps to boost plants' photosynthesis [26].

The bacteria and viruses can be removed from water and other substances using UV LEDs for sterilization [27].

LEDs of certain wavelengths have also been used for light therapy treatment of neonatal jaundice and acne [28].

Other uses for UV LEDs [29], which have a spectrum from 220 nanometers to 395 nanometers, include water/air purification, surface disinfection, glue curing, free-space non-line-of-sight communication, high performance liquid chromatography, UV curing dye printing, phototherapy (295 nanometer Vitamin D, 308 nanometer Excimer lamp, or laser replacement), medical/analytical instrumentation, and DNA absorption.

In electronic circuits, LEDs have also been utilized to serve as a medium-quality voltage reference. In low-voltage regulators, the forward voltage drop (about 1.7 V for a red LED and 1.2V for an IR) can be employed in place of a Zener diode. Above the knee, the I/V curve of red LEDs is the flattest. Nitride-based LEDs are ineffective for this because of their rather steep I/V curve [30].

The exploration with fusing light sources and wall-covering surfaces for interior walls in the form of LED wallpaper has been encouraged by the progressive shrinking of low-voltage lighting technology, such as LEDs and organic light-emitting diodes (OLEDs), appropriate to include into low-thickness materials.

18.9   RESEARCH AND DEVELOPMENT

18.9.1   ORGANIC LIGHT-EMITTING DIODES

An organic compound makes up the electroluminescent substance that makes up the emissive layer of an organic light-emitting diode (OLED). Due to the pi electrons' delocalization upon conjugation throughout all or a portion of the molecule, the organic substance is electrically conductive and serves as an organic semiconductor [31]. The organic materials can be small organic molecules in a crystalline phase or polymers.

OLEDs provide a variety of potential benefits, such as thin, inexpensive displays with excellent contrast and color spectrum, low driving voltage, and a broad viewing angle. Mobile electronic gadgets, including cellphones, digital cameras, lights, and televisions, have visual displays made from OLEDs [32, 33].

18.9.2   PEROVSKITE LEDS

Perovskite LEDs (PLEDs) are the newest members of LED family. PLEDs are established on the semiconductors called perovskites. The capacity of producing light from PLEDs is already equaled to those of the best-performing OLEDs [34]. The most lucrative feature of these PLEDs is their cost-effectiveness. This can be achieved, as it is possible to process them from solution. This low-cost, low-tech technique makes it possible to produce large-area perovskite-based devices at a very cheap cost. Due to the elimination of non-radiative losses, their efficiency is increased. It implies that the recombination mechanisms that don't result in photons are blocked. The coupling issue, which is common for thin-film LEDs, is resolved, as is the issue of balancing charge carrier injection, which raises the external quantum efficiency (EQE). By raising the EQE above 20%, the most recent PLED devices have surpassed the performance barrier [35].

18.10   CONCLUSION

In conclusion, this book chapter has illuminated the captivating world of LEDs and their profound impact on various aspects of our lives and industries. The journey through the realm of LEDs has revealed several key takeaways.

18.10.1   FUNDAMENTAL PRINCIPLES

LEDs operate on the fascinating principles of quantum mechanics, where the emission of light is a result of electron transitions within semiconductor materials. This fundamental understanding underpins the remarkable efficiency and reliability of LEDs.

18.10.2   EVOLUTION AND UBIQUITY

LEDs have evolved from their early days as indicator lights to become ubiquitous in modern society. They have transformed lighting, displays, and countless applications, thanks to their energy efficiency, longevity, and versatility.

18.10.3   ENERGY EFFICIENCY AND SUSTAINABILITY

LEDs have emerged as champions of energy-efficient lighting, contributing significantly to global sustainability efforts by reducing energy consumption and carbon emissions. The transition to LED lighting has had a substantial environmental impact.

18.10.4   DIVERSE APPLICATIONS

LEDs have found applications in a wide array of fields, from general illumination to specialized areas like horticulture lighting, medical devices, and automotive lighting. Their flexibility and adaptability have driven innovation across industries.

18.10.5   ADVANCEMENTS

Ongoing advancements in LED technology continue to reshape the landscape. Innovations in materials, epitaxy techniques, and phosphor technology have improved color quality and spectral characteristics. Integration with smart control systems and the IoT is paving the way for intelligent lighting solutions.

18.10.6   THE FUTURE OF ILLUMINATION

LEDs are not just lighting technology; they represent a paradigm shift in how we perceive and interact with light. As researchers and engineers push the boundaries of what LEDs can achieve, the future promises even more exciting developments in terms of efficiency, aesthetics, and functionality.

In essence, LEDs have not only revolutionized the way we light up our world but have also catalyzed innovation across various disciplines. Their journey from a simple indicator to a transformative technology is a testament to human ingenuity and the ever-evolving quest for more sustainable, efficient, and dynamic lighting solutions. As we look ahead, the possibilities for LEDs seem boundless, promising to continue brightening our lives in ways we can only imagine.

REFERENCES

1. J. Dong, D. Xiong, Applications of light emitting diodes in health care, Annals of Biomedical Engineering, 45 (11) (2017), 2509–2523.
2. D. Misra, J. Brewer, Crystal radio detector [cat's whisker]: The first wireless device, IEEE Circuits and Devices Magazine, 17 (2) (2001), 12–24.
3. O.V. Losev, Luminous carborundum detector and detection effect and oscillations with crystals, Philosophical Magazine, 7th Series, 5 (39) (1928), 1024–1044.
4. H. Toktamiş, P.O. Hama, Thermoluminescencedosimetric properties of silicon carbide (SiC) used in industrial applications, Applied Radiation and Isotopes, 148 (2019), 138–146.
5. N. Zheludev, The life and times of the LED: A 100-year history, Nature Photonics, 1 (4) (2007), 189–192.
6. K. Lehovec, C. AAccardo, E. Jamgochian, Injected light emission of silicon carbide crystals, Physical Review, 83 (3) (1951), 603–607.
7. K. Lehovec, C.A. Accardo, E. Jamgochian, Injected light emission of silicon carbide crystals, Physical Review, 89 (1) (1953), 20–25.
8. R. Braunstein, Radiative transitions in semiconductors, Physical Review, 99 (6) (1955), 1892–1893.
9. S. Sadeghi, G.O. Eren, S. Nizamoglu, Strategies for improving performance, lifetime, and stability in light-emitting diodes using liquid medium, Chemical Physics Reviews, 2(041302) (2021), 1–43.
10. N. Dharmarasu, P.O. Vaccaro, S. Saravanan, J. Zanardi, K. Kubota, N. Saito, High-density light-emitting diodes using A lateral p-n junction on patterned (311)A GaAs substrates, IEICE Electronics Express, 1 (5) (2004), 86–91.
11. D. Yuan, Q. Liu, Photon energy and photon behavior discussions, Energy Reports, 8 (2) (2022), 22–42.
12. M. Barar, Organic light emitting diodes: The need of future, International Journal of Chemical Concepts, 01 (2015), 168–174.
13. S. Bierhuizen, M.R. Krames, G. Harbers, G. Weijers, Performance and trends of high power light emitting diodes, Proceedings of SPIE, 6669 (2007), 66690B (1–12).
14. P. Narra, D.S. Zinger, An effective LED dimming approach, conference record of the 2004 IEEE industry applications conference, 2004, 39th IAS Annual Meeting, 3 (2004), 1671–1676.
15. M.S. Hossen, M.T. Islam, S. Hossain, Design & fabrication of a gravity powered light, ICMERE, 02 (254) (2015), 1–5.
16. R. Abbasinejad, D. Kacprzak, A comprehensive detailed formula for LED degradation and lifetime estimation leading to reduce CO_2 emissions, Cleaner Engineering and Technology, 9 (100518) (2022), 1–17.
17. S.M. Pawson, M.K.-F. Bader, LED lighting increases the ecological impact of light pollution irrespective of color temperature, Ecological Applications, 24 (7) (2014), 1561–1568.
18. A. Balachandran, M. Siva, V. Parthasarathi, S.K. Vasudevan, S.K. Vasudevan, An innovation in the field of Street lighting system with cost and energy efficiency, Indian Journal of Science and Technology, 8 (17) (2015), 1–5.
19. J. Dong, D. Xiong, Applications of light emitting diodes in health care, Annals of Biomedical Engineering, 45 (11) (2017), 2509–2523.
20. M.S. Fudin, K.D. Mynbaev, K.E. Aifantis, H. Lipsanen, V.E. Bougrov, A.E. Romanov, Frequency characteristics of modern LED phosphor materials, Scientific and Technical Journal of Information Technologies, Mechanics and Optics, 14 (6) (2014), 71–76.
21. T. Numai, Fundamentals of semiconductor lasers, Springer Series in Optical Sciences, 93 (2004), 89–186.
22. L.E.M. Matheus, A.B. Vieira, L.F.M. Vieira, M.A.M. Vieira, O. Gnawali, Visible light communication: Concepts, applications and challenges, IEEE Communications Surveys & Tutorials, 21 (4) (2019), 3204–3237.
23. A.V. Sampath, M.L. Reed, C. Moe, G.A. Garrett, E.D. Readinger, W.L. Sarney, H. Shen, M. Wraback, C. Chua, N.M. Johnson, The effects of increasing AlNMole fraction on the performance of AlGaNActive regions containing nanometer scale compositionally inhomogeneities, International Journal of High Speed Electronics and Systems, 19 (01) (2009), 69–76.
24. F.C. Wang, C.W. Tang, B.J. Huang, Multivariable robust control for a Red–Green–Blue LED lighting system, IEEE Transactions on Power Electronics, 25 (2) (2010), 417–428.
25. P.H. Dietz, W.S. Yerazunis, D.L. Leigh, Very low-cost sensing and communication using bidirectional LEDs, Lecture Notes in Computer Science, 2864 (2004), 175–191.
26. G.D. Goins, N.C. Yorio, M.M. Sanwo, C.S. Brown, Photomorphogenesis, photosynthesis, and seed yield of wheat plants grown under red light-emitting diodes (LEDs) with and without supplemental blue lighting, Journal of Experimental Botany, 48 (7) (1997), 1407–1413.

27. M. Mori, A. Hamamoto, A. Takahashi, M. Nakano, N. Wakikawa, S. Tachibana, T. Ikehara, Y. Nakaya, M. Akutagawa, Y. Kinouchi, Development of a new water sterilization device with a 365 nm UV-LED, Medical & Biological Engineering & Computing, 45 (12) (2007), 1237–1241.

28. R.M.L. Savedra, A.M.T. Fonseca, M.M. Silva, R.F. Bianchi, M.F. Siqueira, White LED phototherapy as an improved treatment for neonatal jaundice, The Review of Scientific Instruments, 92 (6) (2021), 064101 (1–3).

29. K. Song, M. Mohseni, F. Taghipour, Application of ultraviolet light-emitting diodes (UV-LEDs) for water disinfection: A review, Water Research, 94 (2016), 341–349.

30. C.K. Wang, Y.Z. Chiou, H.J. Chang, Investigating the efficiency droop of nitride-based blue LEDs with different quantum barrier growth rates, Crystals, 9 (12) (2019), 677 (1–9).

31. J.H. Burroughes, D.D.C. Bradley, A.R. Brown, R.N. Marks, K. MacKay, R.H. Friend, P.L. Burns, A.B. Holmes, Light-emitting diodes based on conjugated polymers, Nature, 347 (6293) (1990), 539–541.

32. T.R. Hebner, C.C. Wu, D. Marcy, M.H. Lu, J.C. Sturm, Ink-jet printing of doped polymers for organic light emitting devices, Applied Physics Letters, 72 (5) (1998), 519–521.

33. J.N. Bardsley, International OLED technology roadmap, IEEE Journal of Selected Topics in Quantum Electronics, 10 (1) (2004), 3–4.

34. D. Di, A.S. Romanov, L. Yang, J.M. Richter, J.P.H. Rivett, S. Jones, T.H. Thomas, M. Abdi Jalebi, R.H. Friend, M. Linnolahti, M. Bochmann, High-performance light-emitting diodes based on carbene-metal-amides, Science, 356 (6334) (2017), 159–163.

35. A. Armin, P. Meredith, LED technology breaks performance barrier, Nature, 562 (7726) (2018), 197–198.

19 Transistors
Advanced Logic Devices (RAM, Memristors, Gate)

Arpita Roy, Karuna Kumari, and Soumya J. Ray

19.1 INTRODUCTION TO MEMORY AND COMPUTING BEYOND MOORE'S LAW

The world has undergone a revolution in recent years due to the exponential rise of computing devices, which has paved the way for the era of the Internet of Things (IoT) and big data. Nowadays, the enormous expansion of internet data requires increasingly quick and scalable memory technology to increase portable device storage data centers and computers. To address these issues, new technical approaches are being investigated. Since the 1960s, the remarkable advancements in computing and information technology have been prepared by the miniaturization of metal-oxide-semiconductor field-effect transistors (MOSFETs) by Moore's law, which states that the number of integrated transistors in a microprocessor chip doubles roughly every two years [1] as shown in Figure 19.1(a). This has led to an exponential rise in the number of devices per square inch on the chip, which leads to the development of digital complementary metal oxide-semiconductor microprocessors (CMOS).

There are some fundamental problems. First, as leakage currents increase, further lowering of the threshold voltages of the MOSFET device is not possible, which makes it difficult to reduce supply voltage and transistor size in digital circuits [2]. As a result of the excessive heating on the chip known as the "heat wall," which varies from 50 to 100 W per cm^2 [2, 3], the huge power consumption of today's CMOS-based microprocessors has also imposed a strict limit on the maximum clock frequency.

In addition to the heat wall, Moore's law is also being tested by a second hard barrier known as the memory wall [4]. In conventional processors, the central processing unit (CPU) operates at a speed that is much faster than that required to reach the memory where the data are stored [3, 5]. The physical separation of CPU and memory in the von Neumann architecture of modern digital computers, sometimes known as the von Neumann bottleneck, is the root of this basic problem [6–9].

On the one hand, transistors are being redesigned to improve their channel electrostatic control and sub-threshold slope, to address scaling challenges taken by an increase in sub-threshold leakage. Besides that, multiple processing cores were combined on the same chip to improve the performance continuously [2, 10]. The great advantage of this parallel computation approach is that although each core on the chip is operated at a frequency lower than the maximum clock rate not to hit the power barrier, the use of multiple cores at the same time enables increased the overall performance. As a result, the systems on chip (SoC) based on the co-integration of CPU, and graphics processing unit (GPU), which typically uses hundreds of cores running in parallel with high memory bandwidths [5], have been prompted by the requirement of higher energy efficiency.

In addition to this, novel concepts have also been explored to overcome memory wall. Figure 19.1(b) shows the memory hierarchy of conventional processors based on von Neumann architecture. The memory in the processor core is fabricated using registers and flip-flops, which are ultra-high-speed components with low density. SRAM acts as a cache memory and DRAM acts as a main memory which is used to store instructions and to process data also. Both instruction/programs

 DOI: 10.1201/9781003450146-19

(a)

(b)

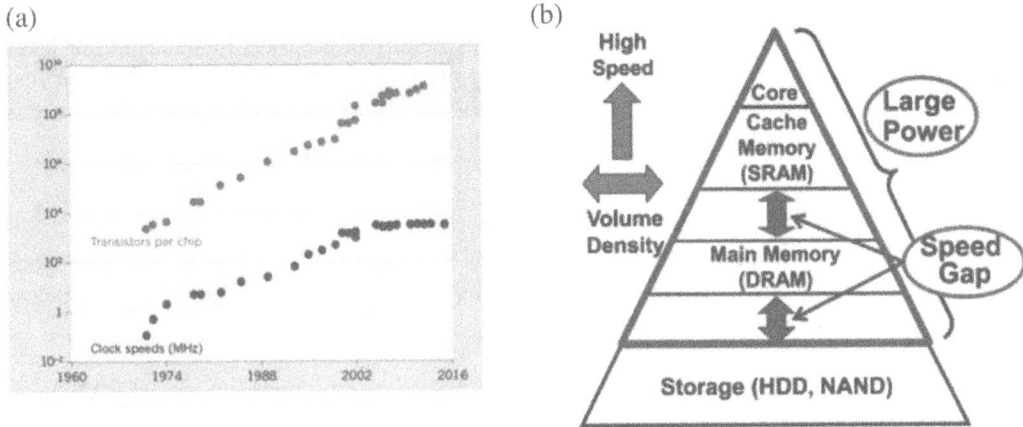

FIGURE 19.1   (a) Scaling trend of the number of transistors per chip and processor operating frequency over the past 50 years and (b) memory hierarchy in computer systems. (Adapted with permission [11]. Copyright [2018].)

and data are stored in mass storage components like hard disc drives (HDDs). When the memory is further away from the CPU, high density activities are required, but high-speed operations are needed from devices located close to the processors.

Major problems in today's data-intensive applications could be summarized as follow:

* Exponentially growing gap between processor and memory in terms of capacity and speed.
* Speed gap between cache memory and main memory.
* Speed gap between main memory and storage memory.
* High-power consumption of the core, the cache, and the main memory, because they are based on volatile memory technologies [11–13].

19.1.1 Storage Class Memory (SCM)

Storage class memory (SCM) has been introduced as a solution to mitigate the memory bottleneck by filling the performance gap that currently exists between the main memory and storage in the memory hierarchy (Figure 19.1[b]). Yet, there is still a performance difference between memory and storage in addition to flash memory's inherent constraints (such as its poor endurance and scaling potential) [11]. On the one hand, emerging SCM classes will be used to match the characteristics of storage types (such as NAND flash) with much better performances; similarly, SCM will also be used to match the characteristics of fast memory types (e.g. DRAM) with non-volatility and low cost. SCMs normally have non-volatile memories, giving them the benefit of zero standby power consumption while preserving the data [14, 15]. The SCM is likely the most significant market for the development of new memory technologies.

19.2   RANDOM ACCESS MEMORY (RAM)

19.2.1 Introduction to RAM

Random access memory, commonly known as RAM [16], stands as the dynamic powerhouse of data manipulation within a computing system. It is the ephemeral canvas where active programs and data reside, providing the dynamic workspace required for a wide range of applications. Unlike long-term storage devices such as hard drives or solid state drives (SSDs), RAM offers a form of volatile memory. This means that data stored in RAM is volatile and is lost when the power is turned off. At its essence, RAM acts as a high-speed bridge between the CPU and the storage devices, enabling swift retrieval and modification of data. This fluidity of access is fundamental for tasks that require real-time processing, such as running applications, browsing the internet, or playing video games. In essence, RAM is the arena where active computations occur, and its significance is initial to the functioning of modern computing systems.

19.2.2 Types of RAM

The types of RAM are shaped by the underlying technology and their suitability for different applications. The schematic diagram of the different types of RAM are shown in Figure 19.2(a-f). Below are some of the prominent types of RAM:

1. *Dynamic RAM (DRAM)*: Dynamic RAM, or DRAM, is the most prevalent type of RAM in modern computing systems. It operates on the principle of capacitive charge storage within integrated circuits. Each memory cell consists of a capacitor and a transistor, representing a binary state. The presence or absence of charge in the capacitor represents the "1" or "0" state, respectively. However, due to the inherent leakage of charge over time, DRAM requires periodic refreshing to maintain data integrity. While DRAM offers high storage density at a relatively low cost, it is characterized by slower access times compared to other types of RAM. This makes it ideal for applications where density is a priority, but speed is less critical.
2. *Static RAM (SRAM)*: Static RAM, or SRAM, operates on a different principle compared to DRAM. It uses bistable flip-flop circuits, which means it retains data as long as power is supplied. This eliminates the need for constant refreshing, resulting in significantly faster access times compared to DRAM. However, SRAM is more expensive to produce and consumes more power, which limits its application in large-scale memory systems. It is often used in cache memory and as registers within the CPU.

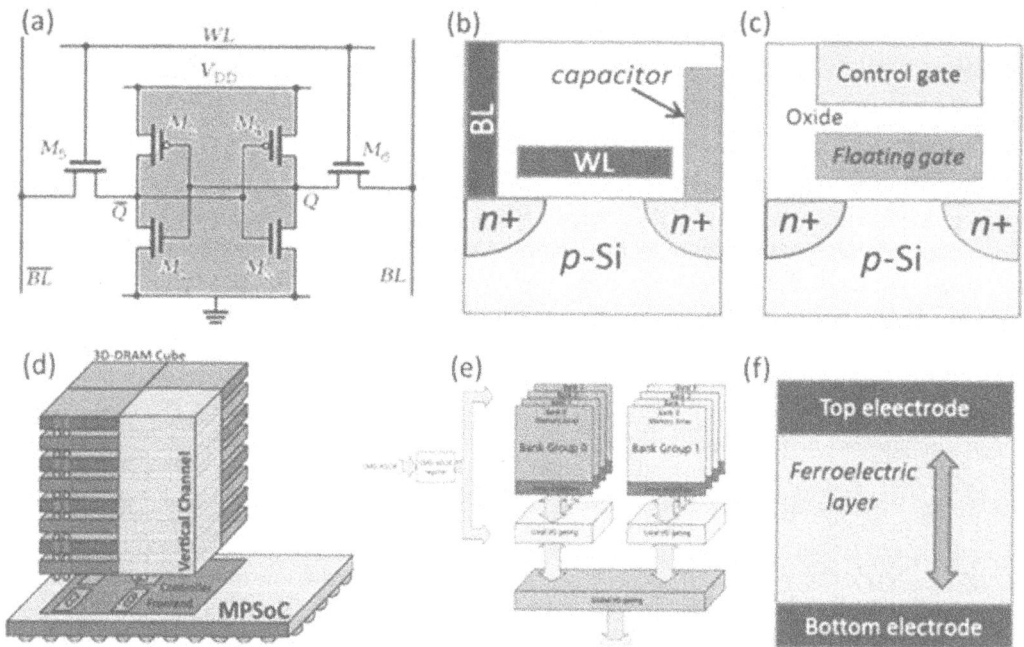

FIGURE 19.2 (a) SRAM architecture, (b) basic Schematic of DRAM, (c) schematic of the flash memory cell, (d) SDRAM architecture, (e) basic architecture of DDR RAM, and (f) basic structure of FeRAM cell. (Adapted with permission from [14]. Copyright [2018].)

3. *Synchronous Dynamic RAM (SDRAM)*: Synchronous dynamic RAM, or SDRAM, represents an evolution of DRAM technology. It synchronizes its operations with the computer's clock speed, allowing for more efficient data access. This synchronous coordination enhances the speed at which data is transferred between the RAM and the CPU. SDRAM has become a standard in many computing systems, including desktops, laptops, and servers, where high-speed data transfer is critical.

4. *Double Data Rate (DDR) RAM*: Double data rate RAM, or DDR RAM, builds upon the foundation of SDRAM. It improves data transfer rates by transmitting data on both the rising and falling edges of the clock signal. This effectively doubles the data transfer rate compared to earlier SDRAM versions. DDR RAM is widely used in modern computing systems, including laptops, desktops, and graphics cards, where high-speed memory access is essential for tasks like gaming and multimedia processing.

5. *Non-Volatile RAM (NVRAM)*: Non-volatile RAM, or NVRAM, represents a departure from traditional RAM in that it retains data even when the system is powered off. This is achieved through various technologies, such as ferroelectric RAM (FeRAM) and magnetoresistive RAM (MRAM). NVRAM combines the speed of RAM with the non-volatility of storage, making it suitable for applications like cache memory and in scenarios where instant data availability is critical, even after a system reboot.

19.2.3   ADVANCEMENTS IN RAM TECHNOLOGY

As technology marches forward, RAM has not remained stagnant. Recent years have witnessed remarkable advancements that push the boundaries of speed, capacity, and energy efficiency:

19.2.3.1   D-Stacked RAM

This innovation involves stacking multiple layers of RAM cells on top of each other. This increases storage capacity within a smaller physical footprint. Additionally, it enhances data transfer rates, as signals can traverse shorter distances within the stacked structure. This technology holds the promise of more powerful and efficient memory solutions.

19.2.3.2   Non-Volatile RAM (NVRAM)

NVRAM [17] technologies have garnered significant attention. They retain data even when the power is turned off, blurring the line between traditional volatile RAM and non-volatile storage. Technologies like MRAM and phase change RAM (PCRAM) hold particular promise, offering potential applications in scenarios where instant data availability is crucial.

19.2.3.3   Hybrid Memory Cube (HMC)

The hybrid memory cube (HMC) represents a leap forward in RAM architecture. It involves vertically stacking memory chips, enabling significantly faster data transfer rates and reduced power consumption compared to traditional DRAM. HMC is a promising solution for high-performance computing applications, where data throughput and energy efficiency are critical considerations. These advancements in RAM technology pave the way for more powerful, energy-efficient, and versatile computing systems, enhancing the capabilities of a wide range of electronic devices.

19.2.3.4   Resistive RAM (ReRAM)

Resistive RAM, or ReRAM, is an emerging technology that shows great interest. It operates on the principle of resistive switching, where the resistance of a material can be altered by applying voltage. This enables ReRAM to function as a non-volatile memory, with the potential for high storage density and fast access times. Researchers are actively exploring ReRAM for applications in both main memory and non-volatile memory.

19.2.4 Recent Research Developments

In recent years, the field of RAM technology has seen a surge of research and innovation. One of the most notable areas of advancement is in the development of novel materials for RAM. Researchers are exploring materials with unique electrical properties, such as ferroelectric materials and 2D materials like graphene, to push the boundaries of RAM performance. Additionally, there is a growing focus on energy efficiency in RAM design. Techniques like near-threshold computing, which operates electronic devices at voltages very close to their threshold voltage, are being investigated to reduce power consumption in RAM modules. Moreover, research efforts are aimed at overcoming the limitations of traditional silicon-based RAM. Emerging technologies like spin-transfer torque RAM (STT-RAM) and domain wall memory are being explored for their potential to offer higher storage densities and lower power consumption compared to conventional RAM technologies.

Furthermore, advancements in memory management algorithms and architectures are optimizing the utilization of RAM in modern computing systems. Techniques like memory compression and intelligent caching strategies are being employed to enhance overall system performance.

19.3 MEMRISTORS

19.3.1 Memristor Fundamentals

The advent of memristors, a term coined from "memory resistor," marks a paradigm shift in electronic components. Conceptualized by Leon Chua in 1971, memristors represent the fourth fundamental two-terminal passive circuit element, alongside resistors, capacitors, and inductors. These devices exhibit a dynamic relationship between resistance and applied voltage, enabling them to "remember" their resistance state based on the history of the electric charge that has passed through them. This unique characteristic grants memristors the ability to retain their resistance state even when the power is turned off. Such a property is invaluable for applications in non-volatile memory storage and neuromorphic computing, where the persistence of information is critical.

19.3.2 Memristor Characteristics and Applications

19.3.2.1 Resistive Switching Behavior

Central to memristor functionality is resistive switching, a phenomenon where the device's resistance state can be altered in response to an applied voltage. This behavior arises from the movement of ions within the memristor's material, leading to a physical change in resistance. Resistive switching [18–23] is the cornerstone of memristor operation, allowing them to function as non-volatile memory elements. By applying specific voltage pulses with defined amplitudes and durations, memristors can transition between low and high resistance states, effectively encoding binary information. The basic schematic diagram of the RRAM device and its corresponding schematic diagram of I-V curve is shown in Figure 19.3(a) and (b) respectively.

19.3.2.2 Applications in Memory

One of the most promising domains for memristor applications is memory technology. Given their non-volatile nature, memristors hold the potential to revolutionize memory storage by offering high-density, energy-efficient, and fast-access memory solutions [24]. Technologies like ReRAM and conductive bridge RAM (CBRAM) are currently under scrutiny for their suitability in various computing applications. In recent years, research has focused on refining memristor materials and device architectures to elevate memory performance. Techniques such as exploring different metal oxides and investigating novel switching mechanisms have shown promise in enhancing the reliability and endurance of memristor-based memory devices.

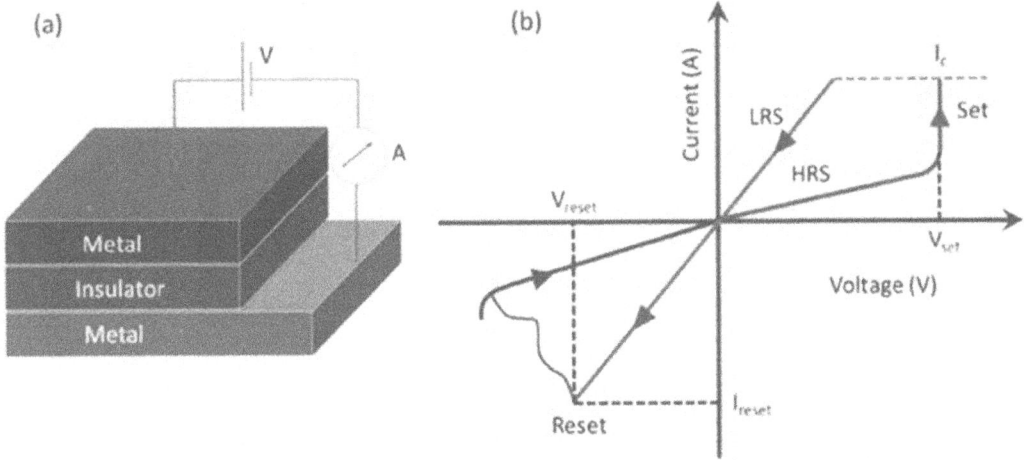

FIGURE 19.3 (a) Schematic diagram of RRAM device in which an insulator is sandwiched between the two metal electrodes. (b) The schematic of the I-V curve obtained in a RRAM device.

19.3.2.2.1 Materials Used for RRAM

A wide array of materials has been explored for resistive switching (RS) applications, encompassing binary oxides, complex metal oxides, carrier-doped manganites with perovskite structure, chalcogenides, polymers, and 2D materials. The pursuit of RS materials has been underway since the 1960s. The development of the RS materials with time is shown in Figure 19.4 [25]. In 1962, Hickmott first

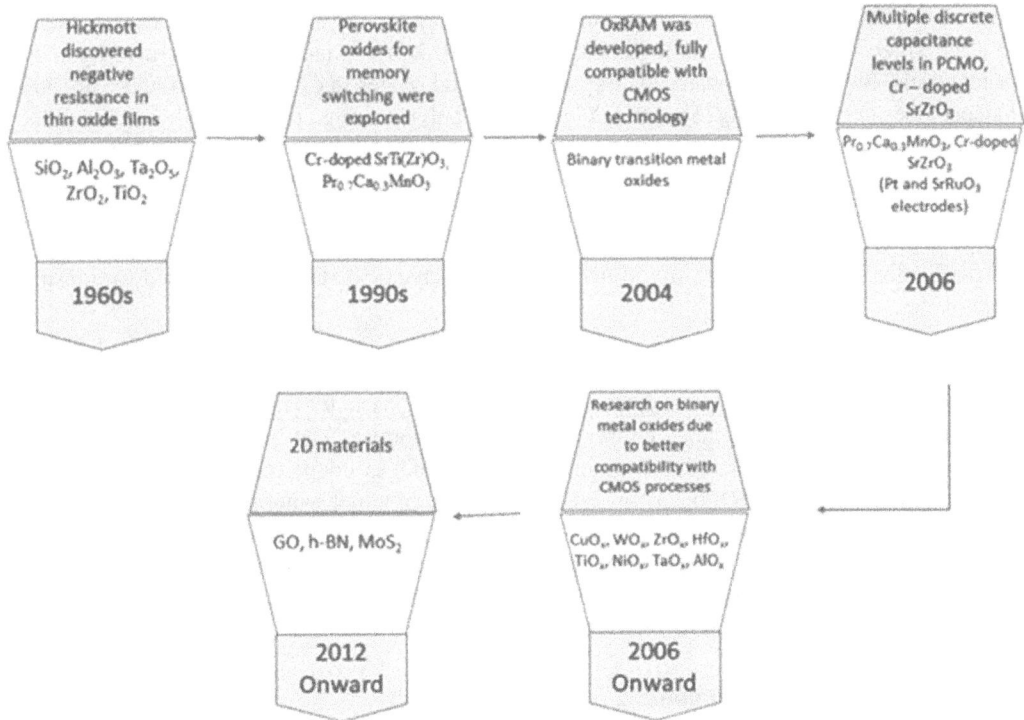

FIGURE 19.4 The development of the RRAM devices and materials in chronological order onwards. (Adapted with permission from [25]. Copyright [2020].)

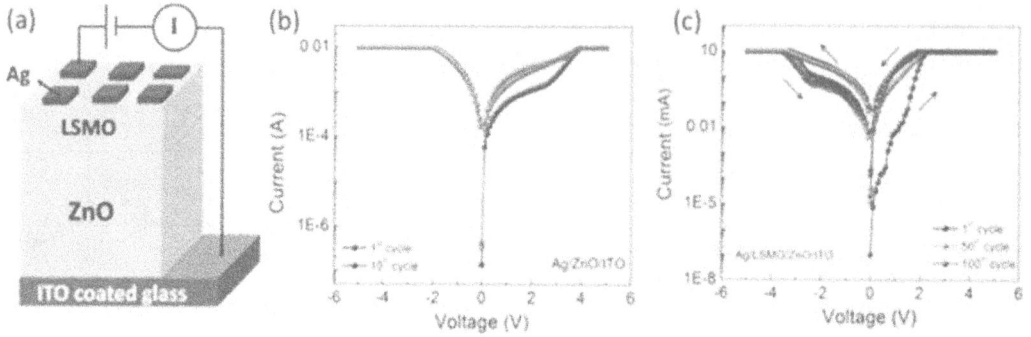

FIGURE 19.5 (a) Schematic diagram of the Ag/LSMO/ZnO device deposited on the ITO substrate. (b) I-V characteristics curve of Ag/ZnO/ITO device for 1st and 10th cycle in semi-logarithmic scale. (c) I-V curve of the Ag/LSMO/ZnO/ITO device for 1st, 50th, and 100th cycle in semi-logarithmic scale. The arrow represents the direction of the current. (Adapted with permission from [26]. (Copyright [2020] Current Applied Physics.)

demonstrated negative resistance in metal-insulator-metal (MIM) structures composed of thin films of SiO2, Ta2O5, Al2O3, TiO2, and ZrO2. Following this, in-depth investigations into the properties of Al2O3, SiO2, and Ta2O5 were conducted, along with the observation of switching properties in NiO. Subsequently, complex oxide materials like pervoskite oxides (e.g., Pr0.7Ca0.3MnO3 or PCMO), Cr-doped SrTi(Zr)O3, and PbZr0.5Ti04 emerged in the 1990s, although their stability for memory-based devices was limited. A significant development occurred in 2004 with the introduction of binary transition metal oxide resistive memory (OxRAM), characterized by stability and compatibility with CMOS technology. Notably, it operated at very low voltages (less than 3 V), marking a pivotal advancement in the field of RS. This breakthrough served as a catalyst for further research into RRAM devices and materials. Subsequently, a range of binary metal oxides, including ZnO, MgO, TiO2, HfO2, Cu2O, Fe2O3, CeO2, Nb2O5, WOx, and MoOx, were investigated as RS materials. Beyond these, materials like ternary metal oxides (e.g., LaAlO3, BiFeO3, SrZrO3), chalcogenides (e.g., GeSx, Ag2S), nitrides (e.g., SiN), polymers (e.g., poly(3,4- ethylenedioxythiophene) polystyrene sulfonate (PEDOT:PSS), poly(o-anthranilic acid) (PARA)), and manganites (e.g., Nd0:7Ca0:3MnO3) have been explored for their potential as RS materials. Additionally, the advent of 2D materials, such as MoS2, hexagonal boron nitride (h-BN), and graphene derivatives (GO, rGO), has offered distinct advantages over traditional binary oxide materials, including high thermal dissipation, enhanced flexibility, transparency, and chemical stability compared to transition metal oxides.

Apart from these, many nanocomposite materials have been utilized as resistive switching materials. For instance, combinations like ZnO/TiO2 can manifest intriguing RS properties. Additionally, nanocomposites consisting of metal oxide nanoparticles (e.g., TiO2, ZnO) embedded in a conductive metal matrix (e.g., copper) have shown promise. The incorporation of materials like graphene oxide (GO) or reduced graphene oxide (rGO) in conjunction with metal nanoparticles further enhances the RS behavior. The introduction of small nanofillers, such as rGO, augments the resistive switching characteristics of manganites. Similarly, the insertion of a thin layer of LSMO between the ZnO and the top electrode amplifies the RS behavior of the ZnO as shown in Figure 19.5(a-c) [26]. Consequently, this approach encompasses various strategies to enhance the performance of oxide-based RRAM devices.

19.3.2.3 Neuromorphic Computing

Memristors play a pivotal role in the advancement of neuromorphic computing [15, 27], a field dedicated to mimic the structure and functionality of the human brain. Their ability to mimic synapse-like behavior positions them as ideal building blocks for artificial neural networks. Memristor-based

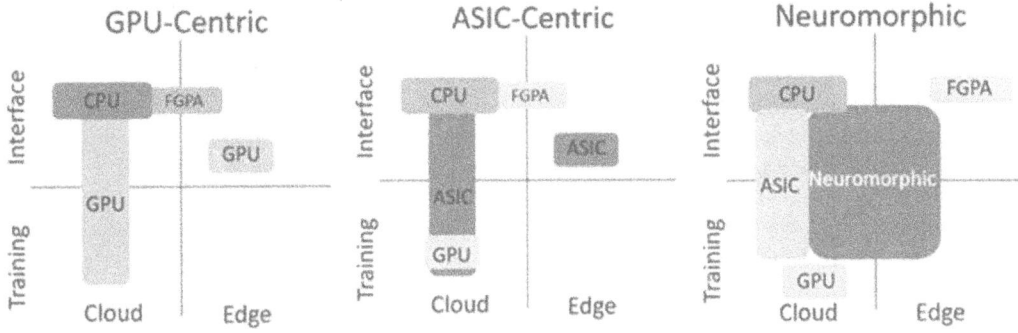

FIGURE 19.6   Evolution of neuromorphic hardware.

neuromorphic systems offer the potential for highly efficient and adaptable computing architectures, capable of tasks like pattern recognition and machine learning. Recent research in this area has focused on designing memristor-based synapses with tunable plasticity, allowing for dynamic adjustments of synaptic weights. Additionally, efforts are underway to integrate memristor-based synapses with spiking neural network architectures, aiming to create more biologically realistic and energy-efficient computing systems.

Neuromorphic computing has been researched over the decades, and recently, there have been significant advancements. As illustrated in Figure 19.6, the most recent improvement can be divided into three main phases. A GPU-centric system, which is primarily used for learning and supports artificial intelligence by using a GPU, is the initial phase. An application-specific integrated circuit (ASIC)-centric system, which is the next stage, is a hot topic of research. An effective and low-power ASIC for machine learning is anticipated to be created by this trend. As a result, numerous semiconductor firms are creating ASIC chips. However, it is anticipated that neuromorphic computing will develop into neuromorphic hardware, enabling ultra-low-power and ultra-high-speed computation to support general-purpose artificial intelligence.

The neuromorphic-centric hardware must be able to handle massive amounts of data in parallel while using incredibly little power. Furthermore, compared to current technology that uses ordinary CMOS components, the neuromorphic semiconductor chip demands a quicker rate of computing. This suggests that creating a novel synaptic device is essential to the development of neuromorphic-centric hardware.

19.3.3   Memristor-Based Logic and Computing

In addition to their applications in memory and neuromorphic computing, memristors can also serve as fundamental elements in logic operations. This convergence of memory and logic opens up new vistas for highly efficient and adaptive computing systems. By incorporating memristors into logic gates, researchers are exploring the potential for memristor-based computing architectures that could outperform traditional CMOS-based systems in specific applications.

Recent strides in memristor-based logic circuits have focused on minimizing power consumption and enhancing computational speed. Strategies such as designing memristor-based logic gates with lower switching energy and exploring new materials with faster switching speeds have demonstrated significant potential in advancing the capabilities of memristor-based computing.

19.3.4   Challenges and Future Directions

Despite the immense potential of memristors, several challenges need to be addressed. Notably, variability poses a significant concern, as ensuring consistent behavior across a large number of memristors can be intricate. Additionally, scalability and fabrication techniques require further research and development to make memristor-based technologies economically viable on a large scale. In the coming years, research in memristor technology is expected to center around overcoming these challenges. Efforts to develop reliable fabrication processes, explore novel materials, and refine memristor-based architectures will be pivotal in realizing the full potential of memristors in memory computing, and neuromorphic applications.

19.3.5 RECENT RESEARCH DEVELOPMENTS IN MEMRISTORS

In recent years, the field of memristor technology has witnessed a surge of research and innovation, driving advancements in several key areas.

19.3.5.1 Multifunctional Memristors

Researchers have made significant strides in developing memristors with multifunctional capabilities. These devices can perform a range of functions beyond memory storage, including logic operations and signal processing. By harnessing the versatile properties of memristors, scientists are working towards creating highly integrated and adaptable computing systems that can excel in diverse applications.

19.3.5.2 Memristor-Based Synaptic Plasticity

Advancements in understanding synaptic plasticity, the ability of synapses to strengthen or weaken over time, have led to the development of memristor-based synapses with tunable plasticity. This enables the creation of artificial neural networks that can undergo dynamic adjustments in synaptic weights, mimicking the learning processes observed in biological systems. Figure 19.7 depicts the behavior of organic memristors with synaptic plasticity for the application of neuromorphic computing [27].

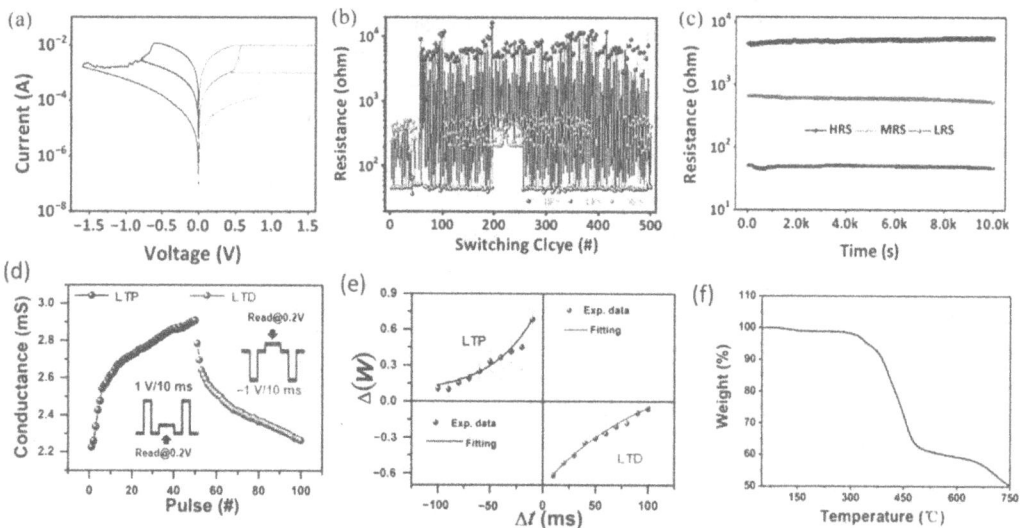

FIGURE 19.7 (a) Electrical parameters of the memristor. (a) Multistate I-V characteristic curve of memristor. (b) The device endurance and (c) retention at room temperature. (d) Long-term synaptic plasticity of the Ta/EV(ClO4)2/BTPA-F/Pt memristor. (e) Schematic illustration of the spike-timing-dependent plasticity properties of the device. (f) Synaptic weight retention performance of the device in response to temperature change. (Adapted with permission from [27]. Copyright [2023] Nanomaterials.)

19.3.5.3 Three-Dimensional Memristor Integration

Recent developments have focused on stacking multiple layers of memristor cells in three-dimensional architectures. This approach increases memory density while minimizing the physical footprint, leading to more compact and energy-efficient memory solutions. Three-dimensional memristor integration holds promise for addressing the growing demand for high-capacity memory in modern computing systems.

Thus, the emergence of memristors has ushered in a new era of possibilities in electronic componentry. With their unique resistive switching behavior, memristors are poised to revolutionize memory technology, neuromorphic computing, and logic operations. Ongoing research efforts are tackling challenges related to variability, scalability, and fabrication techniques, paving the way for memristor-based technologies to become integral components of future computing systems. Recent research developments in memristor technology are expanding the boundaries of what is possible. From multifunctional memristors to non-volatile neuromorphic memory, these advancements are driving innovation in a wide range of applications. As researchers continue to push the boundaries of memristor technology, we can anticipate even more transformative developments in the years to come.

19.4 GATE TRANSISTORS

19.4.1 Overview of Gates in Logic Circuits

In digital electronics, logic gates are fundamental building blocks that process binary information, performing logical operations based on Boolean algebra. These operations include AND, OR, and NOT, and they form the backbone of all computational tasks, from basic arithmetic to complex algorithms. Gate transistors are pivotal components in modern semiconductor technology, serving as the building blocks of digital integrated circuits. They play a fundamental role in controlling the flow of electronic signals and executing logic operations. Over the years, gate transistors have evolved to meet the demands of miniaturization and power efficiency, from the conventional CMOS transistors to cutting-edge fin field-effect transistors (FinFETs) and tunnel field-effect transistors (TFETs).

19.4.2 Advanced Gate Technologies

Traditional CMOS technology has long been the workhorse of digital electronics. However, with the increasing demand for faster and more energy-efficient devices, advanced gate technologies have emerged.

19.4.2.1 FinFETs (Fin Field-Effect Transistors)

FinFETs represent a significant advancement over traditional planar CMOS transistors. Instead of a flat channel, FinFETs use a three-dimensional "fin" structure that extends above the substrate. This design provides better control over current flow and reduces leakage, resulting in improved performance and energy efficiency.

19.4.2.2 Tunnel Field-Effect Transistors (TFETs)

TFETs operate based on a different principle compared to conventional transistors. Instead of relying solely on the electric field to control current flow, TFETs utilize quantum tunneling through a barrier. This enables them to achieve lower sub-threshold swing, making them highly suitable for low-power applications.

19.4.3 Quantum Gates

In the realm of quantum computing, gate operations take on a whole new level of complexity. Quantum gates manipulate quantum bits (qubits), which can exist in multiple states simultaneously (superposition) and become entangled, allowing for computations that classical logic gates cannot perform.

19.4.3.1   Quantum Logic Gates

Quantum logic gates, such as Controlled NOT (CNOT) and Hadamard gates, form the basis of quantum algorithms. They enable operations on qubits, allowing for the creation and manipulation of quantum states.

19.4.3.2   Quantum Supremacy

The concept of quantum supremacy refers to the point at which quantum computers can perform tasks that are fundamentally impossible for classical computers, signaling a major milestone in the field of quantum computing.

19.4.4   Gate Transistor Architecture and Operation

Gate transistors typically comprise a gate, source, drain, and channel region. The gate electrode plays a pivotal role in controlling the flow of current between the source and drain regions. By applying a voltage to the gate, an electric field is induced, either attracting or repelling charge carriers in the channel, thereby allowing or blocking current flow between the source and drain.

19.4.5   Recent Research Developments in Gate Transistors

In recent years, gate transistor technology has experienced a surge in research and development efforts, leading to several notable advancements.

19.4.5.1   Gate-All-Around (GAA) Transistors

Gate-all-around (GAA) transistors represent a monumental leap in transistor architecture. In GAA transistors, the gate completely surrounds the channel, offering even greater control over current flow. This design allows for increased transistor density and improved electrostatic control, paving the way for continued miniaturization and enhanced performance. Recent studies have focused on refining fabrication techniques to maximize the benefits of GAA transistors, aiming for unprecedented levels of integration and energy efficiency.

19.4.5.2   Low-Power Variants

Researchers have made substantial progress in developing low-power gate transistors to address the burgeoning demand for energy-efficient electronics. Novel materials and fabrication techniques are being explored to reduce leakage currents and enhance overall power efficiency. Techniques such as high-K metal gate (HKMG) technology and advanced dielectric materials have emerged as key enablers for achieving these objectives. Recent studies have shown promising results in significantly reducing power consumption without compromising performance.

19.4.5.3   Beyond Silicon: Alternative Materials

As the limitations of silicon-based transistors become more apparent, researchers are venturing into the realm of alternative materials for transistor fabrication. Compound semiconductors like gallium nitride (GaN) and silicon carbide (SiC) offer unique electrical properties that can enable higher-speed operation and higher breakdown voltages. Additionally, 2D materials such as graphene and transition metal dichalcogenides (TMDs) hold tremendous promise for future transistor technology. Recent experiments with these materials have demonstrated encouraging results, opening up new avenues for next-generation transistors.

19.4.5.4 Quantum Tunneling Transistors

Quantum tunneling transistors leverage the principles of quantum mechanics to achieve unprecedented levels of energy efficiency. By exploiting the quantum tunneling effect, these transistors can operate at extremely low voltages, reducing power consumption significantly. This technology holds immense potential for applications in ultra-low-power integrated circuits and energy-efficient computing systems. Recent breakthroughs in the design and fabrication of quantum tunneling transistors have shown promising results, indicating a potential paradigm shift in semiconductor technology.

19.4.6 FUTURE DIRECTIONS AND CHALLENGES

While recent research developments in gate transistors have propelled the field forward, several challenges remain. Ensuring reliability and consistency in the fabrication of advanced transistor technologies, especially those involving alternative materials, is a critical consideration. Additionally, addressing the thermal challenges associated with high transistor densities is paramount to prevent overheating in densely packed integrated circuits.

In the years ahead, research in gate transistor technology is poised to focus on overcoming these challenges. Continued exploration of novel materials, innovative transistor architectures, and advanced fabrication techniques will drive further advancements in semiconductor technology. Additionally, the integration of gate transistors into emerging fields such as quantum computing and neuromorphic computing holds the potential to revolutionize computing paradigms.

19.5 CONCLUSION

In the ever-evolving landscape of electronics, this chapter has taken us on a journey through the heart of cutting-edge technology. We began with the foundational transistor, a device that revolutionized computing and catalyzed the digital age. From its early iterations as a bipolar junction transistor (BJT) to the contemporary era of field-effect transistors (FETs), we witnessed the relentless march of miniaturization, fueled by the prescient vision encapsulated in Moore's law. Transistors, these minuscule sentinels, have indelibly shaped our world. They have empowered us to harness computational power beyond the wildest dreams of their progenitors. Moore's law, that prophetic roadmap, has been the compass guiding us through decades of exponential growth, ushering in an era where processing power once deemed fantastical now resides in our pockets. Yet, our exploration did not end with transistors alone. We ventured into a realm of advanced logic devices, each bearing its own distinct imprint on the technological tapestry. RAM, the dynamic powerhouse of data manipulation, emerged as the fulcrum of real-time computation. Memristors, with their capacity to remember resistance states, beckon us toward a future where memory storage transcends the confines of volatility. The promise of non-volatile memory, coupled with its potential in neuromorphic computing, stands as a testament to the transformative power of this nascent technology. Gate transistors, from the conventional CMOS to the vanguard of FinFETs and TFETs, herald a new era of speed, power efficiency, and integration. As we confront the challenges posed by diminishing transistor sizes, these innovations rise to the occasion, offering solutions that bridge the gap between theoretical limits and real-world applications. Central to this journey has been an acknowledgment of the materials that constitute this electronic tapestry. Silicon, with its exceptional semiconductive properties, remains the cornerstone of modern semiconductor technology. However, as we approach the limits of miniaturization, alternative materials such as compound semiconductors and innovative fabrication techniques have risen to prominence. They offer not just incremental improvements, but the potential for paradigm-shifting advances in performance and energy efficiency. In this epoch of ceaseless innovation, we find ourselves on the cusp of a new frontier. The convergence of transistors, RAM, memristors, and advanced gate technologies propels us toward a future bound only by the horizons of imagination.

Together, these components herald a new age of computing prowess, where memory and logic intertwine seamlessly, and where quantum leaps in speed, efficiency, and functionality become the new norm. As we look to the future, challenges such as reliability in fabrication processes, thermal management, and the integration of emerging technologies like quantum computing and neuromorphic computing remain at the forefront. However, the dynamism and ingenuity of the semiconductor community continue to propel us forward. In conclusion, this chapter underscores the pivotal role that RAM, memristors, and gate transistors play in shaping the future of computing. Their evolution, marked by relentless innovation and interdisciplinary collaboration, heralds a future where computing capabilities are limited only by our imagination. It is through the continued exploration and advancement of these advanced logic devices that we embark on a journey towards more powerful, efficient, and versatile computing systems.

REFERENCES

1. G. E. Moore. "Cramming more components onto integrated circuits." Electronics, vol. 38, pp. 114–117, 1965.
2. M. Horowitz. "Computing's energy problem (and what we can do about it)." IEEE Int. SolidState Circuits Conference (ISSCC), pp. 10–14, 2014.
3. P. A. Merolla, J. V. Arthur, R. Alvarez-Icaza, A. S. Cassidy, J. Sawada, F. Akopyan, B. L. Jackson, N. Imam, C. Guo, Y. Nakamura, B. Brezzo, I. Vo, S. K. Esser, R. Appuswamy, B. Taba, A. Amir, M. D. Flickner, W. P. Risk, R. Manohar, and D. S. Modha. "A million spiking- neuron integrated circuit with a scalable communication network and interface." Science, vol. 345, no. 6197, pp. 668–673, 2014.
4. W. A. Wulf, and S. A. McKee. "Hitting the memory wall: Implications of the obvious." ACM SIGARCH Computer Architecture News, vol. 23, no. 1, pp. 20–24, 1995.
5. G. Indiveri, and S.-C. Liu. "Memory and information processing in neuromorphic systems." Proceedings of IEEE, vol. 103, no. 8, pp. 1379–1397, 2015.
6. R. S. Williams. "What's next?" Computing in Science and Engineering, vol. 19, no. 2, pp. 7–13, 2017.
7. S. Salahuddin, K. Ni, and S. Datta. "The era of hyper-scaling in electronics." Nature Electronics, vol. 1, pp. 442–450, 2018.
8. J. M. Shalf, and R. Leland. "Computing beyond Moore's law." Computer, vol. 48, no. 12, pp. 14–23, 2015.
9. M. Di Ventra, and Y. V. Pershin. "The parallel approach." Nature Physics, vol. 9, pp. 200–202, 2013.
10. M. M. Waldrop. "The chips are down for Moore's law." Nature, vol. 530, no. 7589, pp. 144–147, 2016.
11. M. Alayan. "Investigation of HfO2 based Resistive Random Access Memory (RRAM): Characterization and modeling of cell reliability and novel access device." Diss. Université Grenoble Alpes, 2018.
12. T. Endoh. "Nonvolatile logic and memory devices based on spintronics." Proceedings IEEE International Symposium on Circuits and Systems, vol. 2015–July, pp. 13–16, 2015.
13. T. Endoh, T. Ohsawa, H. Koike, T. Hanyu, and H. Ohno. "Restructuring of memory hierarchy in computing system with spintronics-based technologies." Dig. Tech. Pap. – Symp. VLSI Technol., pp. 89–90, 2012.
14. S. S. Ulhas. "Defect engineering in HfO2/TiN-based resistive random access memory (RRAM) devices by reactive molecular beam epitaxy." Diss. Universitäts-und Landesbibliothek Darmstadt, 2018.
15. J. Park. "Neuromorphic computing using emerging synaptic devices: A retrospective summary and an outlook." Electronics, vol. 9, no. 9, p. 1414, 2020.
16. K. Kumari, A. Kumar, D. Kotnees, J. Balakrishnan, A. D. Thakur, and S. J. Ray. "Structural and resistive switching behaviour in lanthanum strontium manganite-reduced graphene oxide nanocomposite system." Journal of Alloys and Compounds, vol. 815, p. 152213, 2020.
17. A. Kumar, K. Kumari, A. D. Thakur, and S. J. Ray. "Graphene mediated resistive switching and thermoelectric behavior in lanthanum cobaltate." Journal of Applied Physics, vol. 127, p. 235103, 2020.
18. K. Kumari, A. Kumar, A. D. Thakur, and S. J. Ray. "Charge transport and resistive switching in a 2D hybrid interface." Materials Research Bulletin, vol. 139, p. 111195, 2021.
19. K. Karuna, S. Majumder, A. D. Thakur, and S. J. Ray. "Temperature-dependent resistive switching behaviour of an oxide memristor." Materials Letters, vol. 303, p. 130451, 2021.
20. K. Kumari, A. D. Thakur, and S. J. Ray. "The effect of graphene and reduced graphene oxide on the resistive switching behavior of La0.7Ba0.3MnO3." Materials Today Communications, vol. 26, p. 102040, 2021.

21. S. Majumder, K. Kumari, and S. J. Ray. "Pulsed voltage induced resistive switching behavior of copper iodide and La0.7Sr0.3MnO3 nanocomposites." Materials Letters, vol. 302, p. 130339, 2021.

22. K. Kumari, A. D. Thakur, and S. J Ray. "Structural, resistive switching and charge transport behaviour of (1-x) La0.7Sr0.3MnO3(x)ZnO composite system." Applied Physics A, vol. 128, no. 11, p. 992, 2022.

23. K. Kumari, S. J. Ray, and A. D. Thakur. "Resistive switching phenomena: A probe for the tracing of secondary phase in manganite." Applied Physics A, vol. 128, no. 5, p. 430, 2022.

24. K. Karmakar, A. Roy, S. Dhibar, S. J. Ray, and B. Saha. "Instantaneous gelation of a self-healable wide band gap semiconducting supramolecular mg(II) metallohydrogel: An efficient nonvolatile memory design with supreme endurance." ACS Applied Electronic Materials, vol. 5, pp. 3340–3349, 2023.

25. V. Gupta, S. Kapur, S. Saurabh, and A. Grover. "Resistive random access memory: A review of device challenges." IETE Technical Review, vol. 37, no. 4, pp. 377–390, 2020.

26. K. Kumari, S. Kar, A. D. Thakur, and S. J. Ray. "Role of an oxide interface in a resistive switch." Current Applied Physics, vol. 35, pp. 16–23, 2022.

27. J. Zeng, X. Chen, S. Liu, Q. Chen, and G. Liu. "Organic memristor with synaptic plasticity for neuromorphic computing applications." Nanomaterials, vol. 13, no. 5, p. 803, 2023.

20 Non-Volatile Memory Devices

Neeraj Mehta

20.1 INTRODUCTION

The two types of conventional mainstream semiconductor memory are volatile and non-volatile. The random access memory (RAM) can be either volatile or non-volatile [1–4]. However, static and dynamic RAM are examples of volatile memories that lose their saved data when the power is turned off. Although they are user-friendly and efficient, data loss is a significant drawback. Non-volatile memory (NVM) will continue to contain the previously recorded data even when the power source is turned off. Thus, in the event of a power outage, NVM does not lose any stored data since read-only memory (ROM) technology serves as the foundation for all commonly used NVMs.

They can save the information in either memory. However, both memories differ greatly from one another (Table 20.1).

The emphasis of computer system development and research has recently switched to semiconductor memory advances that blur the distinction between memory (slow, expensive, volatile) and storage (quick, affordable, non-volatile), as well as complete the hierarchical arrangement of memory and storage currently in place.

TABLE 20.1
Differences between VMs and NVMs

S. No.	Volatile Memory (VM)	Non-Volatile Memory (NVM)
i	Volatile memory only retains its information when the system is turned on. The content deletes itself when the system is turned off.	Another type of memory that retains data even when the system is powered down is non-volatile memory.
ii	Data and information are temporarily stored in volatile memory. Random access memory (RAM) is the ideal instance of volatile memory.	The data is permanently stored in non-volatile memory. Read-only memory (ROM) is the best instance of non-volatile memory.
iii	This memory performs better than non-volatile memory in terms of speed.	The performance of this memory is slower than volatile memory.
iv	Volatile memory transmits data more easily than non-volatile memory.	In this memory, data transport is challenging.
v	In volatile memory, read and write operations are carried out simultaneously.	In non-volatile memory, only read operations are carried out.
vi	Storage capacity is lower than that of non-volatile memory.	Storing capacity is higher than that of volatile memory.
vii	The cost per unit size is high for this type of memory.	The cost per unit size is low for this type of memory compared to volatile memory.
viii	The functioning of the computer system is affected by volatile memory.	ROM does not affect the system's performance. But it significantly affects the computers' storage capacity.
ix	Data in the volatile memory is directly accessible to the processor.	The non-volatile memory is not directly accessible by the processor.
x	The information related to the actions that the CPU is currently carrying out is stored in volatile memory, or RAM.	Any type of data is permanently stored in non-volatile memory.

 DOI: 10.1201/9781003450146-20

The persistent storage of certain data is a requirement for almost every electronic device. The advent of flash memory and the rise in popularity of battery-powered electronic appliances, primarily mobile phones, have been two important developments that have fueled the remarkable expansion of the NVM market. With the growing use of new data-centered devices like digital still cameras, MP3 players, and other personal digital assistants, a distinct trend in the NVM industry started to emerge in the early years of the new century.

The non-volatile memory devices (NVMDs) are read/write electronic data storage components that retain data even after being powered off [1–5]. They typically consist of magnetic disc drives and certain semiconductor chips. The semiconductor-based NVM devices are essential pieces of every aspect of the information age, from storage units in massive cloud databases to portable personal devices, and they account for one of the largest portions of the $400 billion semiconductor market today [6, 7]. NVM is growing more and more in demand as computer technology advances, and consumers' demands for higher read and write speeds and decreasing power usage are driving these changes.

The floating gate MOS transistor has served as the main building block for the development of NVMs for more than 40 years [8]. Non-volatile semiconductor memory has continuously grown in importance in our daily lives [9]. For instance, the railway exploits millions of commuters each day, having electronically erasable and programmable read-only memory at railway passes to expand the reliability of processing components for railway electronic security arrangements. In addition, non-volatile semiconductor memories are utilized in the majority of household electronics (e.g., cellular phones, digital cameras, audio players, universal-serial-bus [USB] memories) as well as in cars with large-capacity flash memories.

The need for data storage and retrieval has increased dramatically in the digital era. Data-intensive applications have permeated every aspect of our daily lives, from personal computers to cell phones. The creation of NVMDs has been essential in meeting this growing demand for storage. NVM, in contrast to conventional volatile memory, offers permanent and dependable storage solutions by maintaining data even when the power is switched off. This chapter explores the significance of NVM devices, their types, and their applications.

20.2 EVOLUTION OF NON-VOLATILE MEMORY DEVICES: HISTORICAL OVERVIEW

NVMDs have undergone significant historical evolution, progressing from early technologies such as magnetic core memory to modern flash memory and emerging technologies like resistive random-access memory (RRAM) and phase-change memory (PCM) [10]. In this section, we will discuss only the major milestones in the historical evolution of semiconductor-based NVMDs.

C.T. (Tom) Sah played a significant role as a pioneer in the development of semiconductor NVM memory. During the late 1960s and early 1970s, Sah conducted groundbreaking research on the fundamental principles of charge trapping and retention in semiconductor materials [10, 11]. It was Sah [10, 11] who noticed the floating gate memory effects in 1961 but did not publish his findings at that time. The discovery of the floating gate memory effects by Sah was an important observation in the development of NVM technologies.

Beginning in the 1960s, two fundamental strategies for designing semiconductor NVM cells were first studied and reported in the 1960s. For the first time, the potential for NVM in semiconductors was acknowledged at Bell Labs when Kahng and Sze demonstrated a prototype electrically erasable-programmable read-only memory by proposing a design of a floating gate to create an NVMD [12]. Initially, a floating gate cell was designed that stores charges on an electrode without being wired into an external circuit, to put it as simply as possible. The potentiality of NVMDs based on metal-oxide semiconductors was also recognized during this period. Since then, semiconductor memory has contributed greatly to the revolutionary development of digital electronics. This was the basis of the floating gate avalanche-injection metal-oxide-semiconductor (FAMOS). The invention of FAMOS was a significant milestone in NVM technology.

Dov Frohman-Bentchkowsky was an engineer and executive who made significant contributions during his time at Intel Corporation. Frohman-Bentchkowsky joined Intel in 1969 and became one of the early employees of the company [13, 14]. He is credited with inventing the concept of erasable programmable read-only memory (EPROM) while working at Intel. In 1971, Frohman-Bentchkowsky [13, 14] was part of the team of engineers at Intel Corporation who created it. In order to produce a dense, quick-access NVM, he developed the Intel 2708 N-MOS EPROM in 1974 using a novel method that used channel injection and a single dual-layer polysilicon cell. The device could be quickly erased for reprogramming by shining UV light through a window on the box, and it didn't need power to keep its contents.

NVM technology significantly advanced with the creation of EPROM. The advanced version of EPROM was electrically erasable programmable read-only memory (EEPROM). It was invented in 1978. It was developed by a team of engineers at Intel Corporation, including G. Perlegos [15, 16]. EEPROM was used as an innovative alternative to EPROM and programmable read-only memory (PROM).

The development of flash memory is attributed to Dr. Fujio Masuoka [17, 18], who created it while working for Toshiba in the 1980s. He was a gifted student who excelled in maths and had outstanding abilities at the elementary and high school levels by the age of three. As a result, he had a very inquisitive nature and was willing to experiment by going down unknown roads. He worked on DRAM as a development engineer and salesman early in his career at Toshiba. He developed the concept of flash memory while working at Toshiba Corporation in the early 1980s. In 1984, Masuoka et al. filed a patent for his invention, called "Semiconductor memory device and method for manufacturing the same," which described the basic principles and structure of flash memory [18]. According to reports, Shoji Ariizumi, a coworker of Masuoka's, came up with the term "flash" because the act of deleting all the information from a chip made of semiconductors reminded him of a camera's flash. At the IEEE 1984 Integrated Electronics Devices Meeting in San Jose, California, Dr. Masuoka presented the innovation [19]. In the early 1980s, Toshiba created a one-terabyte flash memory using EEPROM, and in 1985, it was released on the market. The details of the pioneers of NVM for the first few decades are illustrated in Figure 20.1.

In 1984, Masuoka and associates introduced NOT-OR (NOR) flash, and then in 1987, at the IEEE International Electron Devices Meeting (IEDM) in San Francisco, they introduced NOT-AND (NAND) flash [20, 21]. In 1987, Toshiba introduced NAND flash memory on a commercial scale [22]. In 1988, Intel Corporation unveiled the first NOR-type flash chip for commercial use 1988 after realizing the invention's enormous potential. In 1989, Samsung and Toshiba introduced

FIGURE 20.1 Schematic showing the historical growth of the NVM technology by different scientists making noteworthy contributions to its development.

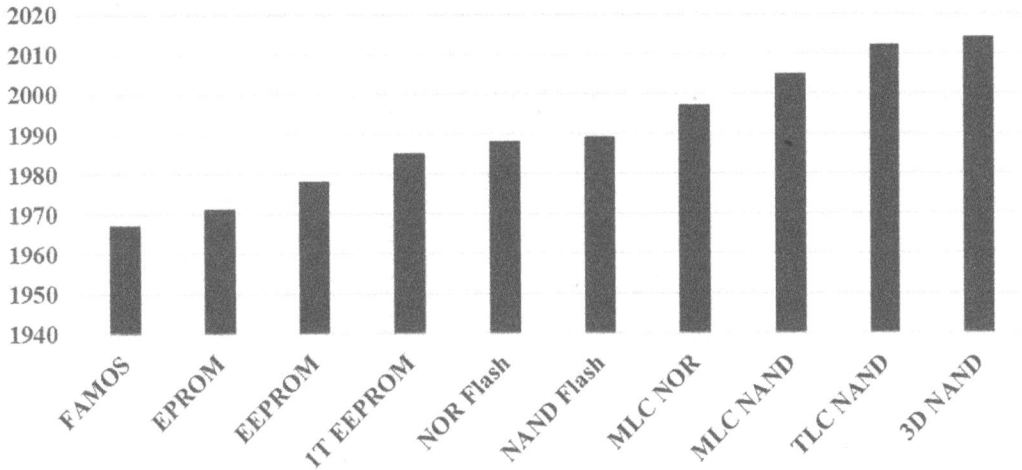

FIGURE 20.2   Timeline showing the evolution of different NVM devices.

the NAND flash. In comparison to NOR flash, it boasted ten times the endurance, faster erase and write speeds, higher density, and lower cost per bit.

Later, in 1997, Intel developed a multi-level cell (MLC) flash, which was a new type of NAND flash memory. More than one bit could be stored in each cell of this two-bit MLC NOR flash. MLC NAND flash memory was first introduced by Toshiba Corporation in 2005. MLC is a type of NAND flash memory that allows multiple bits to be stored in a single memory cell, typically two bits per cell. Triple-level cell (TLC) NAND flash memory was introduced by Samsung in 2012. TLC NAND is a type of NAND flash memory that allows storing three bits of data per cell, hence the "triple-level" designation. It builds upon the MLC technology, which stores two bits per cell. Samsung created 3D NAND memory, sometimes referred to as vertical NAND or V-NAND, in 2014. An important turning point in the development of NVM technology was the creation of 3D NAND. The historical evolution of NVM devices is shown in Figure 20.2.

20.3   TYPES OF NON-VOLATILE MEMORY DEVICES

In the present time, there are various types of NVM devices based on the technology used and the nature of the semiconducting materials. We will discuss some widely used NVMDs in this section.

20.3.1   Flash Memory

The most popular NVM technology is flash memory [23–26]. It uses floating gate transistors to store data, making it possible to maintain knowledge even in the absence of power. NAND flash and NOR flash are just two types of flash memory. NAND flash is frequently used in consumer devices like USB drives, SSDs, and memory cards due to its great density and inexpensive price. Contrarily, NOR memory provides quicker read and write rates and is frequently utilized in devices like micro-controllers and firmware storage. Although NOR-based flash has slow write and erase speeds, it features a complete address/data (memory) interface that permits arbitrary access to any location. It can therefore be used to store program code that only has to be updated occasionally.

The construction of flash memory is shown in Figure 20.3. An array of memory that is stacked with many flash cells is part of flash memory architecture. A memory cell serves as the fundamental building block of flash memory. Typically, metal-oxide-semiconductor field-effect transistor (MOSFET) technology is used to build flash memory cells. A fundamental flash memory cell

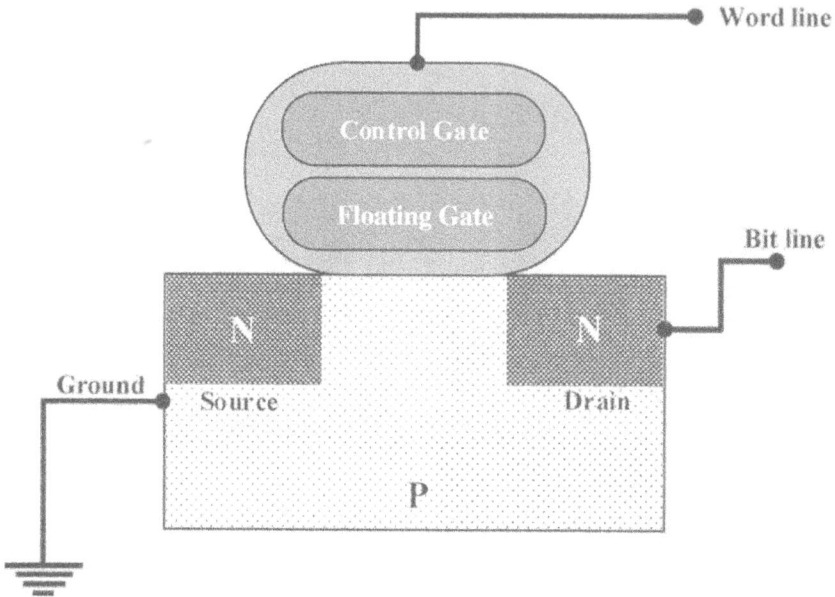

FIGURE 20.3   Basic construction of a typical flash memory.

is made up of a storage transistor having a floating gate and a control gate. The floating gate is an electrically isolated component within the memory cell. It is made of a conductive material, usually polysilicon, and is surrounded by a thin insulating layer known as tunnel oxide. A further insulating layer separates the control gate from the floating gate, which is located above it. The passage of electrical charge to and from the floating gate is managed by the control gate. These gates are separated from other components of the transistor by using a fine sheet of dielectric substance. This separating sheet prevents current from routinely passing through it.

The memory cells must be programmed in order to write data to flash memory. An electric field is produced when a voltage is delivered to the control gate to program a cell. Electrons can tunnel through the tunnel oxide and onto the floating gate thanks to the strong electric field. The floating gate is being charged to represent the binary "1" state. When necessary, flash memory also allows for data erasure. A block, which is a collection of memory cells, gets erased. Usually, a whole block is deleted all at once. The erasing procedure renders the floating gate neutral by removing its charge by putting a greater voltage between the control gate and the substrate. Consequently, the electrons tunnel out of the floating gate and return to their initial state, thus erasing operation is accomplished. Flash memory can be read for data by applying an electrical signal to the control gate and seeing the resulting current flow through the memory cell. The threshold voltage of the transistor is impacted by the existence or absence of charge on the floating gate. The state of the cell (0 or 1) can be ascertained by measuring the current.

Flash memory is organized into pages that are then broken up into blocks. Memory cells often come in groups of many per page. The block structure makes programming and wiping operations efficient. Flash memory, in contrast to RAM, necessitates wiping complete blocks before rewriting them, which might slow down write operations.

The entire flash memory is connected with the word line and bit line. These lines are essential components that enable the reading and writing of data. In a flash memory array, the word line is a control line that is used to choose a particular row, or "word," of memory cells. It is connected to the gates of a row of memory cells [27]. The memory cells in that row become accessible for reading or writing operations when a certain word line is triggered. The bit line, often referred to as the column line, is in charge of transporting the information kept in the memory cells of a particular column. It attaches to the source or drain terminals of a column of memory cells. The bit line reads the voltage or current level from the chosen memory cells during a read operation, revealing the data contained there.

Flash memory has a limited number of write/erase cycles before it starts to degrade. To mitigate this, wear-leveling algorithms are employed to distribute write cycle consistently across the memory cells, ensuring that no individual cells wear out prematurely [28]. Error correction techniques are also employed to identify and accurate errors that may arise during reading and writing operations.

20.3.2   PHASE CHANGE MEMORY (PCM)

An innovative NVM technique called phase change memory (PCM) uses the special qualities of some materials to store data [29, 30]. Ovshinsky [31] made the initial discovery of PCM in the late 1960s when he saw that a device made of chalcogenide glass layer would flip to a highly conductive mode when a threshold voltage was achieved and would remain in a conductive state until a holding voltage was provided. Binary data is represented by PCM through reversible phase transitions between amorphous and crystalline states [32]. This technology has several benefits, such as quick read and write speeds, high endurance, and scalability. In a variety of fields, including embedded devices and high-performance computing, PCM has the potential to replace or supplement current memory technologies.

The main component of a memory cell is PCM. It is a chalcogenide glass [33], commonly made of the elements germanium, antimony, and tellurium (Ge-Sb-Te) [34]. Amorphous and crystalline are the two distinct phases in which this substance can exist. Data storage is made possible by the change between these phases. The substrate, on which the PCM device is typically constructed, acts as the framework of the entire structure [32, 35]. On the substrate, there is a coating of a bottom electrode layer. This electrode offers electrical contact to the PCM material and helps in the passage of current during read and write operations. The PCM material is deposited on top of the bottom electrode. Above the PCM material is a heater element. The heat source during the PCM's phase transition is often a thin strip of conductive material. Generally, tungsten or titanium nitride is used as conductive material. The heating element is electrically isolated from the higher components by having an insulating layer placed on top of it. Heat dissipation and inappropriate electrical contact are prevented by this layer. Above the layer of insulating material, the top electrode is located. It acts as the device's electrical contact and permits the flow of current during read-and-write operations.

Between the top electrode and the heater, the phase-change material is traversed by the electrical current. A programmed zone is produced by current crowding at the heater to phase-change material contact, as seen by the mushroom-shaped borderline. Electrical pulses are used to train and read PCM cells, and as a result, the temperature of the phase-change layer changes. The PCM is amorphized into the high-resistance reset state using a short high-temperature pulse, and it is crystallized into the low-resistance set state using a long-duration pulse of a medium temperature. The construction of PCM is shown in Figure 20.4.

The mushroom-shaped design of PCM devices has many benefits [36]. The form and placement of the heater element help to concentrate the heat produced during the phase transition in a small area, minimizing the effect on surrounding cells and consuming less power. The architecture makes it easier to integrate more memory cells in a given area, which enhances scalability.

20.3.3   RESISTIVE RANDOM-ACCESS MEMORY (RERAM)

Another favorable NVM technology is resistive random access memory (ReRAM), also referred to as memristor-based memory [37, 38]. To store information, its working principle is based on varying the resistance of a solid-state substance, often a metal oxide. Fast access times, great endurance, low power consumption, and scalability are all characteristics of ReRAM. These features make

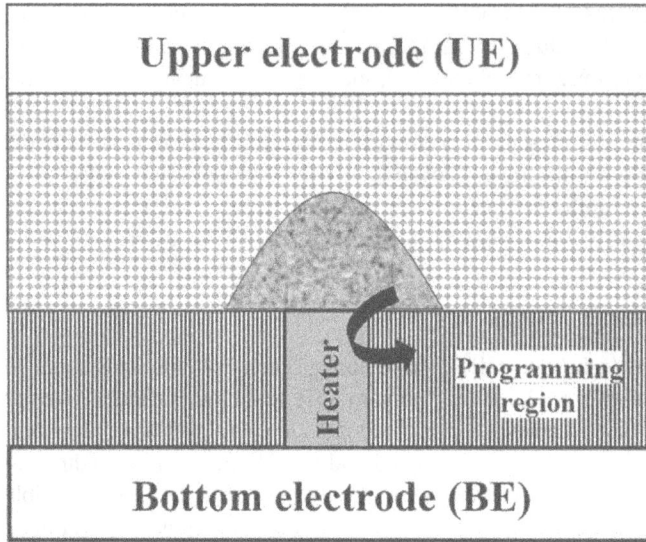

FIGURE 20.4   Basic construction of a typical PCM.

ReRAM suitable for use in sophisticated data storage systems, edge computing, and artificial intelligence [2].

The working of ReRAM and PCM are comparable in certain ways. PCM entails producing enough Joule heating to cause phase transformation between amorphous and crystalline phases. ReRAM, on the other hand, makes use of oxygen vacancies, which are the defects created in a thin oxide layer. These oxygen vacancies represent the sites of the oxygen-free oxide bonds. The charging and drifting of them is possible by applying an external electric field. The construction of ReRAM is shown in Figure 20.5.

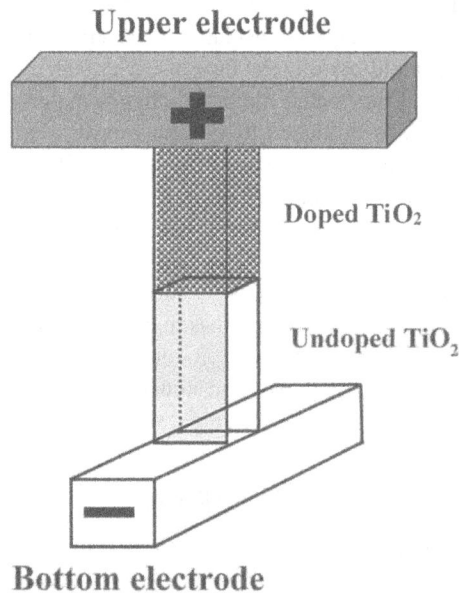

FIGURE 20.5   Basic construction of a typical ReRAM.

Typically, a substrate serves as the base upon which the ReRAM device is constructed. On the substrate, a bottom electrode layer is deposited that allows the flow of current during read and write operations and acts as the ReRAM device's electrical contact. Titanium nitride (TiN), and platinum (Pt) are often-used materials for bottom electrodes. The main element of a metal-oxide-based ReRAM is the metal-oxide layer. It is a metal-oxide thin layer that demonstrates resistive switching behavior, which is frequently made of titanium dioxide (TiO_2). The thin layer of TiO_2 consists of two regions. The lower region is highly resistive, since it is made of undoped TiO_2 with no oxygen vacancy. The upper region is conductive, and it consists of doped TiO_2 with plentiful oxygen vacancies. When the applied voltage is positive, the charged oxygen vacancies in the doped layer are attracted to the pure TiO_2 layer by the electric field. Due to this, there is a variation in the length of the doped region followed by a change in the overall resistivity. The low resistivity ions predominate in this situation. Consequently, the device offers minimal resistance because of the extension of the doped zone to its entire length. In contrast, when the applied voltage is negative, it exhibits the maximum resistance when the undoped area covers the entire length.

20.4 APPLICATIONS

NVM devices have revolutionized data storage and computing in numerous ways [2, 39]. In this section, we will discuss some potential applications of NVMDs. NVMs are widely used in consumer electronics and enterprise storage. Flash memory has established itself as the main component of portable electronics like smartphones, tablets, and digital cameras because it provides great storage capacity in a small package [40]. The most outstanding instances of enterprise storage are NVMDs [41]. Although their performance is steadily improving, the changes are significant and useful. For instance, due to their superior performance, lower power consumption, and less danger of mechanical failure, conventional hard disc drives (HDDs) have been replaced by SSDs that utilize flash memory technology in many data centers.

Many cloud-based analytics and artificial intelligence applications require rapid data processing and analysis [42]. NVM's low latency and high bandwidth characteristics enable faster data ingestion and real-time processing, enabling more efficient and timely insights. Embedded systems, including Internet of Things (IoT) devices, automotive electronics, and industrial automation, utilize NVM technologies like PCM and RRAM. They offer dependable and affordable storage solutions. NVM can be used as a cache for frequently accessed data. This is known as "storage-class memory" or "persistent memory." By using NVM as a cache, embedded systems can dramatically accelerate data access, leading to improved overall system performance.

By facilitating effective virtual machine migration, accelerating data access speed, and lowering latency, NVMDs play a crucial part in cloud computing architecture. NVM has a noteworthy role in enhancing the efficiency, performance, and reliability of cloud computing environments. NVM provides much quicker read and write speeds compared to traditional spinning hard drives. This results in quicker data access, reducing latency in cloud applications. This is especially useful for applications like securities trading platforms, entertainment servers, and media delivery networks that need real-time data processing. In cloud environments, virtualization is a key technology that enables the efficient sharing of physical resources among multiple virtual machines. NVM can enhance the performance of virtualization by providing VMs with quicker access to their data and reduced I/O latency (i.e., waiting time), providing more responsive and efficient virtualized workloads. NVM provides data persistence, meaning that data stored in NVM is not lost when power is turned off. This attribute is crucial for cloud services that require high levels of data availability, such as databases, file systems, and distributed storage systems.

Figure 20.6 summarizes the unique applications of NVMs in different devices and related fields.

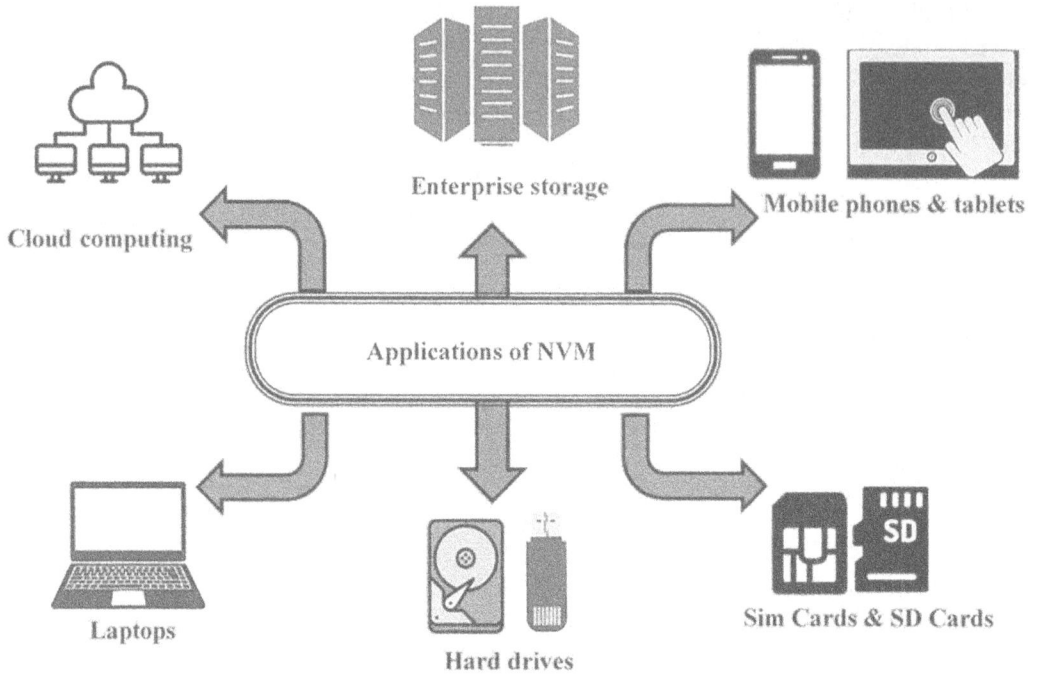

FIGURE 20.6   A glimpse of exclusive applications of NVMs in diverse fields.

20.5   PRESENT STATUS AND FUTURE ASPECTS

Compared to volatile memory, NVM uses user-configurable tools and does not need to be refreshed on a regular basis. Thus, consumer electronics employ NVM extensively due to its inexpensive cost and low power requirements [43]. NVM is, therefore, frequently employed for long-term consistent storage or secondary storage, and it is quite common in the digital media industry. In the race to produce universal memory, NVM technologies are receiving a lot of interest from the semiconductor industry.

The development of very dense storage media, such as HDDs, digital video discs (DVDs), Blu-ray, and flash memories, is being driven by the high demand for digital data storage. For instance, flash memory SSDs have replaced several HDDs with smaller capacities due to the rapid reduction of flash memories [44].

Most non-volatile memories are expected to be based on standard flash technologies for many years to come, which utilize multi-level cells and transistor-based charge storage memory cells to store multiple logic bits in a single physical cell.

According to expectations, the current floating gate, flash technology-based NVM mainstream will serve as the standard technology for the foreseeable future [45]. NVM systems have completely changed the way that data is stored since they provide enduring and dependable solutions. Continuous reductions in the price and electrical power demand of integrated circuits have fueled the market's spectacular growth for electronic portables. The way people interact with digital information has changed as a result of advancements in technologies like flash memory, magnetic storage, PCM, and RRAM, which provide quicker access, larger capacity, and better energy efficiency [46].

The North American, Asia-Pacific, Latin American, European, Middle Eastern, and African regions make up the majority of the NVM market worldwide. The Asia-Pacific region is predicted to dominate the worldwide NVM market. Presently, North America and Asia-Pacific are the two largest NVM markets. The Middle East, Africa, and Latin America could be some of the market's other significant geographical areas. Over the course of the next five to seven years, the presence of

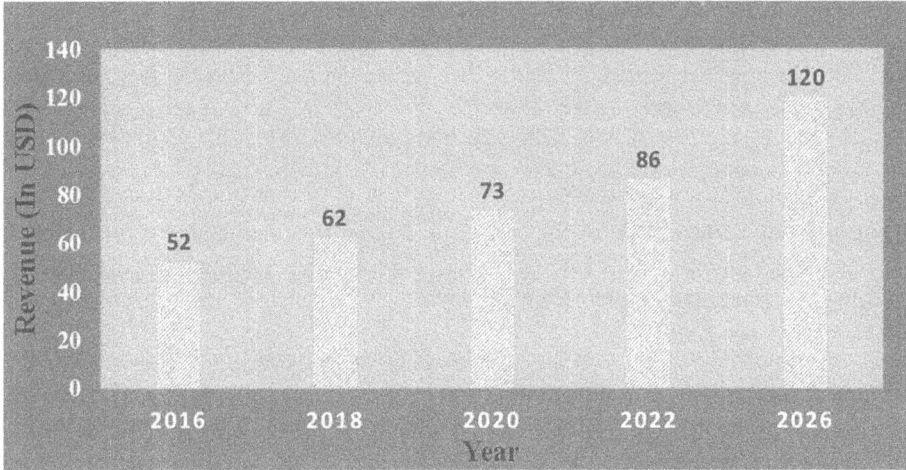

FIGURE 20.7   Plot showing the market size of NVM devices between 2016 to 2026.

major market players in nations like China, India, Japan, and South Korea is also expected to have a favorable effect on the local market. The market is expected to increase as next generation NVMs are being used in different electronic items (e.g., laptops, AI gadgets, mobile phones) to improve the consumer experience. According to a global assessment of the NVM market trend, its market value is anticipated to expand at a compound annual growth rate of around 8.7 between the mid-2010s and the mid-2020s, with starting and final market values of USD 120 billion and USD 52.3 billion, respectively (see Figure 20.7).

The future of NVM holds several exciting possibilities as technology continues to advance [47]. There are various key future aspects and trends to consider. Storage-class memory is a category of memory that falls between conventional volatile memory (such as DRAM) and NVM (such as NAND flash). Samsung's Z-NAND and Intel's Optane are two examples of SCM. SCM provides persistence, low latency, and high-speed access. It might make it possible to create databases and applications in innovative ways and speed up different cloud computing workloads. As these technologies develop, they may alter the NVM market and provide even better cloud computing and data storage solutions [48]. In-memory computing eliminates the need to fetch data from slower storage tiers by storing and processing data directly in high-speed, NVM. Workloads involving analytics, AI, and data-intensive applications are all considerably accelerated by this method.

PCM was successfully introduced to the market as Storage Class Memory in 2017 by Intel and Micron Technologies [49]. The GeSbTe alloy has a structural phase transition, which results in a significant difference in electric resistance between the amorphous and crystalline phases, and it is the basis for the switching process used by PCM. An outlook on the atomic layer deposition of chalcogenides for next-generation phase change memory was put forth by Lee et al. [50]. Their findings on atomic layer deposition (ALD) of GeSbTe (GST) alloys represent a current research effort to broaden the application of the ALD technology beyond the usual GST alloys.

In memory devices, computing is anticipated to increase in popularity in cloud environments as a result of developments in NVM technology, resulting in even faster data processing and analysis. Instead of processing data in a centralized cloud data center, edge computing processes data closer to the data source. NVM is the best option for edge computing scenarios because of its speed and persistence combination, where low latency and data robustness are essential. Real-time decision-making is facilitated by NVM, which also enables IoT devices, driverless vehicles, and industrial automation. With their heavy reliance on data processing, AI and machine learning workloads greatly benefit from NVM's quick read and write rates. Large dataset support and task acceleration provided by NVM could result in more effective AI model deployment in cloud environments in the near future.

The persistent nature of NVM can also contribute to increased security. Sensitive information might be safely stored on encrypted NVM, and as the memory is non-volatile, the encryption will hold even after the power is turned off. Better data protection and adherence to privacy laws may result from this. Future systems might employ hybrid memory architectures that combine different memory technologies, such as DRAM, SCM, and traditional NAND flash, to optimize performance and cost. Adaptive memory architectures that intelligently allocate data to the most appropriate memory tier could become more prevalent.

The properties of NVM can influence how hardware and software are created. To fully utilize the advantages of NVM, cross-stack interaction between software, operating systems, and hardware platforms may become more crucial, allowing developers to construct applications that are more effective and responsive.

With the rise in popularity of several consumer gadgets, like computers, flash drives, and optical discs, it is anticipated that the demand for NVMs will increase dramatically. NVM devices will become even more important as technology advances, opening the door for groundbreaking uses in fields like edge computing and artificial intelligence, among others. Deep neural networks can be trained using enterprise data centers, flash storage, and cloud storage thanks to the rising popularity of machine learning applications, IoT devices, and artificial intelligence that need high throughput and low latency. Thus, the market trend of NVM devices indicates that demand will increase even more as a result of the expanding size and quantity of data centers.

ACKNOWLEDGMENT

Neeraj Mehta is thankful to his university for providing an incentive grant under Institutes of Eminence (IOE) scheme (Dev. Scheme No. 6031).

REFERENCES

1. P. Cappelletti, (2015), Non-volatile memory evolution and revolution. IEEE International Electron Devices Meeting (IEDM), Washington, DC, USA, pp. 10.1.1–10.1.4, https://doi.org/10.1109/IEDM.2015.7409666
2. R. Waser, Resistive non-volatile memory devices. Microelectronic Engineering, 86 (2009) 1925–1928.
3. J. S. Meena, S. M. Sze, U. Chand, T.-Y. Tseng, Overview of emerging nonvolatile memory technologies. Nanoscale Research Letters, 9 (2014) 526. https://doi.org/10.1186/1556-276X-9-526
4. Y. Fujisaki, Review of emerging new solid-state non-volatile memories. Japanese Journal of Applied Physics, 52 (2013) 040001.
5. J.-S. Lee, Progress in non-volatile memory devices based on nanostructured materials and nanofabrication. Journal of Materials Chemistry, 21 (2011) 14097–14112. https://doi.org/10.1039/C1JM11050K
6. E. A. Sack, D. A. Laws, Westinghouse: Microcircuit Pioneer from molecular electronics to ICs. IEEE Annals of the History of Computing, 34 (2012) 74–82.
7. M. T. Ghoneim, M. M. Hussain, Review on physically flexible nonvolatile memory for internet of everything electronics. Electronics, 4 (2015) 424–479.
8. A. Fazio, Flash memory scaling. MRS Bulletin, 29 (2004) 814–817. https://doi.org/10.1557/mrs2004.233
9. W. Banerjee, Challenges and applications of emerging nonvolatile memory devices. Electronics, 9 (2020) 1029. https://doi.org/10.3390/electronics9061029
10. C. T. Sah, A new semiconductor tetrode, the surface-potential controlled transistor. Proceedings of the IRE, 49 (1961) 1625. https://doi.org/10.1109/JRPROC.1961.287763
11. C. T. Sah, Evolution of the MOS transistor: From conception to VLSI. Proceedings of the IEEE, 76 (1988) 1295. https://doi.org/10.1109/5.16328
12. D. Kahng, S. M. Sze, A floating gate and its application to memory devices. Bell System Technical Journal, 46 (1967) 1288–1295. https://doi.org/10.1002/j.1538-7305.1967.tb01738.x
13. D. Frohman-Bentchkowsky, (1972), Integrated MNOS memory organization, US Patent 3641512A.
14. D. Frohman-Bentchkowsky, A fully decoded 2048-bit electrically programmable FAMOS read-only memory. IEEE Journal of Solid-State Circuits, 6 (1971) 301–306. https://doi.org/10.1109/JSSC.1971.1050191

15. A. Gupta, T.-L. Chiu, M. Chang, A. Renninger, G. Perlegos, (1982), A 5V-only 16K EEPROM utilizing oxynitride dielectrics and EPROM redundancy. IEEE International Solid-State Circuits Conference. Digest of Technical Papers, San Francisco, CA, USA, pp. 184–185, https://doi.org/10.1109/ISSCC.1982.1156369

16. G. Perlegos, S. Pathak, A. Renninger, W. Johnson, M. Holler, J. Skupnak, M. Reitsma, G. Kuhn, (1980), A 64K EPROM using scaled MOS technology. EEE International Solid-State Circuits Conference. Digest of Technical Papers, San Francisco, CA, USA, pp. 142–143, https://doi.org/10.1109/ISSCC.1980.1156041

17. F. Masuoka, Field effect semiconductor memory apparatus with a floating gate, US Patent, 3825945 (1974).

18. F. Masuoka, S. Horii, T. Tanigami, T. Yokoyama, Semiconductor memory device and manufacturing method for the same, US Patent, 7304343B2 (1984).

19. F. Masuoka, M. Asano, H. Iwahashi, T. Komuro, S. Tanaka, (1984), A new Flash EEPROM cell using triple polysilicon technology. IEEE Tech. Dig. IEDM, pp. 464–467, https://doi.org/10.1109/IEDM.1984.190752

20. F. Masuoka, M. Asano, H. Iwahashi, T. Komuro, S. Tanaka, (1985), A 256K Flash EEPROM using triple polysilicon technology. IEEE ISSCC. pp. 168–169, https://doi.org/10.1109/ISSCC.1985.1156798

21. H. Takato, K. Sunouchi, N. Okabe, A. Nitayama, K. Hieda, F. Horiguchi, F. Masuoka, (1988), High-performance CMOS surrounding gate transistor (SGT) for ultra high-density LSIs," Technical Digest., International Electron Devices Meeting, San Francisco, CA, USA, pp. 222–225, https://doi.org/10.1109/IEDM.1988.32796

22. F. Masuoka, M. Momodomi, Y. Iwata, R. Shirota, (1987), New ultra high-density EPROM and flash EEPROM with NAND structure cell. International Electron Devices Meeting, Washington, DC, USA, pp. 552–555, https://doi.org/10.1109/IEDM.1987.191485

23. M. Gill, (1996), Flash memories: A review. Proceedings of Nonvolatile Memory Technology Conference, Albuquerque, NM, USA, p. 142, https://doi.org/10.1109/NVMT.1996.534690

24. F. Masuoka, T. Endoh, (1995), Flash memories, their status and trends. Proceedings of 4th International Conference on Solid-State and IC Technology, Beijing, China, pp. 128–132, https://doi.org/10.1109/ICSICT.1995.499651

25. K. Rajkanan, (1996), Flash memory technology: A review. IEEE International Workshop on Memory Technology, Design, and Testing, Singapore, pp. 47–48, https://doi.org/10.1109/MTDT.1996.782491.

26. P. Olivo, E. Zanoni, (1999). Flash Memories: An Overview. In: Flash Memories. Springer, Boston, MA. https://doi.org/10.1007/978-1-4615-5015-0_1

27. P. Pavan, R. Bez, P. Olivo, E. Zanoni, (1997), Flash memory cells-an overview. Proceedings of the IEEE, 85 (1997) 1248–1271. https://doi.org/10.1109/5.622505

28. J. Ranaweera, I. Kalastirsky, E. Gulersen, W. T. Ng, C. A. T. Salama, A novel programming method for high speed, low voltage flash E^2PROM cells. Solid-State Electronics, 39 (1996) 981–89. https://doi.org/10.1016/0038-1101(96)00002-0

29. J. C. Bernede, Materials for erasable optical disks. Materials Chemistry and Physics, 32 (1992) 189–195. https://doi.org/10.1016/0254-0584(92)90276-E

30. N. Akahira, N. Yamada, K. Kimura, M. Takao, (1988), Recent advances in erasable phase-change optical disks. Proc. SPIE 0899, Optical Storage Technology and Applications, https://doi.org/10.1117/12.944624

31. R. Ovshinsky, Reversible electrical switching phenomena in disordered structures. Physical Review Letters, 21 (1968) 1450–1453. https://doi.org/10.1103/PhysRevLett.21.1450

32. O. Zilberberg, S. Weiss, S. Toledo, Phase-change memory: An architectural perspective. ACM Computing Surveys, 45 (2013) 1–33. https://doi.org/10.1145/2480741.2480746

33. K. Tanaka, Photoinduced processes in chalcogenide glasses. Current Opinion in Solid State and Materials Science, 1 (1996) 567–571. https://doi.org/10.1016/S1359-0286(96)80074-X

34. A. L. Lacaita, Phase change memories: State-of-the-art, challenges and perspectives. Solid-State Electronics, 50 (2006) 24–31. https://doi.org/10.1016/j.sse.2005.10.046

35. W. Zhang, R. Mazzarello, M. Wuttig, E. Ma, Designing crystallization in phase-change materials for universal memory and neuro-inspired computing. Nature Reviews Materials, 4 (2019) 150–168. https://doi.org/10.1038/s41578-018-0076-x

36. D. Ielmini, A. L. Lacaita, Phase change materials in non-volatile storage. Materials Today, 14 (2011) 600–607. https://doi.org/10.1016/S1369-7021(11)70301-7

37. Y. T. Li, S. B. Long, Q. Liu, H. B. Lu, S. Liu, M. Liu, An overview of resistive random access memory devices. Chinese Science Bulletin, 56 (2011) 3072. https://doi.org/10.1007/s11434-011-4671-0

38. S. Hamdioui, H. Aziza, G. C. Sirakoulis, (2014), Memristor based memories: Technology, design and test. 9th IEEE International Conference on Design & Technology of Integrated Systems in Nanoscale Era (DTIS), Santorini, Greece, pp. 1–7, https://doi.org/10.1109/DTIS.2014.6850647

39. A. Chen, A review of emerging non-volatile memory (NVM) technologies and applications. Solid-State Electronics, 125 (2016) 25–38. https://doi.org/10.1016/j.sse.2016.07.006
40. G. Haas, V. Leis, What modern NVMe storage can do, and how to exploit it: High-performance I/O for high-performance storage engines. Proceedings of the VLDB Endowment, 16 (2023) 2090–2102. https://doi.org/10.14778/3598581.3598584
41. S.-W. Lee, B. Moon, C. Park, J.-M. Kim, S.-W. Kim, (2009), Advances in flash memory SSD technology for enterprise database applications. SIGMOD '09: Proceedings of the 2009 ACM SIGMOD International Conference on Management of data, pp. 863–870, https://doi.org/10.1145/1559845.1559937
42. I. Awoyelu, T. Omodunb, J. Udo, Bridging the gap in modern computing infrastructures: Issues and challenges of data warehousing and cloud computing. Computer and Information Science, 7 (2014) 33–40. http://dx.doi.org/10.5539/cis.v7n1p33
43. R. Bez, A. Pirovano, Non-volatile memory technologies: Emerging concepts and new materials. Materials Science in Semiconductor Processing, 7 (2004) 349–355. https://doi.org/10.1016/j.mssp.2004.09.127
44. N. R. Mielke, R. E. Frickey, I. Kalastirsky, M. Quan, D. Ustinov, V. J. Vasudevan, Reliability of solid-state drives based on NAND flash memory. Reliability of solid-state drives based on NAND flash memory. Proceedings of the IEEE, 105 (2017) 1725–1750. https://doi.org/10.1109/JPROC.2017.2725738
45. S. Kargar, F. Nawab, Challenges and future directions for energy, latency, and lifetime improvements in NVMs. Distrib Parallel Databases, (2022). https://doi.org/10.1007/s10619-022-07421-x
46. S. Agarwal, H. K. Kapoor, Improving the lifetime of non-volatile cache by write restriction. IEEE Transactions on Computers, 68 (2019) 1297–1312. https://doi.org/10.1109/TC.2019.2892424
47. K. Ishimaru, (2019), Future of non-volatile memory from storage to computing, IEEE International Electron Devices Meeting (IEDM), San Francisco, CA, USA, pp. 1.3.1–1.3.6, https://doi.org/10.1109/IEDM19573.2019.8993609
48. M. Si, H.-Y. Cheng, T. Ando, G. Hu, P. Ye, Overview and outlook of emerging non-volatile memories. MRS Bulletin, 46 (2021) 946–958. https://doi.org/10.1557/s43577-021-00204-2
49. W. Zhang, E. Ma, Unveiling the structural origin to control resistance drift in phase-change memory materials. Materials Today, 41 (2020) 156–176. https://doi.org/10.1016/j.mattod.2020.07.016
50. Y. K. Lee, C. Yoo, W. Kim, J. Jeon, C. Hwang, Atomic layer deposition of chalcogenides for next-generation phase change memory. Journal of Materials Chemistry C, 9 (2021) 3708–3725. https://doi.org/10.1039/D1TC00186H

21 Sensors Based on Semiconductors

*Alper Durmaz, İbrahim M. Kahyaoğlu,
Erdi C. Aytar, and Selcan Karakuş*

21.1 INTRODUCTION

The size-dependent properties (optical, electrical, thermal, and catalytic) of nanoparticles, which are highly valued in nanotechnology, serve as electrochemical labels. Their size is described by a variety of techniques and is controlled by synthesis; however, because of their high surface energy, instability may occur. Applications in environmental engineering, biomedicine, IT, aviation, and food analysis are among them. Electrochemical sensors with metal and semiconductor nanoparticles (SNPs) exhibit tremendous promise. For detection, metal and SNPs can be bonded to electrodes by electropolymerization, covalent bonding, electrodeposition, or physical adsorption. By adding conductive nanoparticles to composite electrodes, adsorption is improved and interactions with substrates are made possible. Nanoparticles (NPs) can be arranged to produce controllable active regions in nanoelectrode assemblies. The irreversibility of redox processes is aided by the catalytic qualities of metal and SNPs, which lower potential needs in electrochemical analysis. These NPs are used in electrochemical sensors and nanodevices for investigation of bacteria and cells, protein interactions, DNA hybridization, and small molecule detection. Medical diagnosis is revolutionized by DNA analysis, and metal and SNPs have great promise to improve chemical and biological detection in this domain. Electrochemical DNA sensors take advantage of these NPs' large surface area to optimize DNA immobilization and detection capabilities. These developments boost the effectiveness and dependability of DNA analysis by improving sequence-specific DNA detection. Further investigation is necessary to investigate more potent surface immobilization strategies. SNPs and metal are essential for improving DNA immobilization and advancing the creation of sensitive DNA sensors. Colloidal gold enhances DNA immobilization and lowers detection limits. Effective DNA labels that strengthen electrical detection signals are metal and SNPs. Electrochemical DNA sensors are a prominent use for NPs of gold and silver. Using actual saliva samples, a recently created portable, label-free photoelectrochemical gene sensor effectively detects SARS-CoV-2 infection. Utilizing Arduino and 3D printing to reduce in size, it offers a practical and economical substitute for conventional RT-PCR and electrochemical methods. Nazari-Vanani et al. developed a DNA-based genosensor for detecting Leishmania infantum genomic DNA. This genosensor utilizes cadmium sulfide nanolayers and specific single-stranded DNA sequences, eliminating the need for PCR or labeling. Impedance measurements during DNA target hybridization with cadmium sulfide nanolayer-attached probes are conducted without external stimuli. The genosensor can detect complementary DNA strands in the concentration range of 1.0×10^{-14} to 1.0×10^{-6} mol L^{-1}. They successfully detected Leishmania infantum DNA at concentrations ranging from 5–50 ng μL^{-1}, with a detection limit of 1.2 ng μL^{-1} [1]. An essential part of electrochemical studies is the electrocatalysis of organic molecule detection by metal and SNPs. Zinc oxide (ZnO), tin oxide (SnO_2), tungsten trioxide (WO_3), titanium dioxide (TiO_2), and niobium pentoxide (Nb_2O_5) are examples of semiconductor-based gas sensors that are easy to produce, affordable, and have excellent performance. They have strong sensitivity, quick response, and recuperation times. Metal sulfides such as bismuth sulfide (Bi_2S_3) and molybdenum disulfide (MoS_2) are common semiconductor hydrogen sensors. Because an electrochemical gas sensor needs a solid electrode to be in contact

DOI: 10.1201/9781003450146-21

with a liquid electrolyte phase, its design is difficult. These sensors typically use metal and SNPs as conductive materials, with the expectation that SMOs' detecting capabilities will be improved by the addition of small-sized quantum dots (QDs). Plant communication is greatly aided by volatile organic compounds (VOCs), which are released by crops for a variety of reasons. Electronic noses that use SMOs gas sensors are frequently used to identify VOCs quickly and accurately in crops, which helps with disease and insect pest analysis. MOS E-noses, like SnO_2 and WO_3, address sensitivity and selectivity challenges using a "selective odor measurement using a multisensory array" approach [2]. MOS gas sensors find extensive use across diverse scientific research domains, offering several advantages, including affordability, high reliability, lightweight design, stability, ease of integration, sensitivity, and rapid response times. In nature, insects receive information through their antennae. MOS E-noses, when combined with enrichment and analysis techniques, are capable of detecting insect pheromones. Some studies have utilized MOS E-noses to detect key components of European grapevine moth (*Lobesia botrana*) pheromones. However, the sensitivity and selectivity of MOS gas sensors alone for insect pheromone detection are relatively limited. Nevertheless, the "selective odor measurement using a multisensory array" approach, employing MOS gas sensors like SnO_2 and WO_3, effectively addresses this limitation. This system combines various sensors to enhance selectivity against different gases and VOCs. [3]. Sensors offer precise gas detection with unique response patterns, enhancing odor perception and recognition. MOS gas sensors find diverse applications in agriculture, food industry, environmental monitoring, and medical diagnostics. In agriculture, they aid in plant variety selection, insecticide testing, and plant health analysis. In the food industry, they assess flavor, aroma, and quality. Their rapid response and high sensitivity enable early pathogen detection and spoilage signs [4]. In Table 21.1, some examples of 2D materials and their applications in VOCs sensors are presented.

Mohamed et al. synthesized undoped and Fe-doped TiO_2 nanostructures using the VLS growth technique in an alumina tube furnace with Pt-coated Si and quartz substrates. Undoped TiO_2 nanostructures featured densely grown nanowires (200–400 nm diameter, >12 μm length), indicating VLS growth. In contrast, Fe-doped TiO_2 exhibited agglomerated NPs (160–450 nm). Both materials demonstrated enhanced photocatalytic activity, suggesting potential applications in optical devices and organic pollutant degradation [7]. Lan et al. proposed a Schottky-contacted single ZnO micro/nanowire sensor for protein kinase activity detection. ZnO, combined with functional molecules, forms planar van der Waals heterojunctions, enabling the study of phosphorylated peptide sensors through the piezotronic effect. These sensors can specifically identify phosphorylated peptides. Experimental results showed significant relative current changes with increasing compressive strain under positive bias for different phosphorylated peptide concentrations. This study highlights a promising protein kinase sensor using a Schottky-contacted single ZnO micro/nanowire enhanced by the piezotronic effect. Understanding electrostatic impacts on charge transfer in nanowire/molecule heterojunctions

TABLE 21.1
Some Examples of 2D Materials and Their Applications in VOC Sensors

Material	Analyte	References
$SnO_2/LaFeO_3$	Formic acid gas	[5]
SMOs	Methane	[6]
TiO_2 nanostructures	Organic pollutants	[7]
ZnO micro/nanowire	Protein kinase activity	[8]
PANI/ $Ti_3C_2T_x$	Ethanol	[9]
rGO-SnO_2	Ethanol	[10]
Graphene/ZnO	Acetone and formaldehyde	[11]

will aid future van der Waals heterojunction detector and transistor designs [8]. Butturini et al. monitored dissolved methane in aquatic ecosystems to enhance carbon cycle understanding. They evaluated a MOS methane sensor encapsulated in a hydrophobic membrane for gas diffusion. Results showed a detection limit of 0.1–0.2 μmol/L and stable response (18.5–28°C). Elevated sulfur concentrations affected sensor response at low methane levels, suggesting the monitoring of dissolved sulfite levels could mitigate these effects [6]. In another study, Xia et al. developed a method to detect formic acid gas (HCOOH) using SnO_2 QDs combined with $LaFeO_3$. SnO_2 QDs improved the porous structure of LaFeO3, aiding gas diffusion. $LaFeO_3$'s detection properties were studied with varying SnO_2 QD doping levels, increasing sensitivity to HCOOH. A $LaFeO_3$-based sensor with 2.5% SnO_2 QD doping achieved a low 1 ppm detection limit and a strong response (31.5–100 ppm at 210°C). Hetero-connections between SnO_2 QDs and $LaFeO_3$ enhanced HCOOH detection, showcasing the advantages of QD sensitization in semiconductor gas sensors [5].

In another study, Li et al. discovered that lung cancer patients and healthy people had similar breath patterns. Principal component analysis (PCA) and linear discriminant analysis (LDA) worked together to improve the diagnosis and separate the two groups. With their ability to detect numerous gases, MOS sensor arrays show potential, pending improvements in manufacturing processes, novel materials, and environmental conditions. Gas recognition rates are improved by optimized algorithms beyond PCA and LDA, highlighting the promise of MOS gas sensor arrays in lung cancer diagnosis [12].

21.2 GAS SENSORS BASED ON METAL-ORGANIC FRAMEWORK DERIVED SEMICONDUCTORS

Metal-organic frameworks, or MOFs, are adaptable structures with modifiable void spaces that find use in biomedical applications. Strong fluorescence characteristics allow them to improve sensing capacities and enable tracking in living things. In recent years, MOFs have drawn a lot of interest as cutting-edge materials.

21.2.1 SYNTHESIS OF MOFS/SEMICONDUCTORS

For rapid, on-site ion measurement and molecular detection, MOFs and semiconductors are combined in ion-sensitive field-effect transistors (ISFETs) (Figure 21.1). Motivated by their extraordinary features as third-generation materials, researchers are examining the expanding uses of MOFs in numerous domains, despite their huge band gaps, particularly in sensing [13].

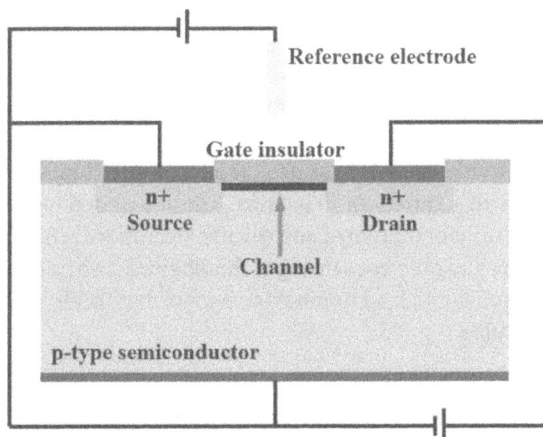

FIGURE 21.1 Schematic diagram of an ISFET sensor.

Several methods are available for synthesizing MOF/semiconductor composites. Shi et al. created a selective daylight CO2 sensor by synthesizing sulfide semiconductors within a titanium-based MOF using ultrasonication and thermal treatment [14]. Wang et al. achieved strong antibacterial properties against *E. coli* and *S. aureus* bacteria under visible light by encapsulating semiconductor QDs within MOF structures through ultrasonication [15]. Montañez et al. incorporated copper-based MOF films into thermally grown silica layers, using a layer-by-layer synthesis approach up to 10 nm in size, to produce highly efficient capacitors operating at low voltages [16].

21.2.2 MOF-Derived Semiconductor Sensors

MOF-supported semiconductor sensors, leveraging MOFs' large surface area, find applications in fields like food, cosmetics, environment, and pharmaceuticals. Chen et al. synthesized a Ni-MOF to create a semiconductor dopamine sensor [17]. This sensor, labeled as 0.7–310.2 µM, offers high sensitivity over a wide range and a low detection limit. Yang et al. developed a Ti-MOF for NO_2 gas detection [18], and their Ti-MOF semiconductor sensor responded well in the 50–200 ppm NO_2 concentration range. Zhang et al. used Cu-MOFs to create semiconductor-enhanced luminescent sensors for KRAS gene detection [19]. This Cu-MOF sensor demonstrates exceptional sensitivity from 0.1 pM to 1 nM. MOF-derived semiconductor sensors, harnessing MOFs' porous structure, large surface area, and semiconductor properties, are expanding their applications. Future target compounds are well-suited to MOFs' adaptable pore structure. MOF-based sensors are being researched further, offering faster, more advanced, molecule-specific, extremely sensitive, recyclable, and portable sensors with improved features.

21.3 NANOSTRUCTURED SEMICONDUCTORS SENSORS

Nanostructured semiconductors, designed at the nanoscale, exhibit unique physical, chemical, and electrical characteristics, enabling exceptional sensitivity in sensing applications. With precise control over size, shape, and composition, they detect analytes at low concentrations, maximizing performance. The increased surface area excels in ultrasensitive sensing platforms, responding to minute variations in surroundings. Enhanced surface chemistry improves selective sensing, ensuring preferential interaction with target analytes for reduced interference. Intrinsic quantum confinement guarantees quick reaction and recovery times, facilitating rapid analyte detection and baseline restoration. The small size allows effective analyte adsorption and diffusion, ensuring swift response kinetics. These sensors offer the benefits of portability and energy efficiency with reduced power requirements.

Running at room temperature lowers power requirements and makes tiny, low-power sensor devices possible by doing away with the need for energy-intensive thermal control systems. The cost-effectiveness, lightweight design, and high sensitivity of nanostructured MO gas sensors are highly appreciated; methods such as particle size reduction are used to improve sensing capabilities. Due to their large bandgaps and chemical stability, semiconductor MOs have drawn interest in the field of gas detection. For effective harmful gas detection, nanostructured semiconductors are being actively researched and used in a variety of industries. These semiconductors have high conductivity, customizable surface termination groups, and diverse operational functionalities. A new era of extremely sensitive, selective, and efficient sensing technologies is ushered in by their amazing sensing properties. Future research could lead to innovative tools that tackle many scientific, industrial, and biomedical problems [20].

21.3.1 Fabrication for Nanostructured Semiconductor Sensors

Fabrication techniques for nanostructured semiconductor sensors are crucial for precise synthesis and assembly of semiconductor materials at the nanoscale. These techniques allow customization

TABLE 21.2

Synthesis Method for Nanostructured Semiconductor Sensors

Methods	Description
Chemical vapor deposition (CVD) method	Produces nanostructures with the aid of various gas mixtures on a heated substrate.
Physical vapor deposition (PVD) method	Deposition of semiconductor using sputtering and evaporation.
Sol-gel method	Polycondensation and hydrolysis reactions.
Hydrothermal method	Crystallizing substances from hot, high-pressure solutions.
Electrochemical deposition method	To deposit metal layer by layer on selected conductive substrates.
Template-assisted method	To make nanofibers within a porous membrane or layer.
Laser ablation method	Process of removing atoms from a solid using an intense laser beam, either thermally or non-thermally.
Lithography method	Using an electron beam or light.

of size, shape, composition, and surface properties, which are vital for sensing performance. Novel fabrication approaches for these sensors include various methods (Table 21.2)

Through regulated precursor gas reactions, chemical vapor method (CVD) is frequently used to precisely generate semiconductor films or nanowires layer by layer on substrates. It creates a variety of nanostructures in the semiconductor industry (0D, 1D, and 2D). It is difficult to achieve efficiency in air pressure CVD; different growth parameters must be carefully considered since they affect the final nanomaterials. It is difficult to design an effective atmospheric pressure CVD system; factors including reactor size, substrate alignment, gas flow, precursors, temperature, duration, and flow rate must be taken into consideration [21].

Physical vapor method (PVD) allows semiconductor material to be deposited using physical processes, similar to sputtering and evaporation. This makes it perfect for producing thin films or nanostructures with outstanding uniformity and adherence. The ideal PVD technique for thin film sensing layers is electron beam evaporation, which has benefits for batch production, well-made layered structures, and reasonably priced, uniformly doped alternatives. Electron beam evaporation is a preferred method for accurate and dependable thin film sensor layer deposition in photovoltaic diode technology due to its uniform doping and flexibility [22].

The sol-gel method synthesizes nanostructured materials by converting a precursor solution (sol) into a solid gel and then heat-treating it to form the desired semiconductor structure. In a study by Zhang et al. (2023), zinc oxide nanocrystals (ZnO NCs) were synthesized on a foam-like gallium nitride (F-GaN) substrate using this method to create a gas-sensitive material for an H_2S gas sensor. The composite structure of ZnO NCs/F-GaN, when used for detecting 50 ppm of H2S at 220°C, showed remarkably fast response and recovery times of 78 and 31 seconds, respectively. This improved sensing response is mainly attributed to the synergistic effect of combining ZnO NCs and F-GaN. The researchers proposed a new catalytic mechanism involving F-GaN in gas sensing, expanding the applications of GaN-based SMOs. The gas-sensitive material created using the sol-gel method with F-GaN has clear advantages, making it suitable for practical use. This approach also enhances our understanding of gas sensing. Combining the sol-gel method with F-GaN not only improves sensing but also opens the door to oxide/GaN-based sensors, offering a promising approach for gas sensor fabrication [23]. By cultivating semiconductor nanoparticles in hot, pressured water, hydrothermal techniques yield well-defined nanostructures. The synthesis process affects the quality and properties of nanomaterials and is essential for crystal growth. Microwave-assisted hydrothermal synthesis (MAH) is a technique used by Ortega and associates [24] that yields high-quality nanostructures with exact control over size, composition, phase, and form. When ZnO concentration hits supersaturation in MAH synthesis, crystal nucleation begins because $Zn(OH)_4^{2-}$, a growth unit for ZnO nanoparticles, has been dehydrated. ZnO nanocrystals develop more quickly thanks to the MAH technique, which also reduces certain surface areas and facet shapes. Chemical leakage into the surrounding environment is avoided when operating inside a sealed autoclave. This approach provides quick heating, quick reaction times, and high yields for the synthesis of ceramic nanostructures while being economical, ecologically benign, and energy efficient. In conclusion, the MAH process provides a productive and ecologically sustainable way to create superior ceramic nanostructures with exact control over their properties [24].

Electrochemical deposition method, including electrodeposition and electroless deposition, controls the growth of semiconductor nanostructures on conductive substrates through electrochemical reactions. Çepni and Öznülüer Özer used electrodeposition to synthesize $In(OH)_3$ films in their 2022 study. They achieved this through a two-step electrochemical process that increases pH near the electrode and prompts the deposition of indium hydroxide from reduced indium ions. This method is cost-effective, simple, and applicable under ambient conditions, making it a green and convenient alternative to harsh chemical processes. It eliminates the need for high temperature/pressure environments, enabling direct material formation on the substrate surface [25].

Template-assisted method utilizes templates with predefined nanopores to guide the growth of semiconductor nanostructures. Alinauskas et al. employed this method in their 2017 study for SnO_2 nanomaterial. They used polycarbonate membranes as templates, subjecting them to a ten-step immersion process in a sol, with gentle wiping between steps to aid solution ingress into the pores. The impregnated membranes were placed on various substrates and left to mature for five hours, followed by annealing at 550°C for five hours. In some cases, an additional annealing step at 800°C for five hours was carried out. The high porosity of these materials makes them suitable for sensing ions or molecules, and this method holds promise for creating various functional nanomaterials [26].

Laser ablation method uses a laser beam to create SNPs or nanowires. In 2023, Dikovska et al. produced ZnO/Zn_2TiO_4 composite nanostructures through laser ablation in air. They used pulsed laser deposition (PLD) with different $ZnO-TiO_2$ targets. A nanosecond Nd:YAG laser operating at 1064 nm with specific parameters was employed. The ablated material was deposited onto SiO_2 substrates. This research introduces a laser-based method for controlled nanoscale material deposition. It successfully fabricated high-purity ZnO/Zn_2TiO_4 composites without chemicals, offering cost-effectiveness and conventional technology use without requiring a high vacuum environment [27].

Photolithography, electron beam lithography, and nanoimprint lithography create precise nanostructures in semiconductor materials on substrates. In a 2023 study, Rani et al. used two-photon polymerization (TPL) with a femto-second laser integrated with an inverted microscope. They drop-casted two-photon active functional (TPAF) resin on a glass coverslip, spin-coated it, and exposed it to the laser while systematically moving it on a motorized stage to create micro/nanostructures. This demonstrates a laser-based method for micro/nanostructure fabrication with precision. This study utilized an internally developed TPL system to create precise micro/nanostructures within a TPAF resin. It involved precise laser focusing, controlled substrate traversal, and removal of unexposed resin using dimethylformamide (DMF) [28].

In recent years, enhancing the sensitivity of nanostructured semiconductor sensors has become essential for applications in chemical analysis, environmental monitoring, and gas sensing, utilizing materials like MOs (e.g., ZnO, SnO_2) or semiconducting nanoparticles. Several techniques are employed to improve their performance:

- By increasing interactions with target analytes, surface functionalization—achieved by adding chemical groups or molecules to the sensor's surface—enhances sensitivity and selectivity. The addition of alterations such as organic ligands or metal nanoparticles enhances the sensor's sensitivity to substances even further.
- Doping is a useful technique that modifies the electrical structure of a material to increase its receptivity to analytes. It involves inserting intentional impurities, or dopants, into the semiconductor lattice [10].

- Careful temperature control maximizes sensing performance.
- Signal-to-noise ratio and accuracy improve with machine learning and noise reduction.
- Selective coatings filter unwanted interferents for precise analyte targeting.
- Energy band engineering enhances semiconductor sensor performance by modifying the material's energy band structure, increasing charge carrier mobility and sensitivity to chemicals.
- Optical enhancement techniques like surface plasmon resonance and photonic crystal structures improve sensor optical characteristics, enhancing overall detection capabilities.

21.3.2 Nanostructured Semiconductor Sensors in Clinical Diagnosis and Environmental Monitoring

The identification of biomarkers for diseases such as prostate cancer and heart disease has been revolutionized by high-sensitivity nanostructured semiconductor sensors, which are crucial for medical diagnostics. Early disease identification is made possible by these sensors, which is essential for raising survival rates, especially in cases of infectious diseases. Their quick on-site tests are helpful in scenarios when resources are few since they guarantee accurate and timely disease detection for appropriate treatment. Lahcen et al.'s 2021 study developed a POC nanostructured molecularly imprinted polymer (MIP) sensor for breast cancer biomarker monitoring. By optimizing parameters and incorporating gold nanoparticles (AuNS), they increased the sensor's surface area, enhancing Her-2 protein adsorption. This led to the successful LSG-AuNS-MIP sensor, demonstrating high sensitivity and selectivity in detecting Her-2 in human serum samples. Integrated into a compact POC device, it showcases promise for extended disease detection applications due to its stability, flexibility, and ease of production and modification as a diagnostic tool [29]. In 2021, Foroughi et al. successfully synthesized three-dimensional cubic europium (Eu^{3+}) and cuprous oxide (Cu_2O) nanostructures resembling clover-like face-centered patterns (Eu^{3+}/Cu_2O CLFNs). The study aimed to create an effective electrochemical sensor for nevirapine ($C_{15}H_{14}N_4O$) detection. The hydrothermal process yielded a significant quantity of high-quality 3D cubic Eu^{3+}/Cu_2O CLFNs, extensively characterized for electrochemical properties. The modified electrode exhibited enhanced nevirapine oxidation peak currents, improved electron transfer, effective catalytic activity in nevirapine detection, and reduced oxidation overpotential. Under optimized conditions, it demonstrated excellent reproducibility, a low LOD, and high sensitivity, providing an innovative approach for sensitive nevirapine determination and monitoring using a modified electrode configuration [30].

Nanostructured semiconductor sensors are essential equipment for monitoring the quality of the air and water since they are capable of accurately detecting contaminants in the water as well as dangerous gases including carbon dioxide (CO_2), carbon monoxide (CO), sulfur dioxide (SO_2), and nitrogen oxides (NO_x). It is impossible to exaggerate how important they are in determining the extent of environmental pollution and supporting mitigation measures. The atmosphere has recently been severely contaminated by a variety of toxic gases, including SO_2, NH_3, NO_2, N_2O, and H_2S, among others. Especially, NO_2, primarily emitted from activities like fossil fuel combustion and nuclear power plants, is one of the most hazardous environmental gases. Even at trace levels in the air, NO_2 can cause serious health issues, including throat irritation, asthma, cardiac problems, headaches, pulmonary edema, and, in extreme cases, death. Mathankumar developed novel Ag-doped WO_3 for gas sensing, with exceptional sensitivity to NO_2 at an optimal temperature of 150°C. The Ag-2 sample showed a remarkable fivefold increase in gas response compared to pure WO_3. First-principal simulations revealed a 60% stronger binding between NO_2 and Ag-doped WO_3 compared to H_2S and NH_3. Highly sensitive NO_2 sensors are essential for tracking and reducing health risks associated with NO_2 pollution. Theoretical predictions that match actual results present a promising direction in this regard [31]. To protect the safety of water sources, nanostructured sensors are essential for monitoring water quality and detecting contaminants such as organic pollutants, heavy metals, and microbiological diseases. Magro et al. created extremely sensitive gold-interdigitated electrode-based nanostructured thin film sensors in mineral water by employing ZnO and TiO$_2$. At a precision of 1×10^{-16} M, these sensors demonstrated potential as dependable monitoring tools in a variety of water matrices by reliably differentiating antibiotic concentrations in river water [32]. To prevent foodborne illnesses, nanostructured semiconductor sensors are crucial for detecting infections and deterioration signs in food. For instance, Ahmadian-Fard-Fini et al. mixed ethylene diamine with extracts of lemon, grapefruit, or turmeric to make extremely luminous carbon dots. These carbon dots offered simplicity, sensitivity, and specificity in bacterial detection, making them an effective and sensitive approach for quantitatively detecting *E. coli* germs [33]. In conclusion, these sensors, essential for identifying biomarkers, supporting early disease diagnosis, guaranteeing accurate drug monitoring, and aiding in pollution reduction and environmental evaluation, prove indispensable in crucial sectors due to their adaptability and exceptional performance.

21.4 APPLICATIONS OF INORGANIC SEMICONDUCTORS IN PHOTOELECTROCHEMICAL SENSORS

In vitro diagnostics and medical research depend heavily on biochemical sensors, and there is growing demand for better performance. Utilizing heterogeneous nanostructures, photoelectrochemical sensors are effective tools for detecting chemical and biochemical molecules because they excel at light absorption and electron transmission. This section thoroughly examines how they can be used to diagnose both inorganic and organic molecules in vitro, assisting in the early detection and management of diseases. It offers advances in this developing discipline by revealing new directions for experts.

One-dimensional (1D) semiconductor nanomaterials have garnered significant attention due to their unique structural characteristics, expansive specific surface areas, remarkable electronic, optical, and thermal properties, and exceptional conductivity, making them ideal for photoelectric applications. Among various fabrication methods, electrospinning stands out for its ability to precisely control morphologies, such as nanofibers, nanotubes, and more. It also allows for aligned structures, high porosity, and tunable mechanical properties. These qualities hold promise across fields like photoelectric devices, catalysis, energy storage, sensors, and capacitors. Ultraviolet photodetectors (UV PDs), capable of converting light into electrical signals, find extensive use in environmental monitoring, optical communication, and biological sensing, driving advancements in nanoelectronics and nano-optoelectronics. The exceptional properties of 1D semiconductor nanomaterials, combined with versatile techniques like electrospinning, contribute to their potential in various applications, including their role in UV PDs and the broader field of nano-optoelectronics [34]. A molecularly imprinted photoelectrochemical (PEC) sensor for chlortetracycline (CTC) detection was presented in 2023 by Zhang et al. on a ZnO-MWCNTs-NH$_2$ substrate; they used a special mixture of 5,10,15,20-tetrakis(4-carboxylphenyl) porphyrin (TCPP) and poly(o-phenylene-diamine) (PoPD). With a linear detection range from 0.5 nmol/L to 10 mol/L and an impressively low detection limit of 0.17 nmol/L, this sensor demonstrated exceptional sensitivity. Furthermore, it successfully identified CTC in genuine samples, such as milk, pork, and lake water, with recoveries varying from 86.4% to 112.0%. This useful instrument shows creativity in improving conventional photoactive materials and potential for identifying a range of environmental toxins [35].

21.5 APPLICATIONS OF NANOWIRES AND 2D SEMICONDUCTOR SENSORS

Nanotubes are ultra-thin structures with diameters in the range of tens of nanometers or less and limitless length. When these tubes reach scales where quantum mechanical effects come into play, they are often referred to as quantum wires. Various types of nanotubes exist, including superconducting, metallic, semiconductor, and insulating varieties. Molecular nanotubes are built from repeated organic or inorganic molecular units. Several techniques, such as chemical etching, chemical vapor deposition, and electrospinning, have been developed to synthesize nanotubes. These materials are categorized as 1 D due to their extremely high aspect ratio, often exceeding 1000. Nanotubes exhibit unique properties not found in bulk or three-dimensional materials because the confinement of electrons within them leads to distinct energy levels, differing from traditional energy levels. 1D semiconductor nanowires have gained significant attention in chemical sensing applications, particularly MOs nanowires, renowned for their stability and high surface-to-volume ratio. Increased surface area enhances gas molecule adsorption, improving sensing responses. Nanowire size influences electronic properties, influenced by analyte molecule adsorption and desorption processes. Ongoing research explores alternative materials and structures. Advances in 2D nanomaterial preparation have facilitated functional devices with varying thicknesses. Integrating 2D structures like MOFs, graphene, and transition metal dichalcogenides into chemical sensors shows promise, with their large surface area, electronic properties, defects, and surface states significantly impacting sensing capabilities. Novel semiconductor nanostructures offer opportunities for enhanced gas sensing systems, addressing traditional sensor limitations, enabling miniaturization, and enhancing functionality. Recent efforts have focused on developing complex structures based on both 1D and 2D materials.

Si nanowires form utilizing CVD through the vapor-liquid-solid (VLS) mechanism in the "bottom-up" procedure. They are deposited onto a silicon substrate while suspended in an ethanol solution. After applying a photoresist and patterning metal electrodes using the lift-off technique, the procedure is finished with passivation and surface modification procedures. The "top-down" strategy, in contrast, enables exquisite nanowire control and is CMOS compatible. It entails defining terminals and patterning microscale electrodes on a silicon-on-insulator (SOI) wafer by low-density boron or phosphorus doping. With the aid of electron beam lithography, nanoscale Si nanowires are created, and metal contact terminals are created. Larger diameter nanowires with regulated orientation are made possible by the top-down technique, which is excellent for quick, mass production that is compatible with CMOS technology. The bottom-up strategy, albeit more popular, might produce nanowires with a lower diameter and a random orientation. Although alignment steps cannot be used with CMOS, they can improve orientation. Regarding the characteristics of nanowires, each approach has specific benefits and drawbacks.

Monitoring and diagnosing diabetes through breath acetone levels offers a convenient and non-invasive approach. Effective materials for this purpose should exhibit high acetone response, selectivity, low sensitivity to humidity, and the ability to work at room temperature. Chemical-resistant gas sensors, known for their simplicity, hold potential for breath acetone detection. Some semiconductor materials respond to airborne acetone, causing resistance changes. However, many require high operating temperatures, leading to issues like increased power consumption and complex design. In a study by Rudie et al. (2022), 2D MXene-based nanocomposites were investigated for their potential in chemical-resistant sensors. Most showed high sensitivity and selectivity to acetone. Among them, the 1D/2D CrWO/Ti3C2 nanocomposite exhibited the best acetone detection response, with sensitivity nine times higher than KWO nanowires. The sensor also demonstrated excellent selectivity for acetone over other common breath vapors, suggesting its potential in diabetes management devices [36].

Using a CVD reactor and PbI_2 precursor growth, Ghoshal et al. (2019) produced $(C_4H_9NH_3)2PbI_4$ perovskite nanowires and nanoplatelets. SEM (ZEISS SUPRA 55) and AFM (multi-mode) were used to characterize the structures, and optical measurements were made with the use of lasers and a confocal microscope setup. It is a noteworthy accomplishment that this study used Low-temperature vacuum-assisted chemical vapor deposition (LTCVD) to demonstrate regulated and template-free development of 2D perovskite nanowires. Outstanding optical characteristics, homogeneous thickness, and compositions were all present in the nanowires. They displayed one of the greatest polarization ratios for perovskites, 0.73. With a photo-current anisotropy ratio of 3.62, a hybrid device made of $(C_4H_9NH_3)2PbI_4$/graphene/hybrid also served as a polarization photodetector. These results highlight perovskite nanowires' potential for use in optoelectronic systems [37].

21.6 APPLICATIONS OF SEMICONDUCTOR-BASED HEAVY METAL SENSORS

Detecting heavy metal pollution is crucial for the environment and human health. Methods for this purpose fall into traditional techniques (gravimetry, volumetry, colorimetry), instrumental methods (atomic absorption spectroscopy, inductively coupled plasma-optical emission spectroscopy, inductively coupled plasma-mass spectrometry), and electrochemical methods (voltammetry, amperometry, impedance). Despite their pros and cons, they all share a common limitation: the need for laboratory-based measurements [38]. Modern sensor technology allows for accurate on-site measurements, including analysis of heavy metals. A very sensitive arsenic sensor with a 1 nM detection limit was created by Zhou et al. employing semiconductor black phosphorus functionalized with dithiothreitol and gold nanoparticles [39]. With a sensitivity of 40 mV/100 ppm, César et al. created an electrolyte-insulator-semiconductor (EIS) structure for Pb^+ ion detection in water [40]. A thymine-Hg^{2+}-thymine base pair with MoS_2 nanolayers was used by Wang et al. to build an electrochemical sensor for Hg detection that operates in the range of 0.01–4 g/L in the presence of Hg^{2+} ions [41]. The electrical activity of metals has led to an increase in the usage of electrochemical sensors for the detection of heavy metals. In order to improve sensor electrochemical properties, semiconductors are essential, and this role will probably continue to be played until other high-tech materials overtake them. The possibility of increasingly more sensitive sensors becoming available as semiconductor technology develops highlights the significance of semiconductor research in this area.

21.7 APPLICATIONS OF POLYMER SEMICONDUCTOR-BASED FLEXIBLE SENSORS

Throughout history, clothing has been essential for human survival, adapting to changing needs and technology. It helps individuals cope with climate and is vital for social reasons. As technology advances, the clothing industry evolves, integrating conductive fabrics and sensors into textiles. These textile innovations have applications in healthcare, communication, security, transportation, energy, and more.

Conductive polymers have greatly impacted the development of conductive textile products due to their flexibility and reasonable conductivity. Polymers like polyaniline (PANI), polypyrrole (PPy), polythiophene (PTh), and poly(3,4-ethylenedioxythiophene) (PEDOT) are commonly used and can be chemically modified to enhance their conductivity [42]. Semiconductor sensors using conductive polymers have a wide range of applications. For example, a CH/PVP sensor with a ZnO film showed high sensitivity to hydrogen gas [43], and perovskite nanocrystals integrated into a conductive polymeric matrix resulted in a semiconductor NO_2 gas sensor with a strong response [44]. These studies highlight the significance of polymer semiconductor sensors in wearable technology and gas sensor applications. As industry advances and environmental concerns grow, the demand for these sensors is expected to increase.

21.8 ULTRA-SENSIVITE AND SELECTIVE 2D HYBRID SEMICONDUCTOR BIOSENSOR

Nanoelectronic developments are addressing issues with semiconductor devices and material technology. Organic electronic transistors (OCTs) perform best at low frequencies and poorly at high frequencies. There are few organic semiconductor possibilities for entirely organic circuits, and incorporating them into tiny devices is challenging. Surface effects, stiffness, and limited ion permeability are problems that inorganic semiconductors must deal with. Graphene and other 1D and 2D layered semiconductors are suitable for bioelectronics because of their mobility, flexibility, and inertness. The carrier transfer and analytical capabilities of hybrid transistors, which combine organic and inorganic semiconductors, are improved. Utilizing the compatibility of organic and

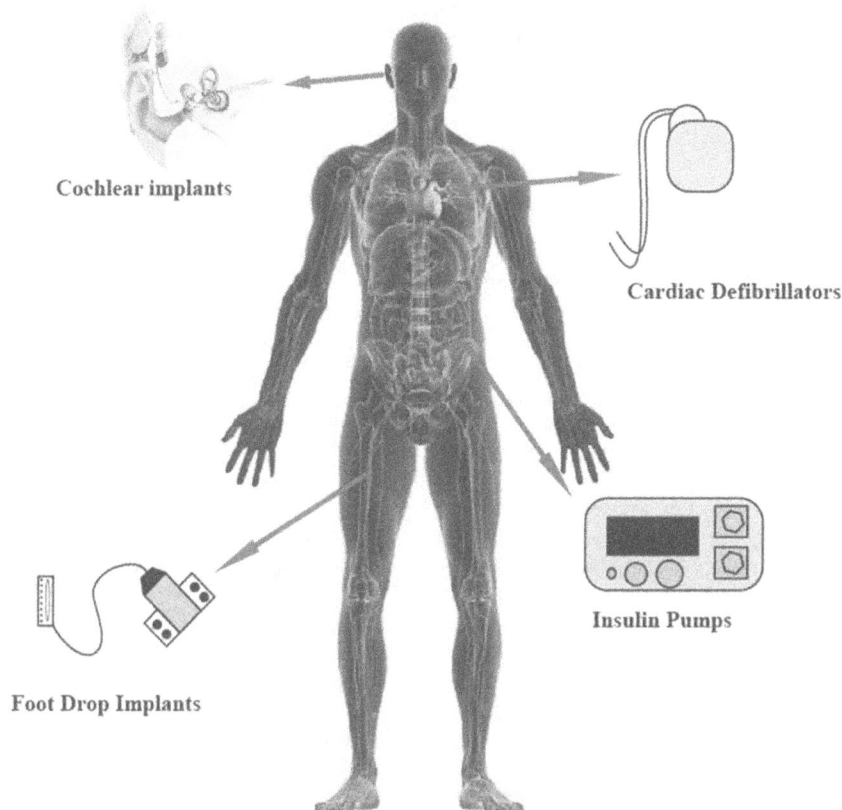

FIGURE 21.2 The biomedical applications of semiconductor sensors.

inorganic elements, their production is simple. The development of high-performance biosensors using hybrid organic-inorganic materials is made possible by these developments in nanoelectronics. Some sensors used in the biomedical field are shown in Figure 21.2.

Li et al. developed a $BiVO_4$/2D-C_3N_4/DNA aptamer PEC sensor for microcystin-LR (MC-LR) detection, achieving record-breaking sensitivity. The 2D-C3N4 material enhanced performance by shortening hole transfer distances and immobilizing DNA aptamers [45]. Ji et al. (2023) designed an InSe-FET biosensor for rapid CA125 antigen detection in clinical samples. It exhibited high sensitivity and selectivity without interference from other molecules, making it a promising tool for medical diagnostics [46]. In conclusion, the development of ultra-sensitive and selective 2D hybrid semiconductor biosensors represents a remarkable advancement in the field of bioelectronics.

21.9 FUTURE BIOMEDICAL APPLICATIONS OF SEMICONDUCTOR SENSORS

Future biomedical applications of semiconductor sensors will revolutionize healthcare by facilitating rapid, accurate diagnosis and individualized therapy. In point-of-care devices, these sensors will be widely utilized to provide fast analyses and lessen the need of centralized laboratories. Remote patient monitoring will be made possible by implantable sensors that track vital signs and recurring conditions. Wearables with semiconductor sensors will monitor health parameters continually, assisting in the early diagnosis and treatment of ailments. These sensors will be essential for early cancer detection, neurological research, the tracking of infectious diseases, and intelligent medication delivery systems. When paired with artificial intelligence, they will produce massive amounts of healthcare data for more precise diagnostics and treatment recommendations, ushering in a new era of precision healthcare excellence. They will also promote regenerative medicine and genomics research.

REFERENCES

1. R. Nazari-Vanani, H. Heli, and N. Sattarahmady, "An impedimetric genosensor for Leishmania infantum based on electrodeposited cadmium sulfide nanosheets," Talanta **217**, 121080 (2020).
2. L. Cheng, Q. H. Meng, A. J. Lilienthal, and P. F. Qi, "Development of compact electronic noses: a review," Meas Sci Technol **32**, 062002 (2021).
3. C. Wehrenfennig, M. Schott, T. Gasch, D. Meixner, R. A. Düring, A. Vilcinskas, and C. D. Kohl, "An approach to sense pheromone concentration by pre-concentration and gas sensors," Physica Status Solidi (a) **210**, 932 (2013).
4. Z. Zheng, and C. Zhang, "Electronic noses based on metal oxide semiconductor sensors for detecting crop diseases and insect pests," Comput Electron Agric **197**, 106988 (2022).
5. Z. Xia, C. Zheng, J. Hu, Q. Yuan, C. Zhang, J. Zhang, L. He, H. Gao, L. Jin, X. Chu, and F. Meng, "Synthesis of SnO_2 quantum dot sensitized $LaFeO_3$ for conductometric formic acid gas sensors," Sens Actuators B Chem **379**, 133198 (2023).
6. A. Butturini, and J. Fonollosa, "Use of metal oxide semiconductor sensors to measure methane in aquatic ecosystems in the presence of cross-interfering compounds," Limnol Oceanogr Methods **20**, 710 (2022).
7. S. H. Mohamed, M. El-Hagary, and S. Althoyaib, "Growth of undoped and Fe doped TiO_2 nanostructures and their optical and photocatalytic properties," Appl Phys A Mater Sci Process **111**, 1207 (2013).
8. F. Lan, Y. Chen, J. Zhu, Q. Lu, C. Jiang, S. Hao, X. Cao, N. Wang, and Z. L. Wang, "Piezotronically enhanced detection of protein kinases at ZnO micro/nanowire heterojunctions," Nano Energy **69**, 104330 (2020).
9. L. Zhao, K. Wang, W. Wei, L. Wang, and W. Han, "High-performance flexible sensing devices based on polyaniline/MXene nanocomposites," InfoMat **1**, 407 (2019).
10. X. Xu, W. Liu, S. Wang, X. Wang, Y. Chen, G. Zhang, S. Ma, and S. Pei, "Design of high-sensitivity ethanol sensor based on Pr-doped SnO_2 hollow beaded tubular nanostructure," Vacuum **189**, 110244 (2021).
11. L. Liu, D. Zhang, Q. Zhang, X. Chen, G. Xu, Y. Lu, and Q. Liu, Liu et al., "Smartphone-based sensing system using ZnO and graphene modified electrodes for VOCs detection," Biosens Bioelectron **93**, 94 (2017).
12. G. Li, X. Zhu, J. Liu, S. Li, and X. Liu, "Metal oxide semiconductor gas sensors for lung cancer diagnosis," Chemosensors **11**, 251 (2023).
13. N. Liu, J. Zhang, Y. Wang, Q. Zhu, C. Wang, X. Zhang, J. Duan, B. Hou, and J. Sheng, "Combination of metal-organic framework with Ag-based semiconductor enhanced photocatalytic antibacterial performance under visible-light," Colloids Surf A Physicochem Eng Asp **644**, 128813 (2022).
14. H. Shi, J. Du, H. Wang, Z. Jia, D. Jin, J. Cao, J. Hou, and X. Guo, "Engineering unsaturated coordination of conductive TiO_x clusters derived from Metal–Organic–Framework incorporated into hollow semiconductor for highly selective CO_2 photoreduction," Chem Eng J **440**, 135735 (2022).
15. M. Wang, L. Nian, Y. Cheng, B. Yuan, S. Cheng, and C. Cao, "Encapsulation of colloidal semiconductor quantum dots into metal-organic frameworks for enhanced antibacterial activity through interfacial electron transfer," Chem Eng J **426**, 130832 (2021).
16. L. M. Montañez, K. Müller, L. Heinke, and H. J. Osten, "Integration of thin film of metal-organic frameworks in metal-insulator-semiconductor capacitor structures," Microporous Mesoporous Mater **265**, 185 (2018).
17. C. Chen, J. Ren, P. Zhao, J. Zhang, Y. Hu, and J. Fei, "A novel dopamine electrochemical sensor based on a β-cyclodextrin/Ni-MOF/glassy carbon electrode," Microchem J **194**, 109328 (2023).
18. C. R. Yang, P. W. Cheng, and S. F. Tseng, "Highly responsive and selective NO2 gas sensors based on titanium metal organic framework (Ti-MOF) with pyromellitic acid," Sens Actuators A Phys **354**, 114301 (2023).
19. X. Zhang, P. Wang, Z. Liang, W. Zhong, and Q. Ma, "A novel Cu-MOFs nanosheet/BiVO4 nanorod-based ECL sensor for colorectal cancer diagnosis," Talanta **266**, 124952 (2024).

20. N. P. Zaretskiy, L. I. Menshikov, and A. A. Vasiliev, "On the origin of sensing properties of the nano-structured layers of semiconducting metal oxide materials," Sens Actuators B Chem **170**, 148 (2012).

21. I. A. Ahmad, and Y. H. Mohammed, "Synthesis of ZnO nanowires by thermal chemical vapor deposition technique: Role of oxygen flow rate," Micro Nanostructures **181**, 207628 (2023).

22. K. GangaReddy, and M. V. R. Reddy, "Physical vapour deposition of Zn2+ doped NiO nanostructured thin films for enhanced selective and sensitive ammonia sensing," Mater Sci Semicond Process **154**, 107198 (2023).

23. Z. Zhang, L. Nie, Q. Zhou, Z. D. Song, and G. bo Pan, "Chemiresistive H2S gas sensors based on composites of ZnO nanocrystals and foam-like GaN fabricated by photoelectrochemical etching and a sol-gel method," Sens Actuators B Chem **393**, 134148 (2023).

24. P. P. Ortega, C. C. Silva, M. A. Ramirez, G. Biasotto, C. R. Foschini, and A. Z. Simões, "Multifunctional environmental applications of ZnO nanostructures synthesized by the microwave-assisted hydrothermal technique," Appl Surf Sci **542**, 148723 (2021).

25. E. Çepni, and T. Öznülüer Özer, "Electrochemical deposition of Indium(III) hydroxide nanostructures for novel battery-like capacitive materials," J Energy Storage **45**, 103678 (2022).

26. L. Alinauskas, E. Brooke, A. Regoutz, A. Katelnikovas, R. Raudonis, S. Yitzchaik, D. J. Payne, and E. Garskaite, "Nanostructuring of SnO2 via solution-based and hard template assisted method," Thin Solid Films **626**, 38 (2017).

27. A. O. Dikovska, R. G. Nikov, G. V. Avdeev, G. B. Atanasova, T. Dilova, D. B. Karashanova, and N. N. Nedyalkov, "ZnO/Zn2TiO4 composite nanostructures produced by laser ablation in air," Physica E Low Dimens Syst Nanostruct **150**, 115707 (2023).

28. S. Rani, R. K. Das, A. Jaiswal, G. P. Singh, A. Palwe, S. Saxena, and S. Shukla, "4D nanoprinted sensor for facile organo-arsenic detection: A two-photon lithography-based approach," Chem Eng J **454**, 140130 (2023).

29. A. Lahcen, S. Rauf, A. Aljedaibi, J. I. de Oliveira Filho, T. Beduk, V. Mani, H. N. Alshareef, and K. N. Salama, "Laser-scribed graphene sensor based on gold nanostructures and molecularly imprinted polymers: Application for Her-2 cancer biomarker detection," Sens Actuators B Chem **347**, 130556 (2021).

30. M. M. Foroughi, S. Jahani, Z. Aramesh-Boroujeni, M. Rostaminasab Dolatabad, and K. Shahbazkhani, "Synthesis of 3D cubic of Eu3+/Cu2O with clover-like faces nanostructures and their application as an electrochemical sensor for determination of antiretroviral drug nevirapine," Ceram Int **47**, 19727 (2021).

31. G. Mathankumar, S. Harish, M. K. Mohan, P. Bharathi, S. K. Kannan, J. Archana, and M. Navaneethan, "Enhanced selectivity and ultra-fast detection of NO2 gas sensor via Ag modified WO3 nanostructures for gas sensing applications," Sens Actuators B Chem **381**, 133374 (2023).

32. C. Magro, T. Moura, J. Dionísio, P. A. Ribeiro, M. Raposo, and S. Sério, "Nanostructured metal oxide sensors for antibiotic monitoring in mineral and river water," Nanomaterials **12**, 1858 (2022).

33. S. Ahmadian-Fard-Fini, M. Salavati-Niasari, and D. Ghanbari, "Hydrothermal green synthesis of magnetic Fe3O4-carbon dots by lemon and grape fruit extracts and as a photoluminescence sensor for detecting of *E. coli* bacteria," Spectrochim Acta A Mol Biomol Spectrosc **203**, 481 (2018).

34. Z. Li, Y. Hou, Y. Ma, F. Zhai, and M. K. Joshi, "Recent advances in one-dimensional electrospun semiconductor nanostructures for UV photodetector applications: A review," J Alloys Compd **948**, 169718 (2023).

35. X. Zhang, T. Li, X. Gao, J. Lin, B. Zeng, C. Gong, and F. Zhao, "Porphyrin and molecularly imprinted polymer double sensitized cathode photoelectrochemical sensor for chlortetracycline," Microchem J **193**, 109094 (2023).

36. A. Rudie, A. M. Schornack, Q. Wu, Q. Zhang, and D. Wang, "Two-dimensional Ti3C2 MXene-based novel nanocomposites for breath sensors for early detection of diabetes mellitus," Biosensors **12**, 332 (2022).

37. D. Ghoshal, T. Wang, H.-Z. Tsai, S.-W. Chang, M. Crommie, N. Koratkar, S.-F. Shi, D. Ghoshal, T. Wang, S. Shi, H. Tsai, S. Chang, M. Crommie, and N. Koratkar, "Catalyst-free and morphology-controlled growth of 2D perovskite nanowires for polarized light detection," Adv Opt Mater **7**, 1900039 (2019).

38. A. Nigam, N. Sharma, S. Tripathy, and M. Kumar, "Development of semiconductor based heavy metal ion sensors for water analysis: A review," Sens Actuators A Phys **330**, 112879 (2021).

39. G. Zhou, H. Pu, J. Chang, X. Sui, S. Mao, and J. Chen, "Real-time electronic sensor based on black phosphorus/Au NPs/DTT hybrid structure: Application in arsenic detection," Sens Actuators B Chem **257**, 214 (2018).

40. R. R. César, A. D. Barros, R. O. Nascimento, O. L. Alves, I. Doi, J. A. Diniz, and J. W. Swart, "Electrolyte-Insulator-Semiconductor Structure for Pb+ Detecting," Procedia Eng **87**, 188 (2014).

41. R. Wang, C. Y. Xiong, Y. Xie, M. J. Han, Y. H. Xu, C. Bian, and S. H. Xia, "Electrochemical sensor based on MoS2 nanosheets and DNA hybridization for trace mercury detection," Chin J Anal Chem **50**, 100066 (2022).

42. D. Lv, W. Shen, W. Chen, Y. Wang, R. Tan, M. Zhao, and W. Song, "Emerging poly(aniline co-pyrrole) nanocomposites by in-situ polymerized for high-performance flexible ammonia sensor," Sens Actuators A Phys **349**, 114078 (2023).

43. R. Kumar, H. Rahman, S. Ranwa, A. Kumar, and G. Kumar, "Development of cost effective metal oxide semiconductor based gas sensor over flexible chitosan/PVP blended polymeric substrate," Carbohydr Polym **239**, 116213 (2020).

44. D. Jang, H. Jin, M. Kim, and Y. Don Park, "Polymeric interfacial engineering approach to perovskite-functionalized organic transistor-type gas sensors," Chem Eng J **473**, 145482 (2023).

45. Y. Li, Y. Bu, F. Jiang, X. Dai, and J. P. Ao, "Fabrication of ultra-sensitive photoelectrochemical aptamer biosensor: Based on semiconductor/DNA interfacial multifunctional reconciliation via 2D-C3N4," Biosens Bioelectron **150**, 111903 (2020).

46. H. Ji, Z. Wang, S. Wang, C. Wang, K. Zhang, Y. Zhang, and L. Han, "Highly stable InSe-FET biosensor for ultra-sensitive detection of breast cancer biomarker CA125," Biosensors (Basel) **13**, 193 (2023).

22 Semiconductor-based Ferroelectrics

Rijith Sreenivasan, Akhila Muhammed,
and Sumi V. Sasidharan Nair

22.1 INTRODUCTION

22.1.1 FERROELECTRICS–DEFINITION

The switching behavior of polarizations in a material, known as ferroelectricity, has been a significant subject in the field of material science. The history of ferroelectrics can be traced back to the discovery of ferroelectricity in Rochelle salt by Valasek in 1920, despite the initial report of Rochelle salt occurring almost three centuries earlier [1]. Valasek's observation of a polarization-electric field hysteresis loop in Rochelle salt crystal when subjected to an external electric field unveiled the defining characteristic of ferroelectrics, marking the beginning of a new era in this field [2]. For a material to be considered ferroelectric, it must possess a crystalline structure falling within one of the ten polar point groups: (C_1), (C_2), (Cs), (C_{2v}), $(C4)$, (C_{4v}), (C_3), (C_{3v}), $6(C_6)$, or (C_{6v}). Additionally, its spontaneous polarizations' orientation must be changeable or reversible in reaction to an external electric field, provided the temperature is below the Curie temperature (Tc). As depicted in Figure 22.1, Rochelle salt demonstrates a monoclinic polar space group $(C_2, 2)$ within the temperature range of 255 K to 296 K. During this specific temperature range, a hysteresis loop between polarization (P) and electric field (E), confirming the presence of ferroelectricity in Rochelle salt, can be observed [3]. This hysteresis loop signifies the occurrence of a phase transition. A ferroelectric material is typically characterized as having an intrinsic lattice polarization, denoted as P, which can be reversed by applying an external electric field, represented by E, greater than the coercive field, E_c. In other words, a ferroelectric material must exhibit a hysteresis loop between polarization (P) and electric field (E) when subjected to an applied electric field. Ferroelectrics usually exhibit a phase-transition temperature, T_0, above which they become paraelectric, although some may not follow this pattern and instead melt before reaching that state. It's worth noting that all ferroelectrics are also pyroelectric, and conversely, all pyroelectrics are piezoelectric. However, the reverse is not true, as pyroelectric ZnO does not possess ferroelectric properties. Therefore, ferroelectric materials lack a center of symmetry and cannot exist in glasses. Some ferroelectrics exhibit center of symmetry during their paraelectric phase, such as $BaTiO_3$, while others, like KH_2PO_4, do not possess this property. It is worth noting that most ferroelectric families are composed of non-oxide materials, although oxides are the most extensively studied due to their durability and wide range of practical applications.

22.2 CLASSIFICATIONS OF FERROELECTRICS

Ferroelectrics can be classified into two distinct categories based on their phenomenological properties. The first category encompasses ferroelectrics that have the ability to polarize along a single axis, either in the up or down direction. They are piezoelectric even in the non-polar state. The remarkable characteristic exhibited by ferroelectrics belonging to this class is the reversible alteration of their spontaneous polarization in response to mechanical stress, specifically shear stress. Additionally, these ferroelectrics demonstrate a highly significant dependency on both

DOI: 10.1201/9781003450146-22

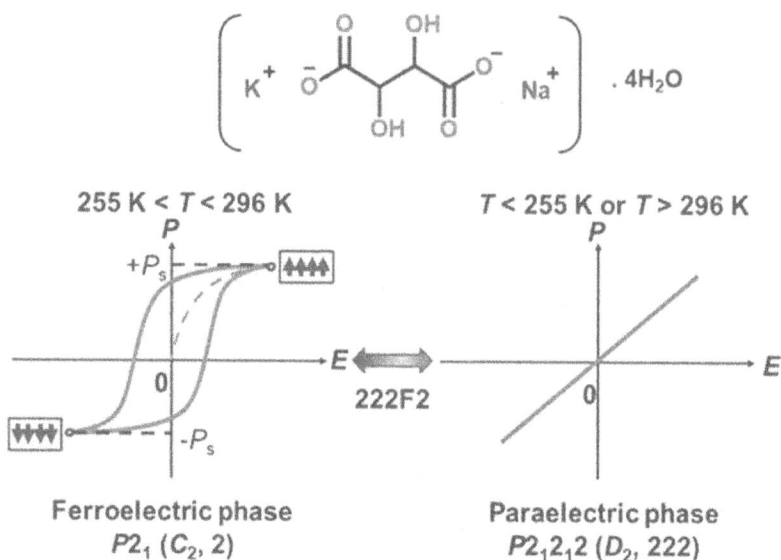

FIGURE 22.1 The P-E curves depicting the conceptual relationship between polarization and electric field for the ferroelectric and paraelectric phases of Rochelle salt. Polarizations are visually represented by dark-blue arrows. The point group of each phase is indicated through the use of Schönflies and international symbols, which are presented in parentheses. (Adapted with permission from [4], Copyright [2020], American Chemical Society.)

the electrical boundary conditions and temperature when it comes to their electromechanical behavior.

The second classification of ferroelectrics pertains to a group of substances referred to as perovskites, named after one of its members, $CaTiO_3$. These ferroelectrics share a common chemical formula, ABO_3, where A represents a divalent or monovalent metal, and B represents a tetravalent or pentavalent metal. In a state of complete cubic symmetry, the A atoms occupy the corners of the cube, the B atoms reside at the body-centers, and the O atoms are situated at the face-centers. Among the various perovskites, $BaTiO_3$ has received the most extensive research attention. As the temperature decreases, $BaTiO_3$ undergoes a sequential transition, traversing through all these structural phases. Since the discovery of Rochelle salt, research on ferroelectrics has predominantly focused on ferroelectric ceramics such as $BaTiO_3$ (BTO) and Pb (Zr, Ti) O_3 (PZT) [5, 6]. These inorganic ferroelectrics are widely recognized for their exceptional performance characteristics, including high polarization (Ps) and a high piezoelectric coefficient (d_{33}).

Molecular ferroelectrics possess distinct properties compared to conventional inorganic ferroelectrics. These properties include being environmentally friendly, lightweight, easy to process, and exhibiting structural diversity, mechanical flexibility, low acoustical impedance, and occasionally homochirality [7–9]. These advantages position molecular ferroelectrics as promising alternatives to traditional inorganic ferroelectrics.

22.3 THE PHENOMENOLOGICAL BEHAVIOR OF FERROELECTRICS– THEORY OF MOLECULAR FERROELECTRICS

These design theories are developed based on a chemical perspective. In essence, by employing specific chemical concepts and methodologies, it becomes possible to systematically convert centrosymmetric materials into multifunctional ferroelectric materials. Furthermore, these theories enable the initial establishment of effective approaches to control and enhance ferroelectricity and

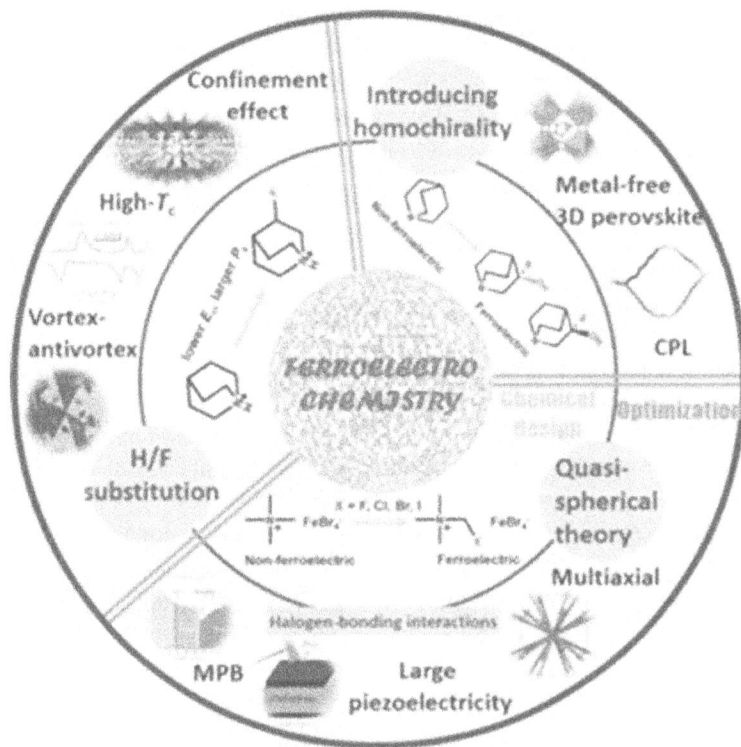

FIGURE 22.2 An illustrative diagram depicting the concept of ferroelectrochemistry is presented, which focuses on the purposeful design and optimization of molecular ferroelectrics using phenomenological theories. (Adapted with permission [4], Copyright [2020], American Chemical Society.)

piezoelectricity within molecular systems. From this point of view three chemical design approaches offer valuable insights into the purposeful design of molecular ferroelectrics. These approaches encompass modifications applied to spherical molecules, the incorporation of homochirality, and the substitution of hydrogen (H) or fluorine (F). These strategies, discussed in previous studies [10, 11], enable the attainment of five enantiomorphic point groups (C_1, C_2, C_4, C_3, C_6) through the introduction of homochirality. Additionally, other polar point groups (C_s, C_{2v}, C_{4v}, C_{3v}, C_{6v}) can be achieved by tailoring the molecular structure. This concept focuses on the intentional design and optimization of molecular ferroelectrics, guided by phenomenological theories (as illustrated in Figure 22.2), with the objective of enhancing their performance.

22.3.1 Quasi-spherical Theory

Spherical molecules, including [Me_4N] $^+$, dabco (1,4-diazabicyclo [2.2.2] octane), and quinuclidine, possessing low rotational energy barriers, hold great promise for constructing structural phase transitions. For instance, [Me_4N]$FeCl_4$ exhibits four phase transitions at temperatures of 295 K, 309 K, 344 K, and 384 K, sequentially transitioning from the space groups Pbcm, Pma2, Amm2, Cmcm to Pm$\bar{3}$m [12] and this was reported as an example of utilizing spherical molecules. Furthermore, the family of spherical molecules has been expanded to include phosphonium-based [$Me4P$]$^+$ cation. An example, [$(CH_3)_4P$]$CdCl_3$, displays a paraelectric-to-ferroelectric phase transition at 348 K with Aizu notation 6/mF6 [13]. It not only exhibits high-Tc ferroelectricity but also demonstrates luminescence upon doping Sb^{3+} into its structure. Additionally, materials incorporating 3-pyrrolidinium as the organic component, such as (3-pyrrolidinium) $CdBr_3$ and (3-pyrrolidinium) $MnCl_3$, have been reported to showcase intriguing ferroelectric properties. Nevertheless, the occurrence of ferroelectric-type phase transitions is not always observed in spherical molecules due to their inherent high symmetry, which often favors the formation of centrosymmetric structures that are incompatible with ferroelectricity [14–16]. Recognizing this challenge, significant efforts have been dedicated to developing targeted design approaches over the years. Eventually, a breakthrough was achieved with the introduction of the quasi-spherical theory.

The quasi-spherical theory is a phenomenological approach based on the Curie symmetry principle that involves making subtle modifications to spherical molecules [10]. By introducing a dipole moment, this theory aims to achieve lower-symmetric ferroelectric phases. By decreasing the molecular symmetry while maintaining the quasi-spherical shape, a ferroelectric phase transition can still occur. This transition leads to an increase in the symmetry of the paraelectric phase. The enhancement in polarization is beneficial for data storage in a ceramic-like form [17]. At high temperatures, tetrahedral or octahedral anions with high symmetry may appear spherical due to their rotational motion. However, as the temperature decreases, these anions undergo a transition to a non-spherical shape. According to the Landau theory, the symmetry of the ordered ferroelectric phase is a subset of the higher-symmetry paraelectric phase in the setting of ferroelectric materials. This suggests that the symmetries of the disordered paraelectric phase and the ordered ferroelectric phase have a "group-subgroup relationship" or a "parent-child relationship." The ferroelectric material exhibits uniaxial ferroelectricity when the symmetry of the ordered ferroelectric phase is the greatest non-isomorphic subgroup of the parent symmetry.

22.3.2 Chemical Strategies for Achieving Homochirality in Ferroelectrics

The identification of ferroelectricity in Rochelle salt marked a significant milestone in the development of ferroelectrics. While the association between homochirality and ferroelectricity originated from Rochelle salt, the significance of homochirality in the generation and utilization of ferroelectricity went unnoticed. In 2010, the discovery of two homochiral organic molecules, namely (R)-2-methylpiperazine-1,4-diamine [18] and (R)-2-methylpiperazinium bis(dichloroiodate) [19], brought forth the possibility of them being potential ferroelectrics. This breakthrough introduced a novel approach for the purposeful design of molecular ferroelectrics [20]. For a material to exhibit ferroelectricity, it needs to have a crystalline structure belonging to one of the polar point groups. Out of the 11 chiral point groups, five of them (C_1, C_2, C_3, C_4, and C_6) possess the necessary polarity that allows for ferroelectric behavior [21]. Based on the group-subgroup relationship between the low-temperature ferroelectric phase and the high-temperature paraelectric phase, Aizu [21] identified 22 out of the 88 different types of full ferroelectric phase transitions as chiral-to-chiral transition (Table 22.1).

TABLE 22.1

Variations of Full Ferroelectric Phase Transitions that Exhibit Chiral-to-Chiral Transformation

Crystal System	Aizu Notation of Phase Transition
Monoclinic	2F1
Orthorhombic	222F1; 222F2
Tetragonal	4F1; 422F1; 422F2; 422F4
Trigonal	3F1; 32F1; 32F2; 32F3
Hexagonal	6F1; 622F1; 622F2; 622F6
Cubic	23F1; 23F2; 23F3; 432F1; 432F2; 432F4; 432F

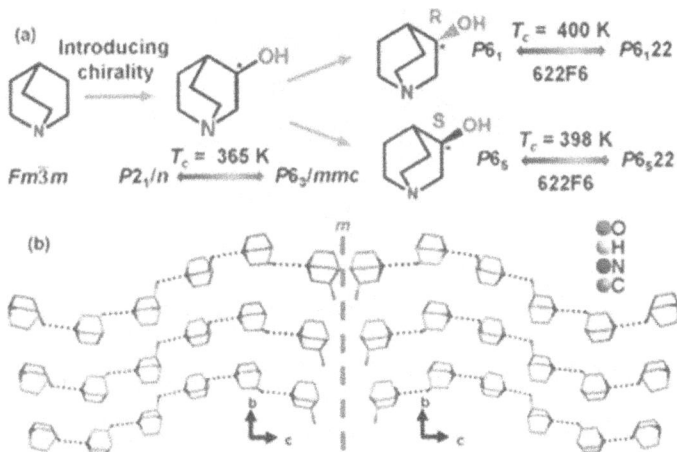

FIGURE 22.3 (a) Scheme of acquiring 3-quinuclidional by introducing homochirality. (b) Packing views of crystal structures of 22, with H atoms omitted for clarity. (Adapted with permission [4], Copyright [2020], American Chemical Society.)

22.3.2.1 Incorporating Homochirality in the Fabrication of Single-Component Ferroelectrics

The question has emerged whether it is possible to achieve ferroelectricity by simultaneously introducing homochirality and leveraging its spherical geometry. A remarkable study on this subject reveal that the addition of a hydroxyl group to the quinuclidine molecule successfully yields quasi-spherical molecules, specifically (R)-and (S)-3-quinuclidinol [22], by reducing symmetry. This introduction of homochirality in the molecule results in the acquisition of ferroelectric properties. The presence of homochirality facilitates the crystallization of molecules into polar-chiral point groups, satisfying the fundamental prerequisite for ferroelectricity. In both homochiral materials, ferroelectric-type phase transitions occur at a relatively high temperature of approximately 400 K, leading to the formation of chiral-nonpolar space groups. Specifically, the materials transition to space groups P6122 (D6, 622) and P6522 (D6, 622) in their respective paraelectric phases, as denoted by the Aizu notation 622F6. As depicted in Figure 22.3, the homochiral crystals exhibit crystallization in the anticipated enantiomorphic space groups. Specifically, they crystallize in the ferroelectric phase as P61 (C6, 6) and P65 (C6, 6) for the respective materials [23]. In contrast, the crystal of (Rac)-3-quinuclidinol [24] exhibits crystallization in the centrosymmetric space group P21/n (C2h, 2/m). This observation implies that the presence of homochirality facilitates the easier crystallization of molecules in polar-chiral point groups, meeting the fundamental requirement for ferroelectricity. Another example of a single-component system is the homochiral compound (4aR,8aR)-2,3-di(thiophen-2-yl)-decahydroquinoxaline. Although there were no indications of phase transitions, its potential ferroelectricity was confirmed through the observation of a P-E hysteresis loop. In the historical context, the discoveries of homochiral ferroelectrics have primarily involved multi-component organic amine and metal coordination compounds [25]. Single-component homochiral ferroelectrics with high curie temperature (Tc) are extremely uncommon.

22.4 2D FERROELECTRICS

The key characteristic of ferroelectrics is their ability to maintain two or more stable polar states with a moderate energy gap. These states can be interchanged by applying an electric field. Thanks to dedicated research efforts, significant advancements have been made in comprehending the underlying mechanisms of ferroelectricity, accurately quantifying ferroelectric performance, and successfully implementing diverse applications based on ferroelectric materials. The advancements in quantum-mechanics-based first-principles calculations have empowered researchers to delve into the fundamental factors driving ferroelectric polarization. Furthermore, these calculations have provided insights into the impact of ferroelectric polarization on crucial properties like electronic structures and phonon dispersion [26]. Based on symmetry considerations, the Landau-Ginzburg-Devonshire (LGD) theory provides a useful approach to characterize and understand the ferroelectric performances exhibited by materials [27]. Through the utilization of these theoretical tools, it becomes feasible to determine the underlying source of ferroelectric polarization. One such example is the soft phonon-mode theory, which suggests that the phonon frequency is directly related to the disparity between long-range forces (e.g., dipole-dipole interaction) and short-range forces (e.g., Pauli repulsion). This theory provides valuable insights into the mechanisms governing ferroelectric polarization [28].

Over the past few decades, there has been a rapid advancement in micro/nanoelectromechanical systems (M/NEMS) driven by the need for intelligent components with micro/nano-scale dimensions and low energy consumption. These systems have created a demand for the development of diverse components, including sensors, transducers, and actuators, that can operate efficiently in the micro/nano-scale domain [29]. 2D ferroelectrics are emerging as a prominent area of research, particularly due to their potential as functional materials for M/NEMS. Their unique properties and characteristics make them an attractive research focus in the field. As a result, 2D ferroelectrics are gaining significant attention as a promising area of study (Figure 22.4).

After extensive research, numerous 2D ferroelectrics have been discovered, each exhibiting unique and notable properties. For instance, 2D In_2Se_3 demonstrates the phenomenon of locking between in-plane and out-of-plane ferroelectric polarization [31]. In addition, the odd-even layer effect has been observed in few-layer SnS [32]. These distinctive properties are rarely observed in 3D ferroelectric materials. Furthermore, 2D ferroelectrics have a diverse set of material properties,

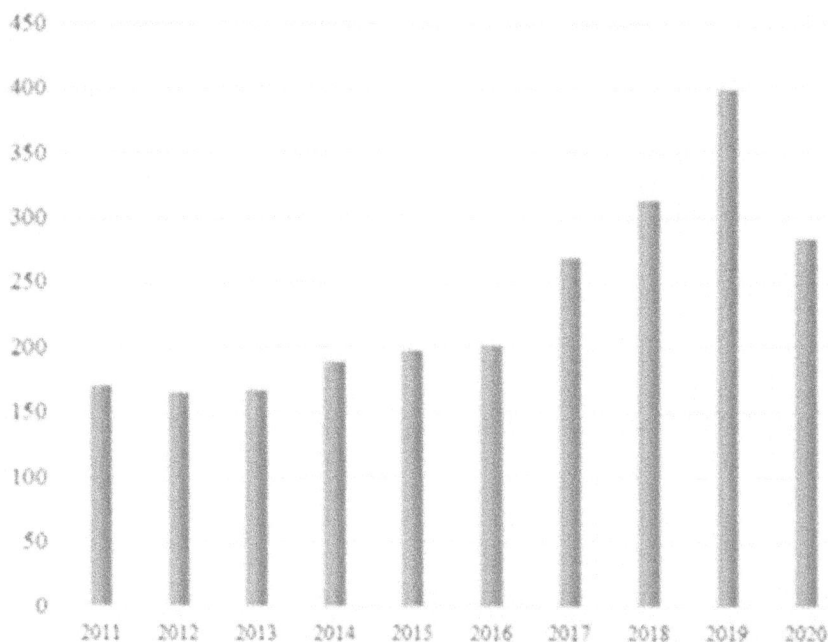

FIGURE 22.4 The number of publications on 2D ferroelectrics per year. (Data from database "webofknowledge.com, "2D ferroelectric/two-dimensional ferroelectric." Adapted with permission [30], Copyright [2021], Wiley-VCH GmbH.)

including well-known phenomena such as the piezoelectric effect, pyroelectric effect, and bulk photovoltaic effect (BPVE), as well as newly found phenomena such as valley polarization and spin polarization. These distinct features enable the creation of intelligent structures with a wide range of functions, such as 2D ferroelectric-based energy harvesters [33], tunnel junctions [34], field-effect transistors [35], and photodetectors [36]. Many of these nanodevices based on 2D ferroelectrics are expected to operate exceptionally well. The ferroelectric field-effect transistor, based on the 2D semiconductor In_2Se_3, demonstrates a substantial memory window, with an on/off ratio surpassing 10^8, achieving a peak on-current of $862\ \mu A\ \mu m^{-1}$, all while operating at a low supply voltage.

22.4.1 THEORY OF 2D FERROELECTRICS

First-principles calculations have indeed been widely employed in the field of materials science to gain insights into the fundamental microscopic mechanisms governing material behavior. These calculations are based on the principles of quantum mechanics and provide a detailed understanding of the electronic structure and properties of materials. When it comes to 2D ferroelectrics, first-principles calculations play a crucial role in several aspects. One important application is the determination of stable crystal structures. By calculating the total energy of different crystal structures, researchers can identify the most energetically favorable configuration, which corresponds to the stable crystal structure. This information is essential for predicting the behavior and properties of 2D ferroelectrics.

Another significant application is quantifying the ferroelectric polarization. Ferroelectric materials exhibit a spontaneous electric polarization that can be reversed by the application of an external electric field. First-principles calculations can provide accurate calculations of the polarization, allowing researchers to quantify and understand this important property in 2D ferroelectrics. Furthermore, first-principles calculations aid in exploring the underlying ferroelectric mechanisms in 2D materials.

22.4.1.1 Determining Stable Crystal Structures

Non-centrosymmetry is indeed a crucial condition for ferroelectric polarization, and identifying the stable crystal structures is a key step in studying these materials. In the context of first-principles calculations for 2D ferroelectrics, there are several methods used to determine stable crystal structures. These include: (1) Experimental verification: In some cases, the stable crystal structure of a 2D ferroelectric material is determined through experimental techniques, such as in the case of 2D SnTe [37]. (2) Analogy to homologous 2D compounds: Homologous 2D compounds with known crystal structures can serve as a reference to predict the stable crystal structure of a new 2D ferroelectric material. For example, the crystal structure of an SbN monolayer can be inferred based on the crystal structure of related compounds. (3) Structural optimization algorithms: Optimization algorithms, often based on energy minimization techniques, can be employed to determine the stable crystal structure of a 2D ferroelectric material. These algorithms search for the configuration that minimizes the total energy of the system, considering various structural parameters. The $LiAlTe_2$ monolayer is an example where structural optimization was used to determine the stable crystal structure [38]. (4) Analogy to stable 3D counterparts: In some cases, the stable crystal structure of a 2D ferroelectric material can be inferred by analogy to its stable 3D counterpart, especially if the 3D structure has a small cleavage energy. The WO_2Cl_2 monolayer is an example where the stable crystal structure is derived from its stable 3D counterpart [39].

It is worthwhile to noted that 2D materials can have multiple stable states, and each of these states could potentially exhibit ferroelectric behavior. The ground state and other stable states can correspond to local energy minima. For example, in the case of MoS_2 monolayer, the ground state $2H$-MoS_2 is non-ferroelectric, while the stable $d1T$-MoS_2 phase is predicted to be ferroelectric [40]. Considering the existence of multiple stable states in various 2D materials, there is still a vast potential for the discovery of new 2D ferroelectrics. Exploring different stable states and their ferroelectric properties can lead to the identification of novel 2D materials with desirable characteristics. To assess the dynamic stability of a 2D crystal structure, phonon spectrum analysis based on first-principles calculations can be employed. Dynamically unstable crystal structures often exhibit one or more phonon branches with negative frequencies. Analyzing the phonon spectrum provides valuable information about the stability and vibrational properties of the crystal structure.

22.4.1.2   Measuring the Magnitude of Ferroelectric Polarization

Ferroelectric polarization can be understood as the concentration of electric dipole moments. However, for a long time, there was no clear and universally accepted definition for the polarization of bulk materials [41]. The simplest model, known as the Clausius-Mossotti (CM) model, defined polarization as the average density of dipole moments within a given unit cell, assuming localized point charges without any overlap. However, this definition was limited to pure ionic crystals and was not applicable to most materials, as the charge density in real materials changes continuously in a periodic manner. To address the oversimplification of the CM model, several definitions based on the continuous charge density, $\rho(r)$ have been established. These definitions encompass both differential and integral forms. The integral form of polarization within a chosen unit cell (V_{cell}) is expressed as $P = (1/V_{cell}) \int \rho(r)P(r)\, dr$, where $\rho(r)$ represents the charge density and $P(r)$ represents the local polarization at position r. The most reliable definition of polarization, which addresses the limitations of static charge distribution, was proposed. It introduces the concept of effective polarization (ΔP), which is associated with the flow of charge ($j(r, t)$). This definition is expressed as $\Delta P = (1/V_{cell}) \int\int j(r, t)\, dr\, dt$, where the integral is taken over the volume (V_{cell}) and time (t). This definition has overcome the inherent shortcomings of static charge distribution and has become the foundation of modern polarization theory. In the framework of quantum mechanics, this definition has been further developed and expanded [42]. In the case of 2D ferroelectric materials, the accurate quantification of in-plane polarization typically relies on the modern theory of polarization. However, for the out-of-plane polarization, especially in monolayer samples, the integral form based on charge density is usually accurate enough. This is because the absence of periodic structures in the out-of-plane direction eliminates the uncertainty associated with polarization calculations.

22.4.1.3   Ferroelectric Polarization Mechanisms

The mechanism behind ferroelectric polarization is often complex, involving the interplay of multiple factors. One widely used approach to investigate the driving force for ferroelectric polarization is the soft-mode theory [43]. This theory assumes that the frequency of an optical phonon is determined by the difference between short-range repulsion and long-range forces. The short-range repulsion tends to favor a non-polar lattice with high symmetry, while the long-range force favors a polar lattice with low symmetry. When these forces balance each other, the phonon frequency becomes zero, resulting in the stabilization of the corresponding phonon mode known as the "soft mode." The occurrence of the soft phonon mode at the center or edges of the momentum space is often associated with polar or non-polar modes, respectively. The soft-mode theory provides valuable insights into the driving force and behavior of ferroelectric polarization. Studying the phonon spectrum can enhance our comprehension of how atomic displacement contributes to ferroelectric polarization. Nevertheless, additional computations are necessary to determine the underlying factors that induce these atom displacements. Various methods have been validated as effective for this objective. In this context, we provide a concise overview of the calculations involving projected density of states (PDOS), interatomic force constants, and born effective charges.

22.4.1.4   Phase Transition

First-principles calculations, which do not rely on any prior information, have certain limitations. Due to the extensive computational requirements, they are best suited for structures with fewer than hundreds of atoms and are most effective at absolute zero temperature. However, during phase transitions, it is crucial to consider heat fluctuations. Therefore, more practical methods or theories need to be employed, such as ab initio molecular dynamics, first-principles effective Hamiltonians, and the LGD theory. Among these methods and theories, the LGD theory has been successfully utilized to examine the phase transition in bulk ferroelectrics from a phenomenological perspective [44]. The LGD theory operates on the principle that a system with different symmetries cannot undergo a smooth transition between two phases. Consequently, the potential energy of the system can be expressed as a Taylor expansion using order parameters, where only terms that satisfy the higher symmetry are retained. Within the framework of the LGD theory, the potential energy of a proper ferroelectric at the phase transition is represented as a power series expansion of polarization.

$$E = \sum_i^1 \left[(T - Tc) \frac{A}{2} P_i^2 + \frac{B}{4} P_i^4 + \frac{C}{6} P_i^6 \right] + \frac{C}{6} \sum_{i,j} (P_i - P_j)^2$$

where A, B, and C represent material parameters, T is the temperature, and the summations are performed over all unit cells (i) and their nearest neighbor unit cells (j).

The LGD theory can be extended to incorporate electronic polarization, allowing for the interpretation of the electronic origin within its framework [45] Furthermore, the LGD theory can also consider the influence of external fields, providing insights into their effects on ferroelectric performance. In the case of bulk materials, the extended LGD theory has successfully revealed the impacts of piezoelectric, flexoelectric, and electric effects on ferroelectric phase transitions and the formation of domain walls. Investigating these effects is crucial for understanding 2D ferroelectrics. For instance, by including the electric effect in the potential energy, it becomes possible to elucidate the electronic improper ferroelectric origin in monolayer $MoTe_2$. To determine the unknown parameters in the LGD theory, fitting the results of first-principles calculations with an effective Hamiltonian can be employed. This approach establishes a link between theoretical methods and enables the investigation of 2D ferroelectrics on a larger scale. While theoretical calculations can make predictions, experimental verification remains essential for confirming the existence of 2D ferroelectricity. Although accurately measuring or detecting ferroelectric polarization in 2D materials is challenging due to their weak response and reduced dimensions, a few studies have successfully measured and detected ferroelectric polarization using both direct and indirect measurement techniques.

22.5 APPLICATIONS

22.5.1 CAPACITORS

Ferroelectric capacitors differ from conventional capacitors in that they utilize ferroelectric materials instead of dielectrics. The polarization and direction of the ferroelectric material within the capacitor play a crucial role. With their remarkable ability to retain charge, ferroelectric capacitors have proven to be highly valuable in applications such as ferroelectric random access memory (FeRAM) and other non-volatile memory technologies. In electronics, these capacitors are used in integrated circuits and microelectronic devices, offering efficient energy storage and enabling high-density data storage capabilities. Perovskite materials, known for their exceptional dielectric constants of up to 30,000, have emerged as powerful components in ferroelectric capacitors. This unique characteristic enhances their capacitance and makes them suitable for various applications. In memory applications, ferroelectric capacitors serve as data storage elements. They store the logical value of the data, which can be read by applying an electric field. During the read operation, an electric field is employed to measure the amount of charge required to switch the memory cell to its opposite state. This reveals its previous state. However, it is imperative to note that the application of an electric field during the read operation erases the current memory state. This requires an additional write operation to restore the desired bit value. Consequently, this constraint poses a limitation to using single ferroelectrics as memory elements, as each read cycle must be followed by a write cycle. Additionally, ferroelectric capacitors have a high write cycle limit, though not infinite [46].

22.5.2 NONVOLATILE MEMORY

Nonvolatile memory in ferroelectrics has revolutionized technologies for data storage and memory. As an alternative to conventional nonvolatile memory technologies, ferroelectric materials provide enhanced performance, low power consumption, and excellent data retention. Ferroelectric RAM (FeRAM) is random access memory similar in construction to DRAM, but it uses a ferroelectric layer instead of a dielectric layer. FeRAM surpasses traditional memory technologies by exploiting ferroelectric materials' inherent polarization characteristics. Within a FeRAM cell, the polarization state of the ferroelectric capacitor represents stored data, providing a high-density, nonvolatile memory solution. FeRAM is an emerging alternative to nonvolatile random access memory technologies, which offers advantages over flash memory. FeRAM's advantages over flash include lower power usage, faster write performance, and a much-increased maximum number of write-erase cycles (exceeding 1016 for 3.3V devices). It also exhibits low power consumption due to the nonvolatile nature of ferroelectric capacitors, eliminating the need for constant power supply during data retention. FeRAM is ideal for portable and energy-efficient electronic devices [47].

Ferroelectric RAM was proposed by an MIT graduate student, Dudley Allen Buck, in his master's thesis, titled "Ferroelectrics for Digital Information Storage and Switching," published in 1952 [48]. FeRAM development began in the late 1980s. In the early 1990s, development was based at NASA's Jet Propulsion Laboratory to improve readout methods, such as introduction of a nondestructive readout using UV pulses. A fabless semiconductor company, Ramtron, has developed much of the current FeRAM technology, and one of the major licensees is Fujitsu. Fujitsu operates the largest semiconductor foundry production line with FeRAM capabilities. They manufactured stand-alone FeRAMs and embedded FeRAMs based on Ramtron's technology between 1999 and 2010. Texas Instruments collaborated with Ramtron to develop FeRAM test chips in a modified 130-nm process.

22.5.3 THERMISTOR

The thermistor is a temperature sensor achieved by calibrating the material resistance as a function of temperature. Ferroelectric materials that undergo the ferroelectric-paraelectric transition exhibit variation in electrical conductivity as a function of temperature. This electrical behavior allows ferroelectric materials to be applied not only as a positive temperature coefficient thermistor but also as a negative temperature coefficient thermistor. Ferroelectric oxides with a perovskite structure are applied as thermistors. The thermistors' resistivity is modulated by metal cations distributed in different interstices or oxygen vacancies. For example, polycrystalline n-type $BaTiO_3$ exhibits a positive temperature coefficient of resistivity behavior that results from a temperature dependent, Schottky-type grain boundary potential barrier due to adsorbed oxygen at the grain boundaries. When the $BaTiO_3$ is sintered in air, a defect reaction occurs at the grain boundary and produces a neutral Ba vacancy. Then, it may be ionized by an electron introduced by the donor dopant. In other words, this capturing of a free electron by the neutral Ba vacancy at the grain boundary during the ferroelectric phase transition is responsible for the positive temperature coefficient of resistivity behavior [49].

22.5.4 FERROELECTRIC THIN FILMS IN ELECTRONICS

Ferroelectric thin films have opened up new possibilities in electronics. By depositing ferroelectric materials as thin films on substrates, ferroelectricity integration into existing technologies becomes feasible. Ferroelectric thin films have been employed in non-volatile memory devices, such as FeFETs and ferroelectric capacitors, offering high-density data storage, low power consumption, and improved performance. This integration has the potential to revolutionize electronics, enabling advanced devices with enhanced functionalities.

22.5.5   FERROELECTRIC TUNNEL JUNCTIONS

Ferroelectric tunnel junctions (FTJs) have emerged as a cutting-edge area of research in the field of nanoelectronics. These structures combine the unique properties of ferroelectric materials with tunnel barriers, enabling precise control over electrical conductivity through polarization switching. FTJs offer exciting opportunities for the development of innovative electronic devices, such as non-volatile memories, logic devices, and spintronics. The engineering of FTJs involves optimizing ferroelectric materials, interface properties, and tunnel barrier characteristics to achieve high-performance, low-power devices. Ongoing research in this field aims to unlock the full potential of FTJs and pave the way for advanced electronic technologies with enhanced functionality and efficiency.

22.5.6   FERROELECTRIC PHOTOVOLTAICS

Ferroelectric photovoltaics represent a compelling frontier in renewable energy research. These photovoltaic devices leverage ferroelectric materials' unique properties to convert solar energy into electrical power. Ferroelectric materials offer advantages such as high dielectric constants and excellent photovoltaic response, enabling efficient energy conversion. By harnessing ferroelectrics' inherent polarization and band gap engineering capabilities, researchers strive to enhance solar cell efficiency and stability. Ferroelectric photovoltaics hold promises for sustainable energy generation, offering a pathway toward environmentally friendly power sources. Ongoing research aims to further optimize and advance the performance of ferroelectric photovoltaic devices to realize their full potential in the renewable energy landscape [50].

22.5.7   FERROELECTRIC-BASED ENERGY STORAGE

Ferroelectric-based energy storage has emerged as a promising field, addressing the growing demand for efficient and sustainable energy storage solutions. By harnessing the unique properties of ferroelectric materials, such as their piezoelectric and pyroelectric characteristics, energy can be harvested from mechanical vibrations and temperature gradients. The utilization of ferroelectrics in energy storage offers several advantages. These materials provide high energy density, fast charge-discharge rates, and long cycle life, making them attractive for various applications. Ferroelectric capacitors and devices store electrical energy electrostatically, offering potential for compact and high-performance energy storage systems. In the future, ferroelectric-based energy storage holds immense scope. As research progresses, efforts are focused on enhancing ferroelectric materials' performance, improving their energy storage capacity, and optimizing device engineering. Furthermore, advancements in nanoscale fabrication techniques and the exploration of novel ferroelectric materials are expected to unlock new opportunities in energy storage.

22.6   CONCLUSION AND PROSPECTIVE

To conclude, we summarized the advances in molecular ferroelectrics in terms of the targeted chemical design approaches. Meanwhile, detailed information of phase transitions and symmetry breaking are also included. In this perspective, we first reviewed the chemical design idea of symmetry lowering, namely the quasi-spherical theory, to modulate ferroelectricity at the molecular level of shape, symmetry, and specific interactions by tailoring the high symmetric cations. This strategy has been successfully applied to [Me4N]⁺, dabco, and quinuclidine. The reveal of the connection between ferroelectricity and homochirality is also an important step in the development of molecular ferroelectrics. As mentioned above, the introduction of homochirality is a unique design strategy when constructing molecular ferroelectrics.

Ferroelectric materials are widely known for their electronic, optical, thermal, mechanical, and magnetic properties, as well as the complex interplay between these properties. However, our understanding of 2D ferroelectrics is still limited. While extensive research has been conducted on the flexoelectric effect in 3D ferroelectrics over the past decades, the potential enhancement of this effect at the nanoscale due to increased strain gradient remains largely unexplored in 2D ferroelectrics. Therefore, 2D ferroelectrics present an ideal platform for investigating the significant flexoelectric effect and its influence on other properties like the bulk photovoltaic effect. Nevertheless, detecting and quantifying the small ferroelectric responses in 2D ferroelectrics pose significant challenges. It may be necessary to introduce additional direct or indirect precision measurements and employ a combination of multiple methods to ensure accurate characterization. The ultimate goal of exploring 2D ferroelectrics is to design intelligent components for nanoelectromechanical systems (NEMS). Consequently, it is crucial to continuously explore and discover more 2D ferroelectric materials with desirable properties. Simultaneously, the study of novel structures based on 2D ferroelectrics, such as sensors, actuators, transducers, and energy harvesters, is of utmost importance.

In summary, research on 2D ferroelectrics is still in its early stages. Further exploration is required not only to unravel the diverse underlying mechanisms but also to investigate numerous material properties. This endeavor necessitates interdisciplinary collaboration spanning physics, chemistry, mechanics, engineering, materials science, and mathematics. Significant progress in the exploration of 2D ferroelectrics can be achieved by concurrently conducting research studies in all these areas.

REFERENCES

1. J. Valasek, Piezo-electric and allied phenomena in Rochelle salt, Phys. Rev. 17 (1921) 475.
2. J. Valasek, Properties of Rochelle salt related to the piezo-electric effect, Phys. Rev. 20 (1922) 639.
3. F. Mo, R.H. Mathiesen, J.A. Beukes, K.M. Vu, Rochelle salt–a structural reinvestigation with improved tools. I. The high-temperature paraelectric phase at 308 K, IUCrJ. 2 (2015) 19–28.
4. H.-Y. Liu, H.-Y. Zhang, X.-G. Chen, R.-G. Xiong, Molecular design principles for ferroelectrics: Ferroelectrochemistry, J. Am. Chem. Soc. 142 (2020) 15205–15218.
5. R. Clarke, A.M. Glazer, Critical phenomena in ferroelectric crystals of lead zirconate titanate, Ferroelectrics. 14 (1976) 695–697.
6. H.D. Chen, K.R. Udayakumar, L.E. Cross, J.J. Bernstein, L.C. Niles, Dielectric, ferroelectric, and piezoelectric properties of lead zirconate titanate thick films on silicon substrates, J. Appl. Phys. 77 (1995) 3349–3353.
7. S. Horiuchi, Y. Tokura, Organic ferroelectrics, Nat. Mater. 7 (2008) 357–366.
8. W. Zhang, R.-G. Xiong, Ferroelectric metal–organic frameworks, Chem. Rev. 112 (2012) 1163–1195.
9. W. Li, Z. Wang, F. Deschler, S. Gao, R.H. Friend, A.K. Cheetham, Chemically diverse and multifunctional hybrid organic–inorganic perovskites, Nat. Rev. Mater. 2 (2017) 1–18.
10. H.-Y. Zhang, Y.-Y. Tang, P.-P. Shi, R.-G. Xiong, Toward the targeted design of molecular ferroelectrics: Modifying molecular symmetries and homochirality, Acc. Chem. Res. 52 (2019) 1928–1938.
11. Y.-L. Liu, S.-Q. Lu, Y.-Y. Tang, X.-G. Chen, J.-X. Gao, H.-J. Li, R.-G. Xiong, Fluorination observed T c increase of 110 K is challenging the hydrogen–deuterium isotope effect, Chem. Commun. 55 (2019) 10007–10010.
12. J. Harada, N. Yoneyama, S. Yokokura, Y. Takahashi, A. Miura, N. Kitamura, T. Inabe, Ferroelectricity and piezoelectricity in free-standing polycrystalline films of plastic crystals, J. Am. Chem. Soc. 140 (2018) 346–354.
13. L. Zhou, P.-P. Shi, X.-M. Liu, J.-C. Feng, Q. Ye, Y.-F. Yao, D.-W. Fu, P.-F. Li, Y.-M. You, Y. Zhang, An above-room-temperature phosphonium-based molecular ferroelectric perovskite,[(CH3) 4P] CdCl3, with Sb3+-doped luminescence, NPG Asia Mater. 11 (2019) 15.
14. M. Paściak, M. Wołcyrz, A. Pietraszko, Structural origin of the x-ray diffuse scattering in (CH 3) 4 NCdCl 3 and related compounds, Phys. Rev. B. 78 (2008) 24114.
15. B. Morosin, Crystal structure of tetramethylammonium cadmium chloride, acta, Crystallogr. Sect. B Struct. Crystallogr. Cryst. Chem. 28 (1972) 2303–2305.
16. B. Morosin, E.J. Graber, Crystal structure of tetramethylammonium manganese (II) chloride, Acta Crystallogr. 23 (1967) 766–770.
17. P.-P. Shi, Y.-Y. Tang, P.-F. Li, H.-Y. Ye, R.-G. Xiong, De novo discovery of [Hdabco] BF4 molecular ferroelectric thin film for nonvolatile low-voltage memories, J. Am. Chem. Soc. 139 (2017) 1319–1324.
18. P.-F. Li, W.-Q. Liao, Y.-Y. Tang, W. Qiao, D. Zhao, Y. Ai, Y.-F. Yao, R.-G. Xiong, Organic enantiomeric high-T c ferroelectrics, Proc. Natl. Acad. Sci. 116 (2019) 5878–5885.

19. L. Chen, G. Han, H. Ye, R. Xiong, A new chiral quinoxaline derivative with ferroelectric property, Chinese J. Chem. 28 (2010) 1799–1802.

20. T. Zhang, L.-Z. Chen, M. Gou, Y.-H. Li, D.-W. Fu, R.-G. Xiong, Ferroelectric homochiral organic molecular Crystals, Cryst, Growth Des. 10 (2010) 1025–1027.

21. K. Aizu, Possible species of "ferroelastic" crystals and of simultaneously ferroelectric and ferroelastic crystals, J. Phys. Soc. Japan. 27 (1969) 387–396.

22. P.-F. Li, Y.-Y. Tang, Z.-X. Wang, H.-Y. Ye, Y.-M. You, R.-G. Xiong, Anomalously rotary polarization discovered in homochiral organic ferroelectrics, Nat. Commun. 7 (2016) 13635.

23. C. Yang, W. Chen, Y. Ding, J. Wang, Y. Rao, W. Liao, Y. Tang, P. Li, Z. Wang, R. Xiong, The first 2D homochiral lead iodide perovskite ferroelectrics:[R-and S-1-(4-chlorophenyl) ethylammonium] 2PbI4, Adv. Mater. 31 (2019) 1808088.

24. H.-Y. Ye, Y.-Y. Tang, P.-F. Li, W.-Q. Liao, J.-X. Gao, X.-N. Hua, H. Cai, P.-P. Shi, Y.-M. You, R.-G. Xiong, Metal-free three-dimensional perovskite ferroelectrics, Science. 361 (2018) 151–155.

25. E. Bousquet, J. Junquera, P. Ghosez, First-principles study of competing ferroelectric and antiferroelectric instabilities in BaTiO 3/BaO superlattices, Phys. Rev. B. 82 (2010) 45426.

26. J. Zhu, X. Yao, Y. Bai, C. Jin, X. Zhang, Q. Zheng, Z. Xu, L. Chen, S. Wang, Y. Liu, Anomalous polarization enhancement in a vdW ferroelectric material under pressure. 14 (2023) 4301.

27. W. Cochran, Crystal stability and the theory of ferroelectricity, Adv. Phys. 9 (1960) 387–423.

28. R. Frisenda, E. Navarro-Moratalla, P. Gant, D. PÚrez De Lara, P. Jarillo-Herrero, R.V. Gorbachev, A. Castellanos-Gomez, Recent progress in the assembly of nanodevices and van der Waals heterostructures by deterministic placement of 2D materials, Chem. Soc. Rev. 47 (2018) 53.

29. J. Xiao, H. Zhu, Y. Wang, W. Feng, Y. Hu, A. Dasgupta, Y. Han, Y. Wang, D.A. Muller, L.W. Martin, Intrinsic two-dimensional ferroelectricity with dipole locking, Phys. Rev. Lett. 120 (2018) 227601.

30. L. Qi, S. Ruan, Y. Zeng, Review on recent developments in 2D ferroelectrics: Theories and applications, Adv. Mater. 33 (2021) 2005098.

31. Y. Bao, P. Song, Y. Liu, Z. Chen, M. Zhu, I. Abdelwahab, J. Su, W. Fu, X. Chi, W. Yu, Gate-tunable in-plane ferroelectricity in few-layer SnS, Nano Lett. 19 (2019) 5109–5117.

32. F. Xue, J. Zhang, W. Hu, W.-T. Hsu, A. Han, S.-F. Leung, J.-K. Huang, Y. Wan, S. Liu, J. Zhang, Multidirection piezoelectricity in mono-and multilayered hexagonal α-In2Se3, ACS Nano. 12 (2018) 4976–4983.

33. L. Kang, P. Jiang, H. Hao, Y. Zhou, X. Zheng, L. Zhang, Z. Zeng, Giant tunneling electroresistance in two-dimensional ferroelectric tunnel junctions with out-of-plane ferroelectric polarization, Phys. Rev. B. 101 (2020) 14105.

34. M. Si, A.K. Saha, S. Gao, G. Qiu, J. Qin, Y. Duan, J. Jian, C. Niu, H. Wang, W. Wu, A ferroelectric semiconductor field-effect transistor, Nat. Electron. 2 (2019) 580–586.

35. C. Chang, W. Chen, Y. Chen, Y. Chen, Y. Chen, F. Ding, C. Fan, H.J. Fan, Z. Fan, C. Gong, Recent progress on two-dimensional materials, Acta Phys. Chim. Sin. 37 (2021) 2108017.

36. K. Liu, J. Lu, S. Picozzi, L. Bellaiche, H. Xiang, Intrinsic origin of enhancement of ferroelectricity in SnTe ultrathin films, Phys. Rev. Lett. 121 (2018) 27601.

37. C. Liu, W. Wan, J. Ma, W. Guo, Y. Yao, Robust ferroelectricity in two-dimensional SbN and BiP, Nanoscale. 10 (2018) 7984–7990.

38. Z. Liu, Y. Sun, D.J. Singh, L. Zhang, Switchable out-of-plane polarization in 2D LiAlTe2, Adv. Electron. Mater. 5 (2019) 1900089.

39. L.-F. Lin, Y. Zhang, A. Moreo, E. Dagotto, S. Dong, Frustrated dipole order induces noncollinear proper ferrielectricity in two dimensions, Phys. Rev. Lett. 123 (2019) 67601.

40. G.-B. Liu, D. Xiao, Y. Yao, X. Xu, W. Yao, Electronic structures and theoretical modelling of two-dimensional group-VIB transition metal dichalcogenides, Chem. Soc. Rev. 44 (2015) 2643–2663.

41. K.M. Rabe, P. Ghosez, First-principles studies of ferroelectric oxides, in: Phys. Ferroelectr. A Mod. Perspect., Springer, 2007: pp. 117–174.

42. R. Resta, M. Posternak, A. Baldereschi, Towards a quantum theory of polarization in ferroelectrics: The case of KNbO 3, Phys. Rev. Lett. 70 (1993) 1010.

43. A.A. Sirenko, C. Bernhard, A. Golnik, A.M. Clark, J. Hao, W. Si, X.X. Xi, Soft-mode hardening in SrTiO3 thin films, Nature. 404 (2000) 373–376.

44. P. Chandra, P.B. Littlewood, A Landau primer for ferroelectrics, in: Phys. Ferroelectr. A Mod. Perspect., Springer, 2007: pp. 69–116.

45. A. Chanana, U.V. Waghmare, Prediction of coupled electronic and phononic ferroelectricity in strained 2D h-NbN: First-principles theoretical analysis, Phys. Rev. Lett. 123 (2019) 37601.

46. J.P.B. Silva, M.B. Silva, M.J.S. Oliveira, T. Weingärtner, K.C. Sekhar, M. Pereira, M. Gomes, High-performance Ferroelectric–Dielectric multilayered thin films for energy storage capacitors, Adv. Funct. Mater. 29 (2019) 1–8.

47. H. Li, R. Wang, S. Han, Y. Zhou, Ferroelectric polymers for non-volatile memory devices: A review, Polym. Int. 69 (2020) 533–544.

48. D.A. Buck, Ferroelectrics for digital information storage and switching. Master's thesis, Massachusetts Inst of Tech, Report R212 (1952).

49. H.K. Muchenedi, M.L.N.M. Mohan, Fabrication of ferroelectric liquid crystal thermistor, IEEE Trans. Electron Devices. 67 (2020) 5063–5068.

50. X. Han, Y. Ji, Y. Yang, Ferroelectric photovoltaic materials and devices, Adv. Funct. Mater. 32 (2022) 2109625.

23 Role of Semiconductors in Energy Devices

*Navid Nasajpour Esfahani, Amir Koohbor,
Hamid Garmestani, and Steven Y. Liang*

23.1 INTRODUCTION

The history of semiconductors dates back to the late 19th and 20th centuries, when many devices such as radios, radars, televisions, and telephones were developed using the concept of electronic vacuum tubes. This technology was then referred to as solid-state semiconductors. During these years, researchers made significant progress in making semiconductors less expensive, more reliable, lighter, and smaller. Since silicon is neither a good conductor nor a good insulator, it has been considered one of the main semiconductor materials due to its availability. Silicon possesses a face-centered cubic crystal structure. The energy diagram of an intrinsic semiconductor, like ideal silicon, is explained by the band gap concept, which is based on the Fermi level. The Fermi level is an energy level situated between the conduction band and the valence band and is also known as the forbidden region. The crystal lattice remains stable without an additional energy source, but when external energy is applied to the lattice, electrons tend to move from the valence band to the conduction band. In contrast to intrinsic semiconductors, extrinsic semiconductors contain impurities or dopants that can modify specific properties depending on the dopants used. In general, there are two main types of semiconductors known as n-type and p-type. In p-type semiconductors, "p" stands for positive, and these materials are created by adding trivalent impurities or donors, such as boron, indium, and gallium. For n-type materials, which have a negative charge, they contain pentavalent impurities or acceptors, such as arsenic, phosphorus, and antimony. Various solid-state semiconductors can be created by connecting an n-zone material to a p-zone material, a configuration known as p-n junctions. P-N junctions can be found in different semiconductor devices, including junction field-effect transistors (JFETs), bipolar junction transistors (BJTs), and diodes. The difference between p-type and n-type materials is illustrated in Figure 23.1 [1], providing a fundamental understanding of semiconductors. In this chapter, we will describe the applications of semiconductor materials in the field of energy based on current research.

23.2 ELECTRICAL POWER GENERATION

The basic device that can be created by the p-n junction is called a diode. Within the p-n junction, there is a region where the p and n materials meet, and this region lacks electrons and electron holes. This area is referred to as the depletion region, as depicted in Figure 23.2 [1].

As a result, the p and n materials should adjust their energy levels at the p-n junction (Figure 23.3) [1].

Diodes are known as the simplest p-n junctions. They permit current to flow easily in one direction but prevent current flow in the opposite direction [1]. The behavior of the p-n junction can be quantify by using Shockey equation as:

$$I = I_s(e^{\frac{V_D q}{nkT}} - 1) \tag{23.1}$$

where I is a diode current, I_s reverse saturation current, V_D voltage (across the diode), q charge of electron, (1.6×10^{-19}), n is quality factor (between 1-2), k Boltzmann constant, and T is temperature

DOI: 10.1201/9781003450146-23

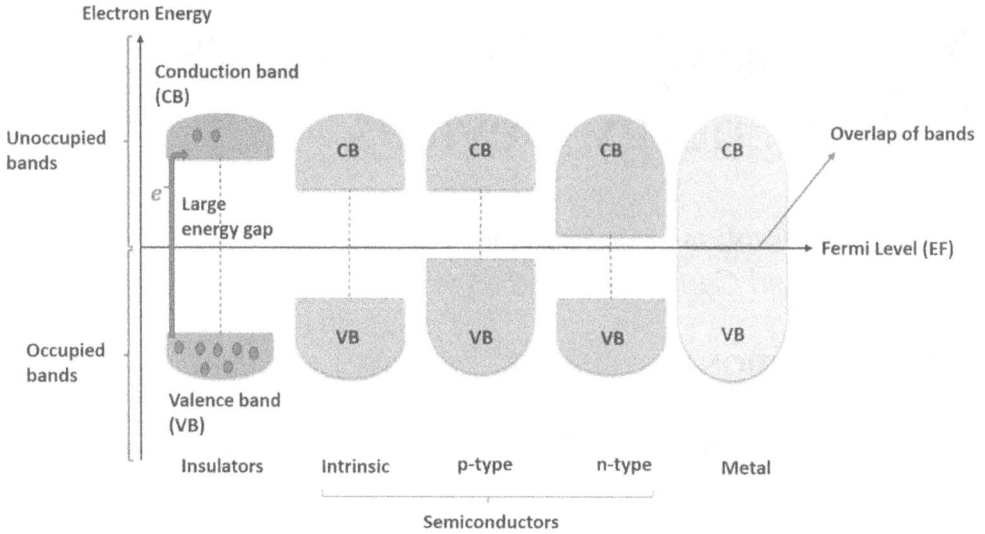

FIGURE 23.1 The difference between the p-type and n-type materials. (Adapted with permission [2], Copyright [2023], Elsevier.)

[1]. The orientation of a p-n junction can occur in only two ways based on the voltage source. The first way is called forward bias, in which electrons are transferred from the negative terminal of the battery into the n-type material. When they enter the depletion region, the supplied potential must be high enough to facilitate the diffusion of electrons into the p-type material. Finally, they move to the positive terminal of the battery from the p-type material. The second way is called reverse bias. In reverse bias, the voltage source is reversed, resulting in a change in the behavior of the p-n junction. Different types of diodes can be observed in Figure 23.4 [1].

The behavior of Zener diodes is similar to that of ordinary diodes when they are forward-biased. However, Zener diodes are typically used in a reverse-bias situation. The advantage of these diodes is to provide stable voltage to prevent breakdown. Another type of diode is called a light-emitting diode (LED), which can produce light based on the received electrical input. This is quite different from photodiodes, as exposure to light can generate a current in photodiodes. However, both types of these devices can operate in the visible spectrum of humans but can also be used outside this range in ultraviolet (UV) and infrared (IR). Filament-based light sources have been replaced by LEDs in various applications due to their high efficiency, physical robustness, availability in different colors, and small size. The design of energy transition in LEDs is intended to emit radiation at short wavelengths, and the color of the LED depends on the materials used and the voltage drop. The operation of photodiodes can be divided into two modes: photovoltaic and photoconductive. In the photovoltaic mode, the photodiode acts as a source of voltage, and this mode is used in solar cells. In the second mode, which is the photoconductive mode, the photodiode acts like a current source. Next is the

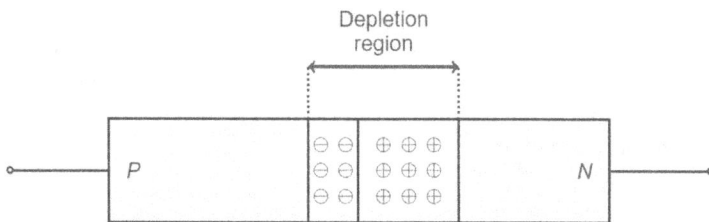

FIGURE 23.2 The schematic of the p-n junction and depletion region. (Adapted with permission [1], Copyright (2022), Springer.)

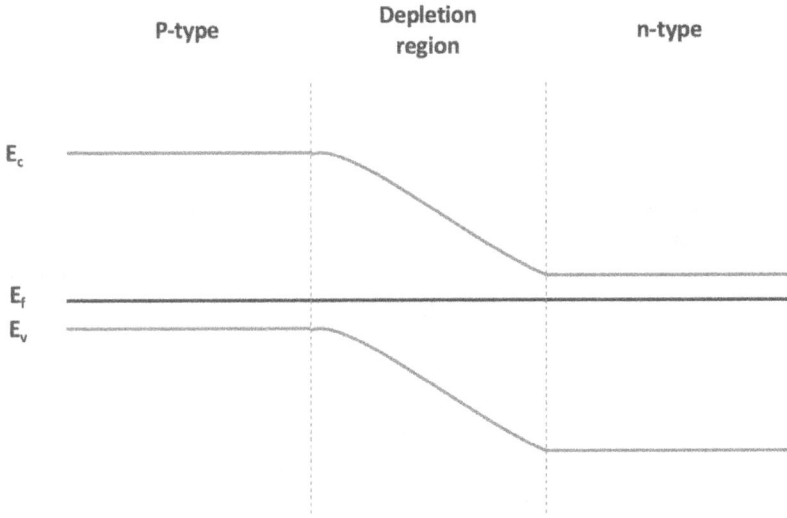

FIGURE 23.3 The energy bonds of p-n junction. (Adapted with permission from Reference [1], Copyright (2022), Springer.)

Schottky diode, which relies on semiconductor-to-metal contact, unlike traditional diodes. It offers two advantages compared to traditional diodes: fast switching times and low turn-on voltages.

Finally, there's the Varactor diode, a special diode used for controlling capacitance. Next, the conversion mechanism of sunlight into electricity is explained. In photovoltaic systems, the electric potential between two materials is harnessed by exposing a proton to the electron-hole pair. This technology in solar cells involves the use of p-n junctions. In both cases, the main goal is to collect the charge carriers as electric current [3]. Cadmium sulfide (CdS) is one of the most important common materials in the application of solar cells. CdS nanowires are formed on gold-coated substrates through the thermal decomposition of $Cd(S_2CNProp_2)_2$ and $Cd(S_2CNEt_2)_2$ in a chemical vapor deposition (CVD) process at low pressure [4, 5]. Tin oxide is also another important material in the application of solar cells. Tin oxide is an n-type semiconductor and has been used in solar cell application by many researchers [3]. In addition to these materials, researchers have developed tandem axial silicon nanowires for photovoltaic applications. The voltage values of p-i-n axial and radial structures are similar to each other, even though their junction geometries are different. This surprising comparison exists between coaxial and p-i-n axial silicon nanowires. The Fermi level in silicon nanowires is equal to the electrochemical potential of the redox pair, causing a bending of the

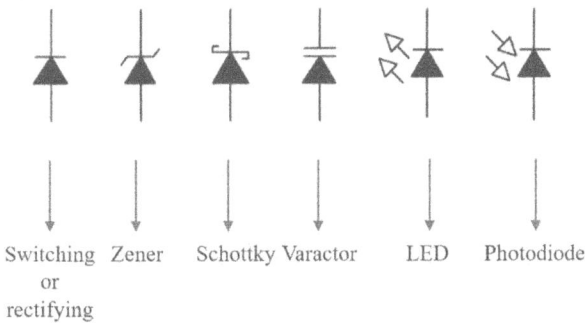

FIGURE 23.4 Different types of diodes. (Adapted with permission from Reference [1], Copyright (2022), Springer.)

energy bands and creating an imbalanced charge region where the segregation of photogenerated holes and electrons occurs [6].

23.3   ENERGY STORAGE SYSTEMS

Supercapacitors, like batteries, are well-known electrochemical energy storage devices. This technology typically stores electrical energy, which is gathered from the movement of electrons within conductive materials such as gold, copper, aluminum, etc., and converts it from its chemical form into energy when needed. Since the late 1800s, the storage of electrical energy at the interface between an electrolyte and a solid has been known. This phenomenon is known as an electrical double layer. In the 1950s, the development of supercapacitors (SCs) began, and Standard Oil of Ohio was reported to have invented the first electronic device that used double layer charge storage in 1966. The company patented it in 1971 [7]. In electric double-layer capacitors (EDLCs), no reduction-oxidation reaction is required; in other words, no substances gain or lose electrons. Energy is stored in such a manner that charge storage is achieved through the dissociation (the breaking of molecules or compounds into smaller pieces such as ions and atoms) and adsorption (the transfer of molecules from a fluid bulk to a solid surface) of the electrolyte ions on the surface of the metal-based electrode materials. Because of this characteristic, EDLCs exhibit very good resilience and cycling performance over millions of cycles. To achieve the high capacitance required for EDLCs, the electrode surface area and the thickness of the Helmholtz layer (double layer) are significant determining factors. Porous carbon materials, due to their large surface area, are often used as electrode materials for ion adsorption. Using the model of double-layer capacitance, it can be concluded that when polarization affects the interface between the electrode and electrolyte, the division of charges is observed [8]. In addition to the use of semiconductors in these devices, researchers have investigated various materials for semiconductors in energy storage applications. Silicon carbide (SiC) is one of the most well-known semiconductor materials and a primary silicon compound. This material has a highly adjustable and wide bandgap, very high electron mobility, and stable chemical properties, which has led to significant interest among researchers [9]. Silicon-based SCs have a very high energy density. Not only that, but they are also known to be very promising energy storage devices because they are compatible with modern devices [10]. Since the semiconductor properties are very compatible and the technology to process silicon has advanced, not only do the SiC-based electrodes benefit, but the method for preparation of high-performance silicon carbide-based electrodes resembles that of only silicon-based electrodes. Additionally, silicon itself is very reactive compared to SiC, which doesn't react with other chemicals as much, so using SiC in capacitors is a better option because it has benefits such as working very well in a broad range of temperatures. SiC is known to have many different structures, high compatibility, which makes it easier to work with other chemicals, and a relatively large surface area. Due to these characteristics, when SiC is in different shapes, it becomes very good for making electrodes in devices, batteries, and catalysts [11]. To ensure that SiC performs at its best, extensive research has been conducted, and many researchers are enthusiastic about its utilization. This text briefly discusses the research that has been conducted and the applications of SiC as electrodes in supercapacitors. John P. Alper and his colleagues utilized CVD, a technique that others also employ, to grow SiC nanowire (3C-SiC) films doped with nitrogen. These small wires, which are excellent for storing electrical energy, exhibited a capacitance of 240 mFcm^{-2}. Furthermore, when these wires were submerged in water, they could be used as components in a supercapacitor. Despite being used approximately 20,000 times, they retained 95% of their performance. However, it was discovered that if these wires were used to store and release electricity at rates exceeding 200 millivolts per second, the wires began to resist the flow of electricity, resulting in reduced performance. Researchers are currently seeking solutions to this issue and believe that improving the connection between these tiny wires and their substrate will enhance performance when used at high speeds [12]. Utilizing a unique method that involves a chemical process and low temperature to carve out these tiny wires, Roya and her team also produced SiC nanowires. When these nanowires, which are highly stretched and thin structures with nanoscale diameters, undergo passivation to reduce their reactivity and enhance their stability, their capacitance value reaches 1.7 mFcm^{-2}. This can be compared to electrode materials composed of carbon-based materials in micro-supercapacitor electrodes. Furthermore, they maintain a retention rate of 95% or higher after undergoing 1000 charge/discharge cycles. Moreover, these created wires can connect with the surface beneath them through physical contact, eliminating the need for additional components to collect electricity. This is another reason why SiC is highly suitable for manufacturing energy storage devices. Carbides serve as a base upon which electrodes can grow and reform. This uniqueness lies in the fact that nothing is required to hold the electrode together and keep it intact. Additionally, SiC is an excellent conductor of electricity [13]. Carbon fiber fabrics have a surface capacitance of 23 mFcm^{-2}, on which Gu and the rest of the researchers fabricated the SiC nanowires. The capacitance did not increase significantly even after about 100,000 cycles were conducted at room temperature. The SiC nanowires exhibit excellent thermal stability, as evidenced by the fact that even when the experiment was conducted at approximately 60°C, both the carbon fabric bases and the shape of the SiC nanowires remained largely unchanged [14]. Chen and his team produced thin layers of 3C-SiC nanowire films on special graphite-based paper using an extraordinary method that reduces specific materials. These wires are not only very thin and long but also braided together to form a net-like structure. This structure is highly advantageous in applications such as batteries due to its large surface area. The optimal capacitance is 25.6 mFcm^{-2}, and as a result, even after 2000 cycles, it maintains 100% capacity retention [15]. By altering the ratio of silicon and SiO$_2$, commonly known as silica, Liang and other researchers employed a catalyst CVD-free method to produce a large quantity of SiC nanowires with various shapes, structures, and sizes on carbon fiber fabrics. This particular material can hold a substantial amount of electrical energy, approximately 46.7 mFcm^{-2}, on its surface. Even the best carbon-based supercapacitor electrodes cannot match this capacity. Remarkably, even after more than 1000 uses, the material continues to function exceptionally well and does not lose its ability to store energy. Therefore, to maintain its long-term performance, using SiC nanowires on carbon fiber fabrics is an excellent choice [16]. Furthermore, a team of researchers led by Sanger closely examined how well SiC nanocauliflower, a tiny structure, functions as a material for electrodes in supercapacitors. They utilized a unique technique called DC magnetron co-sputtering to deposit SiC nanocrystals onto a surface composed of porous alumina and covered with silver. As a result of this process, the SiC nanocrystals were capable of holding up to 300 Fg^{-1}. The supercapacitor they created is an excellent device for storing electrical charges, with a capacity of up to 188 Fg^{-1}. Remarkably, even after approximately 30,000 cycles of use, the supercapacitor still retains slightly more than 97% of its performance, demonstrating its durability [17]. SiC nanocrystal based devices exhibit excellent electrochemical performance and effectiveness, making them highly promising for energy storage applications as supercapacitor electrodes. The studies mentioned above have revealed that to achieve a high surface capacitance of 46.7 mFcm^{-2}, these silicon electrodes need to be constructed and produced on carbon cloth. However, directly integrating these carbon-based electrodes with current silicon-based electronic components, such as transistors and computer chips, can be quite challenging. This could lead to unsatisfactory results, as observed in attempts to use nanocrystalline SiC films, which achieved a lower surface capacitance of 0.07 mFcm^{-2} on the silicon electrodes compared to SiC films [16]. The capacitive performance of the SiC electrode can be further enhanced by increasing the electrode's surface area. Liu and his colleagues employed a very special process known as laser chemical vapor deposition, which created highly precise patterns of graphene mixed with SiC on flat silicon surfaces and very small silicon wires. These patterns exhibited a very strong crystal structure. The graphene mixed with SiC and silicon nanowire nanomatrix can also store a significant amount of electric charge on its surface, approximately 3.2 mFcm^{-2}. Furthermore, its charge storage capacity improved even further after 10,000 cycles of use, increasing to 115%. This demonstrates its long-term stability and effectiveness [18]. In summary, supercapacitors are vital in electrochemical energy storage, and SiC-based electrodes show great promise in advancing the performance and applicability of these energy storage devices. Researchers continue to explore methods to optimize SiC electrodes for various applications, contributing to the development of efficient and durable energy storage solutions.

23.4 ELECTRONIC AND OPTOELECTRICAL DEVICES

The use of 2D materials has emerged as a highly promising choice for channel materials in transistors. Additionally, the short conduction channels in 2D semiconductors result from their natural atomic thickness, leading to rapid switching performance, while their good flexibility ensures suitability for wearable devices. Although many researchers have made significant efforts to develop transistors based on 2D materials for electronic applications, it remains challenging to bridge the gaps between theoretical and practical considerations. For instance, the theoretical room temperature electron mobility of MoS_2 as a 2D semiconductor exceeds $400 \ cm^2V^{-1}s^{-1}$, while experimentally it is below $200 \ cm^2V^{-1}s^{-1}$ [19]. By using a technique called the real-space projector-augmented wave (PAW), the calculations were simplified. By doing this, the size of the hexagonal unit that's making up the material was found to be 3.14 Å. Also, to confirm that the layers did not meddle with each other during the computer simulation, a large distance of 10 Å was kept between them. Moreover, another unique method, which involves a 11×11 K-point sampling of the Brillouin zone, was used in order to map and plot the material's properties. The size of the band gaps is very important in semiconductors, but typically the density functional theory within the local density approximation (DFT-LDA) downplays it. Even though it downplays the size, it still gives a very accurate description of, for example, the scattering of individual bands, the effective masses, and the differences in energy between valleys. It is good to know that the lowest point in the conduction band is actually along the way between the K points in the Brillouin zone and not the K or K' points. This is suggested by the most up-to-date calculations using a technique called GW quasiparticle calculations. The reason for this method is because it is better at explaining band structures in semiconductors. In Figure 23.5, you can see the results of our calculations with DFT-LDA. These results indicate that the material has a direct gap, which means that it requires 1.8 eV of energy to move an electron from one state to another at a specific point called the K point. The

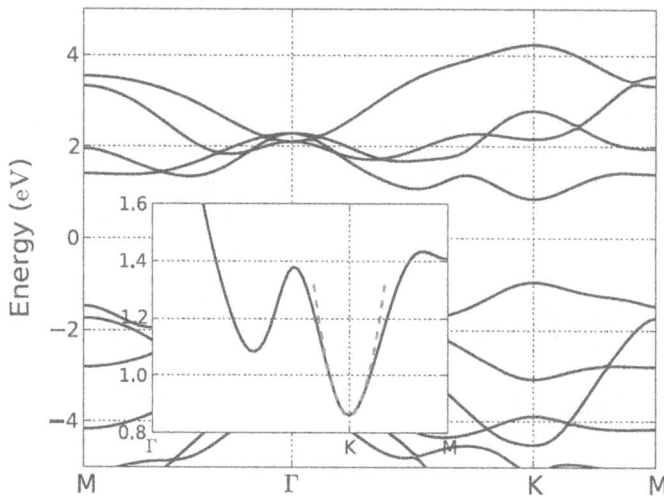

FIGURE 23.5 Band structure of single-layer MoS_2. The conduction band valley at The K point, within the energy range pertinent to low-field mobility, is accurately represented by a parabolic band (indicated by the dashed line) with an effective mass of 0.48 times the electron rest mass (m_e). (Adapted with permission [19], Copyright (2012), Arxiv.)

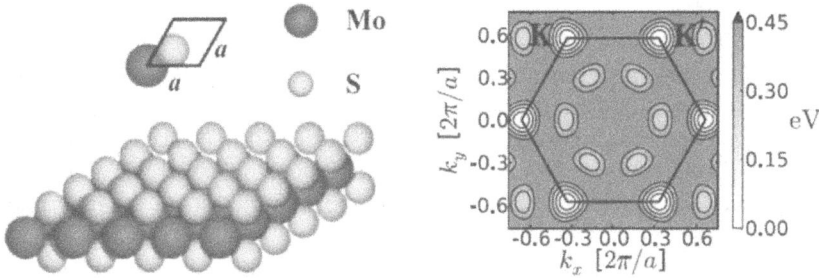

FIGURE 23.6 The atomic arrangement and conduction band characteristics of a single-layer MoS$_2$. On the left, you can see the primitive unit cell and the arrangement of molybdenum and sulfur atoms in shifted hexagonal layers. On the right, there's a contour plot illustrating the conduction band's lowest-lying valleys, as computed using DFT-LDA within the hexagonal Brillouin zone of single-layer MoS$_2$. In the case of n-type MoS$_2$, the low-field mobility is influenced by the properties of carriers within the K and K' valleys, which are energetically distinct from the secondary valleys inside the Brillouin zone. (Adapted with permission [19], Copyright [2012], Arxiv.)

small picture in the graph shows a close look of the conduction band, the bottom part of the energy levels, along the Γ-K-M path [19].

The conduction band looks like a perfect curve, almost like a parabola in the K valley. Although there's a point of about 200 milli-electron volts (meV) between the and the K points, it is not relevant for the behavior of the material when applying a low-electric field. On the right side in Figure 23.6, there's a plot that displays the lowest part of the conduction band's shape in the 2D Brillouin zone. In this zone, the K and K' valleys can be found, and they are filled with electrons in n-type MoS$_2$. They can be located at the corner of the hexagonal region of the zone. These parts of the band consist of specific natures which are parabolic and isotropic. So therefore, if an interest takes place in the movement of electrons in a low-electric field, by describing these parts of the band thoroughly while thinking of them as simple parabolic shapes, one could achieve their answer, and also by using $\varepsilon k = \frac{\hbar^2 k^2}{2m^*}$, with an electron mass (m^*) of 0.48 m$_e$, and k being measured with respect to the points K and K' in the Brillouin zone. The carriers with their 2D nature have a consistent measure of the number of states they can occupy, which is revealed as ρ_0. ρ_0 is decided by two factors, one being the spin of the particles, which g_s is 2, and the other being the K, K'–valley degeneracy, $g_v = 2$. The overall formula is given by $\rho_0 = \frac{g_s g_v m^*}{2\pi \hbar^2}$ [19].

Here are many states that are available, which emerge from the conduction band's mass in the K and K' valleys. These masses are very effective, and they cause distribution of carriers which do not overlap. This takes place only if the density of the carrier is not extremely high. In a single layer of MoS$_2$, three phonon branches are associated with the physical movement of the material. These are the acoustic branches. Another six branches are associated with its optical properties. These are the optimal branches. Together, nine branches can occur. Between the three acoustic branches, the behavior of the branch that involves out-of-plane flexural motion is in such a way that the frequency is raised quadratically as q, which is the size of the vibration, approaches zero. As for the remaining two acoustic branches, also known as transverse acoustic (TA) and longitudinal acoustic (LA), they have a frequency that is dependent on the speed of the vibrations inside the material (C_λ), also known as the in-plane sound velocity. The equation $\omega_{q\lambda} = C_\lambda q$ explains this. In this case, C_λ is 6.5 × 10^3 m/s for the LA and 4.4 × 10^3 m/s for the TA. There is a gap in the vibrations, which completely separates the optical and acoustic branches, even at visually similar points at the zone boundary. The two lowest optical branches are unique and do not significantly interact with electric charges. Consequently, they are not relevant for this study. The transverse optical (TO) and longitudinal optical (LO) branches that follow have an energy level of 48 meV. In these two branches, the atoms of sulfur (S) and molybdenum (Mo) move in opposite directions. In materials exhibiting observable polarization, the LO-TO splitting between these two modes occurs due to an interaction between the polarization and the atomic structure resulting from vibrations, particularly when considering super long-wavelength vibrations. A comprehensive understanding of Born effective charges is crucial to explain this phenomenon based on fundamental principles [19, 20]. Nonetheless, there is no regular pattern in the 2D materials in the vertical direction through the layers. As a result, the LO-TO splitting effect is not observable. Therefore, for the purposes of this study, the interaction with the extensive polarization caused by the LO mode will not be considered. The almost flat phonon at around 50 meV is referred to as the homopolar mode. It is commonly found in layered materials. In the homopolar mode, the thickness of the layers changes due to vibration. Unlike the Mo layer, which remains stationary, the sulfur layers move in opposite directions. This vibration results in a change in the potential energy (PE) stored in the material and has been shown to cause significant deformation in bulk MoS_2 [21]. The quality of a metal-semiconductor junction is assessed by the Schottky barrier height (SBH), where the energy barrier is higher when the SBH is higher. The SBH is described as the energy difference between the semiconductor's band edge and the metal's work function, a concept known as the Schottky-Mott rule. This implies that altering the contact can be achieved by using different metals with an appropriate work function. However, the selection of a different metal for a typical top contact may have a limited impact due to an unavoidable Fermi-level pinning effect (FLPE). This phenomenon can be attributed to the poor interface states between the semiconductor and the metal electrode [22].

23.5   SENSORS

While sensors are not typically categorized as energy devices, they have the capability to directly measure important parameters of energy devices. Furthermore, they can be used in various other applications. For example, at the center of the nervous system, the human brain constantly receives external information, processes it, and initiates body movements, whether they are chemically or physically based [23]. We can't naturally sense UV rays and ultrasound, which are pieces of information in our environment. Therefore, special tools such as sensors and detectors have been created to help us better understand our surroundings in a more precise way. This section focuses on semiconducting polymers, which are a unique type of material used in devices for sensing. It explores objects like humidity sensors, photoelectric detectors (PDs), and gas sensors, which can help gather important information to ensure the safety and comfort of our living spaces. Visible light is the portion of the light spectrum that is visible to the human eye. It constitutes only a small fraction (0.036%) of the entire range of light wavelengths, which extends from 10 to 106 nm. Photoelectric Detection refers to unique devices that function somewhat like human eyes but for electronics. These are small instruments made of semiconductor materials capable of detecting light through the photoelectric effect. The photoelectric effect occurs when light excites electrons, generating an electric signal [24]. PDs are used in many fields such as environmental monitoring, aerospace, health care, information technology, and industrial manufacturing [25]. Gas sensors detect gases that include explosive and toxic analytes, monitor air quality, and oversee the operation of industrial processes. These factors underscore the importance of gas sensors in human society [26]. Chemiresistor sensors are straightforward to operate, cost-effective, easy to produce, and compact. The majority of fiber-based gas sensors fall into this category. In a gas sensor, the active sensing layer exhibits high sensitivity to the gas molecules in its vicinity. Changes in conductivity occur when certain semiconductor polymers undergo reversible reactions with the gas molecules [27]. This makes them perfect candidates for use as active layers in gas sensors. Specifically, polyaniline (PANI) can transition among three different states of oxidation in various chemical vapors: partially oxidized emeraldine form, fully reduced leucoemeraldine form, and fully oxidized pernigraniline form [28]. Many harmful chemical acids can lead to the protonation of the PANI structure (adding a proton to a molecule to form a conjugate acid), which enhances its ability to conduct electricity. On the other hand, NH_3, which is an alkaline gas, or amines that are organic in nature, can facilitate and accelerate the process of removing protons from PANI, known as deprotonation, thereby reducing its conductivity [29]. Furthermore, chemicals like C_{12}, N_2H_4, NO_x, SO_x, and F_2, which can readily exchange electrons with PANI, have the ability to alter the charge of PANI. This results in significant changes in how effectively PANI can conduct electricity [30]. When humidity levels change, certain semiconducting polymers exhibit variations in their electrical conductivity. This property makes them suitable for use as specialized layers in humidity-detecting sensors. The operation of these humidity sensors depends on the interaction between water molecules and the active layers, as well as physical adsorption and minor chemical interactions in real-time humidity monitoring systems [31]. Therefore, active materials need to interact with and attract water so that the sensors work well [32]. Initially, water molecules on the outermost layer adhere to these active materials by forming chemical bonds. As air humidity increases, the heightened moisture causes additional layers of water molecules to adhere to the sensitive layer through physical attraction [33]. In summary, while sensors are not typically considered energy devices, they play a crucial role in measuring key parameters of energy devices and finding applications in various other fields. They enhance our understanding of the environment and enable precise monitoring. Semiconducting polymers, such as PANI, are valuable materials in sensors, including humidity sensors, PDs, and gas sensors. These sensors detect and respond to changes in environmental factors, from light to gas concentrations and humidity levels. The conductivity of these polymers can be modified by various chemical interactions, making them versatile components in sensor technology. This understanding of how sensors operate and interact with their surroundings is essential for developing effective energy devices and improving our overall quality of life.

23.6   INTERNET OF THINGS APPLICATIONS

With the advancement of information technology, people have found themselves increasingly immersed in the world of computers, the internet, and other digital technologies driven by algorithms and data. These technologies heavily rely on the logical operations of hardware. Just as humans have a natural inclination to understand things, they also seek to connect and interact with digital worlds and information in a way that feels intuitive and comprehensible. Textiles serve as a link between people and the world around them. It's more accurate to describe them as a bridge that endures, provides comfort, and offers flexibility [34]. Furthermore, these smart textiles can be employed for information sharing, representing a novel, innovative, and resourceful concept. These distinctive textiles, often referred to as the "second layer" of human skin, possess the ability to mimic an integral part of the human body, comprehending and responding to the needs of both the body and its surrounding environment [34]. Interactive interfaces work like magic. They are capable of performing various tasks, including comprehending signals, translating them into computer language, processing real-time information, retaining that information, providing feedback, and ultimately facilitating our interaction with others [35]. In this section, we discuss four things that these unique fibers can do with information. They can compute, communicate, sense, and display. Semiconducting polymers are the special materials that make up these unique fibers. These textiles and smart fibers produce a vast amount of data very quickly because, as their name suggests, they are very smart and can rapidly detect changes in the surrounding environment and our body [36]. Forwarding signals to another computer and then receiving the processed information can be very frustrating and complex. There are many challenges to address, such as the accuracy and sensitivity of the signals, the speed at which computers perform calculations, and the amount of bandwidth available for data transmission. When fabrics and textiles are capable of processing information like computers, they can rapidly process information in real-time. By integrating small computers into these e-textiles, the need for physical connections and one-way data transfers will no longer be necessary. This creates new possibilities for using textiles to perform a wide range of tasks and communicate with the world in entirely new ways [37]. While there is a fantastic way to connect machines and humans using extended reality technology, incorporating small electronic components like transistors and memristors into these fabrics could enhance creativity, expand imagination, and foster the generation of new ideas [38]. "By integrating fiber electronics into textiles, we can create wearable sensors, displays, and various other electronic components, offering enhanced functionality while maintaining comfort and versatility." In conclusion, the integration of smart textiles and advanced materials into our daily lives has the potential to revolutionize how we interact with the digital world. These textiles, often referred to as the "second layer" of human skin, have the remarkable capability to compute, communicate, sense, and display information, making them versatile tools for various applications. By processing information like computers, these textiles can rapidly respond to changes in our environment and body, eliminating the need for complex data-forwarding processes. While the primary focus of these innovations has been on enhancing human-machine interactions and creativity, they also have significant implications for energy devices. The ability to incorporate small electronic components into textiles creates new possibilities for energy-efficient wearable technologies, energy harvesting from clothing, and improved interfaces for controlling energy devices. These developments not only provide comfort and flexibility but also contribute to the advancement of energy-related applications, bridging the gap between smart textiles and the world of energy devices.

23.7   MEDICINE AND BIOLOGICAL SYSTEMS

In recent years, there has been a growing demand for medical gadgets that can be worn seamlessly in everyday routines. These medical devices offer immediate, personalized healthcare solutions that are cost-effective and provide real-time advice for maintaining a healthy life. To fulfill this crucial role, textiles have emerged as the preferred choice. Thanks to advancements in organic electronics, smart textiles can now collect valuable information. This information can be used to monitor various aspects such as calories burned, posture, sleep quality, stress levels, and cognitive abilities related to thinking, understanding, and learning. Subsequently, this data is transmitted back to individuals through their clothing, assisting them in maintaining overall health, well-being, and even preventing diseases [39]. Semiconducting polymer-based wearable devices possess a unique ability that sets them apart. They are capable of performing electrophysiological sensing and photodetection in a manner that is compatible with living organisms. Furthermore, they can easily detect physiological and chemical changes, which can be quite challenging when using inorganic semiconductors and metals. In this section, we explore how semiconducting polymer-based textiles and fibers are utilized to detect health-related information and aid in medical recovery through wearable technologies. Biopotentials, which refer to the electrical signals in the human body generated by chemical processes at the cellular level, are characterized by their low voltage. The examination and study of these biopotentials are of utmost importance for biological research and biomedical monitoring [40]. An example would be an electromyogram (EMG), which measures muscle activity, and an electrocardiogram (ECG), which records the heart's electrical activity. Both are commonly used to detect issues with the heart's rhythm and muscle problems [41]. Electrocorticography (ECoG) and the electroencephalogram (EEG) are also unique electrical signals in the body. These signals are used to create brain-computer interfaces, which are technologies that enable communication between external devices and the brain. They spread across the skin, since the body conducts the signals, allowing them to be recorded [42]. The complicated link between the neurons that form brain networks (nervous system), is responsible for how humans move, act, and see things [43]. To create computers that can match the speed of the human brain, it's essential to continuously adjust and monitor the brain signals produced. By doing so, we can gain a better understanding of the brain and develop computers that function similarly to the human brain [44]. Furthermore, it is believed that understanding and collecting signals from both the inner and outer parts of the brain can be beneficial in addressing certain diseases, including Parkinson's disease, paralysis, and muscle stiffness [45]. Therefore, clear and detailed methods are necessary to accurately capture signals in terms of both time and location. In other words, advanced electrodes are required

[46]. In summary, recent years have witnessed a growing demand for wearable medical devices that seamlessly integrate into daily life, providing cost-effective, personalized healthcare solutions with real-time advice. Textiles have emerged as a preferred platform for these devices, enabled by advancements in organic electronics. Smart textiles can now gather valuable health-related data, including metrics like calories burned, posture, sleep quality, stress levels, and cognitive abilities. This data is transmitted through clothing, assisting individuals in maintaining overall health and preventing diseases. Semiconducting polymer-based wearable devices possess unique capabilities, excelling in electrophysiological sensing and photodetection compatible with living organisms. They outperform inorganic counterparts in detecting physiological and chemical changes. These innovations in wearable technology and advanced sensors hold promise for energy devices. By integrating health and activity data, energy devices can adapt and optimize energy consumption patterns. This can lead to more efficient and personalized energy management, contributing to a sustainable and efficient energy ecosystem.

23.8 ENVIRONMENTAL REMEDIATION

Recently, the world population has been growing very rapidly, and industries have experienced significant development, resulting in environmental pollution and an energy crisis. This has not only impacted human health, potentially putting it in jeopardy, but has also raised concerns about the sustainability of the environment. Consequently, people have shifted their focus towards inventing technologies aimed at removing pollutants and protecting the environment. Fujishima and his team were the first to introduce the concept of heterogeneous photocatalysis [47]. It was introduced in 1972 when utilizing a photoelectrochemical cell with a single crystal TiO_2 electrode to attempt to split water (H_2O). Heterogeneous photocatalysis is an advanced oxidation process (AOP) and has gradually laid the foundation for environmental remediation [48]. There are several advantages to heterogeneous photocatalytic reactions. Firstly, they utilize sustainable sunlight. Secondly, they operate under less harsh conditions compared to older methods such as membrane separation, adsorption, bioprocesses, and chemical sedimentation. Lastly, they are highly efficient and require minimal energy. To protect the environment and conserve energy, it is essential to use solar energy for pollutant remediation. Due to these advantages, numerous studies on photocatalytic pollutant treatment have been conducted, with many research papers published on this topic in the last 20 years. For instance, in 2020 alone, over 9000 papers on photocatalytic pollutant degradation were published. It's worth noting that in addition to treating pollutants in water, photocatalysis also addresses pollutants in the atmosphere and soil. These pollutants encompass heavy metals, plastic waste, NO_x, persistent organic pollutants (POPs), sulfur-containing fossil fuels, and harmful bacteria. In recent years, researchers have developed various semiconductors, including silver-based or bismuth-based semiconductors, metal oxide semiconductors, metal-free semiconductors, semiconductors made from metal sulfides, unique polymers, and metal-organic frameworks (MOFs). These materials are specifically designed to target different types of environmental pollution. Researchers have also explored ways to enhance the performance of these materials, such as repairing defects, creating heterojunction systems, and incorporating specific substances. Furthermore, researchers have proposed the use of environmental photocatalysis as a process to remediate a wide range of pollutants [49]. However, several challenges hinder the widespread environmental application of this technology. These challenges include inefficient utilization of solar energy, semiconductor costs, and rapid energy loss in photogenerated electron-hole pairs ($e–h^+$). The thermodynamics and kinetics of photocatalysis remain insufficiently understood, a topic often overlooked in previous reviews. Therefore, it is crucial to explore ways to enhance semiconductor materials for environmental remediation, focusing on changing the materials in use. The review begins by discussing the fundamentals of environmental photocatalysis, including its underlying principles, operational mechanisms, associated challenges, and the role of reactive oxidation species (ROS). It further examines various semiconductor materials and techniques that can enhance their effectiveness in environmental cleanup. Ultimately, the review proposes different perspectives and ideas for improving the utilization of these heterogeneous photocatalysts for environmental purposes, with the hope of providing valuable guidance. It's important to note that the efficiency of photocatalysts can be influenced by various factors beyond the migration of electrons and holes (e–h+). These factors include the reactivity of ROS, the specific environmental conditions, the dimensional structures and shapes of the semiconductors used, the migration of other carriers and electrons, the band gap structure, and different kinetic and thermodynamic processes [50]. Environmental pollutants are typically classified into two main groups. The first group comprises pollutants found in water, such as heavy metals, persistent organic pollutants, disease-causing microorganisms, and fossil fuels containing sulfur. The second group consists of pollutants found in the air, including sulfur dioxide, nitrogen oxides, ammonia, and volatile organic compounds. However, disease-causing microorganisms can also be present in various locations, such as laboratory surfaces and operating tables. In addition to these categories, the most common type of solid waste, namely plastic waste, contains microplastics that are often found in bodies of water and the natural environment. These microplastics can be considered a distinct type of pollutant. Consequently, scientists are actively researching ways to efficiently utilize heterogeneous photocatalysis to address pollution-related issues [51]. A lot of research has been conducted in recent decades to explore eco-friendly techniques for removing natural pollutants. One promising method is heterogeneous photocatalysis, which has the potential to harness a significant amount of solar energy. This method relies on photocatalytic reactions triggered by light to break down and remove various pollutants found in nature, including toxic substances, harmful gases, and heavy metals. What's particularly intriguing is that the photocatalysis process can occur in both gases and liquids, making environmental cleanup more straightforward with semiconductor-based technology [52]. The recent growth in global population and industrial development has led to environmental pollution and an energy crisis, raising concerns about both human health and environmental sustainability. In response, innovative technologies like heterogeneous photocatalysis have been developed to address these challenges. Heterogeneous photocatalysis, introduced by Fujishima and his team in 1972, harnesses solar energy for environmental remediation. This technology offers several advantages, such as utilizing sustainable sunlight, operating under less harsh conditions compared to older methods, and being highly efficient while conserving energy. It has been extensively studied, with thousands of research papers published in recent years, highlighting its potential for addressing various pollutants in water, air, and soil. Researchers have developed a range of semiconductor materials tailored to specific environmental pollution challenges, and strategies for enhancing their performance have been explored. However, challenges remain, including inefficient solar energy utilization, semiconductor costs, and rapid energy loss in photocatalytic reactions. The understanding of the thermodynamics and kinetics of photocatalysis is still evolving. To overcome these challenges and advance the application of photocatalysis, changing the materials used is a key focus. Environmental pollutants are typically categorized into waterborne and airborne pollutants, including heavy metals, organic pollutants, microorganisms, and plastic waste. Heterogeneous photocatalysis offers promise for addressing these pollutants efficiently, breaking them down through photocatalytic reactions triggered by light. This versatile technology can be applied in both gas and liquid phases, simplifying environmental cleanup efforts. In summary, heterogeneous photocatalysis represents a promising solution to environmental pollution and sustainability challenges. By harnessing solar energy and utilizing semiconductor materials, it offers a path toward a cleaner and more sustainable future, aligning with the broader goals of energy device development and environmental protection.

23.9 CONCLUSION

In summary, the influence of semiconductor materials and associated technologies reaches far and wide, touching upon numerous critical fields and sectors. Their impact can be observed across energy storage, electronic devices, sensors, Internet of Things (IoT) applications, medicine and biology, and environmental remediation, heralding a new era of possibilities and improvements. Starting with energy storage, semiconductor materials have revolutionized this sector by enhancing the efficiency and performance of energy storage systems. Lithium-ion batteries, for example, have benefited immensely from advances in semiconductor technology, resulting in longer-lasting, more reliable power sources. This has not only improved the reliability of portable electronic devices but also has significant implications for the adoption of renewable energy sources, contributing to a more sustainable energy landscape. In the realm of electronic devices, semiconductor materials have been instrumental in miniaturization, allowing for the creation of smaller, more powerful, and energy-efficient electronic components. This has led to the proliferation of smartphones, tablets, and other compact devices that have become an integral part of our daily lives. Semiconductor advancements continue to drive innovation in this space, promising even more compact and efficient electronics in the future. Sensors, a crucial component of IoT applications, have also benefited from semiconductor materials. These materials have enabled the development of highly sensitive and responsive sensors, facilitating the seamless connection of devices and data sharing in IoT ecosystems. This, in turn, has led to smart homes, cities, and industries, with applications ranging from remote monitoring of health conditions to optimizing energy consumption. In the field of medicine and biology, semiconductor materials have played a pivotal role in the development of cutting-edge diagnostic and therapeutic tools. From semiconductor-based imaging technologies like MRI and CT scans to the miniaturization of medical devices for remote patient monitoring, these materials have improved healthcare outcomes and patient experiences. Additionally, advancements in semiconductor-based lab-on-a-chip technologies have accelerated research in genomics, proteomics, and drug discovery, offering new avenues for understanding and treating diseases. Environmental remediation has also benefited from semiconductor materials. These materials have been harnessed to create efficient and cost-effective solutions for purifying water, air, and soil. Semiconductor-based photocatalysis, for instance, can break down pollutants and contaminants, contributing to cleaner and healthier environments. As we look to the future, the relentless pursuit of semiconductor technology advancement promises to bring even more transformative innovations. These innovations will address the challenges of our rapidly evolving world, from powering electric vehicles with longer-lasting batteries to creating wearables that monitor health in real-time. Furthermore, semiconductor materials will continue to contribute to a brighter and more sustainable future by enabling renewable energy solutions, improving healthcare outcomes, and protecting our environment. Their pervasive influence across various sectors underscores their importance in shaping the modern world and driving progress toward a more technologically advanced and sustainable global society.

REFERENCES

1. V. K. Jain, Semiconductor Devices, in Solid State Physics, Springer International Publishing, 2022: pp. 505–533. doi: 10.1007/978-3-030-96017-9_17.

2. N. Nasajpour-Esfahani, A critical review on intrinsic conducting polymers and their applications, J Ind Eng Chem. 125 (2023) 14–37. https://doi.org/10.1016/j.jiec.2023.05.013

3. M. Grätzel, Photoelectrochemical cells, Nature. 414 (2001) 338–344. https://doi.org/10.1038/35104607

4. Y.J. Hsu, S.Y. Lu, Low temperature growth and dimension- dependent photoluminescence efficiency of semiconductor nanowires, Appl Phys A Mater Sci Process. 81 (2005) 573–578. https://doi.org/10.1007/s00339-004-2714-y

5. C.J. Barrelet, Y. Wu, D.C. Bell, C.M. Lieber, Synthesis of CdS and ZnS nanowires using single-source molecular precursors, J Am Chem Soc. 125 (2003) 11498–11499. https://doi.org/10.1021/ja036990g

6. A.J. Bard, A.B. Bocarsly, F.R.F. Fan, E.G. Walton, M.S. Wrighton, The concept of Fermi level pinning at semiconductor/liquid junctions. Consequences for energy conversion efficiency and selection of useful solution redox couples in solar devices, J Am Chem Soc. 102(11) (1980) 3671–3677. doi: 10.1021/ja00531a001

7. M.J. Penn, R. Baer, D. Walter, Z.Y. Zhao, Y. Wu, H.B. Cao, Electrochemical Capacitors: Challenges and Opportunities for Real-World Applications, John R. Miller and Andrew Burke Electrochem. Soc. Interface 17 (2008) 53. doi: 10.1149/2.F08081IF.

8. H. Helmholtz, Studien über electrische Grenzschichten, Annalen Der Physik Und Chemie. 243 (1879) 337–382. https://doi.org/10.1002/andp.18792430702

9. H. Morkoç, S. Strite, G.B. Gao, M.E. Lin, B. Sverdlov, M. Burns, Large-band-gap SiC, III-V nitride, and II-VI ZnSe-based semiconductor device technologies, J Appl Phys. 76 (1994) 1363–1398. https://doi.org/10.1063/1.358463

10. Z. Liu, Y. Cai, R. Tu, Q. Xu, M. Hu, C. Wang, Q. Sun, B.-W. Li, S. Zhang, C. Wang, T. Goto, L. Zhang, Laser CVD growth of graphene/SiC/Si nano-matrix heterostructure with improved electrochemical capacitance and cycle stability, Carbon N Y. 175 (2021) 377–386. https://doi.org/10.1016/j.carbon.2021.01.004

11. N. Yang, H. Zhuang, R. Hoffmann, W. Smirnov, J. Hees, X. Jiang, C.E. Nebel, Electrochemistry of nanocrystalline 3C silicon carbide films, Chem Eur J. 18 (2012) 6514–6519. https://doi.org/10.1002/chem.201103765

12. J.P. Alper, M.S. Kim, M. Vincent, B. Hsia, V. Radmilovic, C. Carraro, R. Maboudian, Silicon carbide nanowires as highly robust electrodes for micro-supercapacitors, J Power Sources. 230 (2013) 298–302. https://doi.org/10.1016/j.jpowsour.2012.12.085

13. J.P. Alper, M. Vincent, C. Carraro, R. Maboudian, Silicon carbide coated silicon nanowires as robust electrode material for aqueous micro-supercapacitor, Appl Phys Lett. 100(16) (2012) 163901. doi: 10.1063/1.4704187

14. L. Gu, Y. Wang, Y. Fang, R. Lu, J. Sha, Performance characteristics of supercapacitor electrodes made of silicon carbide nanowires grown on carbon fabric, J Power Sources. 243 (2013) 648–653. doi: 10.1016/j.jpowsour.2013.06.050

15. J. Chen, J. Zhang, M. Wang, L. Gao, Y. Li, SiC nanowire film grown on the surface of graphite paper and its electrochemical performance, J Alloys Compd. 605 (2014) 168–172. doi: 10.1016/j.jallcom.2014.03.155

16. J. Liang, J. Lu, P. Gao, W. Guo, H. Xiao, The crystallization and growth of SiC nanowires converted from self-assembly Si nanorods on carbon fabric and their electrochemical capacitance property, J Alloys Compd. 827 (2020) 154168. doi: 10.1016/j.jallcom.2020.154168

17. A. Sanger, A. Kumar, A. Kumar, P.K. Jain, Y.K. Mishra, R. Chandra, Silicon carbide nanocauliflowers for symmetric supercapacitor devices, Ind Eng Chem Res. 55(35) (2016) 9452–9458. doi: 10.1021/acs.iecr.6b02243

18. Z. Liu et al., Laser CVD growth of graphene/SiC/Si nano-matrix heterostructure with improved electrochemical capacitance and cycle stability, Carbon N Y. 175 (2021) 377–386. doi: 10.1016/j.carbon.2021.01.004

19. K. Kaasbjerg, K.S. Thygesen, K.W. Jacobsen, Phonon-limited mobility in n-type single-layer MoS2 from first principles, Phys Rev B Condens Matter Mater Phys. 85 (2012). https://doi.org/10.1103/PhysRevB.85.115317

20. X. Gonze, C. Lee, Dynamical matrices, born effective charges, dielectric permittivity tensors, and interatomic force constants from density-functional perturbation theory, Phys Rev B. 55 (1997) 10355–10368. https://doi.org/10.1103/PhysRevB.55.10355

21. R. Fivaz, E. Mooser, Mobility of charge carriers in semiconducting layer structures, Phys Rev. 163(3) (1967) 743–755. doi: 10.1103/PhysRev.163.743

22. H. Hasegawa, T. Sawada, On the electrical properties of compound semiconductor interfaces in metal/insulator/ semiconductor structures and the possible origin of interface states, Thin Solid Films.103 (1983) 119–140.

23. W. Bialek, Physical limits to sensation and perception, Annu Rev Biophys Chem. 16 (1987) 455–478. https://doi.org/10.1146/annurev.bb.16.060187.002323

24. W. Ouyang, F. Teng, J. He, X. Fang, Enhancing the photoelectric performance of photodetectors based on metal oxide semiconductors by charge-carrier engineering, Adv Funct Mater. 29 (2019). https://doi.org/10.1002/adfm.201807672

25. Y. Zhang, W. Xu, X. Xu, J. Cai, W. Yang, X. Fang, Self-powered dual-color UV–green photodetectors based on SnO2 millimeter wire and microwires/CsPbBr3 particle heterojunctions, J Phys Chem Lett. 10 (2019) 836–841. https://doi.org/10.1021/acs.jpclett.9b00154

26. D. Bonardo, N.L.W. Septiani, F. Amri, S. Humaidi, B. Yuliarto, Review—Recent development of WO3 for toxic gas sensors applications, J Electrochem Soc. 168 (2021) 107502. https://doi.org/10.1149/1945-7111/ac0172

27. G.E. Collins, L.J. Buckley, Conductive polymer-coated fabrics for chemical sensing, Synth Met. 78 (1996) 93–101. https://doi.org/10.1016/0379-6779(96)80108-1

28. X.-R. Zeng, T.-M. Ko, Structures and properties of chemically reduced polyanilines, Polymer (Guildf). 39(5) (1998) 1187–1195. doi: 10.1016/S0032-3861(97)00381-9

29. Y. Zhang, J.J. Kim, D. Chen, H.L. Tuller, G.C. Rutledge, Electrospun polyaniline fibers as highly sensitive room temperature chemiresistive sensors for ammonia and nitrogen dioxide gases, Adv Funct Mater. 24(25) (2014) 4005–4014. doi: 10.1002/adfm.201400185

30. D. Li, J. Huang, R.B. Kaner, Polyaniline nanofibers: A unique polymer nanostructure for versatile applications, Acc Chem Res. 42(1) (2009) 135–145. doi: 10.1021/ar800080n

31. S.-J. Park, J.-Y. Jeon, T.-J. Ha, Wearable humidity sensors based on bar-printed poly(ionic liquid) for real-time humidity monitoring systems, Sens Actuators B Chem. 354 (2022) 131248. doi: 10.1016/j.snb.2021.131248

32. Y. Lu, G. Yang, Y. Shen, H. Yang, K. Xu, Multifunctional flexible humidity sensor systems towards noncontact wearable electronics, Nanomicro Lett. 14(1) (2022) 150. doi: 10.1007/s40820-022-00895-5

33. A. Tripathy et al., Design and development for capacitive humidity sensor applications of lead-free Ca, Mg, Fe, Ti-oxides-based electro-ceramics with improved sensing properties via physisorption, Sensors. 16(7) (2016) 1135. doi: 10.3390/s16071135

34. G. Chen, X. Xiao, X. Zhao, T. Tat, M. Bick, J. Chen, Electronic textiles for wearable point-of-care systems, Chem Rev. 122 (2022) 3259–3291. https://doi.org/10.1021/acs.chemrev.1c00502

35. T. Fernández-Caramés, P. Fraga-Lamas, Towards the internet-of-smart-clothing: A review on IoT wearables and garments for creating intelligent connected e-textiles, Electronics (Basel). 7 (2018) 405. https://doi.org/10.3390/electronics7120405

36. A. Libanori, G. Chen, X. Zhao, Y. Zhou, J. Chen, Smart textiles for personalized healthcare, Nat Electron. 5 (2022) 142–156. https://doi.org/10.1038/s41928-022-00723-z

37. A. Satharasinghe, T. Hughes-Riley, T. Dias, Photodiodes embedded within electronic textiles, Sci Rep. 8 (2018) 16205. https://doi.org/10.1038/s41598-018-34483-8

38. S. Liu, K. Ma, B. Yang, H. Li, X. Tao, Textile electronics for VR/AR applications, Adv Funct Mater. 31 (2021). https://doi.org/10.1002/adfm.202007254

39. Y. Yamada, Textile-integrated polymer optical fibers for healthcare and medical applications, Biomed Phys Eng Express. 6 (2020) 062001. https://doi.org/10.1088/2057-1976/abbf5f

40. J. Oreggioni, A.A. Caputi, F. Silveira, Biopotential Monitoring, in: Encyclopedia of Biomedical Engineering, Elsevier, 2019: pp. 296–304. https://doi.org/10.1016/B978-0-12-801238-3.64161-2

41. W.G. Stevenson, K. Soejima, Recording techniques for clinical electrophysiology, J Cardiovasc Electrophysiol. 16 (2005) 1017–1022. https://doi.org/10.1111/j.1540-8167.2005.50155.x

42. T. Buchner, On the physical nature of biopotentials, their propagation and measurement, Physica A Stat Mech Appl. 525 (2019) 85–95. https://doi.org/10.1016/j.physa.2019.03.056

43. I.B. Dimov, M. Moser, G.G. Malliaras, I. McCulloch, Semiconducting polymers for neural applications, Chem Rev. 122 (2022) 4356–4396. https://doi.org/10.1021/acs.chemrev.1c00685

44. R. Bawa, Advances in Clinical Immunology, Medical Microbiology, COVID-19, and Big Data, Jenny Stanford Publishing, New York, 2021. https://doi.org/10.1201/9781003180432

45. P. Fattahi, G. Yang, G. Kim, M.R. Abidian, A review of organic and inorganic biomaterials for neural interfaces, Adv Mater. 26 (2014) 1846–1885. https://doi.org/10.1002/adma.201304496

46. G. Buzsáki, C.A. Anastassiou, C. Koch, The origin of extracellular fields and currents-EEG, ECoG, LFP and spikes, Nat Rev Neurosci. 13 (2012) 407–420. https://doi.org/10.1038/nrn3241

47. A. Fujishima, K. Honda, Electrochemical photolysis of water at a semiconductor electrode, Nature. 238 (1972) 37–38. https://doi.org/10.1038/238037a0

48. B.P. Mishra, K. Parida, Orienting Z scheme charge transfer in graphitic carbon nitride-based systems for photocatalytic energy and environmental applications, J Mater Chem A Mater. 9 (2021) 10039–10080. https://doi.org/10.1039/D1TA00704A

49. Y. Ren, D. Zeng, W.-J. Ong, Interfacial engineering of graphitic carbon nitride (g-C3N4)-based metal sulfide heterojunction photocatalysts for energy conversion: A review, Chinese J Catal. 40 (2019) 289–319. https://doi.org/10.1016/S1872-2067(19)63293-6

50. Q. Tian et al., Synergetic effects of the interfacial dyadic structure on the interfacial charge transfer between surface-complex and TiO_2, Appl Surf Sci. 496 (2019) 143711. doi: 10.1016/j.apsusc.2019.143711

51. X. Jiao et al., Photocatalytic conversion of waste plastics into C_2 fuels under simulated natural environment conditions, Angewandte Chemie International Edition, 59(36) (2020) 15497–15501. doi: 10.1002/anie.201915766

52. C. Gao, J. Low, R. Long, T. Kong, J. Zhu, Y. Xiong, Heterogeneous single-atom photocatalysts: Fundamentals and applications, Chem Rev. 120(21) (2020) 12175–12216. doi: 10.1021/acs.chemrev.9b00840

24 Role of Semiconductors in Future Flexible Batteries

Joseph C. M., Vinuth Raj T. N., and Priya A. Hoskeri

24.1 INTRODUCTION

A battery is an electrochemical device designed to store energy and deliver electrical power to electronic devices. The appeal of wearable electronics is enhanced by the mechanical flexibility, rollability, and stretchability of these devices. In order to facilitate the seamless integration and self-sufficiency of systems, it is imperative that the devices are equipped with batteries possessing comparable physical attributes. The current redesign efforts are focused on the bulk battery structure, which includes the active materials, electrodes, electrolytes, separators, and packaging. The current need necessitates the adoption of an interdisciplinary approach, wherein chemists, materials scientists, and engineers collaborate closely to address the challenges at hand. Based on the findings of the literature review, it has been identified that there exist two potential methodologies for the development of flexible batteries. The first technique involves substituting inherently rigid materials with pliable and malleable compounds. The second way entails transforming rigid materials into flexible structures [1–5]. In both the scenarios, it is imperative to ensure the consistent functioning of the connections between the components of the battery, even when subjected to repeated deformation. Lithium ion batteries (LIBs) are distinguished in the current state of the art of portable energy storage due to their exceptional attributes, including high energy density, power density, and longevity. However, in the case of flexible electronics, the electrode designs in LIBs experience significant degradation in performance when subjected to physical bending [6]. Flexible LIBs have been successfully developed by meticulous use of structural engineering techniques. In all the futuristic flexible devices, the organic semiconductor is one of the inevitable components for the fabrication of devices with a storing facility.

24.2 SEMICONDUCTOR MATERIALS IN BATTERY TECHNOLOGY

Semiconductors play a significant role in the realm of batteries, notably in advanced battery technologies like LIBs. While semiconductors do not function as a primary energy storage component in batteries, they do play a vital role in improving the performance, safety, and efficiency of various aspects of battery technology. Semiconductors are involved in several mechanisms that contribute to the operation of the batteries. Battery management systems (BMSs) rely on semiconductors that are crucial elements for the purpose of overseeing and regulating the condition, charge level, health status, and general efficacy of batteries. BMSs employ mainly semiconductor devices for the purpose of voltage, current, and temperature measurements. Additionally, these systems regulate the charging and discharging procedures to guarantee the battery's safe and efficient functioning. Semiconductor materials are also employed in the implementation of protection circuits in batteries, with the purpose of averting instances of overcharging, over-discharging, and short circuiting. The functionality of these circuits is dependent on the utilization of semiconductor components such as voltage regulators, voltage detectors, and current sensors. These components serve the purpose of monitoring and regulating the voltage and current levels of the battery, thereby contributing to the overall safety of the battery system.

DOI: 10.1201/9781003450146-24

The utilization of semiconductors in power electronics plays a crucial role in the conversion and inversion of electrical energy in sophisticated battery applications, particularly in hybrid and electric vehicles. This process facilitates the efficient transfer of electrical energy between the battery and the propulsion system. Insulated gate bipolar transistors and metal oxide semiconductor field effect transistors are commonly employed in the realm of energy conversion for better efficiency. Semiconductors play a crucial role in charging and discharging control circuits by effectively managing the current flow throughout the process of charging and discharging a battery. This practice aids in sustaining ideal charging rates and mitigating potential harm caused by excessive charging or discharging. The investigation of specific semiconductor materials for energy storage purposes, such as supercapacitors, is an area of interest in the field of energy storage and conversion. Supercapacitors are known for their high power density and ability to rapidly charge and discharge energy. These have the potential to enhance conventional battery technologies in specific applications.

Semiconductors play a crucial role in the thermal management of batteries, as they are employed to effectively monitor and regulate temperature levels. The use of temperature sensors and control circuits serve the purpose of mitigating the occurrence of overheating, a phenomenon that can result in the deterioration of battery performance and pose potential safety risks. Semiconductors have a significant impact on enhancing the performance, safety, and efficiency of batteries, hence facilitating the advancement of important battery technologies essential for a wide range of applications, including portable devices, electric cars, and renewable energy storage systems.

24.3   ORGANIC SEMICONDUCTING MATERIALS IN BATTERY TECHNOLOGY

The present chapter focuses on the analysis of contemporary and prospective flex circuit batteries, which are expected to contribute significantly to the advancement of energy conservation. The issues associated with enhancing conductivity, cycle life, charge and discharge rates, and improving power ratings are also matters of interest. Electronic devices have been fabricated using organic semiconductors derived from polymers, small molecules, aromatic compounds, and phthalocyanines. The enhancement of the electrical, thermal, mechanical, and optical properties of these materials can be achieved through the use of novel functional organic groups. These materials offer several benefits, including cost effectiveness and simplified preparation methods for use in various devices. Different coating processes such as spin coating and dip coating are employed to fabricate films of organic semiconductors. The simplified preparation procedure of these materials facilitates cost effective device fabrication. In order to accommodate the advancements in new and contemporary technology, there is a pressing need for the development of novel electric energy storage devices, specifically for low-cost batteries. Organic semiconductors have shown great potential as materials for organic batteries. Certain metal phthalocyanines (MPCs) and their doped combinations have exhibited encouraging results in the realm of micropower energy applications. The incorporation of embedded MPCs into nanostructures holds significant potential for researchers to advance in the field of flexible electronics by enabling energy conservation.

In the studies done by Oh et al. [7], the use of carbon nanotube-decorated α-iron oxide particles acting as the anode and lithium iron phosphate acting as the cathode was introduced for the first time, showing potential use as freestanding anodes in flexible lithium polymer batteries. The utilization of fibrous mats composed of these materials as both electrodes and a gel polymer electrolyte enabled the achievement of reduced stress during the bending of the electrochemical pouch cell. The incorporation of bendability, demonstrated via 5000 cycles, has been implemented inside a widely recognized LIB chemical system utilizing lithium cobalt oxide ($LiCoO_2$) and graphite. Zinc ion batteries represent a promising technological advancement due to their comparatively high capacity. The active materials utilized in this particular battery configuration include several advantageous characteristics. These include affordability, plentiful availability, enhanced safety measures, environmental friendliness, and long-term sustainability [8, 9]. According to the findings of Zamarayeva et al. [10], the utilization of a zinc metal anode in conjunction with a manganese oxide cathode, as well as the incorporation of flexible binders such polyvinyl alcohol and polyacrylic acid, has demonstrated the ability to maintain performance even after undergoing bending. Furthermore, interesting results have been reported in recent studies on flexible energy storage with Li/S batteries [11].

LIBs commonly consist of four essential components: a cathode, an anode, an electrolyte, and a separator. The cathode materials encompass lithium iron phosphate ($LiFePO_4$), lithium manganese oxide ($LiMn_2O_4$), $LiCoO_2$, lithium nickel cobalt manganese oxide (often referred to as NMC) ($LiNiMnCoO_2$), lithium nickel cobalt aluminum oxide ($LiNiCoAlO_2$), and lithium titanate. Graphite is the predominant choice for anode materials in the majority of cases. The materials commonly employed as separators in this context are typically microporous films made of polypropylene or polyethylene. The electrolytes commonly employed consist of lithium hexafluorophosphate ($LiPF_6$) dissolved in ethylene carbonate (EC). However, efforts are being made to enhance safety and capacity retention by modifying the electrolytes using additives and developing novel electrolyte materials. Silanes are employed for the purpose of enhancing the stability of the electrolyte. Silyl phosphates, such as tris (trimethylsilyl) phosphate, have been employed as supplementary substances in the ethylene carbonate electrolyte to enhance stability and maintain capacity. Improved electrolytes have been developed in the form of polyether functionalized silanes [12]. Electrolyte additives, specifically aminoalkylsilanes with oligo (ethylene oxide) unit, have been developed for the purpose of enhancing stability and capacity retention [13]. In recent years, several advancements have been made in the field of battery technology, leading to the emergence of various novel batteries. These include conventional batteries, dual ion batteries, mixed ion batteries, aluminum ion/chloroaluminate batteries, and mixed ion batteries [14].

Flexible batteries are commonly designed with planar structures that exhibit many forms such as paper, thin film, and spinel-like structures. The assembly process involves the layer-by-layer arrangement of battery electrode layers, which encompass current collectors and the supported cathode/anode materials, as well as electrolytes, separators, and packaging. Different types of batteries include paper-based (flexible), graphene, laminated thin-film-structured, sodium ion, belt type, lithium, and organic batteries like pyrene-4,5,9,10 and tetraone/Zn batteries. They have exhibited notable characteristics, including high flexibility, lightweight construction, exceptional chemical and electrochemical stability, and a large surface area [15–17]. In recent years, there has been a growing interest in research focused on the utilization of MPCs as electrodes in flexible batteries. The deactivation of MPCs has been observed to exhibit catalytic activity in lithium batteries. Specifically, the discharge energy of $Li/SOCl_2$ batteries, when catalyzed by MPCs, surpassed the performance of the same battery in the absence of these compounds. Therefore, MPCs represent a highly promising category of catalysts, exhibiting certain advantages in comparison to metal and metal oxide catalysts due to their cost effectiveness. The impact of the central metal atom in Li/$SOCl_2$ batteries was investigated through a comprehensive examination of MPCs, where M represents Mn^{2+}, Fe^{2+}, Co^{2+}, Ni2+, and Cu^{2+} [18].

Pertaining to organic semiconductors and compounds, scholars have identified several notable benefits, including a highly theoretical specific capacity, ample availability of raw materials, environmentally favorable characteristics, and robust structural design capabilities. Several studies have indicated that the incorporation of nanostructures such as graphene into organic carbonyl electrode materials leads to a notable increase in surface area, resulting in a greater number of active sites and enhanced battery conductivity. In a study conducted by Mengqian Xu et al. [19], it was observed that the combination of 3,4, 9,10 Perylene Tetra Carboxylic Dianhydride (PTCDA) and sodium salt in an aqueous solution, followed by hydrolysis and ethanol anti-solvent treatment, resulted in the formation metal substituted precursor. The subsequent addition of graphene to this mixture exhibited remarkable electrochemical energy storage properties. This enhancement can be attributed to the increased specific surface area, improved dispersion, greater exposure of the active site, and enhanced electrical conductivity of the composite material. These findings have significant implications for the advancement of high-performance organic electrode materials in LIBs. There is a greater abundance of cathode materials available for LIBs as compared to the anode materials. The researchers are currently investigating the anode materials for future LIBs. Due to their substantial theoretical capacity, large reserves of raw materials and distinctive physicochemical qualities, compounds including zinc (Zn) and manganese (Mn) with selenium (Se) have emerged as prominent anode materials for LIBs. In a study conducted by Jiaoyu Xiao et al. [20], Zn-Mn PTCDA was successfully synthesized using a straightforward hydrothermal reaction. The synthesis of two-dimensional (2D) elliptical leaf-shaped $Zn_{0.697}Mn_{0.303}Se/C$ composites was achieved through direct selenization, using Zn-Mn PTCDA as the precursor. This composite exhibited a significantly high specific surface area of 213.9 m^2g^{-1}. After 110 cycles at a current density of 100 mAg^{-1}, the reversible capacity of the system is measured to be 1005.14 $mAhg^{-1}$. Following 1000 cycles under a high current density of 1 Ag^{-1}, the observed capacity remains favorable at 653.79 $mAhg^{-1}$. The combination of Zn-Mn PTCDA shows potential for further development in the realm of LIBs [20].

A novel organic battery was developed for micro-power energy applications by incorporating a PTCDA organic semiconductor with varying compositions of black carbon (BC) and copper chloride as the electrolyte. The addition of different ratios of BC was performed, and the obtained results were thoroughly examined to identify the optimal composites for PTCDA-BC batteries. The voltage of the battery exhibits an increase from an initial value of 0.17 V to a final value of 0.45 V. Concurrently, the short circuit current, denoted as Isc, undergoes a transition from 12 mA to 25 mA. The battery, which had a PTCDA:BC ratio of 80:20, demonstrated the highest maximum power output, denoted as Pmax, measuring 3.08 mW. The results obtained from the study suggest that batteries utilizing PTCDA:BC composites have potential for application in micro-power energy systems [21].

The ternary composite (PTCDA/CNT/MPc) was utilized as the positive electrode material by Fan et al. [22], while sodium metal served as the negative electrode. The findings indicate that the electrochemical performance of the PTCDA/CNT/MPc composite surpasses that of the pure PTCDA. One notable finding was the congruence between the experimental and theoretical values of the specific capacity of the batteries. At a current density of 10 Ag^{-1}, the PTCDA/CNT/MPc electrode material demonstrates a remarkable capacity of 90.1 $mAhg^{-1}$. The increased conductivity of the CNT/MPc composite carbon material contributes to the enhanced conductivity of PTCDA. Additionally, it expands the transport pathway for ions, hence promoting the adsorption of sodium ions into the composite positive electrode. Hence, the composite electrode demonstrates a notable enhancement in reversible capacity and cycle life. This improvement is attributed to its exceptional intercalation and deintercalation performance for sodium ions, as well as its charge discharge rate performance [22].

24.4 ORGANIC ELECTROCHROMIC MATERIALS

Energy storage devices with a smart function of changing color obtained by incorporating electrochromic materials into battery or supercapacitors electrodes were discussed in an earlier review [23]. In this work, the working principles of supercapacitors, batteries, and electrochromic (EC) devices with a focus on the material candidates for electrochromic energy storages were discussed. The challenges of the integration of EC energy systems for simultaneous realization of electrochromism and energy storage were also discussed [23].

In the literature survey on metal oxides and inorganic EC materials, the achievable EC color is limited to bluish and brownish [24, 25] during the oxidation and reduction phases. Exploring the organic EC semiconductors, which have shown multiple coloration phases for variation in the input voltage in the range of 1–3 V, are gaining importance because of their lower basing voltages and variety of coloration during the forward and reverse bias voltages. In this direction, few MPC thin films have been explored to understand the electrochromic properties. MPCs have a good chemical stability and a visible color change during the oxidation or reduction due to their rich colors. Further, a detailed literature survey indicated that the annealed EC thin films had better stability compared to the non-annealed ones.

In an earlier report, organic semiconductor-doped rechargeable zinc air batteries (RZABs) were prepared and studied for power storage applications [26] and preparation of cobalt- and nitrogen-doped porous carbon derived from phloroglucinol-formaldehyde polymer networks with 2-methyl imidazole and cobalt phthalocyanine as precursors was reported. The CoN-PC-2 catalyst prepared in this study exhibited commendable electro catalytic activity for both oxygen reduction reaction (ORR) and oxygen evolution reaction (OER), evidenced by a half-wave potential of 0.81 V. Moreover, the catalyst demonstrated an outstanding performance in zinc-air batteries, achieving a peak power density of 158 mWcm^{-2} and displaying excellent stability during charge-discharge cycles. The findings from this study aimed to provide valuable insights and guidelines for further research and the development of hierarchical micro-mesoporous carbon materials from polymer networks, facilitating their potential commercialization and widespread deployment in energy storage applications [26].

In one of the works [27] on the study of EC properties of vacuum-evaporated manganese phthalocyanine (MnPc) thin films, MnPc thin films annealed at 423 K showed a good electrochemical reversibility of 84%. Cyclic voltammograms of MnPc thin films deposited at RT and vacuum annealed at different temperatures are shown in Figure 24.1. Optical density (OD) was calculated at

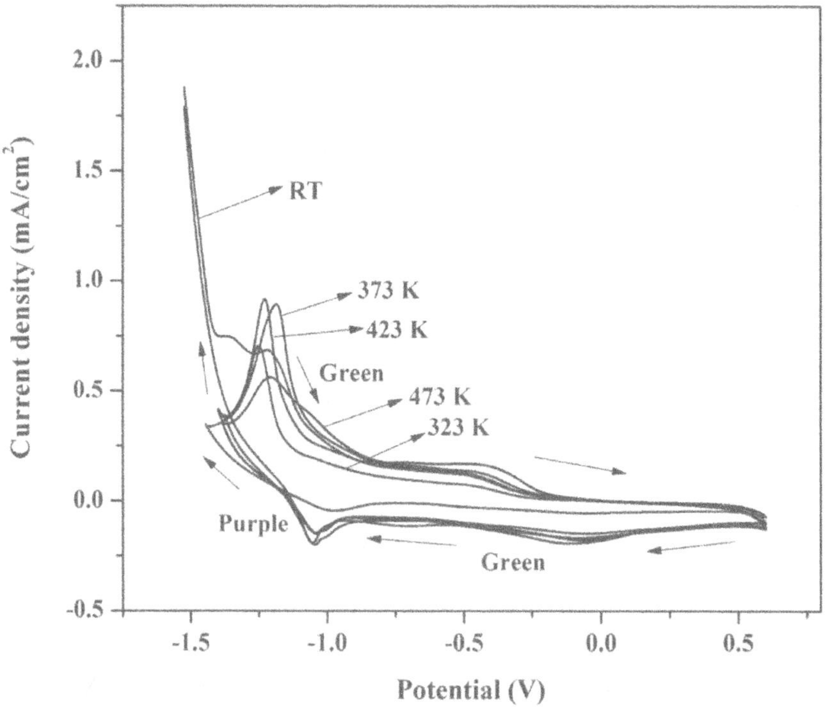

FIGURE 24.1 Cyclic voltammograms of MnPc thin films deposited at RT and vacuum annealed at different temperatures. (Adapted with permission [27]. Copyright (2021) Elsevier Publishers.)

TABLE 24.1
Electrochromic Parameters of MnPc Thin Films for Different Vacuum Annealing Temperatures

Annealing Temperature (K)	Peak Current Density (mA/cm²)		Diffusion Coefficient (× 10⁻¹⁶ cm²/s)		Electrochemical Reversibility (%)	Optical Density Change (ΔOD)	η (cm²/C)
	I_{pc}	I_{pa}	for I_{pc}	for I_{pa}			
RT	0.69	−0.19	1.31	0.100	50	0.103	37
323	0.70	−0.05	1.33	0.0066	56	0.206	57
373	0.89	−0.15	2.16	0.0657	70	0.223	68
423	0.91	−0.20	2.24	0.1136	84	0.249	82
473	0.56	−0.14	0.86	0.0584	79	0.151	45

a wavelength of 550 nm for MnPc thin films vacuum annealed at different temperatures. OD gives the extent of transmission of electromagnetic radiation through a substance. The change in optical density, delta OD (Δ_{OD}), for MnPc thin films during intercalation and deintercalation increased with the increase in annealing temperature. Δ_{OD} was a maximum for the MnPc thin film annealed at 423 K, as shown in Table 24.1. It was observed that the diffusion coefficient for I_{pc} (cathode peak current density) was a maximum with a value of 2.24×10^{-16} cm²/s for the MnPc thin film annealed at 423 K. This can be due to the increase in the crystallinity of the MnPc thin films annealed at 423 K, which led to the free movement of K+ ions between the voids in the crystal lattice. Chowdhury et al. [28] have also reported that the diffusivity of the ions in the crystal increases with the increase in the crystallinity. MnPc thin film annealed at 473K showed a notable decrease in its diffusion coefficient as attributed to the presence of grains of different sizes, which might have hindered the free movement of ions in the lattice. Hence an extension of this work on optimizing the methods of increasing the charge holding capacity of these films might make them eligible candidates as electrodes in battery technology.

24.5 ORGANIC INORGANIC COMPOSITES FOR SUPERCAPACITORS

Recently supercapacitors emerged as one of the promising devices for charge storage due to their high energy and power densities. Supercapacitors store energy through electrochemical reaction, and a study on graphene-doped lanthanum aluminate nanocomposite done using cyclic voltammetry technique was reported [29].

Cyclic Voltammetry studies were done for reduced graphene oxide (RGO)-LaAlO3 nanocomposites at different voltage sweeps on pure RGO and pure $LaAlO_3$ as electrode materials. CV profiles of an RGO-$LaAlO_3$ nanocomposite electrode sample at various scan rates ranging from 2 to 100 mVs⁻¹ and a potential window of −0.3 V to 1.2 V were reported [29]. The consistency of these results demonstrates the outstanding capacitive behavior of this nanocomposite electrode. During the negative scan, aluminum oxide is reduced from Al^{+3} to Al^{+2} and regains its +3 state in the presence of electrolytes during the positive scan, which explains its faradaic behavior. At high scanning rates, the oxidation and reduction peaks of CV curves continuously shifted to higher and lower potentials, causing polarization effect and ohmic resistance of the electrode materials. This suggests that the electrochemical reaction is quasi-reversible and the degree of irreversibility increases as the scan rate of the potential increases.

Figure 24.2 shows a plot of capacitance retention of an asymmetric supercapacitor device (ASD) fabricated using RGO-$LaAlO_3$ nanocomposites. The capacitance retention of ASD fabricated using RGO-$LaAlO_3$ nanocomposites showed a reasonably good storing capacity for a large number of cycles with this combination of organic-inorganic composite materials.

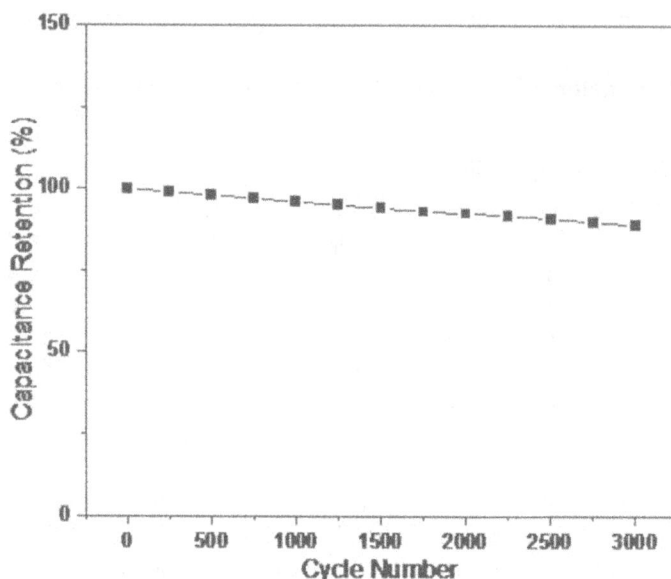

FIGURE 24.2 Plot of capacitance retention of an asymmetric super capacitor device (ASD) fabricated using RGO-LaAlO₃ nanocomposites. (Adapted with permission from [29]. Copyright (2020) Elsevier Publishers.)

24.6 CONCLUSION

In this chapter, the importance of integration of organic/inorganic semiconductors for electrochromic and supercapacitor devices for charge storage applications was discussed. In conclusion, the doping of supercapacitors with organic semiconductors could enhance the specific capacitance, resulting in a high charging discharging rate, a long life cycle, and outstanding power density, making these devices among the most promising candidates for future storage devices.

REFERENCES

1. D. Shavit, The developments of LEDs and SMD Electronics on transparent conductive Polyester film, Vac. Int. 1 (2007) 34–36.
2. S. Mahmud, M. Rahman, Md. Kamruzzaman, O. Md. Ali, S. A. Md. Emon, H. Khatun, R. Md. Ali, Recent advances in lithium-ion battery materials for improved electrochemical performance, a review, Res. Eng. 15 (2022) 100472–100491.
3. G. Qian, X. Liao, Y. Zhu, F. Pan, X. Chen, Y. Yang, Designing flexible lithium-ion batteries by structural engineering, ACS Energy. Lett. 4 (2019) 690–701.
4. J. Ryu, W.-J. Song, S. Lee, S. Choi, S. Park, A game changer: Functional nano/micro materials for smart rechargeable batteries, Adv. Funct. Mater. 30 (2020) 1902499.
5. E. Pomerantseva, F. Bonaccorso, X. Feng, Y. Cui, Y. Gogotsi, Energy storage: The future enabled by nanomaterials, Sci. 366 (2019) 8285.
6. L. de Biasi, B. Schwarz, T. Brezesinski, P. Hartmann, J. Janek, H. Ehrenberg, Chemical, structural and electronic aspects of formation and degradation behavior on different length scales of Ni-rich NCM and Li-rich HE-NCM cathode materials in Li-ion batteries, Adv. Mater. 31 (2022) 1900985.
7. S.H. Oh, O.H. Kwon, Y.C. Kang, J.K. Kim, J.S. Cho, Highly integrated and interconnected CNT hybrid nano fibers decorated with α-iron oxide as freestanding anodes for flexible lithium polymer batteries, J. Mater. Chem. A. 7 (2019) 12480–12488.
8. F. Mo, G. Liang, Q. Meng, Z. Liu, H. Li, J. Fan, C. Zhi, A flexible rechargeable aqueous zinc manganese-dioxide battery working at −20°C, Energy Environ. Sci. 12 (2019) 706–715.
9. H. Li, Z. Liu, G. Liang, H. Yang, H. Yan, M. Zhu, Z. Pei, Q. Xue, Z. Tang, Y. Wang, B. Li, C. Zhi, Waterproof and tailorable elastic rechargeable yarn zinc ion batteries by a cross linked polyacrylamide, electrolyte, ACS Nano. 12 (2018) 3140–3148.

10. A.M. Zamarayeva, A. Jegraj, A. Toor, V.I. Pister, C. Chang, A. Chou, J.W. Evans, A.C. Arias, Electrode composite for flexible zinc–manganese dioxide batteries through in situ polymerization of polymer, hydrogel, Energy Technol. 8 (2020) 1901165.
11. J.H. Kim, Y.H. Lee, S.J. Cho, J.G. Gwon, H.J. Cho, M. Jang, S.Y. Lee, S.Y. Lee, Nanomat Li–S batteries based on all-fibrous cathode/separator assemblies and reinforced Li metal anodes: Towards ultrahigh energy density and flexibility, Energy Environ. Sci. 12 (2019) 177–186.
12. X. Chen, M. Usrey, P.H. Adrian, R. West, R.J. Hamers, Thermal and electrochemical stability of organosilicon electrolytes for lithium-ion batteries, J. Power Sources. 241 (2013) 311–319.
13. J.L. Wang, H. Luo, Y.J. Mai, X.Y. Zhao, L.Z. Zhang, Synthesis of amino alkyl silanes with oligo (ethylene oxide) unit as multifunctional electrolyte additives for lithium-ion batteries, Sci. Chi. Chem. 56 (2013) 739–745.
14. R. Borah, F.R. Hughson, J. Johnston, T. Nann T, On battery materials and methods, Mater. Today Adv. 6 (2020) 100046.
15. X. Dong, L. Chen, X. Su, Y. Wang, Y. Xia, Flexible aqueous lithium-ion battery with high safety and large volumetric energy density, Angew. Chem. 55 (2016) 7474–7477.
16. Z. Guo, J. Li, Y. Xia, C. Chen, F. Wang, A.G. Tamirat, Y. Wang, Y. Xia, L. Wang, S. Feng, A flexible polymer based Li–air battery using a reduced graphene oxide/Li composite anode, J. Mater. Chem. A. 6 (2018) 6022–6032.
17. Z. Guo, Y. Ma, X. Dong, J. Huang, Y. Wang, Y. Xia, An environmentally friendly and flexible aqueous zinc battery using an organic cathode, Angew. Chem. 57 (2018) 11737–11741.
18. Z. Xu, G. Zhang, Z. Cao, J. Zhao, H. Li, Effect of N atoms in the backbone of metal phthalocyanine derivatives on their catalytic activity to lithium battery, J. Mol. Catalysis A: Chem. 318 (2010) 101–105.
19. M. Xu, J. Zhao, J. Chen, K. Chen, Q. Zhang, S. Zhong, Graphene composite 3,4,9,10- perylenetetracarboxylic sodium salts with a honeycomb structure as a high performance anode material for lithium ion batteries, Nanoscale Adv. 3 (2021) 4561–4571.
20. J. Xiao, H. Liu, Y. Lu, L. Zhang, J. Huang, Zn–Mn-PTCDA derived two-dimensional leaf-like $Zn_{0.697}Mn_{0.303}Se/C$ composites as anode materials for high-capacity li-ion batteries, Ceram. Int. B. 47 (2021) 7438–7447.
21. A. Dere, Perylene-3,4,9,10-tetracarboxylic dianhydride (PTCDA) based composites organic battery, Phys. B: Cond. Matter. 547 (2018) 127–133.
22. W. Fan, R. Chua, C. Wang, H. Song, Y. Ding, X. Li, M. Jiang, Q. Li, L. Liu, A. He, Synthesis and characteristic of the ternary composite electrode material PTCDA/CNT@MPc and its electrochemical performance in sodium ion battery, Comp. Part B: Eng. 226 (2021) 109329.
23. P. Yang, P. Sun, W. Mai, Electrochromic energy storage devices, Mater. Tod. 19 (2016) 394–402.
24. T.Y. Yun, X. Li, J. Bae, S.H. Kim, H.C. Moon, Non-volatile li-doped ion gel electrolytes for flexible WO_3 based electrochromic devices, Mater. Des. 162 (2019) 45–51.
25. R. Kumar, D.K. Pathak, A. Chaudhary, Current status of some electrochromic materials and devices: A brief review, J. Phys. D: Appl. Phys. 54 (2021) 503002.
26. Y. Kumar, S. Akula, K.P. Elo, M. Käärik, J. Kozlova, A. Kikas, J. Aruväli, V. Kisand, J. Leis, A. Tamm, K. Tammeveski, Cobalt phthalocyanine doped polymer based electrocatalyst for rechargeable zinc-air batteries, Mater. 16 (2023) 510.
27. B.R. Sridevi, A.H. Priya, C.M. Joseph, Effect of annealing on the optical, structural and electrochromic properties of vacuum evaporated manganese phthalocyanine thin films, Thin Solid Films. 723 (2021) 138584.
28. M. Chowdhury, M.T. Sajjad, V. Savikhin, V. Hergue´, K.B. Sutija, S.D. Oosterhout, M.F. Toney, P. Dubois, A. Ruseckas, I.D.W. Samuel, Tuning crystalline ordering by annealing and additives to study its effect on exciton diffusion in a polyalkylthiophene copolymer, Phys. Chem. Chem. Phys. 19 (2017) 12441–12451.
29. T.N. Vinuth Raj, A.H. Priya, H.B. Muralidhara, C.R. Manjunatha, K. Yogesh Kumar, M.S. Raghu, Facile synthesis of perovskite lanthanum aluminate and its green reduced graphene oxide composite for high performance supercapacitors, J. Electroanalyt. Chem. 858 (2020) 113830.

25 Semiconductor-based Materials for Water-Splitting Applications

Arnet Maria Antony, R. Geetha Balakrishna,
K. Pramoda, and Siddappa A. Patil

25.1 INTRODUCTION

The rise in world population has proportionally led to copious energy demand. The exhaustible sources of fossil fuels are lacking and unable to cope with the escalating energy crisis. The extensive usage of the non-renewable resources in comparison to their formation has drastically depleted these resources. On the other hand, the burning of fossil fuels produces large amounts of carbon dioxide (CO_2), adding to environmental pollution. These concerns fuel the urge for the exploration of new technologies to harness renewable and sustainable resources. One such energy with absolute efficiency and zero residual carbon print is hydrogen (H_2) fuel. The one and only byproduct formed on combustion of H_2 is water, making it the cleanest energy. This drives the interest of researchers in producing H_2 fuel more proficiently. Conventionally, H_2 is produced by methane reformation leading to large amounts of CO_2 emission, impacting the environment. It was Fujishima and Honda who first demonstrated the possibility of water splitting by photo-electrocatalysis for the generation of H_2. They employed an n-type TiO_2 semiconductor as a photoelectrode in an electrochemical cell coupled with a platinum black electrode [1]. This finding was followed by a widespread exploration in the photo-electrocatalytic water splitting. Additionally, the vast availability of water terrestrially and the regeneration of the same on combustion of the H_2 fuel makes water splitting convenient. The process of water splitting can be denoted by two simple half electrode reactions, the hydrogen evolution reaction (HER) and the oxygen evolution reaction (OER) at the cathodic and anodic sites, respectively. In general, these electrode reactions can be provided as

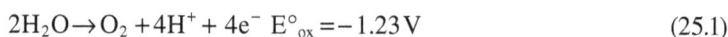

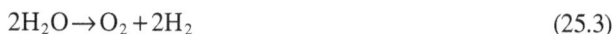

$$2H_2O \rightarrow O_2 + 4H^+ + 4e^- \quad E°_{ox} = -1.23\,V \tag{25.1}$$

$$4H^+ + 4e^- \rightarrow 2H_2 \quad E°_{red} = 0.00\,V \tag{25.2}$$

$$2H_2O \rightarrow O_2 + 2H_2 \tag{25.3}$$

The simplicity of the electrocatalytic reaction for water splitting as seen in the above equations, however, does not simplify the process. The electrolysis of water requires an energy of 237 kJmol^{-1} and a potential of minimum 1.23 V under standard conditions [2]. These threshold values have been achieved by the implementation of the photo-electrochemical approach, replacing the direct electrochemical method by employing the inexhaustive solar energy, promising an eco-friendly method in producing H_2 fuel. The schematic representation of the photo-electrolysis set-up is shown in Figure 25.1. The process kicks off with the absorption of the photons by the semiconductor material, which results in the excitation of the electrons present in the valence band into the conduction band, leading to separation of electrons and holes in the electrode. While some of these electrons and holes, on their way to the surface of the electrode, are lost due to recombination, the others are transferred into the electrolyte system. The H$^+$ ions present take the electrons from the semiconductor

DOI: 10.1201/9781003450146-25

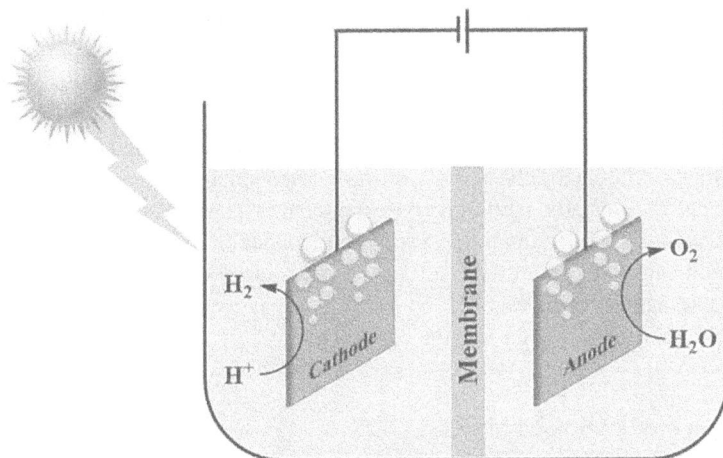

FIGURE 25.1 Schematic representation of a photo-electrochemical cell set-up for the photo-electrolysis of water using a semiconductor electrode. (Image credit, Author.)

surface to bring about the HER at the cathode. Likewise, the anode oxidizes water to bring about the OER generating O_2 [3].

Besides, the H_2 generation depends on the pH of the electrolyte. Both theoretical and experimental analyses indicate that the rate of the HER is boosted under acidic medium. The presence of a higher concentration of protons energetically favors the reduction process. The reduction in an acidic medium can be described as occurring in the following three steps.

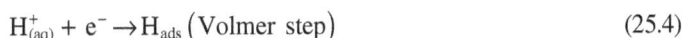

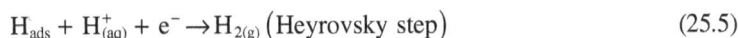

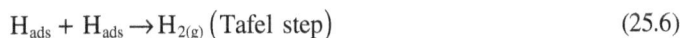

$$H^+_{(aq)} + e^- \rightarrow H_{ads} \ (\text{Volmer step}) \tag{25.4}$$

$$H_{ads} + H^+_{(aq)} + e^- \rightarrow H_{2(g)} \ (\text{Heyrovsky step}) \tag{25.5}$$

$$H_{ads} + H_{ads} \rightarrow H_{2(g)} \ (\text{Tafel step}) \tag{25.6}$$

In the initial step (Equation 25.4), the proton undergoes reduction and gets adsorbed onto the surface of the electrocatalyst. This step is known as the Volmer step. The adsorbed H_{ads} undergoes competition between the Heyrovsky (Equation 25.5) and Tafel (Equation 25.6) steps to form the H_2 molecule. According to the Sabatier's principle, the closer the value of ΔG_H to zero, the better the electrocatalytic activity for HER [4]. To date, the Pt electrode still stands as the best electrode for HER activity in an acidic medium.

Annually, the power of sunlight reaching the surface of the earth is estimated to be 8.6×10^4 TW, which is much greater than the energy requirements [5]. The unavailability of suitable techniques to harvest this renewable energy is the reason for our inability to utilize solar energy productively. Tapping of this solar energy to some extent has been achieved by using suitable semiconductor photoelectrodes to convert the solar energy into H_2 fuel. The ability of the photoelectrode to utilize the light source in yielding high solar to hydrogen efficiency depends on the typical optoelectronic properties of the semiconductor material [6]. Although semiconductor photoelectrodes have been investigated for nearly half a century, there is still a wide berth for improving their efficiency [7–9]. The benchmark for the photoconversion efficiency for water splitting is about 10%, which is the minimum efficiency required for commercialization of the semiconductor photoelectrode. The Pt/C electrode for HER and RuO_2 electrode for OER are the reference standards for analyzing the efficiency of other semiconductor materials [10, 11]. To achieve this threshold value or higher, ideally, the energy level of the conduction band should be higher than the H^+/H_2 redox potential (0.0 V), while the valence band should be lower than the redox potential of O_2/H_2O (1.23 V) [6]. This implies that the semiconductor electrode should at least provide a hole potential and photocurrent density of ~1.6 V and 8.2 mAcm^{-2}, respectively [2]. Also, the photocatalytic efficiency of the photoelectrode depends on the semiconductor band gap, spectrum range of absorption, intensity of the light absorbed, the charge transport through the semiconductor, and the applied bias voltage [12]. Another major factor influencing the activity of the semiconductor is the pH of the electrolytes utilized. The HER prefers an acidic medium, while the OER is accelerated by an alkaline medium. In line with this background understanding, various semiconductor materials have been designed and appropriate modification incorporated to attain the satisfactory optoelectronic properties facilitating the adequate harnessing of light energy to yield electrical energy, while preserving the chemical stability and catalytic function in the acceptable range.

25.2 RESULTS AND DISCUSSION

25.2.1 Metal Oxide-based Semiconductor Materials

As understood, the electrodes used in photoelectrochemical cells for photo-electrolysis are semiconductors with the inherent ability to trap solar photons for water splitting reactions. These semiconductor electrodes are required to sustain for long periods of time with substantially fast reaction rates without any considerable kinetic loss. The fundamental material properties of the photo-electrolytic semiconductors must be tuned to enable said properties for their application in commercial devices. Nevertheless, satisfying the thermodynamic as well as the kinetic requirements of the 4-electron water splitting is a real challenge. The limitation of TiO_2 as a useful semiconductor electrode was its narrow range of the light absorbed in the solar spectrum and its photo-corrosive property [1]. This paved the way to explore other semiconductor materials of the III-V transition metal systems.

The remarkable stability and ease of synthesis of economical metal oxides and the tunability of the band gap with respect to the transition metal character drew the attention of many researchers for their use in photo-electronics [13, 14]. The electronic character of the conduction band of the transition metal monitors the band gap of the semiconductor. In metal oxides, the O 2p constitutes the valence band, while the metal (s, d, or p) forms the conduction band. Metal oxides with metal s character have large band gaps (Al_2O_3 – 8.8 eV, Ga_2O_3 – 4.5 eV), while it is lower for those with metal d character (Fe_2O_3 – 2.2 eV, Cu_2O – 2.1 eV). Moreover, the presence of defects in the metal oxides, like oxygen vacancy, can lead to intrinsic n- or p-type semiconductors enabling their use as photoanodes (TiO_2, WO_3, Fe_2O_3, ZnO) or photocathodes (NiO, CuO, Cu_2O) respectively [15]. Likewise, the mixing of two metals to form ternary metal oxides improves the band structure and optoelectronic properties. The additional metal introduced into the oxide undergoes hybridization with the O 2p orbital, resulting in a slight rise in the energy of the valance band, thereby decreasing the band gap compared to their binary metal oxide counterpart (Figure 25.2) [14]. The extension of this characteristic property to multifunctional photo-electrocatalysts revealed dynamic activity and stability in HER as well as OER [16]. Furthermore, considerable research work has been carried out to design semiconductor materials with anti-corrosive property under milder pH [17].

Although metal oxide semiconductors have potential as photoelectrodes, their photo-corrosive nature, narrow absorption window, rapid recombination of the electron and the hole, electrical resistivity, sluggish performance, and reduced carrier mobility make the use of metal oxides (binary and ternary) in photo-electrolysis limited [18, 19].

The distinctive properties associated with the nanosize exhibited better photoactivity compared to their bulk equivalents. The large surface area of the nanoparticles enables the prompt diffusion of the charge carriers onto the surface with better charge separation, thus facilitating higher photocatalytic activity [20]. Modeling heterojunction with various combinations of

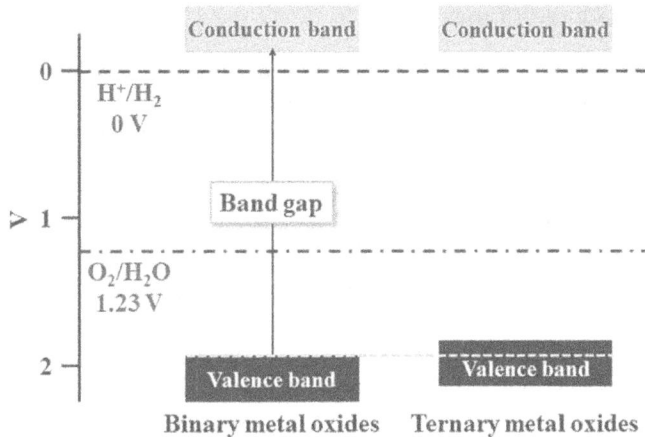

FIGURE 25.2 Illustration of the band gaps in binary and ternary metal oxides. (Image credit, Author.)

nanostructured metal oxides of Ti, Bi, Zn, W, Fe, and Cu has been exhaustively studied for photo-electrocatalytic water splitting owing to the synergistic enhancement of the optoelectronic properties [21, 22]. These heterojunction semiconductors work in broader solar spectrum with improved charge separation amplifying their water splitting ability [23]. Alternatively, doping of the metal oxide with suitable elements like the non-metals C, N, O, and S or the transition metals V, W, and Sn are found to enhance the optical activity and electronic properties when compared to their pure semiconductor materials [24–26]. On doping, additional energy levels are introduced by the dopant assisting in the reduction of the band gap, thereby improving the photocatalytic activity. From Table 25.1, it is evident that the doping of the metal oxides or forming their heterojunction exhibited better photo-electrochemical activity and stability than that of the bare metal oxide irrespective of HER or OER [23]. However, stability of the semiconductor was still a reasonable concern to be dealt with.

TABLE 25.1
Comparison of Selected Metal Oxides as Photocathodes/Photoanodes with Respect to Their Photoelectrochemical Activity and Stability

Photocathode	Electrolyte	Photocurrent Density	Stability (Photocurrent Retention Rate)
Bare Cu_2O thin film	1.0 M Na_2SO_4	5 mA cm^{-2} at 0 V vs RHE	≈0% after 20 min
Cu_2O/C nanowire	0.5 M Na_2SO_4	2.7 mA cm^{-2} at 0 V vs RHE	61.3% after 1000 s
Co_3O_4 nanosheet	0.5 M Na_2S	33.6 μA cm^{-2} at zero bias	–
Co_3O_4/CuO nanosheet	0.5 M Na_2SO_4	6.5 mA cm^{-2} vs SCE at −0.3 V	–
$CuRhO_2$ thin film	1.0 M NaOH	1.0 mA cm^{-2} at −0.9 V vs SCE	70% after 8 h
Rh-doped $SrTiO_3$ thin film	0.1 M K_2SO_4	140 μA cm^{-2} at −0.2 V vs Ag/AgCl	29% after 16 h

Photoanode	Electrolyte	Photocurrent Density	Stability (Photocurrent Retention Rate)
TiO_2 nanowire	0.25 M Na_2S/0.35 M Na_2SO_3	0.5 mA cm^{-2} at −0.2 V vs Ag/AgCl	–
$Ni(OH)_2$/H-TiO_2 nanowire	1 M KOH &1.5 M urea	1.97 mA cm^{-2} at 0.2 V vs Ag/AgCl	97% after 72 h
Bare ZnO nanowire	0.5 M $NaClO_4$	17 μA cm^{-2} at 1.0 V vs Ag/AgCl	–
CN co-doped ZnO nanorod	0.1 M Na_2SO_4	520 μA cm^{-2} at 0.6 V vs Ag/AgCl	–
$ZnWO_4$/WO_3 thin film	0.1 M Na_2SO_4	0.52 mA cm^{-2} at 1.0 V vs Ag/AgCl	–
$CuWO_4$ thin film	0.1 M PBS	0.16 mA cm^{-2} at 0.5 V vs Ag/AgCl	90% after 12 h

25.2.2 Organic Semiconductors

In succession to the thriving exploitation of organic semiconductors in organic photovoltaic devices, these organic semiconductors were investigated for their advantages in water splitting [27–29]. A boost in the photoelectrochemical performance of the inorganic semiconductor resulted in fabricating the former with an organic moiety [30, 31]. This can be attributed to the strong bonding interaction between the transition metal and the organic ligand which passivates the metal oxide layer. Passivation contributed towards the elevation in the durability of the semiconductors when compared to the bare metal oxide. These organic semiconductors exhibit certain advantages such as high charge-transfer transport, tunable band gaps, economical, additional passivation, and solubility in organic solvents [27, 28, 32]. Majorly polymeric compounds with conjugated π-systems, such as poly(p-phenylene), polypyrole, polyaniline (PANI), poly(3-hexylthiophene), etc. (Figure 25.3) are being utilized for water splitting [33, 34]. The extended π-system easily interacts with the photons to initiate the photo-electrocatalysis. Correspondingly, they maintain the charge separation for longer time, decreasing the chances of charge recombination and thereby increasing the catalytic activity.

Following the substantial performance of polymers and copolymers as photocathodes, their performance as photoanodes were investigated. However, they displayed slow kinetics for the oxidation reaction. The retarded activity may be attributed to the low oxidation potential of the organic semiconductors at higher pH [35, 36] and need further venturing into the mechanism. On the other side, carbon nitride was found to have the properties ideal for photoanodes with the ability to achieve the appropriate highest occupied molecular orbital (HOMO) levels for the oxidation of water [37, 38]. Further, the doping of the carbon nitride system substantially enhanced the photocatalytic activity [39].

Likewise, metal organic frames (MOFs), covalent organic frameworks (COFs), and covalent triazine frameworks (CTFs) have also been investigated for their photocatalytic performance (Figure 25.4) [40–42]. Although the organic network easily undergoes electron excitation on irradiation, their poor conductivity restricts the catalytic reaction by the rapid charge recombination. To overcome this bottleneck of a very low carrier mobility, these frameworks can be fabricated into a heterojunction with other moieties like metal nanoparticles, inorganic materials, etc. [43].

25.2.3 Theoretical Calculations and Simulations

Great advancement has been achieved in the theoretical understanding of semiconductor-electrolyte interfaces. Studies have also extended to understand the semiconductor-catalyst-electrolyte interfaces. Conversely, mathematical calculations for the catalyst potential (E_{cat}), current density (J), and potential drop (V), as well as simulations based on first-principles quantum mechanics help to fathom the electronic states, band gap, possible electronic transitions, optical properties, effects of size, and influence of dopants and additives, giving a view into a nearly realistic impression of the mechanistic pathway [44–46]. For instance, Gai et al. [47], analyzed the photoelectrochemical activity of TiO_2 (in the anatase phase) with its mono- and passivated co-doped analogues. The change in the band gap was studied for the pure and doped TiO_2 by performing the calculations for band structure and energy using the frozen-core projector-augmented-wave (PAW) method. The local density

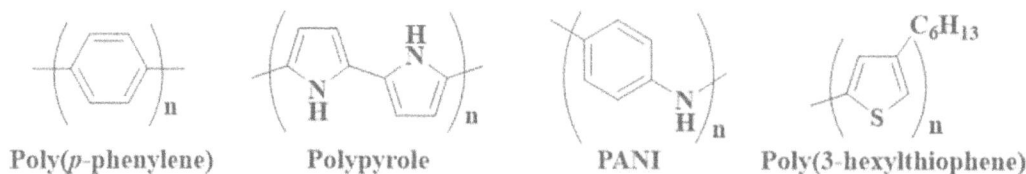

Poly(p-phenylene)   Polypyrole   PANI   Poly(3-hexylthiophene)

FIGURE 25.3 Structures of selected polymer compounds used in organic semiconductors. (Image credit, Author.)

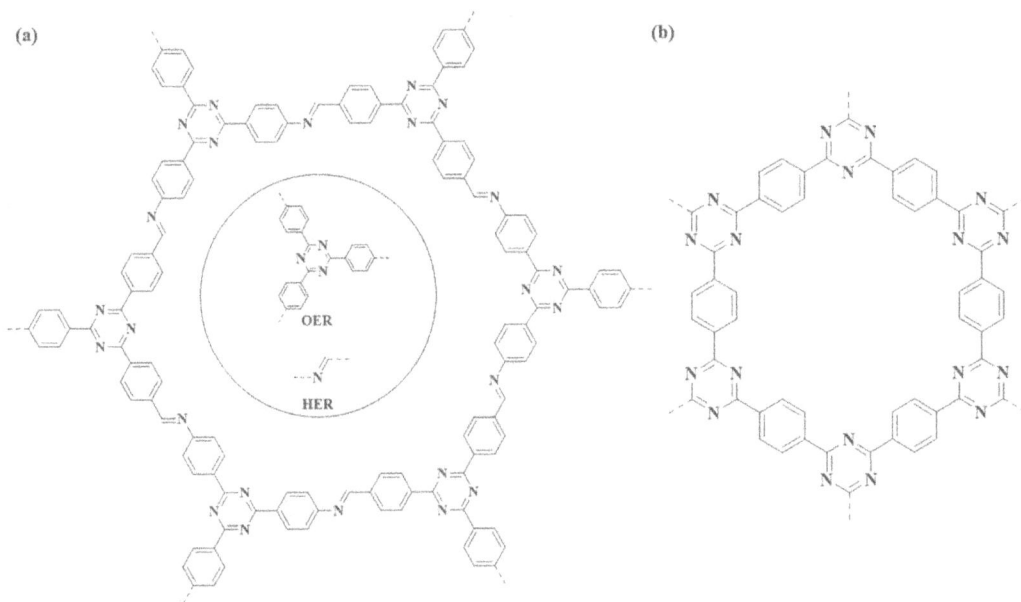

FIGURE 25.4 An example of (a) a COF and (b) a CTF used for the light-driven photocatalytic water splitting. (Image credit, Author.)

approximation (LDA) was implemented as per the Vienna Ab initio Simulation Package (VASP) codes. The calculated shift of the band edges and changes in the band gaps from the donor-acceptor pairs of the co-doped TiO_2 are provided in Table 25.2. From the various co-dopants examined (N, C, V, Nb, Cr, and Mo), the (Mo + C) co-dopant passivated TiO_2 with its reduced band gap to match the visible light region, stood out be the best candidate for the water splitting.

The studies reveal that the evaluation of the theoretical knowledge acquired from simulations can help in understanding the interface physics and mechanistic pathways between the semiconductor electrode, electrocatalyst, and electrolyte. This greatly aids in the design and development of optimal semiconductor materials, providing a guide for an appropriate evaluation of a variety of combinations and permutations of semiconductor materials. Nevertheless, the agreement between the predicted photo-electrocatalytic activity must be assessed in practice for comprehending the semiconductor electrode for real-time applications.

TABLE 25.2
The Effect of Doping with Respect to Pure Anatase TiO_2

Systems	ΔE_v	ΔE_c	ΔE_g
TiO_2 (N + V)	0.38	−0.11	−0.49
TiO_2 (N + Nb)	0.42	0.05	−0.37
TiO_2 (C + Cr)	1.05	−0.31	−1.36
TiO_2 (C + Mo)	1.04	−0.08	−1.12

The values are calculated using the frozen-core PAW method [47]. Positive value is indicative of increase in energy. Adapted with permission [47], Copyright (2023), American Physical Society.

25.3 CONCLUSIONS AND OUTLOOK

Semiconductor materials for the optimal reaping of the light energy to efficiently convert into H_2 fuel has traveled a long way. Starting off from simple conventional metal oxides, a variety of materials irrespective of inorganic systems (binary, ternary oxides, and multifunctional), organic multifunctional systems (MOFs, COFs, and CTFs), nanoparticles, and a range of their permutations and combinations have been immensely explored for the development of semiconductor photoelectrodes for efficient water splitting. Additionally, the synergistic effect on incorporation of electrocatalytic activity coupled with the photoelectrode heightens the efficiency and durability of the semiconductor material. On the other hand, the advancement of computation to realize the mechanistic interaction between the semiconductor, catalyst, and electrolyte interface opens up a new perspective in designing materials with enhanced optoelectronic property. The enlightenment from the existing knowledge and the insights from simulations can lead us in designing and developing promising semiconductor photoelectrodes with exceptional activity.

REFERENCES

1. A. Fujishima, K. Honda, Electrochemical photolysis of water at a semiconductor electrode, Nature 238 (1972) 37–38.
2. M. G. Walter, E. L. Warren, J. R. McKone, S. W. Boettcher, Q. Mi, E. A. Santori, N. S. Lewis, Solar water splitting cells, Chem. Rev 110 (2010) 6446–6473.
3. A. Kudo, Y. Miseki, Heterogeneous photocatalyst materials for water splitting, Chem. Soc. Rev 38 (2009) 253–278.
4. T. R. Cook, D. K. Dogutan, S. Y. Reece, Y. Surendranath, T. S. Teets, D. G. Nocera, Solar energy supply and storage for the legacy and nonlegacy worlds, Chem. Rev 110 (2010) 6474–6502.
5. I. E. Agency, Energy statistics, 2017, International Energy Agency Paris.
6. L. Peter, Fundamental aspects of photoelectrochemical water splitting at semiconductor electrodes, Curr. Opin. Green Sustain. Chem 31 (2021) 100505.
7. S. Chang, X. Huang, C. Y. A. Ong, L. Zhao, L. Li, X. Wang, J. Ding, High loading accessible active sites via designable 3D-printed metal architecture towards promoting electrocatalytic performance, J. Mater. Chem. A 7 (2019) 18338–18347.
8. S. D. Adhikary, A. Tiwari, T. C. Nagaiah, D. Mandal, Stabilization of cobalt-polyoxometalate over poly (ionic liquid) composites for efficient electrocatalytic water oxidation, ACS Appl. Mater. Interfaces 10 (2018) 38872–38879.
9. X. Du, H. Su, X. Zhang, Metal-organic framework-derived M (M = Fe, Ni, Zn and Mo) doped Co9S8 nanoarrays as efficient electrocatalyst for water splitting: The combination of theoretical calculation and experiment, J. Catal 383 (2020) 103–116.
10. A. Kafle, M. Kumar, D. Gupta, T. C. Nagaiah, The activation-free electroless deposition of NiFe over carbon cloth as a self-standing flexible electrode towards overall water splitting, J. Mater. Chem. A 9 (2021) 24299–24307.
11. A. J. Bard, M. A. Fox, Artificial photosynthesis: Solar splitting of water to hydrogen and oxygen, Acc. Chem. Res 28 (1995) 141–145.
12. A. B. Murphy, P. R. Barnes, L. K. Randeniya, I. C. Plumb, I. E. Grey, M. D. Horne, J. A. Glasscock, Efficiency of solar water splitting using semiconductor electrodes, Int. J. Hydrog. Energy 31 (2006) 1999–2017.
13. R. Marschall, L. Wang, Non-metal doping of transition metal oxides for visible-light photocatalysis, Catal. Today 225 (2014) 111–135.
14. A. Walsh, Y. Yan, M. N. Huda, M. M. Al-Jassim, S.-H. Wei, Band edge electronic structure of $BiVO_4$: Elucidating the role of the Bi s and V d orbitals, Chem. Mater 21 (2009) 547–551.
15. Y. Ling, Y. Li, Review of Sn-doped hematite nanostructures for photoelectrochemical water splitting, Part. Part. Syst. Charact 31 (2014) 1113–1121.
16. N. Kumar, K. Naveen, M. Kumar, T. C. Nagaiah, R. Sakla, A. Ghosh, V. Siruguri, S. Sadhukhan, S. Kanungo, A. K. Paul, Multifunctionality exploration of Ca_2FeRuO_6: An efficient trifunctional electrocatalyst toward OER/ORR/HER and photocatalyst for water splitting, ACS Appl. Energy Mater 4 (2021) 1323–1334.

17. B. Kim, T. Kim, K. Lee, J. Li, Recent advances in transition metal phosphide electrocatalysts for water splitting under neutral pH conditions, ChemElectroChem 7 (2020) 3578–3589.
18. M. Forster, R. J. Potter, Y. Ling, Y. Yang, D. R. Klug, Y. Li, A. J. Cowan, Oxygen deficient α-Fe_2O_3 photoelectrodes: A balance between enhanced electrical properties and trap-mediated losses, Chem. Sci 6 (2015) 4009–4016.
19. M. Ziwritsch, n Müller, H. Hempel, T. Unold, F. F. Abdi, R. van de Krol, D. Friedrich, R. Eichberger, Direct time-resolved observation of carrier trapping and polaron conductivity in $BiVO_4$, ACS Energy Lett 1 (2016) 888–894.
20. Z. Zhang, M. F. Hossain, T. Takahashi, Photoelectrochemical water splitting on highly smooth and ordered TiO_2 nanotube arrays for hydrogen generation, Int. J. Hydrog. Energy 35 (2010) 8528–8535.
21. G. Wang, Y. Yang, Y. Ling, H. Wang, X. Lu, Y.-C. Pu, J. Z. Zhang, Y. Tong, Y. Li, An electrochemical method to enhance the performance of metal oxides for photoelectrochemical water oxidation, J. Mater. Chem. A 4 (2016) 2849–2855.
22. J. Luo, L. Steier, M.-K. Son, M. Schreier, M. T. Mayer, M. Grätzel, Cu_2O nanowire photocathodes for efficient and durable solar water splitting, Nano Lett 16 (2016) 1848–1857.
23. Y. Yang, S. Niu, D. Han, T. Liu, G. Wang, Y. Li, Progress in developing metal oxide nanomaterials for photoelectrochemical water splitting, Adv. Energy Mater 7 (2017) 1700555.
24. X. Yang, A. Wolcott, G. Wang, A. Sobo, R. C. Fitzmorris, F. Qian, J. Z. Zhang, Y. Li, Nitrogen-doped ZnO nanowire arrays for photoelectrochemical water splitting, Nano Lett 9 (2009) 2331–2336.
25. M. Xu, P. Da, H. Wu, D. Zhao, G. Zheng, Controlled Sn-doping in TiO_2 nanowire photoanodes with enhanced photoelectrochemical conversion, Nano Lett 12 (2012) 1503–1508.
26. X. Chen, L. Liu, P. Y. Yu, S. S. Mao, Increasing solar absorption for photocatalysis with black hydrogenated titanium dioxide nanocrystals, Science 331 (2011) 746–750.
27. L. Dou, Y. Liu, Z. Hong, G. Li, Y. Yang, Low-bandgap near-IR conjugated polymers/molecules for organic electronics, Chem. Rev 115 (2015) 12633–12665.
28. S. Xiao, Q. Zhang, W. You, Molecular engineering of conjugated polymers for solar cells: An updated report, Adv. Mater 29 (2017) 1601391.
29. J. Hou, O. Inganäs, R. H. Friend, F. Gao, Organic solar cells based on non-fullerene acceptors, Nat. Mater 17 (2018) 119–128.
30. D. Jeon, N. Kim, S. Bae, Y. Han, J. Ryu, WO_3/conducting polymer heterojunction photoanodes for efficient and stable photoelectrochemical water splitting, ACS Appl. Mater. Interfaces 10 (2018) 8036–8044.
31. S. Sharma, S. Singh, N. Khare, Enhanced photosensitization of zinc oxide nanorods using polyaniline for efficient photocatalytic and photoelectrochemical water splitting, Int. J. Hydrog. Energy 41 (2016) 21088–21098.
32. H. Yao, L. Ye, H. Zhang, S. Li, S. Zhang, J. Hou, Molecular design of benzodithiophene-based organic photovoltaic materials, Chem. Rev 116 (2016) 7397–7457.
33. S. Yanagida, A. Kabumoto, K. Mizumoto, C. Pac, K. Yoshino, Poly(p-phenylene)-catalysed photoreduction of water to hydrogen, J. Chem. Soc., Chem. Commun. (1985) 474–475.
34. J. M. Yu, J.-W. Jang, Organic semiconductor-based photoelectrochemical cells for efficient solar-to-chemical conversion, Catalysts 13 (2023) 814.
35. N. S. Lewis, Research opportunities to advance solar energy utilization, Science 351 (2016) aad1920.
36. C. Dai, T. He, L. Zhong, X. Liu, W. Zhen, C. Xue, S. Li, D. Jiang, B. Liu, 2,4,6-Triphenyl-1,3,5-Triazine based covalent organic frameworks for photoelectrochemical H_2 evolution, Adv. Mater. Interfaces 8 (2021) 2002191.
37. Y. Fang, X. Li, X. Wang, Synthesis of polymeric carbon nitride films with adhesive interfaces for solar water splitting devices, ACS Catal 8 (2018) 8774–8780.
38. J. Bian, Q. Li, C. Huang, J. Li, Y. Guo, M. Zaw, R.-Q. Zhang, Thermal vapor condensation of uniform graphitic carbon nitride films with remarkable photocurrent density for photoelectrochemical applications, Nano Energy 15 (2015) 353–361.
39. A. M. Antony, V. Kandathil, M. Kempasiddaiah, R. Shwetharani, R. G. Balakrishna, S. M. El-Bahy, M. M. Hessien, G. A. Mersal, M. M. Ibrahim, S. A. Patil, Graphitic carbon nitride supported palladium nanocatalyst as an efficient and sustainable catalyst for treating environmental contaminants and hydrogen evolution reaction, Colloids Surf. A Physicochem. Eng. Asp 647 (2022) 129116.
40. G. Wang, Y. Liu, B. Huang, X. Qin, X. Zhang, Y. Dai, A novel metal–organic framework based on bismuth and trimesic acid: Synthesis, structure and properties, Dalton Trans 44 (2015) 16238–16241.
41. S. Zhang, G. Cheng, L. Guo, N. Wang, B. Tan, S. Jin, Strong-base-assisted synthesis of a crystalline covalent triazine framework with high hydrophilicity via benzylamine monomer for photocatalytic water splitting, Angew. Chem. Int. Ed 59 (2020) 6007–6014.

42. Y. Wan, L. Wang, H. Xu, X. Wu, J. Yang, A simple molecular design strategy for two-dimensional covalent organic framework capable of visible-light-driven water splitting, J. Am. Chem. Soc 142 (2020) 4508–4516.

43. T. Zhang, W. Lin, Metal–organic frameworks for artificial photosynthesis and photocatalysis, Chem. Soc. Rev 43 (2014) 5982–5993.

44. T. J. Mills, F. Lin, S. W. Boettcher, Theory and simulations of electrocatalyst-coated semiconductor electrodes for solar water splitting, Phys. Rev. Lett 112 (2014) 148304.

45. R. Asahi, T. Morikawa, T. Ohwaki, K. Aoki, Y. Taga, Visible-light photocatalysis in nitrogen-doped titanium oxides, Science 293 (2001) 269–271.

46. A. Govind Rajan, J. M. P. Martirez, E. A. Carter, Why do we use the materials and operating conditions we use for heterogeneous (photo)electrochemical water splitting? ACS Catal 10 (2020) 11177–11234.

47. Y. Gai, J. Li, S.-S. Li, J.-B. Xia, S.-H. Wei, Design of narrow-gap TiO_2: A passivated codoping approach for enhanced photoelectrochemical activity, Phys. Rev. Lett 102 (2009) 036402.

26 Copper Oxide-based Semiconductors for Photo-Assisted Water Splitting

Himanshu S. Sahoo, Debasish Ray,
Sangeeta Ghosh, and Chinmoy Bhattacharya

26.1 INTRODUCTION

The evolutionary growth in modern technology and the enhanced global population extract huge amounts of limited conventional resources such as coal, oils, wood, etc., to meet their energy requirements, damaging the limited stock. The consumption of carbon-based fuels for daily activities causes numerous difficult situations, such as carbon dioxide (CO_2) and other toxic gas emissions into the environment that cause global warming, water pollution, and the destruction of the ecosystem. To avoid such problems, novel advanced technologies for energy generation must be followed. There are many carbon-free energy sources, such as sunlight, biological wastes, hydropower, wind, and nuclear, which have the ability to fulfill energy needs by conserving natural resources. Among the various green energy generation methods, the photoelectrochemical (PEC) splitting of water into H_2 and O_2 is much simpler and more environmentally friendly than the catalytic reformation of hydrocarbons [1, 2]. For the conversion of limitless solar energy into green fuel, PEC solar water splitting is considered a potential technology for sustainable development.

26.1.1 FUNDAMENTAL CONCEPTS INVOLVED IN PEC WATER SPLITTING

Green plants have the capacity to synthesize sugars and oxygen molecules from CO_2 and water molecules utilizing sunlight by a process called photosynthesis. In artificial photosynthesis, hydrogen (H_2) and oxygen (O_2) molecules are produced using a semiconducting material following a similar procedure. PEC devices convert solar energy to chemical energy, the most popular approach combining the electro-catalytic and light-absorbing functions. The process occurs in sunlight through photosensitive materials such as semiconductors. The water splitting experiment using TiO_2 photocatalysts in the presence of ultraviolet irradiation was first explained by Fujishima and Honda in 1972 [3]. In principle, a PEC cell consists of an anode and a cathode immersed in an electrolyte and connected by an external circuit. The key feature is that at least one of the two electrodes should be a semiconductor that can absorb photons and generate e^--h^+ pairs. A p-type semiconductor photocathode reduces the proton (H^+) to H_2, while molecular O_2 is evolved at the counter electrode. On the other hand, the n-type semiconductor photoanode oxidizes water to O_2, and H_2 is produced at the counter electrode. If both electrodes are made up of light-absorbing materials, the cell is called a tandem cell.

When the semiconducting material absorbs light having energy equal to or higher than the band gap, e^--h^+ pairs are generated. The excited electrons in the conduction band (CB) are responsible for the reduction of water ($2H^+ + 2e^- \rightarrow H_2$). In contrast, the holes in the valence band (VB) are involved in the four-electron water oxidation process ($2H_2O + 4h^+ \rightarrow 4H^+ + O_2$). It is also noted that the position of VB must be more positive compared to the oxygen evolution potential.

DOI: 10.1201/9781003450146-26

In summary, the photo-electrolysis of water takes place in three steps;

1. The semiconducting material absorbs the photons with energies higher than the band gap energy, and electron (e^-) and hole (h^+) pairs are formed.
2. The generated electron and hole pairs are separated and migrate to the surface reaction sites.
3. Surface chemical reaction occurs between these carriers with various compounds (e.g., H_2O) to generate H_2 and O_2.

At high pH, the redox reactions can be described as:

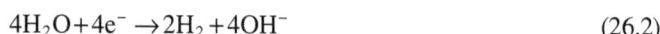

$$4OH^- + 4h^+ \rightarrow 2H_2O + O_2 \tag{26.1}$$

$$4H_2O + 4e^- \rightarrow 2H_2 + 4OH^- \tag{26.2}$$

The overall water-splitting reaction can be written as;

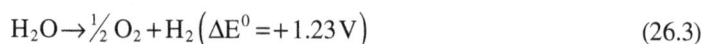

$$H_2O \rightarrow \tfrac{1}{2}O_2 + H_2 \left(\Delta E^0 = +1.23\,V \right) \tag{26.3}$$

where ΔE^0 is the standard reduction potential and is independent of the pH.

Since the oxygen and hydrogen molecules generated in a PEC cell are in gaseous form in two different terminals, there is no chance for the reverse reaction between H_2 and O_2 to form water. It is to be noted that the water-splitting reaction is endothermic and deals with positive free energy changes (ΔG^0: +237 kJ/mol), so this is a non-spontaneous reaction associated with a standard potential difference of 1.23 V. The conversion efficiency and stability of the PEC water splitting process depend on photoelectrodes composed of light-sensitive semiconductor material. Therefore, the selection of suitable photocatalytic materials for PEC water splitting generating H_2 and O_2 as fuel is a primary concern.

26.2 JUSTIFICATION OF CHOOSING CUPROUS OXIDES

Cuprous oxide (Cu_2O)-based semiconductor devices are considered as an alternative to Si-based solar cells. The qualities such as earth abundance, low cost, non- toxicity, and environment-friendly nature of Cu_2O have acquired much interest as photoactive materials in PEC devices. Cu_2O absorbs sunlight in the 300–620 nm range and has direct band gap energy (1.8–2.2 eV) [4]. It has received significant attention as a promising photocatalyst for PEC hydrogen production, although photo-corrosion is the major challenge for regular applications of Cu_2O [5, 6]. The theoretical photocurrent obtainable from the water splitting using Cu_2O as photo-electrode is reported as –14.7 mA/cm² under 1 sun illumination (AM 1.5G) [7]. Experiments show that photo-corrosion is somehow controlled by adding the selective passivation layer on the Cu_2O surface, which limits the direct contact between the photoelectrode and the electrolyte [8]. The best PEC performances of Cu_2O were measured using TiO_2 as a protective layer prepared by the reduction of lactate-stabilized Cu^{2+} ions at pH ~12 [9, 10]. A maximum photocurrent of –7.6 mA/cm² at 0V vs. reversible hydrogen electrode (RHE) was recorded. The position of the Fermi level indicates that it is possible to fabricate an efficient PEC device by coupling with n-type photoelectrodes [11]. The CB of Cu_2O lies 0.7V negative to the HER potential, and the VB lies positive to the oxygen evolution reaction [OER] potential. The flat band potential of Cu_2O was calculated as 0.7 V vs. RHE using the Mott-Schottky (MS) equation [12]. Cu_2O can be used as a promising candidate to fulfill all the requirements as a catalyst for PEC reactions. Due to the small indirect band gap energy (1.2–1.8 eV), the p-type cupric oxide (CuO) is achieving more photocurrent than Cu_2O [13]. It has attracted less attention than Cu_2O because its CB lies more positive than Cu_2O. The CuO can reduce water to molecular hydrogen by forming composites [14]. However, photo-corrosion is the major drawback for both CuO and Cu_2O. Among the different phases of copper oxides, cuprite exhibits the most stable phase with CuO and Cu_3O_4. The Cu_2O unit cell contains a total of six atoms; the copper atoms are arranged in a face centered cubic (fcc) lattice pattern, and the oxygen atoms are occupied at the body-centered cubic lattice. The coordination number of copper in Cu_2O is four.

26.3 COMMON SYNTHESIS METHODS TO PREPARE COPPER OXIDE SEMICONDUCTORS

Researchers are adopting newly developed technologies to synthesize cuprous oxide to improve PEC performances by modulating its properties. Chemical deposition, sputtering, thermal oxidation, anodic oxidation, sol-gel chemistry, radical oxidation, electrodeposition, and solvothermal methods are used to fabricate the p-type Cu_2O thin films [15]. The electrodeposition technique is the most preferred method for depositing Cu_2O on the conducting substrates due to its versatility, low cost, low-temperature requirements, etc. [16].

The photo-corrosion is the major drawback of Cu_2O, so bare Cu_2O is not considered as a productive photo-catalyst for PEC water splitting. The stability and photocurrent density of the Cu_2O semiconductor can be boosted by changing crystal faceting and structure. The crystal facet mainly depends upon the pH of the electrolytic solution. The {111} facet orientation of Cu_2O is dominant at a high pH, whereas the other orientations like {200} are dominant at pH below 10. The orientations, crystal faceting, pH of the bath solution, and temperature should be optimized to enhance the photocurrent and photo-stability [17]. By optimizing the electrochemical solution conditions like pH of the bath, temperature, concentrations, additive reagents, etc., the size, shape, and orientations of the deposited thin films on different substrates can be monitored [18]. The optimized surface area and charge transfer properties with well-formed branched orientations and morphology of Cu_2O thin films can be developed efficiently, leading to solar energy conversion [19, 20]. Tang et al. followed the hydrothermal oxidation route to synthesize strong {110} texture Cu_2O [21]. Maruyama et al. deposited copper (II) oxide thin film by the chemical vapor deposition method [22]. Choi et al. fabricated the Cu_2O thin films on an aluminum substrate via an electrodeposition technique [23]. DC magnetron sputtering method was followed by Reddy et al. for the single-phase Cu_2O thin films with {111} and {200} orientations on the glass substrates [24]. Richardson et al. synthesized Cu_2O thin films on fluorine-doped tin oxide (FTO) by anodic oxidation methods [25]. Zhai et al. and his research group demonstrated an improved electrodeposition method for the deposition of highly crystallized Cu_2O thin films in the presence of citric acid at high pH and medium temperatures [26]. By optimizing the deposition time, the shape of the Cu_2O changes from cubical to tetrahedral to truncated cube and from truncated cube to truncated octahedron. This is ascribed to the interaction of citric ions with {111} facet of Cu_2O. The band gap energies were obtained in the range of 2.21–2.43 eV for the various cuprous oxide thin films, employing UV-Vis absorption spectra. An in situ redox reaction between the metallic copper foil and Cu^{2+} ion can be employed by modifying the copper precursors and anionic groups via hydrothermal strategy for the successful fabrication of rod-shaped arrays, cross-linked octahedra, and truncated octahedra-shaped Cu_2O thin films [27]. The {111} facet rod arrays of Cu_2O possess a large number of copper vacancies, showing the best photocatalytic performances followed by truncated octahedra-shaped {111} and {100} oriented and octahedral {111} facet. Surface reformations occur in rod-shaped arrays, improving photocatalytic activity, while truncated and cross-linked octahedra follow photo-corrosion via oxidation. The most stable photoactivity exhibited by the octahedral and rod-shaped arrays is minimal in the case of truncated octahedra.

26.4 STRATEGIES FOR ENHANCING PEC PERFORMANCES

The formation of a hybrid structure is an alternative way to enhance the PEC performance of copper oxide-based semiconductors. The Cu_2O is made of a hybrid structure with various metal oxide semiconductors, carbon nanostructures, etc. The Schottky barrier formed at the electrode-electrolyte interface suppresses the recombination of charge carriers to increase the photocatalytic efficiency. Different strategies, such as the formation of heterostructures, doping with elements, protective layer formation, and incorporation of co-catalysts, are employed to improve the PEC efficiency.

26.4.1 HETEROSTRUCTURE

Nickel oxide/cuprous oxide nanocomposite was developed by Liu et al. via low-temperature solid reaction method [28, 29]. The NiO/Cu$_2$O composite suspension was spin-coated on the FTO substrate. The presence of Ni, O, and Cu elements is confirmed from the X-ray photoelectron spectra (XPS) analysis of the NiO/Cu$_2$O-6 sample (6% wt. of Cu$_2$O in the sample). The binding energy peaks of the Ni^{2+}, Ni^{3+}, and Cu$^+$ and their satellite peaks confirmed the formation of the Cu$_2$O-modified NiO nanocomposite. The band gap energy of the NiO sample is 3.8 eV, calculated from the Tauc plot. The vacant third 'd' states of Ni present above the valence band is responsible for the absorption in the visible region of the NiO sample. The relative amount of Cu$_2$O present in NiO/Cu$_2$O is directly proportional to the absorption intensity. This indicates that the composite is more favorable for the conversion of solar energy. The bare NiO sample demonstrates a cathodic photocurrent density of –0.3 μA/cm^2. It is noticed that the photocurrent increases after the formation of the NiO/Cu$_2$O junction. The highest photocurrent density reported for the NiO/Cu$_2$O-6 sample (–3 μA/cm^2) is almost ten times that of the bare sample. This increase in photocurrent suggests NiO/Cu$_2$O composite formation, which triggers the charge partition and charge transfer at the electrode/electrolyte interface. The NiO/Cu$_2$O-9 sample exhibited less photocurrent density (–2.25 μA/cm^2), which shows that the excess Cu$_2$O may block the active sites available for reaction. The MS analysis confirmed that the NiO/Cu$_2$O composite shows p-type semi-conductivity. No hydrogen evolution was detected in the case of a bare NiO electrode at –0.0455V vs. RHE. The highest amount of 17.6 μmol hydrogen gas was evolved when a NiO/Cu$_2$O-6 composite was taken as the photocathode. The further Cu$_2$O loading to the NiO composite introduces the recombination sites, which decrease the photoconversion efficiency.

Dubale et al. and his co-workers checked the PEC performance of graphene-modified Cu$_2$O [30]. Cu(OH)$_2$ nanostructures were initially prepared on a Cu mesh electrode by anodization technique in galvanostatic mode. The Cu(OH)$_2$ nanowire arrays/Cu mesh were dipped in GO solution followed by annealing at 500°C in the N$_2$ atmosphere to prepare the graphene/Cu$_2$O nanowires arrays/Cu mesh composite electrode. The Cu(OH)$_2$/Cu mesh electrode dipped in 1.0 mg/mL of GO is denoted as the G-1.0/Cu$_2$O/Cu mesh electrode. A twisted and fractured image was noticed from the scanning electron microscope (SEM) image of Cu(OH)$_2$NWAs (nanowire arrays). However, after graphene incorporation, no fracture was observed. The Cu$_2$O was swaddled into the thin graphene film, confirmed by transmission electron microscope (TEM) images, which suppressed the degradation of the material and enhanced the charge carrier partition with an increase in the PEC activity. The band gap energy of both pure Cu$_2$O/Cu and G-1.0/Cu$_2$O/Cu mesh electrode was 2.03 eV, which was found by extrapolating $(\alpha h v)^2$ vs. hv plot. The linear sweep voltammetry (LSV) under chopped condition of the graphene-modified Cu$_2$O/Cu mesh sample generated a photocurrent density of –4.8 mA/cm^2, which is two-fold higher than the naked Cu$_2$O/Cu mesh electrode (–2.3 mA/cm^2) at 0 V vs. RHE. The photostability of the G-1.0/Cu$_2$O/Cu sample is 83.3% after 20 minutes under light illumination, which is five times more stable than the Cu$_2$O/Cu mesh electrode (14.5%). This indicates that graphene provides excellent behavior in suppressing photo-corrosion of Cu$_2$O and performs as a fast transfer of the charge carriers and high separation rate of electron-hole pairs. The charge transfer is more facile under light illumination due to photoexcitation. The modified G-1.0/Cu$_2$O/Cu mesh electrode possesses lower charge transfer resistance (R$_{ct}$) than the pure Cu$_2$O electrode under dark and light illumination. This indicates that the incorporation of graphene into Cu$_2$O improves the conduction of the photoexcited carriers and hence achieves a better photo-response. This facilitates the electron-hole recombination for better PEC performances. The MS analysis was performed at a fixed frequency of 1 kHz, which is a plot of $1/C^2$ vs. V (potential). The flat-band potential (V_{fb}) carrier density can be calculated by using the MS equation given below:

$$\frac{1}{C^2} = \frac{2}{\varepsilon_0 \varepsilon e N_A A^2} \left(V - V_{fb} - \frac{k_B T}{e} \right) \tag{26.4}$$

Where, C = Differential capacitance

ε_0 = Permittivity of free space
ε = Dielectric constant of semiconductor
e = Electronic charge
N_A = Avogadro's number
A = Cross-sectional area of semiconductor
k_B = Boltzmann constant
T = Absolute temperature

The negative slope of the MS plot indicates that the sample is showing p-type semi-conductivity. Sharma et al. [31] reported a junction of p-type Cu_2O, deposited spray pyrolytically with n-type $SrTiO_3$ for modification in PEC water splitting. The heterojunction composite exhibited remarkable stability and photocorrosion resistivity at applied potentials for constant illumination without any significant loss. The band position of the heterojunction semiconductor, as shown in Figure 26.1, energetically favored the charge separation process and the PEC performance of the materials. The influence of $SrTiO_3$ on Cu_2O was suggested as a blue shift occurred in the absorption edge when thickness increased. The slope of the MS curves determines the value of flat band potential, which is crucial to determine how well the photoelectrode performs in PEC water splitting. The decrease in band bending implied by the positive shift in the V_{fb} relative to Cu_2O improves electron transport.

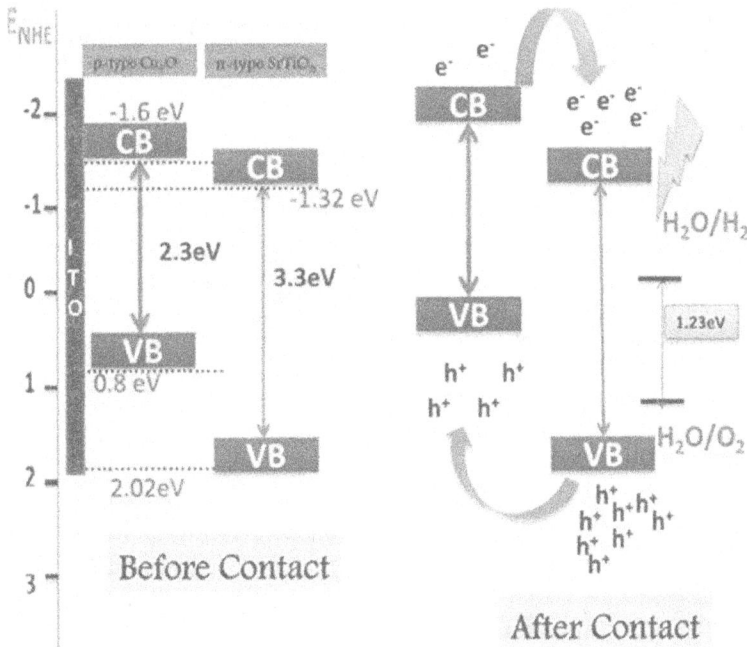

FIGURE 26.1 Energy band diagram of Cu_2O and $SrTiO_3$ before and after the formation of p-n junction. (Adapted with permission [31], Copyright (2014), *The Journal of Physical Chemistry C*.)

A tandem cell of p-type C-modified Cu_2O nanoneedles and n-type TiO_{2-x} nanorods (NRs) is configured for solar-driven water splitting reaction using a Z-scheme by Kaneza et al. [32]. The carbon-modified Cu_2O (C/Cu_2O nano-needles, NNs) electrodes were prepared via the anodization method from KOH solution in galvanostatic mode, followed by soaking in dextrose solution and annealing in an N_2 atmosphere. The n-type TiO_{2-x} nanorods were synthesized via the hydrothermal method. A single cubic Cu_2O phase with a preferred orientation along the {111} orientation was obtained from the X-ray diffraction (XRD) analysis, and the elemental Cu peaks were assigned to the Cu foil substrate. The estimated band gap energy is 2.18 eV for C/Cu_2O NNs and 3.26 eV for TiO_{2-x}NRs, which was confirmed by diffuse reflectance spectra (DRS) analysis. The PEC analyses were performed in a three-electrode setup, taking 0.5 M Na_2SO_4 in 0.1 M KH_2PO_4 at pH 5.0 under light illumination at 1 sun (A.M 1.5, 100 mW/cm²) in N_2-purging. The C/Cu_2O NNs electrode generated a stable photocurrent density of ~0.3 mA/cm² at 0 V vs. RHE (–0.5 V vs. Ag/AgCl), whereas TiO_2 showed only a photocurrent density of 0.06 mA/cm² at 1.23 V vs. RHE (0.8 V vs. Ag/AgCl). In the absence of any redox mediator or any external bias, the PEC performance of C/Cu_2O nanoneedles, when coupled with TiO_{2-x} nanorods in a tandem cell configuration, exhibited a photocurrent density of 0.65 mA/cm². The electrochemical impedance spectra (EIS) analysis revealed that after incorporating carbon in the anodized Cu_2O, the Rct value decreases, implying the facile charge transfer process of charge carriers. The MS analysis generated the positive slope for TiO_{2-x} and the negative slope for C/Cu_2O, confirming the samples' n-type and p-type semiconductivity, respectively.

Jamali et al. [33] reported that the chemical stability of electrodeposited Cu_2O was improved by surface modification with WO_3 and $CuWO_4$. The same synthetic route was explored to carry out these two alterations with different annealing temperatures. After the deposition of $CuWO_4$ on the Cu_2O substrate, the heterostructure sample achieved maximum photocurrent density, i.e., −2.8 mA/cm², which is ~4 times greater than the pure one. Due to the facile charge transportation of bulk Cu_2O to the electrolyte, the composite was exceptionally stable for application in photoelectrochemistry. It was demonstrated that increased light absorption and subsequent photocurrent production in the modified samples, as well as facile charge transfer at the photocathode/electrolyte interface, are mainly responsible for the improved photo-response of the modified composite.

26.4.2 DOPING

Doping with heteroatom is an impressive technique to modulate the electronic structure of the photocatalyst. In tailoring the sample with doping, the query lies in introducing heteroatom with the Cu_2O lattice without hampering its primary growing pattern. This required curbing the in situ growth of hetero nano-composites. Additionally, doping has fascinated the researcher as it has specific qualities for enhancing the PEC application of Cu_2O lattice.

Shyamal et al. employed the electrodeposition technique to fabricate Cu_2O photoelectrodes and lead the modification by introducing Cd^{2+} ions to the same electrolytic bath [34]. For the doping of Cd^{2+} ion in the pure Cu_2O photocathode, the n-type $Cd(OH)_2$ buffer layer was chosen. The Cd^{+2} ion incorporation elevates the PEC activity of the photoelectrode by altering various parameters such as the crystallinity of the synthesized semiconductor, the particle size of the nanomaterials, and facile charge transfer kinetics by preventing the recombination rate of photogenerated charge carriers. The metal deficiency inside the Cu_2O lattice evident by the energy dispersive spectra (EDS) analysis is responsible for the origination of p-type semiconductivity of the Cu_2O photoelectrode, as confirmed by the MS experiment. This deficiency in the metal inside the pure matrix serves as a hole that leads to the p-type property. However, introducing Cd into the semiconductor enhances the copper content along with the enlarged stoichiometric difference between the copper-to-oxygen ratio. The presence of the $Cd(OH)_2$ lattice in the modified samples was demonstrated through the small angle XRD analysis of the optimized 33% Cd-Cu_2O photocatalyst. The analysis revealed the appearance of two new small diffraction peaks at an angle of 29.4° and 35.2° in accordance with the hexagonal $Cd(OH)_2$ with {hkl} plane corresponding to {100} and {101}. The average crystallite size of the photocatalysts was calculated by employing the Debye Scherrer equation: $D = (0.9)/(\lambda \times \beta cos\theta)$, where β: full width at half maxima (FWHM, measured in degree), λ: X-ray wavelength (15.41 nm) and diffraction angle depicted by θ, measured in degrees. The crystalline size of the Cu_2O matrices gradually decreases with the progressive introduction of Cd^{2+} ions into the deposition bath, and optimization has been achieved with about 33% Cd ion introduction to the bath. The successive reduction in crystalline size suggests an enhanced active surface area for the modified thin film photocatalyst, and consequently, it modifies the PEC performance of the samples with the same trend. XPS analysis revealed the simultaneous presence of Cu_2O and Cd^{2+} ions for the optimized photocatalyst. For the bare cuprous oxide Cu_2O (Cu^{+1}), the peaks were reported at the binding energy level of 932.2 eV, whereas the secondary peak appeared at 952.1 eV, corresponding to Cu $2p_{3/2}$ and Cu $2p_{1/2}$, respectively. In comparison, the same peaks were shifted to higher binding energy levels upon the addition of Cd^{2+}-ion. For modified photocatalysts, the peaks were reported at 932.8 eV and 952.7 eV for the Cu level, as mentioned above. Moreover, two new peaks at 405 and 411 eV were reported to correspond to the Cd $3d_{5/2}$ and Cd $3d_{3/2}$, respectively. The PEC activities of the pure and Cd^{2+}-modified Cu_2O matrix were evaluated under periodical illumination of UV-visible light in pH 4.9, using acetate buffer solution in the presence of 0.1(M) Na_2SO_4 as the supporting electrolyte. The photocurrent was reported to be −4.0 mA/cm² for the pure Cu_2O photocatalyst over the metallic copper substrate at an applied external bias of 0.4 V against a normal hydrogen electrode (NHE). Moreover, the photocurrent gradually improved with the successive inclusion of Cd^{2+}-ions into the deposition bath and attained a maximum value for the 33% Cd incorporation in the Cu_2O deposition bath. Further addition of cadmium into the electrolytic bath leads to the suppression of photocurrent generation. The highest photocurrent of −6.1 mA/cm² was reported for 33% Cd^{2+}-incorporated Cu_2O matrix at an applied external bias of 0.4 V vs. NHE, which is ~1.5-fold higher than that of pure Cu_2O matrix. The suppression in the photocurrent for the higher level addition of Cd^{2+} (>35%) was considered due to the excessive distribution of Cd-particles to the vacant site present in the p-type Cu_2O thin film, causing the reduction in the carrier concentration. Furthermore, aggregation of too much $Cd(OH)_2$ layer into the matrix leads to the generation of defects in the crystal lattice, reducing the lifetime of photogenerated charge carriers through recombination at the interfacial region prior to the arrival at the depletion layer. The utmost photo-conversion efficacy of about 2.8% was reported for the pure Cu_2O photocatalyst over the Cu substrate, while the efficiency advances up to 5.8% for the optimized 33% Cd-Cu_2O. The photo-conversion efficiency was obtained using the formula: $\%\eta = (V_{max} \times I_{max})/P_{input}$. The amount of H_2 evolved by the pure and optimized photocatalyst via photocatalytic water reduction was carried out at an applied external bias of 0.0 V against NHE under constant visible light (λ ≥ 390 nm) illumination. The H_2 production rate for a pure Cu_2O photoelectrode was recorded to be 2.0 μmol, which is extended by ~1.5 fold (2.9 μmol) for the 33% Cd incorporated Cu_2O matrix. The amount of H_2 generation and the chronoamperometric (photocurrent vs. time) plots are useful to evaluate the faradaic efficiency. The faradaic efficiency was reported to be 45% and 36% for the optimized (33% Cd) and pure Cu_2O thin films. The incident photon to current conversion efficiency (IPCE) was also documented to be 44% for the bare Cu_2O thin film, which was enhanced almost two times (~87%) for the optimized sample. The impedance spectra were recorded to study the charge transfer kinetics of the thin films at an applied external bias of 0.4 V against NHE. The R_{ct} was significantly reduced by adding a cadmium ion into the electrolytic bath and reaching the lowest value for the 33% Cd-included photocatalyst, resulting in superior charge transfer kinetics at the interfacial region. The p-type nature of these photocatalysts was confirmed by the negative slope of the MS plot.

Hosseini et al. synthesized the C-doped CuO/g-C_3N_4 composite via the microwave-assisted method for PEC water splitting applications [35]. A weak diffraction peak of $2\theta = 27.7°$ is assigned for the {002} plane of g C_3N_4. The mean crystallite size of CuO in C-CuO/CN, CuO/CN, and

CuO was calculated to be 40.9, 16.1, and 20.4 nm, respectively, by applying the Scherrer equation. The XPS scan study of C-CuO/CN shows no elements except carbon, copper, oxygen, and nitrogen. The band gap of the sample is calculated as 1.5 eV from the absorption spectra. Enlarging the crystalline size and carbon content in CuO influences light absorption at higher wavelengths. The photocurrent density exhibited by the CuO/CN was more than two times higher compared to the bare CuO, mainly due to the heterojunction formation. It is also observed that the carbon-doped CuO/CN (C-CuO/CN) demonstrates a photocurrent density of -2.38 mA/cm^2 at an applied bias of 0 V vs. RHE. The photocurrent response of the prepared samples was recorded in the span of 180 seconds at a constant applied bias of 0.125 V against RHE. Anyway, C-CuO/CN exhibited superior PEC performances.

The EIS indicates that the R_{ct} for the photoelectrodes is in the order of 3.48, 7.69, and 18.53 kΩ/cm^2 for the C-CuO/CN composite, CuO/CN, and pure CuO, respectively. The lowest R_{ct} for the C-CuO/CN composite over pure matrix indicates facile charge transfer kinetics at the interfacial region owing to the halting charge recombination rate. The scanty durability of copper-based metal oxides in an aqueous medium limits the potential utilization of these photoelectrodes. The better durability of the C-CuO/CN composite over the other synthesized semiconductors was confirmed by chronoamperometric analysis. The amount of hydrogen evolved by the various photocathodes is determined as 3.91, 2.67, and 2.35 μmol/h-cm^2 for the optimized C-CuO/CN composite, CuO/CN, and pure CuO, respectively.

Shyamal et al. documented the role of Bi introduction into the Cu$_2$O thin film prepared by electrodeposition for the PEC hydrogen generation [36]. The surface morphology as well as the crystalline pattern was influenced by the incorporation of Bi into the Cu$_2$O matrix when the sample was developed either in the presence of the Bi^{+3} ion in the electrolytic bath or previously deposited Bi nanoparticles (NPs) onto the substrate, or the same nanoparticles suspended in the bath. The cubic crystallinity of all the photocathodes was confirmed by XRD analysis. The addition of Bi does not alter the crystalline nature of the materials but instead leads to the broadness of the peak; additionally, the ratio of two major peaks was changed. Consequently, the particle size of the modified sample was in the order of 45 nm, 25 nm, 26 nm, and 18 nm for the pure Cu$_2$O matrix; the Bi ion included Cu$_2$O thin film, Bi nanoparticle-suspended Cu$_2$O photocatalyst, and Bi nanoparticle film/Cu$_2$O semiconductor, respectively. The highest photocurrent of -5.21 mA/cm^2 at an applied external bias of 0.4 V against NHE was reported for the Bi nanoparticle film/Cu$_2$O thin film, followed by -4.9 mA/cm^2 for the Bi nanoparticle-suspended Cu$_2$O photocatalyst, -3.7 mA/cm^2 for the Bi^{+3} ion suspended Cu$_2$O photocatalyst, and -2.6 mA/cm^2 for the pure Cu$_2$O matrix over ITO substrate (ITO/Cu$_2$O), respectively. The band gap energy was reported to vary from 2.1 eV for pure ITO/Cu$_2$O to 1.9 eV for the semiconductor developed on the Bi nanoparticle film, whereas the thickness of the film increases from 2.3 μm for a pure sample to 2.6 μm for ITO/Bi-NPs/Cu$_2$O. The charge transfer kinetics at the electrode/electrolyte interface were evaluated by EIS analysis and reported the lowest R_{ct} of 30.61 Ω for the Bi nanoparticle film/Cu$_2$O, followed by 56.68 Ω for the Bi nanoparticle suspension/Cu$_2$O, 78.84 Ω for the Bi ion/Cu$_2$O, and 128.7 Ω for the pure Cu$_2$O photocatalyst, respectively, as demonstrated in Figure 26.2, The enhancement in photocatalytic activity was due to the combined modification of small crystalline size, high carrier concentration, low charge transfer kinetics, and lower band gap energy upon bismuth inclusion.

Li et al. fabricated the copper oxide thin film over the FTO substrate by employing the electrodeposition technique under constant potentials conditions [37]. To improve the stability of Cu$_2$O, they protected the thin film by covering the Cu$_2$O surface with polyimide (PI) through a simple spin coating method with various levels of PI loading and designated as Cu$_2$O-PI-n, where n stands for the volume of PI loaded. It was reported that the progressive increment in PI loading favors the light absorption capacity of the semiconductors and acquires the maxima for the Cu$_2$O-PI-200 sample. Further increments in PI concentration suppress the light intake, considering the strong light reflection by the dense and glassy surface of the Cu$_2$O-PI-300 thin film. The PI-coated thin films manifest better photo-response than the pure Cu$_2$O. The highest photocurrent density was reported for

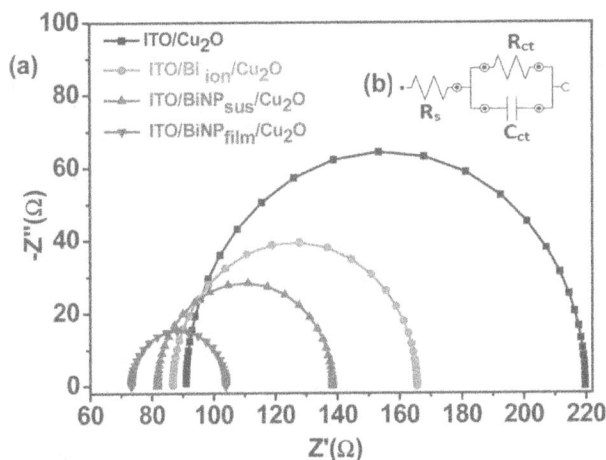

FIGURE 26.2 (a) Nyquist plot of the different Cu_2O thin film electrodes in the presence of 0.1 M Na_2SO_4 electrolytes (pH 4.9) under continuous illumination of 35 mW cm^{-2} at an applied potential of 0.4 V vs. RHE. (b) Equivalent circuit model to analyze the Nyquist plot. (Adapted with permission [36], Copyright (2016), *Journal of Materials Chemistry A.*)

the Cu_2O-PI-200 sample. The reduction in PEC activity of higher loaded (> 200 μl) PI samples resulted from the sluggish conductivity of organic polyimide moiety. To evaluate the durability of the synthesized semiconductors, chronoamperometry was carried out at an applied bias of 0V vs. NHE. From the durability experiment, it was evident that the stability of the optimized sample (Cu_2O-PI-200) has enhanced dramatically with the highest photocurrent of –1.8 mA/cm^2 even after 1200 sec. illumination, whereas the pure semiconductor shows 0.04 mA/cm^2 photocurrent density within the same duration. They also reported that other PI-loaded samples had better photocurrent durability over pure Cu_2O. The protective thin layer leads to inhibition of the photo-corrosion supported by the improved capacitance of the space charge layer and charge recombination of the modified photoelectrodes.

Shyamal et al., for the first time, documented the rare earth Europium (Eu)-modified Cu_2O thin film for the application of the PEC water reduction reaction [38]. They employed a conventional galvanostatic technique to electrodeposit Cu(I)-based metal oxide semiconductors. It was reported that the SEM images of synthesized Cu_2O show the growth of variable-sized cubic lattices upon introducing Eu into the deposition bath. The EDS technique verified the existence of Eu deposits. It was demonstrated that the rare earth introduction does not lead to any primary alteration to the original cubic crystallinity of Cu_2O but only controls the grain size of the particle. This enlargement of the grain size tends to suppress the grain-boundary recombination as it conceals the conducting substrate before the small-grained pure Cu_2O. The photocurrent density progressively improved through the gradual introduction of Eu(III) to the electrodeposition bath and achieved a maximum photocurrent density of –3.2 mA/cm^2 for Cu_2O thin films containing 2.5% Eu^{3+}-ion, while the pure Cu_2O matrix shows only –2.6 mA/cm^2. Beyond the 2.5% Eu^{3+} ion concentration in the electrolytic bath, the PEC performance of the semiconductor again decreases. The reported higher carrier concentration and lower R_{ct} via EIS analysis for the modified sample over the pure one indicates that the cubic Eu nanoparticle acts as a retrieving core for inconvenient impurities, leading to the purification of the Cu_2O crystallite matrix. The higher lifetime of the charge carrier observed through the time-correlated single photon counting (TCSPC) analysis supports the reduction of recombination rate for the photo-generated charge carriers for the doped sample. The enhanced lifetime of the excitons set the right circumstance for electrons in the conduction band to interact with water for the reduction reaction to generate H_2.

26.4.3 CO-CATALYST

Dubale et al. synthesized a stable CuS-modified CuO/Cu$_2$O with the addition of Pt as a co-catalyst. The Cu$_2$O sample was prepared on the FTO substrate via the potentiostatic method. The CuS nanoparticles were deposited by the successive ionic layer adsorption and reaction (SILAR) technique, while the Pt co-catalyst was deposited via sputtering [39]. The emerged XRD diffraction peaks at 36.4^0 and 38.9^0 could be indexed to the {111} crystal planes of Cu$_2$O and CuO, respectively. A maximum photocurrent of −5.4mA/cm^2 was achieved at 0 V vs. RHE in the presence of 1 M Na$_2$SO$_4$ electrolyte (pH 5) for the Cu$_2$O/CuO/CuS heterostructure developed with an optimized nine SILAR cycles for CuS nanoparticles. The photocurrent density of the optimized material is 2.5 times higher than the bare one (−2.2 mA/cm^2). The recombination rate is suppressed due to rapid electron transfer at the surface of the material toward the electrolyte; as a result, the PEC performance is also increased. With a further increase in the number of deposition cycles for the SILAR method, the activity of the material decreases due to the formation of excess CuS over the Cu$_2$O/CuO surface. The chronoamperometric analysis confirmed that the material was relatively stable, and almost 85% of the initial photocurrent was retained after 1 h, as shown in Figure 26.3a2. The co-catalytic activity of CuS for hydrogen evolution was demonstrated as its redox potential is more negative than H$^+$/H$_2$. The incorporation of Pt to CuS, which behaves as a co-catalyst, suggested the facile electron transfer from the photocathode to the electrolyte. As a result, the PEC performance of the materials improves with better stability, as presented in Figure 26.3b2.

Li et al. [40] prepared Cu$_2$O to solve the issues of e$^-$-h$^+$ recombination and photo-corrosion in the presence of any aqueous electrolyte. Amorphous SiO$_x$ was introduced through the dip-coating

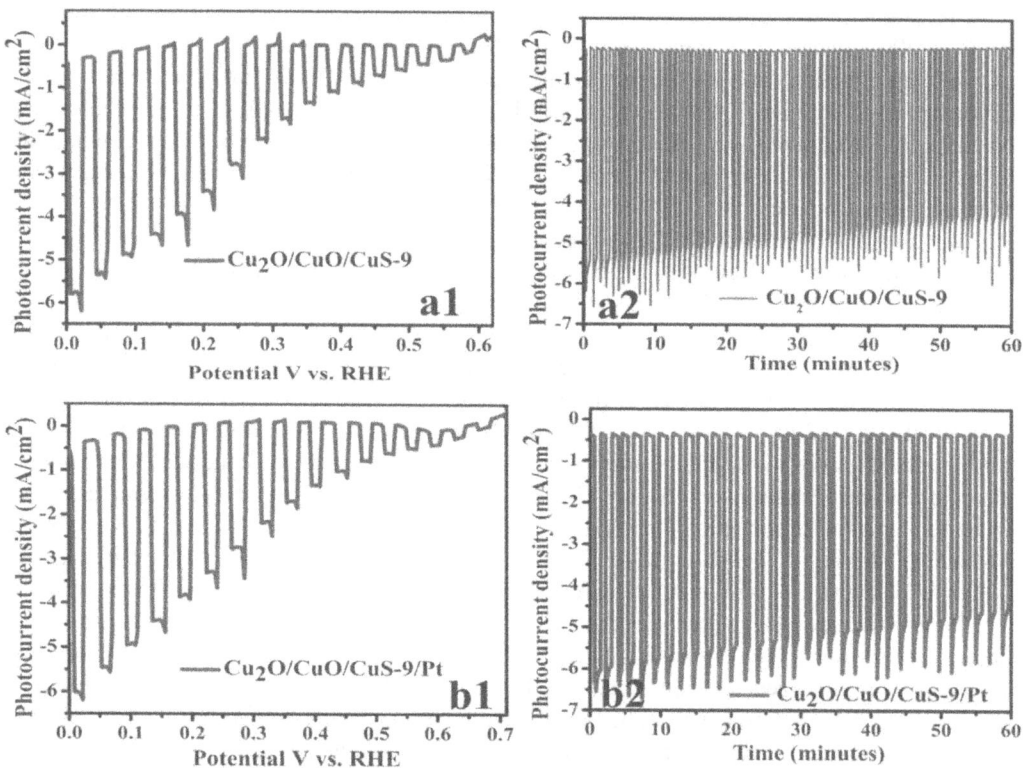

FIGURE 26.3   PEC performances and stability measurement of Cu$_2$O/CuO/CuS-9 (a1 and a2), and Cu$_2$O/CuO/CuS-9/Pt (b1 and b2). (Adapted with permission [39], Copyright (2016), *Journal of Materials Chemistry A*.)

technique as a protective layer on the electrodeposited Cu_2O surface. The PEC water splitting reaction is facilitated for the material compared to the bare Cu_2O, as photocurrent density is three times higher than the pure one. The stability of the compound was also ~3.5 folds higher than pure Cu_2O. Pt catalyst has been decorated to boost the reaction and achieved the maximum photocurrent density to 2.9 mA/cm^2 at 0 V vs. RHE with improved stability, indicating the synergic effect between SiO_x and Pt catalyst. The EIS analysis revealed that the loading of Pt reduced the R_{ct} due to the higher catalytic activity toward the HER. The surface capacitance CPE is increased because of the high charge carrier concentration trapped by the platinum (Pt) catalyst. Cu_2O/SiO_x has a higher charge separation efficiency, confirmed by time-resolved photoluminescence (TRPL), which showed a steeper decay curve and smaller time constants.

26.4.4 PROTECTIVE LAYER FORMATION

Paracchino et al. [9] successfully prepared Cu_2O thin films on FTO substrates via an electrodeposition technique using lactate as a stabilizer with copper sulfate solution at pH12, i.e., in the alkaline medium. An electrodeposited Cu_2O was introduced for solar hydrogen production and was protected from the photocathodic decomposition of water through the formation of a nanolayer of Al:ZnO and TiO_2. The maximum photocurrent of –7.6 mA/cm^2 at 0 V vs. RHE was recorded for the photoelectrode through incorporation of Pt nanoparticles at mild pH, which was almost three times higher than the pure one (−2.4 mA/cm^2 at 0.25 V). The modified protected layers of cuprous oxide ($Cu_2O/ZnO/Al_2O_3$)/TiO_2/Pt) electrode was stable for 20 minutes; however, the bare FTO/Cu_2O sample was degraded within two minutes. The faradic efficiency was calculated at ~100% for the water reduction reaction based on volumetric measurements.

Paracchino et al. also synthesized copper (I) oxide on gold-coated FTO from the basic lactate-stabilized copper sulfate solution, followed by the application of protective layers of n-type oxides (Al-ZnO and TiO_2) on the semiconductor surface by atomic layer deposition (ALD) methods [10]. The Pt co-catalysts were electrodeposited galvanostatically from the aqueous solution containing H_2PtCl_6. The temperature-dependent ALD technique favors a better crystalline structure and proper band alignment, which leads to improved photoactivity. The electrode prepared at 120°C exhibited a photocurrent density of 6.5mA/cm^2 at 0 V/RHE (Figure 26.4a), which was down to 2 mA/cm^2 after 1 hour under the illumination of light, whereas the photoelectrode at 150°C generated 4.5 mA/cm^2 at 0 V/RHE (Figure 26.4a) and that is stable up to 1 hour without any significant change in photocurrent, as demonstrated in Figure 26.4b.

26.4.5 SUBSTRATE MODIFICATION

Shyamal et al. studied the PEC performances of the p-type cuprous oxide thin films electrodeposited on different substrates, i.e., indium doped tin oxide (ITO), Cu-foil, and Al foil from the lactate-stabilized copper sulfate solution in alkaline media by applying a fixed current density of −0.1mA/cm^2 [41]. The Cu_2O adhered better to the Cu and Al substrates than ITO-coated glass due to the better ohmic conductivity. The band gap energy of the Cu_2O was 2.1eV, determined from the UV-visible absorption spectrum analysis. The peak intensity ratio of {111} and {200} peaks is maximum for the Cu substrate, which decreased for Al and ITO substrates, respectively. The photocurrent density measured through the LSV analysis using the Cu_2O deposited on ITO, Al, and Cu foil was −2.6 mA/cm^2, −3.7 mA/cm^2, and −4.6 mA/cm^2, respectively. The electrochemical impedance spectroscopic Nyquist analysis revealed that the diameter of the semicircle follows the order Cu < Al ≪ ITO, indicating the highest R_{ct} value of Cu_2O thin film on ITO, i.e., the least favorable charge transfer process over the surface. The facile charge transfer reaction that occurred on the Cu instead of Al or ITO substrates was explained by the better electronic conductivity of the Cu_2O film matrix on the metallic substrates compared to the ITO-coated glass.

a   b

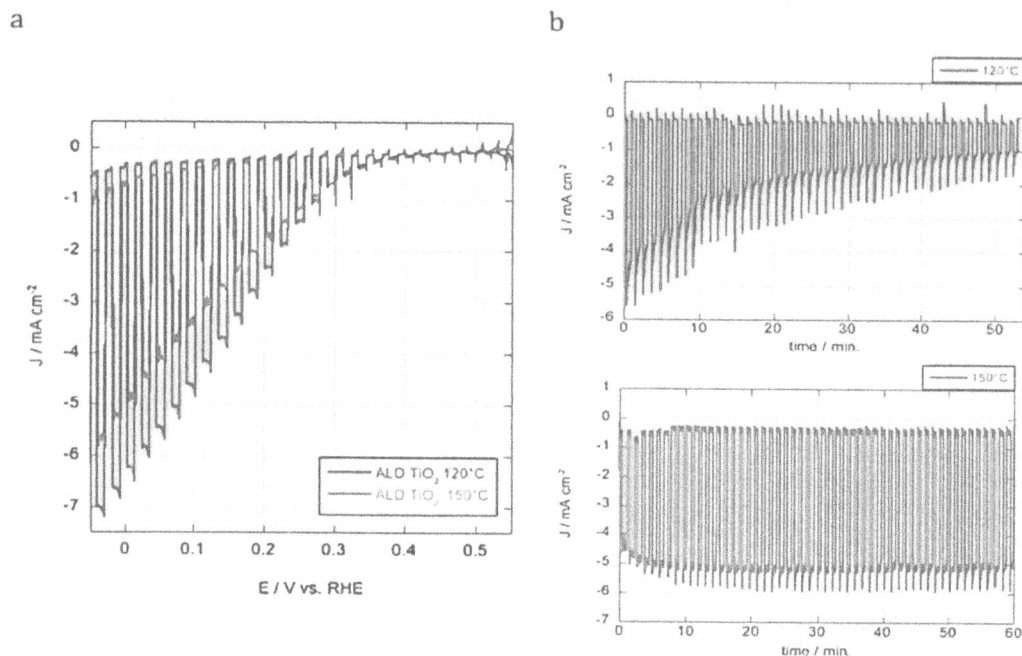

FIGURE 26.4   (a) Cathodic potential sweep under chopped AM 1.5 illumination in Na_2SO_4 0.5 M buffered at pH 5 for Cu_2O/AZO (20 nm)/TiO_2 (10 nm)/Pt photocathodes with TiO_2 top layers deposited at 120°C and 150°C. (b) Stability test at 0 V/RHE in the same electrolyte. Adapted with permission [10], Copyright (2012), Energy & Environmental Science.

26.5   CONCLUSION

In conclusion, copper-based oxide semiconductors are a good photocatalyst candidate for solar energy conversion; still, the limited photostability is a significant drawback for the photocatalytic and PEC properties under solar light illumination. This report discusses different advanced synthetic strategies, such as the formation of heterostructures, doping with different elements, incorporating co-catalysts, etc., to improve the PEC properties and photo-stability of copper-based oxide semiconductor devices. The growth of the protective thin film layers on the Cu_2O effectively enhances the stability by decreasing charge carriers trapping within the film. The primary criterion for incrementing photoactivity is to reduce the recombination of electron and hole pairs. It has been observed that the photo-response is boosted due to the facile transfer of charges at the photoelectrode/electrolyte interfaces by forming the heterostructures with Cu_2O.

REFERENCES

1. Y.K. Hsu, Y.G. Lin, Y.C. Chen, Polarity-dependent photoelectrochemical activity in ZnO nanostructures for solar water splitting, Electrochem. Commun. 13 (2011) 1383–1386.
2. Y.K. Hsu, Y.C. Chen, Y.G. Lin, L.C. Chen, K.H. Chen, Birnessite-type manganese oxides nanosheets with hole acceptor assisted photoelectrochemical activity in response to visible light, J. Mater. Chem. 22 (2012) 2733–2739.
3. A. Fujishima, K. Honda, Electrochemical photolysis of water at a semiconductor electrode, Nature. 238 (1972) 37–38.
4. C.C. Hu, J.N. Nian, H. Teng, Electrodeposited p-type Cu_2O as photocatalyst for H_2 evolution from water reduction in the presence of WO_3, Sol. Energy Mater Sol. Cells. 92 (2008) 1071–1076.
5. P.E. de Jongh, D. Vanmaekelbergh, J.J. Kelly, Cu_2O: A catalyst for the photochemical decomposition of water? Chem. Commun. (1999) 1069–1070. https://doi.org/10.1039/A901232J

6. T. Wang, Z. Luo, C. Li, J. Gong, Controllable fabrication of nanostructured materials for photoelectro-chemical water splitting via atomic layer deposition, Chem. Soc. Rev. 43 (2014) 7469–7484.

7. J.N. Nian, C.C. Hu, H. Teng, Electrodeposited p-type Cu_2O for H_2 evolution from photoelectrolysis of water under visible light illumination, Int. J. Hydrog. Energy. 33 (2008) 2897–2903.

8. D. Kang, T.W. Kim, S.R. Kubota, A.C. Cardiel, H.G. Cha, Electrochemical synthesis of photoelectrodes and catalysts for use in solar water splitting, Chem. Rev. 115 (2015) 12839–12887.

9. A. Paracchino, V. Laporte, K. Sivula, M. Grätzel, E. Thimsen, Highly active oxide photocathode for photoelectrochemical water reduction, Nat. Mater. 10 (2011) 456–461.

10. A. Paracchino, N. Mathews, T. Hisatomi, M. Stefik, S.D. Tilley, M. Graetzel, Ultrathin films on copper(I) oxide water splitting photocathodes: A study on performance and stability, Energy Environ. Sci. 5 (2012) 8673.

11. P.E. de Jongh, D. Vanmaekelbergh, J.J. Kelly, Photoelectrochemistry of electrodeposited Cu_2O, J. Electrochem. Soc. 147 (2000) 486–489.

12. Q. Xu, X. Qian, Y. Qu, T. Hang, P. Zhang, M. Li, L. Gao, Electrodeposition of Cu_2O nanostructure on 3D Cu micro-cone arrays as photocathode for photoelectrochemical water reduction, J. Electrochem. Soc. 163 (2016) H976–H981.

13. C.Y. Chiang, Y. Shin, K. Aroh, S. Ehrman, Copper oxide photocathodes prepared by a solution based process, Int. J. Hydrog. Energy. 37 (2012) 8232–8239.

14. L. Zhang, Y.N. Liu, M. Zhou, J. Yan, Improving photocatalytic hydrogen evolution over CuO/Al_2O_3 by platinum-depositing and CuS-loading, Appl. Surf. Sci. 282 (2013) 531–537.

15. Z. Zang, A. Nakamura, J. Temmyo, Single cuprous oxide films synthesized by radical oxidation at low temperature for PV application, Opt. Express. 27 (2019) 30449.

16. I.S. Brandt, M.A. Tumelero, S. Pelegrini, G. Zangari, A.A. Pasa, Electrodeposition of Cu_2O: Growth, properties, and applications, J. Solid State Electrochem. 21 (2017) 1999–2020.

17. I.V. Bagal, N.R. Chodankar, M.A. Hassan, A. Waseem, M.A. Johar, D.H. Kim, S.W. Ryu, Cu_2O as an emerging photocathode for solar water splitting - A status review, International Journal of Hydrogen Energy. 44 (2019) 21351–21378.

18. C.G. Read, E.M.P. Steinmiller, K.S. Choi, Atomic plane-selective deposition of gold nanoparticles on metal oxide crystals exploiting preferential adsorption of additives, J. Am. Chem. Soc. 131 (2009) 12040–12041.

19. I. Sullivan, B. Zoellner, P.A. Maggard, Copper(I)-based p-type oxides for photoelectrochemical and photovoltaic solar energy conversion, Chem. Mater. 28 (2016) 5999–6016.

20. X.S. Jiang, M. Zhang, S.W. Shi, G. He, X.P. Song, Z.Q. Sun, Microstructure and optical properties of nanocrystalline Cu_2O thin films prepared by electrodeposition, Nanoscale Res. Lett. 9 (2014) 219.

21. Q. Tang, T. Li, X. Chen, D. Yu, Y. Qian, Efficient field emission from well-oriented Cu_2O film, Solid State Commun. 134 (2005) 229–231.

22. T. Maruyama, Copper oxide thin films prepared from copper dipivaloylmethanate and oxygen by chem-ical vapor deposition, Jpn. J. Appl. Phys. 37 (1998) 4099–4102.

23. J. Choi, S.J. Kim, J. Lee, S.C. Nam, J. Kang, J.H. Chang, Controlled growth of Cu_2O particles on a hexagonally nanopatterned aluminium substrate, Nanotechnol. 18 (2007) 215303.

24. A.S. Reddy, S. Uthanna, P.S. Reddy, Properties of dc magnetron sputtered Cu_2O films prepared at dif-ferent sputtering pressures, Appl. Surf. Sci. 253 (2007) 5287–5292.

25. T.J. Richardson, J.L. Slack, M.D. Rubin, Electrochromism in copper oxide thin films, Electrochim. Acta. 46 (2001) 2281–2284.

26. Y.C. Zhai, H.Q. Fan, Q. Li, W. Yan, Morphology evolutions and optical properties of Cu_2O films by an electrochemical deposition on flexible substrate, Appl. Surf. Sci. 258 (2012) 3232–3236.

27. L. Pan, J.J. Zou, T.R. Zhang, S.B. Wang, Z. Li, L. Wang, J. Phys. Chem. C. 118 (2014) 16335–16343.

28. A. Liu, Y. Zhu, K. Li, D. Chu, J. Huang, X. Li, C. Zhang, P. Yang, Y. Du, A high-performance p-type nickel oxide/cuprous oxide nanocomposite with heterojunction as the photocathodic catalyst for water splitting to produce hydrogen, Chem. Phys. Lett. 703 (2018) 56–62.

29. H. Adamu, A.J. McCue, R.S.F. Taylor, H.G. Manyar, J.A. Anderson, Simultaneous photocatalytic removal of nitrate and oxalic acid over Cu_2O/TiO_2 and Cu_2O/TiO_2-AC composites, Appl. Catal. B: Environ. 217 (2017) 181–191.

30. A.A. Dubale, W.N. Su, A.G. Tamirat, C.J. Pan, B.A. Aragaw, H.M. Chen, C.H. Chen, B.J. Hwang, The synergetic effect of graphene on Cu_2O nanowire arrays as a highly efficient hydrogen evolution photo-cathode in water splitting, J. Mater. Chem. A. 2 (2014) 18383–18397.

31. D. Sharma, S. Upadhyay, V.R. Satsangi, R. Shrivastav, U.V. Waghmare, S. Dass, Improved photoelec-trochemical water splitting performance of $Cu_2O/SrTiO_3$ heterojunction photoelectrode, J. Phys. Chem. C. 118 (2014) 25320–25329.

32. N. Kaneza, P.S. Shinde, Y. Ma, S. Pan, Photoelectrochemical study of carbon-modified p-type Cu_2O nanoneedles and n-type TiO_{2-x} nanorods for Z-scheme solar water splitting in a tandem cell configuration, RSC Adv. 9 (2019) 13576–13585.

33. S. Jamali, A. Moshaii, Improving photo-stability and charge transport properties of Cu_2O/CuO for photo-electrochemical water splitting using alternate layers of WO_3 or $CuWO_4$ produced by the same route, Appl. Surf. Sci. 419 (2017) 269–276.

34. S. Shyamal, A. Maity, A.K. Satpati, C. Bhattacharya, Amplification of PEC hydrogen production through synergistic modification of Cu_2O using cadmium as buffer layer and dopant, Appl. Catal. B. 246 (2019) 111–119.

35. S.M. Hosseini, R.S. Moakhar, F. Soleimani, S.K. Sadrnezhaad, S.M. Panah, R. Katal, A. Seza, N. Ghane, S. Ramakrishna, One-pot microwave synthesis of hierarchical C-doped CuO dandelions/g-C_3N_4 nanocomposite with enhanced photostability for photoelectrochemical water splitting, Appl. Surf. Sci. 530 (2020) 147271.

36. S. Shyamal, P. Hajra, H. Mandal, A. Bera, D. Sariket, A.K. Satpati, S. Kundu, C. Bhattacharya, Benign role of bi on an electrodeposited Cu_2O semiconductor towards photo-assisted H_2 generation from water, J. Mater. Chem. A. 4 (2016) 9244–9252.

37. J. Li, W. Li, G. Deng, Y. Qin, H. Wang, Y. Wang, S. Xue, Polyimide stabilized Cu_2O photocathode for efficient PEC water reduction, Ionics. 29 (2023) 685–693.

38. S. Shyamal, P. Hajra, H. Mandal, A. Bera, D. Sariket, A.K. Satpati, M.V. Malashchonak, A.V. Mazanik, O.V. Korolik, A.I. Kulak, E.V. Skorb, A. Maity, E.A. Streltsov, C. Bhattacharya, Eu modified Cu_2O thin films: Significant enhancement in efficiency of photoelectrochemical processes through suppression of charge carrier recombination, J. Chem. Eng. 335 (2018) 676–684.

39. A.A. Dubale, A.G. Tamirat, H.M. Chen, T.A. Berhe, C.J. Pan, W.N. Su, B.J. Hwang, A highly stable CuS and CuS-Pt modified Cu_2O/CuO heterostructure as an efficient photocathode for the hydrogen evolution reaction, J. Mater. Chem. A. 4 (2016) 2205–2216.

40. W. Li, H. Wang, Z. Sun, Q. Wu, S. Xue, Si-doped Cu_2O/SiO_x composites for efficient photoelectrochemical water reduction, J. Power Sources. 492 (2021) 229667.

41. S. Shyamal, P. Hajra, H. Mandal, J.K. Singh, A.K. Satpati, S. Pande, C. Bhattacharya, Effect of substrates on the photoelectrochemical reduction of water over cathodically electrodeposited p-type Cu_2O thin films, ACS Appl. Mater. Interfaces. 7 (2015) 18344–18352.

27 Role of Semiconductor Materials in Wastewater Treatment

Arpita Paul Chowdhury, K. S. Anantharaju,
Subhajit Das, and K. Keshavamurthy

27.1 INTRODUCTION

The main key component among natural resources for living beings on earth is water. The contamination in water bodies is due to pollutants from chemical waste, herbicides, insecticides, heavy metals, food industries, volatile organic compounds, etc. Organic and inorganic pollutants are mainly discharged from industrial effluents and sewage into the water bodies. The pollutant concentration is measured in terms of milligrams of the substance per liter of water (mg/L) or parts per million (ppm) [1]. The toxic inorganic and organic molecules are endangering the health of humans and the environment, and the quality of fresh water has been severely affected, which indirectly leads to a modification and alteration of all the components of freshwater. The serious impact on environment is soil degradation, leading to lower crop yields, marine life destruction, and heavy metals consumption through food chains, causing adverse consequences to humans and animals [2]. These emerging contaminants possess very complex molecular structures and are very hard to detect and remove. The available wastewater treatment is inefficiently designed and so complete removal of contaminants is not possible. For this purpose, different technologies such as physical, chemical, and biological methods are employed [3]. New water remediation technologies are increasing. The remediation techniques, such as adsorption, advanced oxidations, chemical precipitation, membrane process, and ion exchange processes, have drawbacks. For example, the use of membrane technology is costly and energy-intensive, which limits its use in industry. The chemical precipitation requires more chemicals and produces more sludge and secondary product formation [4]. So, the scientific community is focused on alternative solutions which are sustainable and eco-friendly water purification methods. In this context, semiconductor photocatalysts have been a promising emerging method for wastewater treatment. The process is considered eco-friendly and operates under mild reaction conditions. Amazing reactivity is observed for decomposing the pollutants even with trace quantity and no secondary byproduct generation [5]. Fujishima and Honda in 1972 reported TiO_2 electrodes that can produce hydrogen in aqueous medium under sunlight [6]. Since then, this research area has been intensively pursued over the past decades and some significant advancements have been achieved. As the central component of the photocatalytic system, semiconductor photocatalysts have been attracting most of the research focus. In the wastewater treatment process, the photocatalytic process is an attractive method due to the attainment of complete mineralization of the pollutant under mild conditions of temperature and pressure. This is due to the generation of photogenerated electrons/holes in semiconductor materials, which provides efficient oxidation/reduction performance for pollutant degradation. However, some drawbacks exist in single semiconductor photocatalysts; for example, fast recombination rate of electron/hole, quantum yield decrease, and low visible light absorption. Thus, the photocatalytic properties of a semiconductor can be controlled through the design of the new band gap structure to improve the efficiency of these nanomaterials. In this chapter, recent advances in the application of doped semiconductor catalysts and heterostructures that have been recently

DOI: 10.1201/9781003450146-27

developed for wastewater treatment are reviewed. Also, mechanistic insight into photocatalytic reaction is discussed.

27.2 SEMICONDUCTOR PHOTOCATALYST

27.2.1 Mechanism of Semiconductor Photocatalysts

Mostly, semiconductors are used as a model photocatalyst for the essential methods of photocatalysis, owed to its simple accessibility, low cost, high stability, non-toxic nature, and suitable band positions. The major advantages of semiconductor photocatalysis are that it offers a good substitute for energy-intensive treatment methods and can use renewable and pollution-free solar energy [7]. In a semiconductor, firstly, on the illumination of light, an electron is excited from the valence band (VB) into the conduction band (CB), leaving a hole behind. Then the excited electrons/holes migrate to the surface of the catalyst. The chemical potential of +0.5 to 1.5 V versus the normal hydrogen electrode (NHE) should be possessed by the electrons in the CB to exhibit a strong reduction capacity. Similarly, the holes in the VB should have a chemical potential of +1.0 to +3.5 to exhibit a strong oxidative potential. The mechanism of photocatalysis (Figure 27.1) by a semiconductor is represented as:

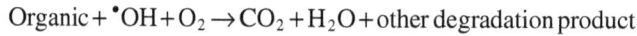

$$Photocatalysts + h\nu \rightarrow e^- + h^+$$

$$h^+ + H_2O \rightarrow H^+ + OH^\bullet$$

$$h^+ + OH^- \rightarrow OH^\bullet$$

$$e^- + O_2 \rightarrow O_2^-$$

$$2e^- + O_2 + 2H^+ \rightarrow H_2O_2$$

$$e^- + H_2O_2 \rightarrow OH^\bullet + OH^-$$

$$Organic + {}^\bullet OH + O_2 \rightarrow CO_2 + H_2O + other\ degradation\ product$$

The major challenges for the pure semiconductor photocatalysts are their large band gap and insufficient sunlight utilization. Yongquan Quab et al. [8] in their research stated that for a stable

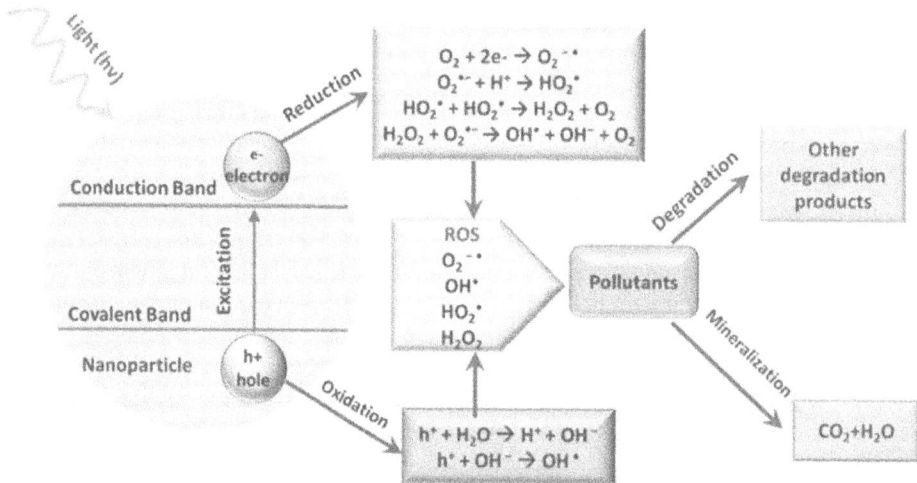

FIGURE 27.1 Schematic illustration of the photocatalytic process. (Adapted with permission [7], Copyright [2022], Elsevier.)

semiconductor catalyst, the semiconductor must first have sufficient band gap to allow effective absorption of light under the visible region. Second, there should be an efficient charge separation. Hence, the band gap engineering of semiconductors is one of the most important targets in photocatalyst design. The efforts on the design and fabrication of heterojunctions are still ongoing. Based on the interface of the different materials, the heterojunction photocatalysts are categorized into four types: (1) semiconductor–semiconductor heterojunction; (2) semiconductor–metal heterojunction; (3) semiconductor–carbon heterojunction; and (4) multicomponent heterojunction.

Among these, in semiconductor/semiconductor heterojunction, the synergistic effect among the semiconductors enhances the photocatalytic activity, thus showing superior new characteristics and performance. The charge may migrate from one component, typically the semiconductor absorbing irradiation, to the other component, with proper band-edge positions in semiconductor-based composites. Reasonably, a composite photocatalyst system, composed of one semiconductor as the light absorber and other components to enhance the photocatalytic processes, is supposed to be an effective prototype to overcome the above shortcomings. Moreover, the second or third component in the composites may also provide active centers for the activation of reactants or the subsequent reduction/oxidation reactions, working as a co-catalyst.

Doping is another alternative and effective way to extend the light absorption to the visible light region of a single semiconductor. The doped ions introduce an additional energy level into the band structure, which can be used to trap electrons or holes in separate carriers from the bands, thus allowing more carriers to successfully diffuse to the surface. Thus, doping modifies the large band-gap of semiconductor and extends the optical property for visible light harvesting and improving the catalytic activity.

27.2.2 BASICS OF HETEROJUNCTION

The heterojunction is defined as the interface between two different semiconductors with unequal band structure. The internal structure of semiconductors comprises some band alignments which act as the basis for the formation of heterostructures. Based on the alignment of energy levels, heterostructures/heterojunction can be categorized into four types: (1) straddling type, (2) staggered type, (3) those with a broken gap, and (4) direct Z scheme (Figure 27.2) [9].

In straddling type heterojunctions (type I), semiconductor A has a wider energy band gap than B. This results in the separation of charge carriers, but recombination is still possible since both electrons and holes are accumulated on the same semiconductor. The type II heterojunction staggered types are formed when the VB and CB of semiconductor A are higher and lower than those of B. The spatial separation formed prevents the charge. The broken gap (type-III heterojunction) is

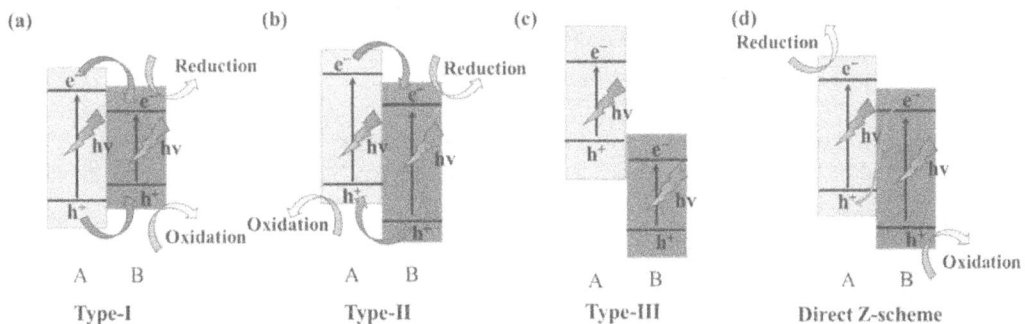

FIGURE 27.2 Schematic illustration of types of heterojunction: (a) straddling gap, (b) staggered gap, (c) broken gap, and (d) Z-Scheme. (Adapted with permission [9]. Copyright [2022], MDPI, Basel, Switzerland, distributed under a Creative Commons Attribution License 4.0 [CC BY].)

similar to the staggered type. The direct Z-scheme heterojunction and type-II heterojunction have similar band structures, but their charge-carrier migration mechanism is different.

27.3 SYNTHESIS TECHNIQUE

27.3.1 HYDROTHERMAL AND SOLVOTHERMAL METHOD

Hydrothermal synthesis is a method where single crystals depend on the solubility of minerals in hot water under high pressure. Here, a steel pressure vessel called an autoclave is used [10]. High temperature (130–250°C) and high vapor pressure (0.3–4 MPa) are controlled. Generally, hydrothermal treatment is followed to prepare iron oxide nanostructures. The solvothermal method is also similar to the hydrothermal method where instead of using aqueous solvent, the nonaqueous solution is preferred.

Recently, Shahzadi et al. [11] synthesized Fe-Doped cadmium oxide (CdO) by hydrothermal method. The catalysts were evaluated toward the photodegradation aqueous solution of methylene blue under 400 W mercury lamp. The cubic structure of Fe and CdO as obtained from XRD with the crystallite size on doping varied from 31 to 45 nm. Azmoon et al. [12] fabricated binary composite MIL 101(Cr)/Fe_3O_4-SiO_2 for the removal of oil contaminants from oilfields. Regmi et al. [13] fabricated Ni-doped $BiVO_4$ semiconductors in a Teflon-lined vessel by microwave hydrothermal treatment. Here, Ni-doping in the V sites is confirmed by the XRD. The researcher observed efficient degradation of ibuprofen with 80% degradation within 90 minutes. Kumar et al. [14] reported a KPCN/GO/$ZnFe_2O_4$ heterostructured photocatalyst. The dye degradation efficiency for rhodamine B, methylene blue, and tetracycline was examined. Around 96% efficiency for rhodamine B was observed under visible light within 30 minutes and 87% for tetracycline within 60 minutes irradiation. Thus, heterojunctions formed improved charge separation and degradation efficiency. Boron-doped graphene oxide/copper sulfide nanocomposite was reported by Farhan and his group [15]. The solutions in the autoclave were kept for 24 hours at 180°C in an oven resulting in a black precipitate. The catalyst can effectively degrade methylene blue under sunlight. Meanwhile, Farid et al. [16] synthesized Ce-doped MoO_3 photoreduction of rhodamine B. Around 91.83% degradation was achieved. Bi_3NbO_7/Bi_2MoO_6 (BNO/BMO) heterojunction was prepared by Cui et al. [17] for photocatalytic degradation of tetracycline and also found that crystallinity increases with BMO content, which indicates the enhanced crystal stability. Figure 27.3 illustrates the formation of Bi_3NbO_7/Bi_2MoO_6 heterojunction.

27.3.2 COPRECIPITATION METHOD

The coprecipitation method is one of the widely used techniques for the preparation of semiconducting material. Here, metal ions are precipitated together by stirring the mixture. Finally washing and drying leads to a high-quality product [18]. Recently, co-doped ZnO/CN nanocomposite and single component (g-C_3N_4 and ZnO) were synthesized by coprecipitation technique. The SEM image revealed a 3D flower-like structure. The optical band gap value of the co-doped ZnO/CN catalyst decreased to 2.23 eV, and this might be due to the synergistic effects among the heterojunction formed in between the co-doped ZnO and CN samples. The photocatalytic treatment for reactive red 120 (RR 120) was studied. Around 250 mL of 20 mg/L RR 120 effectively degraded with 100% efficiency under to solar light irradiation [19]. Meanwhile., Gogoi et al. [20] fabricated ternary quaternary heterojunction Ag/Ag_3PO_4-BiOBr-C_3N_4 photocatalyst for the remediation for RR 120 dye from aqueous medium. The visible light absorption was enhanced in the composite due to the surface plasmon resonance effect of Ag nanoparticles. The degradation efficiency (92.6%) was achieved with the removal rate 0.042 min^{-1}. The charge transfer mechanism and fast interfacial charge transport was further confirmed by the EIS Nyquist plot. Tho et al. [21] fabricated a p-n heterojunction Bi_2S_3/$ZnCo_2O_4$ composite. Here, by varying the weights of the $Bi(NO_3)_3$

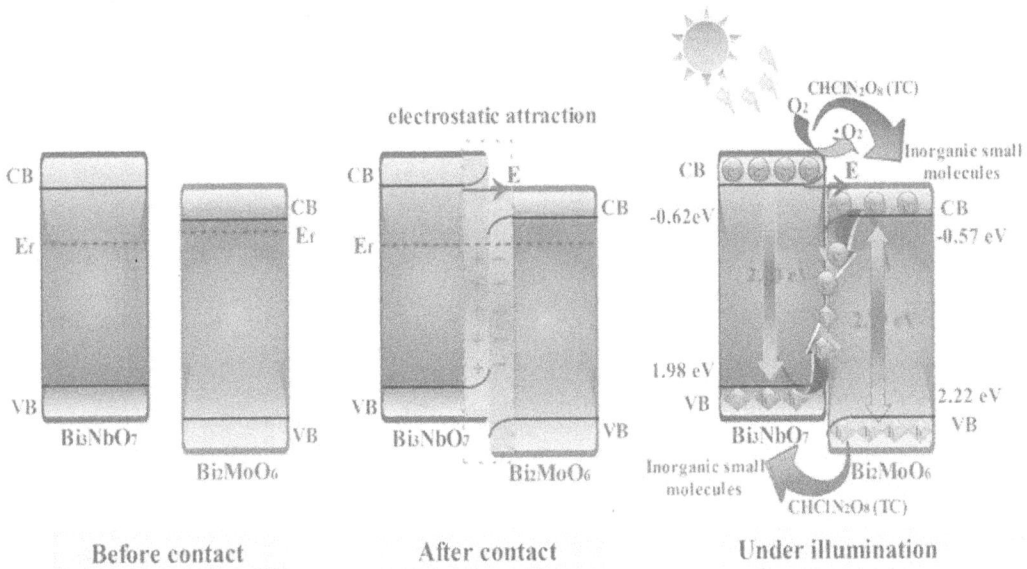

FIGURE 27.3 Heterojunction formed between Bi_3NbO_7 and Bi_2MoO_6 samples after light illumination. (Adapted with permission from reference [17], Copyright [2023], Elsevier.)

and Na_2S precursor, different heterogeneous $xBi_2S_3/ZnCo_2O_4$ were synthesized by coprecipitation method. The heterojunction photocatalyst for decoloration of indigo carmine was evaluated. Mott-Schottky plots proved a heterojunction formed between n-Bi_2S_3 and p-$ZnCo_2O_4$. Furthermore, the investigation of the photocurrent response indicated that the $Bi_2S_3/ZnCo_2O_4$ composite displayed an enhanced response, which was respectively 4.6 and 7.3 times (4.76 μA cm^{-2}) greater than that of the pure Bi_2S_3 and $ZnCo_2O_4$ samples. Highest photocatalyst efficacy (92.1%) was observed with 12 wt% Bi_2S_3 loading for 40 mg L^{-1} indigo carmine dye solution under visible light illumination within 90 minutes. Further FESEM images showed that Bi_2S_3 are precipitated and wrapped on the surface of $ZnCo_2O_4$. The composite proved good recyclability for environmental application. Ahmed and his group prepared CdS/AgI binary composite. The AgI were well incorporated in CdS structure as indicated in XRD spectra. In the precipitation procedure the raw materials structure was unaffected. The organic pollutants, methyl orange and tetracycline hydrochloride were evaluated [22]. $Bi_{3.64}Mo_{0.36}O_{6.55}$ nanospheres fabricated and then nitrogen-doped carbon dots (NCDs) were anchored. The binary composite then enhances the photocatalytic activity for the rhodamine B and bisphenol A degradation. The composite with 0.25% exhibited excellent catalytic activity. This can be attributed to the synergistic effect between the NCDs and $Bi_{3.64}Mo_{0.36}O_{6.55}$ and improved light harvesting efficiency [23]. Chowdhury et al. [24] fabricated a ternary heterojunction of a $BiOCl$-Cu_2CoSnS_4-TiO_2 photocatalyst. The Cu_2CoSnS_4 and TiO_2 nanoparticles are well dispersed on the BiOCl surface. The ternary heterojunction exhibits excellent photocatalytic activity for direct blue 71 degradation. The kinetic constants for the photocatalytic activity showed that the $BiOCl$-Cu_2CoSnS_4-TiO_2-1 photocatalyst is three times higher than that of bare BiOCl, Cu_2CoSnS_4, and TiO_2 particles.

27.3.3 SOL-GEL METHOD

The sol-gel process is a bottom-up synthesis method where after a number of irreversible chemical reactions, a final product is obtained. During the reaction, the homogeneous starting molecules called "sol" become an infinite, heavy, three-dimensional molecule called "gel." Figure 27.4 illustrates the sol-gel process from precursor to aerogel and products derived from them [25].

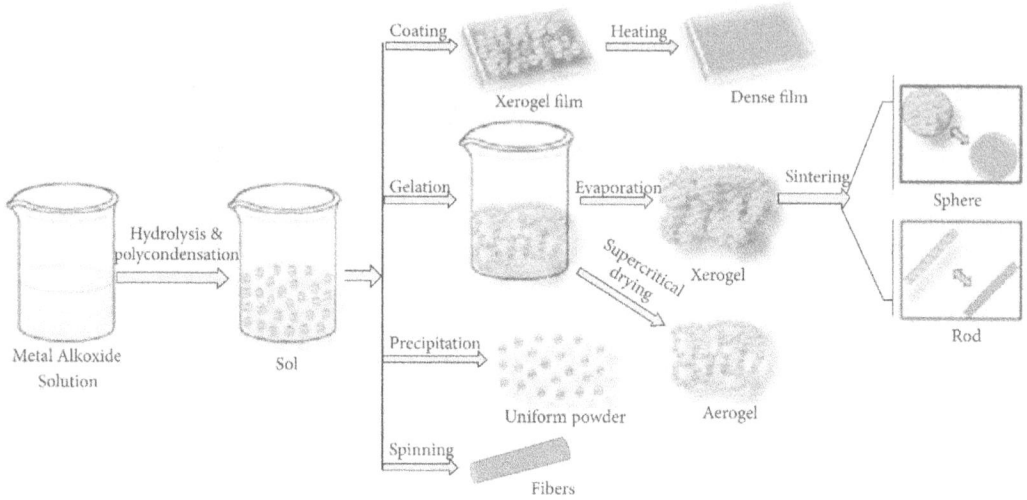

FIGURE 27.4 Overview of a sol-gel technique. (Adapted with permission [25], Copyright @Dmitry Bokov et al. [2021], Hindawi, distributed under a Creative Commons Attribution License.)

Sohaib et al. [26] in their study synthesized pure and doped Mo ZnO-NPs by a facile chemical sol-gel method. This is a solid indication that the Mo^{3+} ions have completely penetrated the wurtzite crystalline lattice of ZnO to create a solid solution and have a high-grade crystalline structure. Hakimi-Tehrani et al. [27] reported a Z-scheme $g\text{-}C_3N_4/WO_3$ nano-photocatalyst. Here, 15 wt% of WO_3 showed photodegradation around four times higher than that of pure $g\text{-}C_3N_4$. The g- C_3N_4 peak from XRD data revealed interlayer structure with a plane filling. So, the interplanar distance of $g\text{-}C_3N_4$ layers increased, as there is a shift in angle observed in the $g\text{-}C_3N_4/WO_3$ nanocomposite. Therefore, it can affirmed that WO_3 nanoparticles are diffused in the g- C_3N_4 interlayer. Alsolami et al. [28] fabricated a two-dimensional (2D) CeO_2 and further synthesized NiS/CeO_2 heterojunctions. The obtained 2D CeO_2 was coupled with narrow band gap NiS for heterojunctions formation. Mixed-phase of NiS/CeO_2 was obtained by the sol-gel synthesis technique as confirmed by XRD, TEM, and XPS characterization. The estimated band gap of the formed heterojunction is 2.07 eV with 9% NiS loading. Ciprofloxacin (CIPF), a model pollutant, photooxidation was performed with the prepared samples. The 12% NiS/CeO_2 showed the best catalytic activity with complete mineralization. The CIPF degradation showed a photoreaction rate of 0.0554 min^{-1}. The superior NiS/ CeO_2 catalyst performance is due to surface structure, good light harvesting ability, and charge carrier separation of the heterojunction formed. The, recyclability experiments proved that the material is photostable, and after five cycles 96.5% photoactivity is achieved. Meanwhile, Belghiti and his group reported the synthesis of a series of ZnO partially coated with different nonmetric layers based on CeO_2, Y_2O_3, or PdO for remediation of charged organic contaminants treatment, using a photocatalysis process. Here, the photodegradation of two organic pollutants was evaluated, which are sulfamethazine (SMZ) and basic yellow 28 (BY28). The photodegradation under UV light for BY28 and SMZ is obtained with complete degradation efficiency. Also, the photocatalyst under visible light degraded BY28 (94%) in 120 minutes and SMZ (60%) in 180 minutes [29]. Recently, Alfryyan et al. [30] synthesized Cd-substituted $NiCoPrFe_2O_4@CNTs$ by sol-gel method and investigated the photocatalytic property. The degradation percentage obtained with different composites for methyl violet degradation are 55.46%, 78.29%, and 85.96%. Among all the prepared catalysts, composition 3 showed the rate constant of 0.012 min^{-1}. This was higher compared to other synthesized photocatalysts. $CuAl_2O_4$-modified $BaSnO_3$ nanoplatelets were synthesized by Tashkandi et al. [31]. The authors claimed $CuAl_2O_4/BaSnO_3$ heterostructure formation by sol-gel-based technique. Mesostructured surfaces and narrow pore diameters of 7.94 nm are obtained. The $CuAl_2O_4$ impregnation facilitated the light absorption capacity. This is due to a decrease in band gap to 2.66 eV with 15.0–20.0 wt% $CuAl_2O_4$. The catalyst promoted the visible-light mineralization of atrazine herbicide. The 2.0 gL^{-1} dose within 40 minutes degraded atrazine herbicide. Basaleh and his group studied the photoreduction of Hg (II) ions under visible illumination. The 1.5wt% PtO/Bi_2WO_6 catalyst proved to be efficient in degrading Hg (II) ions. Around 100% reduction is achieved within 60 minutes of light illumination. The kinetic rate obtained is 0.062 min^{-1}, and approximately six times higher compared to pristine Bi_2WO_6 (0.012 min^{-1}). The incorporation of 1.5wt% PtO enhanced the efficiency of the catalyst with the developed heterojunctions for efficient removal of Hg(II) ions. Also increased light harvesting and decrease in the band gap boosted the charge carrier separation and mobility of the photoexcited carriers [32].

27.3.4 MICROWAVE-ASSISTED METHOD

Microwave assisted synthesis is considered a rapid and facile technique and uses an alternative energy input source. Currently, preparation of organic/inorganic nanostructured materials is of great importance and significance for future rapid development [33]. Vishnu et al. reported Ni-doped magnesium ferrite nanoparticles. The catalyst was prepared using fuel as *Tamarindus indica* seeds extract by microwave. The elements nitrogen, carbon, and oxygen present in seed extract act as strong fuel/reducing agent's essential during the reaction process [34]. The synthesized nanoparticles as confirmed by XRD are spinel cubic with average crystalline size of 17–18 nm. Photodegradation of the two dyes, methylene blue (MB) and rhodamine B (RB), were evaluated. Around 98.1% of MB dye and 97.9% of RB dye were degraded in 0–120 min. In this work initial catalyst dosage was 50 mg with initial dye concentration of 5 ppm. The pH studies further showed that photodegradation efficiency was higher in neutral condition and basic condition. Meanwhile Gorli et al. [35] synthesized Zr-doped TiO_2 as a benign photocatalyst for Bismark brown red dye pollutant. In the experimental setup for photocatalytic degradation, a high pressure 400 W (35000 lumen) metal halide lamp (Osram, India) is utilized as the visible radiation source and the distance between the beaker and radiation source with the UV blocking was maintained at 20 centimeters (Oriel, No. 51472). The results demonstrated that the anatase phase of TiO_2 is not influenced by the presence of dopant. The photocatalyst exhibited a remarkable degradation rate of 99% in 50 minutes. Meanwhile, Zr-doped CeO_2 and activated carbon (AC)–zirconium-doped CeO_2 were reported by Sivakumari and his group [36]. The XRD data revealed that small crystalline size obtained in the present work confirms the good quality and crystallinity. The XRD pattern of bare CeO_2, Zr-doped CeO_2, and AC-Zr-doped CeO_2 clearly showed good crystalline structure, and crystalline phases of composites exhibit good photocatalytic activity toward rhodamine 6G dye in an aqueous solution in the presence of bare UV irradiation. Cortaza et al. [37] fabricated a type II ZnSe/ZnO heterostructure. Rhodamine B and methylene blue dyes of 10 ppm concentration, and 4-Nitrophenol (4-NP) of 18 ppm concentration were degraded and compared using bare ZnO particles and ZnSe/ZnO heterostructures. Initially 100 ml of the dye solution is taken in the presence of 32 mg of the catalyst. Before irradiation, the adsorption-desorption experiment was ensured for maintaining the equilibrium between the surface of the materials and the contaminant solution. In the photocatalytic degradation process, 500W fully reflective solar simulator was used for 240 minutes where the reactor was placed 10 cm below the light source. Meanwhile Gafry et al. [38] prepared Ag-doped ZnO nanostructures. The doped catalyst improved the band gap energy due to the silver doping (Ag). Phenol was chosen as a model pollutant for the catalytic study. The photocatalytic performance showed that the 0.5% Ag-doped ZnO NRs are effective in removing phenol from the contaminated water under solar light irradiation. Further, the photocatalytic kinetics showed that 0.5% Ag–ZnO catalyst is seven times higher compared to 0% Ag-ZnO NRs. The degradation rate reached 98.5% after 5 hours for phenol removal under solar light irradiation. The photocatalytic degradation of tetracycline hydrochloride was carried out in Cheng et al. [39] by a $MoS_2/BiVO_4$ heterojunction synthesized by assisted microwave. The synthesized $MoS_2/BiVO_4$ heterojunction as proved by the characterization analysis was found spherical structure with dimensions in the nano range. The degradation of tetracycline hydrochloride was investigated under visible light irradiation. The photocatalyst $MoS_2/BiVO_4$ heterojunction improved the photocatalytic performance. When compared with bare $BiVO_4$ and MoS_2, the degradation rate achieved is 93.7% in 90 minutes. The tetracycline of 5 mg/L concentration was degraded effectively with $MoS_2/BiVO_4$ at 5 wt% (MB5) mass ratio. Ucker et al. [40] simultaneously reported $CaTiO_3$–ZnS heterostructure. The photodegradation tests on rhodamine B dye were investigated in the presence of the prepared samples. The data revealed that the heterostructure which formed led to effective photocatalytic degradation. The degradation efficiency for $CaTiO_3$ achieved is 50.5% within 30 minutes of irradiation. However, the degradation efficiency reached 95.7% over the binary $CaTiO_3$–ZnS photocatalyst after 180 minutes of irradiation. This clearly indicates that ZnS coating serves in increasing the degree of discoloration over time. Further active radicals responsible for catalytic process are confirmed by the scavenger tests, which revealed that holes were the main active species in the catalytic process. Recently, microwave synthesis was adopted by Li et al. [41] for the synthesis of $Ag/CNQDs/g-C_3N_4$. The XRD data revealed that the introduction of CNQDs and Ag NPs in the matrix of $g-C_3N_4$, did not destroy the chemical structure of the $g-C_3N_4$. The surface plasmon resonance effect of Ag NPs, and $g-C_3N_4$ showed the excellent photocatalytic property. As a result, the optimized $Ag/CNQDs/g-C_3N_4$ composites exhibit excellent photocatalytic degradation of norfloxacin under visible light irradiation. The norfloxacin removal rate was reached almost 100% within 120 min, and the quasi-first-order kinetic constant (k) value was 0.04233 min^{-1}, which was more than 4 times higher than that of $g-C_3N_4$.

27.3.5 ULTRASONICATION AND MICROEMULSION TECHNIQUE

The ultrasonication method is based on sonication, and sonication time varies depending on the chemical constituent used. This method is a highly influential approach for the nanoparticle synthesis. In the ultrasonication method, the ultrasonic sound wave frequency (>20 kHz) is applied to the solution for homogeneous dispersion of nanoparticles into the base fluid as it decreases the cluster formation of nanoparticles by disturbing the intermolecular force. Recently, Murugalakshmi et al. [42] synthesized In_2S_3/Nd_2O_3 heterojunction for sulfasalazine degradation under visible light and for Cr reduction. The Cr (VI) reduction efficiency was 95.23%, and SSZ degradation was 96.19% within 35 and 80 minutes, respectively. The catalyst showed greater efficiency compared to pristine In_2S_3 and Nd_2O_3. The enhanced efficiency of the In_2S_3/10 wt% Nd_2O_3 p-n heterojunction may be due to intimate interfacial contact and proper band that facilitates the better charge carrier separation and also transfer between In_2S_3 and Nd_2O_3 photocatalyst. The catalyst was photostable and highly recyclable after the fifth cycle and only 4% reduction in degradation for Cr (VI) and 7% for SSZ is observed. Manikandan et al. [43] prepared a ternary composite $WO_3@g-C_3N_4@MWCNT$ for photocatalytic tetracycline degradation. The degradation achieved is 79.54% within 120 minutes. The percentage degradation observed is higher than the binary $WO_3@g-C_3N_4$ composite and bare components. Another work reported by Chellanpandi et al. [44] is fabrication of NiO on MK30 surface material using a *Carrisa edulis* fruit extract. Under visible light irradiation, the NiO/MK30 composite was employed to study the photocatalytic degradation activity of methylene blue and tetracycline.

The microemulsions discovery attained an increase significance both in basic research and in different industrial fields. The unique properties, such as large interfacial area, thermodynamic stability, ultra-low interfacial tension, and the ability to solubilize otherwise immiscible liquids paved attention. Microemulsions are obtained by simple mixing of the constituents without the requirement of a specific requirement process. The three kinds of microemulsions are water dispersed in oil (w/o), oil dispersed in water (o/w), and bicontinuous. In 2013, Li and his group reported TiO_2-coated carbon nanotubes prepared by a micro-emulsion method [45]. In the reaction process mixture of distilled water, cyclohexanol and hexadecyl trimethyl ammonium bromide were refluxed together at 25°C for 1 hour using a magnetic stirrer. For the photocatalytic investigation, methylene blue dye was evaluated in aqueous suspension. Maqbool et al. [46] prepared Co–La-doped dysprosium chromite by microemulsion route. The prepared catalyst was used to evaluate crystal violet dye degradation. The experiments were performed under sunlight irradiation. The catalyst showed 70% dye degradation in 90 minutes; this may be due to proper band alignment.

27.3.6 Chemical Vapor Deposition

The chemical vapor deposition is generally used for commercial powder and thin film manufacturing. Recently, nitrogen-doped ZnO was prepared by Hanif and his coworkers for rhodamine B dye [47]. The efficiency achieved for doped ZnO is around 96.90% while for bare ZnO efficiency is 62.95%. It is observed that nitrogen-doped ZnO is 154% times more efficient than undoped ZnO. The reaction took place under visible-light irradiation for 100 min. The average crystallite size is 23.96 and 21.94 nm, for ZnO and nitrogen doped ZnO, respectively. The results showed successful incorporation of nitrogen in ZnO lattice. Iron-loaded carbon black was prepared via chemical vapor deposition by Ahmed et al. [48] for rhodamine elimination. The iron-loaded carbon black was synthesized using chemical vapor deposition under N_2 pyrolysis at a temperature of 800°C. The highest degradation efficiency of around 91.88% was observed at a slightly acidic pH of 4.8. The kinetic study showed the rate constant of 0.0294 min^{-1}. MoS_2 was vertically grown on g-C_3N_4 nanosheets by chemical vapor deposition to prepare nanocomposites by Wang et al. [49]. With a large surface area of 545.2 m^2g^{-1} and a total pore volume of 1.7 cm^3g^{-1}, the sample revealed fast and large adsorption capacity for tetracycline hydrochloride.

27.4 CONCLUSION

In this chapter, the basics of semiconductor photocatalyst are highlighted. The different synthesis techniques and applications of doped semiconductors and heterostructured composites are discussed. The tuning of the band gap and the lattice mismatch between the components are important factors for designing a photocatalyst. The synthesis methods, mainly hydrothermal and solvothermal method, coprecipitation method, sol-gel method, microwave-assisted method, ultrasonication and microemulsion technique, and chemical vapor deposition method are elaborated. Despite the numerous achievements to date, many loopholes still need to be solved such as structural design and more controlled synthesis techniques. These will control the band gap energy of the system required for efficient and complete degradation of the organic pollutant present in water and can be further combined with the simulation predicting method. Thus, future studies are needed to focus on following areas:

1. Alternate and optimized treatment systems for higher efficiency.
2. Large scale testing system procedures should be enabled.
3. The disposal of reused photocatalysts should be considered.
4. Focus should be on low-level pollutant elimination.

REFERENCES

1. J.J. Peirce, R.F. Weiner, P.A. Vesilind, Measurement of Water Quality, in: Environmental Pollution and Control, Elsevier, 1998: pp. 57–76.
2. R. Sankaran, P.L. Show, C.-W. Ooi, T.C. Ling, C. Shu-Jen, S.-Y. Chen, Y.-K. Chang, Feasibility assessment of removal of heavy metals and soluble microbial products from aqueous solutions using eggshell wastes, Clean Technol Environ Policy. 22 (2020) 773–786.
3. N.Y. Donkadokula, A.K. Kola, I. Naz, D. Saroj, A review on advanced physico-chemical and biological textile dye wastewater treatment techniques, Rev Environ Sci Biotechnol. 19 (2020) 543–560.
4. S.S. Imam, R. Adnan, N.H. Mohd Kaus, The photocatalytic potential of BiOBr for wastewater treatment: A mini-review, J Environ Chem Eng. 9 (2021) 105404.
5. J. Hong, K.-H. Cho, V. Presser, X. Su, Recent advances in wastewater treatment using semiconductor photocatalysts, Curr Opin Green Sustain Chem. 36 (2022) 100644.

6. W. Wang, M.O. Tade, Z. Shao, Research progress of perovskite materials in photocatalysis- and photovoltaics-related energy conversion and environmental treatment, Chem. Soc. Rev., 44 (2015) 5371–5402. https://doi.org/10.1039/C5CS00113G

7. Z. Hu, C. Zhang, Y. Zhang, Y. Gu, Y. An, BiOCl-based photocatalysts: Synthesis methods, structure, property, application, and perspective, Inorg. Chem. Commun., 138 (2022) 109277.

8. Y. Qu, X. Duan, Progress, challenge and perspective of heterogeneous photocatalysts, Chem Soc Rev. 42 (2013) 2568–2580.

9. X. Chen, C. Zhao, H. Wu, Y. Shi, C. Chen, X. Zhou, Two-dimensional ZnS/SnS2 heterojunction as a direct Z-scheme photocatalyst for overall water splitting: A DFT study, Materials. 15 (2022) 3786.

10. B.P. Kafle, Introduction to Nanomaterials and Application of UV–Visible Spectroscopy for Their Characterization, in: Chemical Analysis and Material Characterization by Spectrophotometry, Elsevier, 2020: pp. 147–198.

11. I. Shahzadi, M. Aqeel, A. Haider, S. Naz, M. Imran, W. Nabgan, A. Al-Shanini, A. Shahzadi, T. Alshahrani, M. Ikram, Hydrothermal synthesis of Fe-doped cadmium oxide showed bactericidal behavior and highly efficient visible light photocatalysis, ACS Omega. 8 (2023) 30681–30693.

12. P. Azmoon, M. Farhadian, A. Pendashteh, S. Tangestaninejad, Adsorption and photocatalytic degradation of oilfield produced water by visible-light driven superhydrophobic composite of MIL-101(Cr)/Fe_3O_4-SiO_2: Synthesis, characterization and optimization, Appl Surf Sci. 613 (2023) 155972.

13. C. Regmi, Y.K. Kshetri, T.-H. Kim, R.P. Pandey, S.K. Ray, S.W. Lee, Fabrication of Ni-doped $BiVO_4$ semiconductors with enhanced visible-light photocatalytic performances for wastewater treatment, Appl Surf Sci. 413 (2017) 253–265.

14. R. Kumar, A. Sudhaik, P. Raizada, V.-H. Nguyen, Q. Van Le, T. Ahamad, S. Thakur, C.M. Hussain, P. Singh, Integrating K and P co-doped g-C_3N_4 with $ZnFe_2O_4$ and graphene oxide for S-scheme-based enhanced adsorption coupled photocatalytic real wastewater treatment, Chemosphere. 337 (2023) 139267.

15. A. Farhan, M. Zahid, N. Tahir, A. Mansha, M. Yaseen, G. Mustafa, M.A. Alamir, I.M. Alarifi, I. shahid, Investigation of boron-doped graphene oxide anchored with copper sulphide flowers as visible light active photocatalyst for methylene blue degradation, Sci Rep. 13 (2023) 9497.

16. M.T. Farid, S. Aman, N. Ahmad, H.M. Abo-Dief, A.K. Alanazi, R.Y. Khosa, M.Z. Ansari, Z.M. El-Bahy, Ce-Doped MoO_3 photocatalyst as an environmental purifier for removal of noxious rhodamine B organic pollutant in wastewater, Energy Technology. 11 (2023) 2300116

17. B. Cui, W. Leng, X. Wang, Y. Wang, J. Wang, Y. Hu, Y. Du, Enhanced visible-light photocatalytic activity of S-scheme Bi_3NbO_7/Bi_2MoO_6 heterojunction composite photocatalyst, Vacuum. 217 (2023) 112589.

18. T. Athar, Smart Precursors for Smart Nanoparticles, in: Emerging Nanotechnologies for Manufacturing, Elsevier, 2015: pp. 444–538.

19. H. saed kariem Alawamleh, A.H. Amin, A.M. Ali, B.A. Alreda, A.A. Lagum, C. Pecho, N. Taqi, H.M. Salman, M. Fawzi Nassar, Solar light driven enhanced photocatalytic treatment of azo dye Contaminated water based on Co-doped ZnO/g-C_3N_4 nanocomposite, Chemosphere. 335 (2023) 139104.

20. H.P. Gogoi, G. Bisoi, P. Barman, A. Dehingia, S. Das, A.P. Chowdhury, Highly efficient and recyclable quaternary Ag/Ag_3PO_4–BiOBr–C_3N_4 composite fabrication for efficient solar-driven photocatalytic performance for anionic pollutant in an aqueous medium and mechanism insights, Opt Mater (Amst). 138 (2023) 113712.

21. N.T. Mai Tho, N. Van Cuong, V.H. Luu Thi, N.Q. Thang, P.H. Dang, A novel n–p heterojunction Bi_2S_3/$ZnCo_2O_4$ photocatalyst for boosting visible-light-driven photocatalytic performance toward indigo carmine, RSC Adv. 13 (2023) 16248–16259.

22. I. Ahmad, M. Muneer, A.S. Khder, S.A. Ahmed, Novel type-II heterojunction binary composite (CdS/AgI) with outstanding visible light-driven photocatalytic performances toward methyl Orange and tetracycline hydrochloride, ACS Omega. 8 (2023) 22708–22720.

23. W. Sun, S. Yang, Y. Liu, C. Shi, W. Shi, X. Lin, F. Guo, Y. Hong, Fabricating nitrogen-doped carbon dots (NCDs) on Bi3.64Mo0.36O6.55 nanospheres: A nanoheterostructure for enhanced photocatalytic performance for water purification, J Phys Chem Solids. 159 (2021) 110283.

24. A.P. Chowdhury, K.S. Anantharaju, S. Umare, S. Dhar, Facile fabrication of binary BiOCl-Cu_2CoSnS_4 and ternary BiOCl-Cu_2CoSnS_4-TiO_2 heterojunction nano photocatalyst for efficient sunlight-driven removal of direct blue 71 in an aqueous medium, Colloids Surf A Physicochem Eng Asp. 652 (2022) 129841.

25. D. Bokov, A. Turki Jalil, S. Chupradit, W. Suksatan, M. Javed Ansari, I.H. Shewael, G.H. Valiev, E. Kianfar, Nanomaterial by sol-gel method: Synthesis and application, Adv Mater Sci Eng. 2021 (2021) 1–21.

26. M. Sohaib, T. Iqbal, S. Afsheen, M.B. Tahir, A. Masood, M. Rafique, K.N. Riaz, M.A. Sayed, A.F. Abd El-Rehim, A.M. Ali, Novel sol–gel synthesis of Mo-doped ZnO-NPs for photo-catalytic waste water treatment using the RhB dye as a Model pollutant, Environ Dev Sustain. 25 (2023) 11583–11598.

27. M.J. Hakimi-Tehrani, S.A. Hassanzadeh-Tabrizi, N. Koupaei, A. Saffar, M. Rafiei, Synthesis of Z-scheme g-C_3N_4/WO_3 nano-photocatalyst with superior antibacterial characteristics for wastewater treatment, J Solgel Sci Technol. 105 (2023) 212–219.

28. E.S. Alsolami, I.A. Mkhalid, A. Shawky, M.A. Hussein, Sol–gel assisted growth of nanostructured NiS/CeO_2 p-n heterojunctions for fast photooxidation of ciprofloxacin antibiotic under visible light, Appl Nanosci. 13 (2023). 6445–6455.

29. M. Belghiti, L. El Mersly, K. Tanji, K. Belkodia, I. Lamsayety, K. Ouzaouit, H. Faqir, I. Benzakour, S. Rafqah, A. Outzourhit, Sol-gel combined mechano-thermal synthesis of Y_2O_3, CeO_2, and PdO partially coated ZnO for sulfamethazine and basic yellow 28 photodegradation under UV and visible light, Opt Mater (Amst). 136 (2023) 113458.

30. N. Alfryyan, M. Ikram, A. Manzoor, A. Jamil, Z.A. Alrowaili, M.S. Al-Buriahi, A. Irshad, M.I. Din, Synthesis of Cd-substituted $NiCoPrFe_2O_4$@CNTs via sol-gel method: Investigating the structural and photocatalytic properties, Physica B Condens Matter. 660 (2023) 414885.

31. N.Y. Tashkandi, A. Shawky, Enhanced visible-light photocatalytic remediation of atrazine over $CuAl_2O_4$-modified $BaSnO_3$ nanoplatelets synthesized by sol-gel route, J Alloys Compd. 968 (2023) 171826.

32. A.S. Basaleh, S.I. El-Hout, Sol-gel synthesis of photoactive PtO/Bi_2WO_6 nanocomposites for improved photoreduction of Hg (II) ions under visible illumination, Molecular Catalysis. 547 (2023) 113413.

33. R.K. Singh, R. Kumar, D.P. Singh, R. Savu, S.A. Moshkalev, Progress in microwave-assisted synthesis of quantum dots (graphene/carbon/semiconducting) for bioapplications: A review, Mater Today Chem. 12 (2019) 282–314.

34. V.G.S. Singh, N. Kaul, P.C. Ramamurthy, T. Naik, R. Viswanath, V. Kumar, H.S. Bhojya Naik, P. A, A.K. H A, J. Singh, N.A. Khan, Green synthesis of nickel-doped magnesium ferrite nanoparticles via combustion for facile microwave-assisted optical and photocatalytic applications, Environ Res. 235 (2023) 116598.

35. G. Divya, G. Jaishree, T. Sivarao, K.V.D. Lakshmi, Microwave assisted sol–gel approach for Zr doped TiO_2 as a benign photocatalyst for Bismark brown red dye pollutant, RSC Adv. 13 (2023) 8692–8705.

36. G. Sivakumari, M. Rajarajan, S. Senthilvelan, Microwave-assisted synthesis and characterization of activated carbon–zirconium-incorporated CeO_2 nanocomposites for photocatalytic and antimicrobial activity, Res Chem Intermed. 49 (2023) 3539–3561.

37. M. Arellano-Cortaza, E. Ramírez-Morales, S.J. Castillo, L. Lartundo-Rojas, I.Z.- Torres, E.M.L. Alejandro, L. Rojas-Blanco, Microwave-assisted hydrothermal synthesis of type II ZnSe/ZnO hetero-structures as photocatalysts for wastewater treatment, Ceram Int. 49 (2023) 24027–24037.

38. S.S.A. Al Ghafry, H. Al Shidhani, B. Al Farsi, R.G.S. Sofin, A.S. Al-Hosni, Z. Alsharji, J. Al-Sabahi, M.Z. Al-Abri, The photocatalytic degradation of phenol under solar irradiation using microwave-assisted Ag-doped ZnO nanostructures, Opt Mater (Amst). 135 (2023) 113272.

39. C. Cheng, Q. Shi, W. Zhu, Y. Zhang, W. Su, Z. Lu, J. Yan, K. Chen, Q. Wang, J. Li, Microwave-assisted synthesis of MoS_2/$BiVO_4$ heterojunction for photocatalytic degradation of tetracycline hydrochloride, Nanomaterials. 13 (2023) 1522.

40. C.L. Ücker, S.R. Almeida, R.G. Cantoneiro, L.O. Diehl, S. Cava, M.L. Moreira, E. Longo, C.W. Raubach, Study of $CaTiO_3$–ZnS heterostructure obtained by microwave-assisted solvothermal synthesis and its application in photocatalysis, J Phys Chem Solids. 172 (2023) 111050.

41. C. Li, T. Sun, G. Yi, D. Zhang, Y. Zhang, X. Lin, J. Liu, Z. Shi, Q. Lin, Microwave-assisted method syn-thesis of Ag/CNQDs/g-C_3N_4 with excellent photocatalytic activity for the degradation of norfloxacin, Colloids Surf A Physicochem Eng Asp. 662 (2023) 131001.

42. M. Murugalakshmi, K. Saravanakumar, C.M. Park, V. Muthuraj, Efficient photocatalytic degradation of sulfasalazine and reduction of hexavalent chromium over robust In_2S_3/Nd_2O_3 heterojunction under visible light, J Water Process Eng. 45 (2022) 102492.

43. V.S. Manikandan, S. Harish, J. Archana, M. Navaneethan, Fabrication of novel hybrid Z-Scheme WO_3@g-C_3N_4@MWCNT nanostructure for photocatalytic degradation of tetracycline and the evalua-tion of antimicrobial activity, Chemosphere. 287 (2022) 132050.

44. T. Chellapandi, G. Madhumitha, Facile synthesis and characterization of Carrisa edulis fruit extract capped NiO on MK30 surface material for photocatalytic behavior against organic pollutants, Mater Lett. 330 (2023) 133215.

45. Y. Li, L. Li, C. Li, W. Chen, M. Zeng, Carbon nanotube/titania composites prepared by a micro-emulsion method exhibiting improved photocatalytic activity, Appl Catal A Gen. 427–428 (2012) 1–7.

46. H. Maqbool, I. Bibi, Z. Nazeer, F. Majid, S. Ata, Q. Raza, M. Iqbal, Y. Slimani, M.I. Khan, M. Fatima, Effect of dopant on ferroelectric, dielectric and photocatalytic properties of Co–La-doped dysprosium chromite prepared via microemulsion route, Ceram Int. 48 (2022) 31763–31772.

47. M. Hanif, Y. Kim, S. Ameen, H. Kim, L. Kwac, Boosting the visible light photocatalytic activity of ZnO through the incorporation of N-doped for wastewater treatment, Coatings. 12 (2022) 579.

48. H.R. Ahmed, K.H. Hama Aziz, N.N.M. Agha, F.S. Mustafa, S.J. Hinder, Iron-loaded carbon black prepared *via* chemical vapor deposition as an efficient peroxydisulfate activator for the removal of rhodamine B from water, RSC Adv. 13 (2023) 26252–26266.

49. C. Wang, W. Shi, K. Zhu, X. Luan, P. Yang, Chemical vapor deposition growth of MoS_2 on g-C_3N_4 nanosheets for efficient removal of tetracycline hydrochloride, Langmuir. 38 (2022) 5934–5942.

28 Future of the Semiconductor Industry

*Sujit Mukherjee, Debmalya Pal,
Arunava Bhattacharyya, and Subhasis Roy*

28.1 INTRODUCTION

The semiconductor industry serves as a foundational pillar of modern technological progress, providing the backbone for a wide range of electronic devices that have revolutionized our lifestyles, work environments, and modes of communication [1]. From smartphones and laptops to artificial intelligence (AI) and renewable energy systems, semiconductors form the bedrock upon which these innovations are built. As we gaze into the future, it is evident that the semiconductor industry is poised to continue its impressive evolution, driven by a synergy of technological breakthroughs, market dynamics, and societal demands. In the present era of rapid technological advancement, the semiconductor sector encounters a dual landscape of unprecedented challenges and promising opportunities. The insatiable demand for electronic components that are smaller, faster, and more energy-efficient is spurred by emerging trends like the Internet of Things (IoT), 5G connectivity, augmented reality, and advanced AI [2]. To cater to these trends, the industry must forge innovative solutions that stretch existing manufacturing processes' boundaries while fostering novel design, materials, and system integration approaches. However, as the pursuit of enhanced performance and efficiency intensifies, the semiconductor field also faces considerable hurdles. The miniaturization of transistors, as envisioned by Moore's law [3], is reaching tangible limits, making it progressively difficult to sustain the historical pace of shrinking chip dimensions and escalating transistor counts. This prompts inquiries about the sustainability of the exponential computing power growth that characterizes the industry. Moreover, the industry must adeptly navigate intricate geopolitical dynamics, vulnerabilities within supply chains, and concerns about the environmental impact of semiconductor device production and disposal [4, 5]. Given the vital role of semiconductors in critical infrastructure and fundamental technologies, ensuring an uninterrupted supply chain and mitigating potential disruptions become critical responsibilities for industry stakeholders and policymakers. A convergence of factors will sculpt the trajectory for the semiconductor domain. These include pioneering advances in materials like carbon-based transistors, strides in quantum computing, the evolution of 3D integration techniques [6], and the creation of neuromorphic computing models inspired by the human brain. Furthermore, an amplified emphasis on sustainability and ethically responsible manufacturing practices is poised to shape the trajectory of the industry, catalyzing the exploration of environmentally friendly fabrication methods and the development of more energy-efficient devices. In this exploration of the future course of the semiconductor sector, it becomes abundantly clear that the path ahead is one characterized by perpetual adaptation and innovation. As various stakeholders collaborate to surmount challenges and capitalize on opportunities, the semiconductor industry is positioned not just to mold the technological landscape but also to carve out the parameters of advancement in the digital age. The dawn of the Third Industrial Revolution, often called the digital revolution, has orchestrated today's highly automated and processor-driven world [7]. This revolution, which commenced toward the close of the previous century, was distinguished by the emergence of electronics, telecommunications, and computational prowess [8]. Semiconductors constitute an integrated circuit or microchip [9]. Functionally, a semiconductor conducts energy between two materials while blocking the flow to others.

DOI: 10.1201/9781003450146-28

359

Semiconductors play a pivotal role in electronics, exhibiting properties akin to conductors and insulators. In the contemporary world, most electronic components encompass a semiconductor as their fundamental element. Common devices, from medical equipment and watches to cell phones, computers, and laptops, rely on semiconductors. Integrating semiconductors in electronics contributes to heightened speed, efficiency, and affordability. This fundamental notion gave birth to the semiconductor chip, the powerhouse that fuels nearly all digital devices. Around the inception of the digital revolution, Gordon Moore emerged as a visionary who foresaw the ascent of computing power [10]. His prophetic insight, now known as Moore's law, posited that the transistor count on a single microchip would double every two years, resulting in exponential increments in computing capacity while concurrently halving energy costs. Consequently, transistors per microchip have surged from 10 to 30 billion US dollar, ushering in new technological progress [11]. In the present context, AI stands poised to act as the impetus for a fresh, decade-spanning surge within the semiconductor industry. Forecasts predict that the market for AI-enabled semiconductors will burgeon from its current $6 billion in revenue to exceed $30 billion by 2022, boasting a compound annual growth rate (CAGR) of approximately 50% [12]. The trajectory of investments in semiconductor technology will determine the industry's dimensions, growth pace, and the accessibility and benefits it confers.

28.2   BRIEF ABOUT THE CURRENT SCENARIO OF THE SEMICONDUCTOR INDUSTRY

The semiconductor organization is currently in a state of flux. The international chip scarcity that began in 2020 is starting to ease, but the call for semiconductors stays sturdy.

Many elements contribute to this: the ongoing growth of IoT, the rise of electric cars (EVs), and the full-scale adoption of 5G generation. Due to this robust demand, the semiconductor industry is anticipated to develop similarly in the coming years. Market share of semiconductors in various sectors is shown in Figure 28.1. This semiconductor industry faced several significant developments and challenges, including supply chain constraints. The industry was grappling with supply chain disruptions caused by factors such as the COVID-19 pandemic, geopolitical tensions, and natural disasters. These disruptions led to shortages of critical components and increased lead times.

a. *Increased Demand*: The growing demand for semiconductors was driven by various factors, including the expansion of 5G networks, the rise of AI and machine learning, IoT, and advancements in automotive technology.

FIGURE 28.1   Market share of semiconductors in various sectors.

b. *Technological Advancements*: The industry continued to push the boundaries of technology by developing smaller and more power-efficient chips and exploring new materials and manufacturing techniques, such as extreme ultraviolet (EUV) lithography.

c. *Geopolitical Considerations*: Geopolitical tensions were impacting the semiconductor industry, with issues related to trade restrictions, export controls, and intellectual property rights affecting global supply chains and collaborations.

d. *Investments and Acquisitions*: Various companies invested significantly in semiconductor research, development, and manufacturing capacity. Additionally, there were notable mergers and acquisitions within the industry.

e. *Environmental Concerns*: The industry was also focusing on sustainability and reducing its environmental impact, with efforts to develop more energy-efficient technologies and reduce the use of hazardous materials.

28.2.1 CURRENT WORLDWIDE SEMICONDUCTOR MARKETPLACE

In 2021, the global semiconductor industry was expected to be worth $429.5 billion [13]. Asia Pacific dominates the worldwide semiconductor industry, retaining first-class market share in 2021. Broadcom Inc. (US), Intel Corporation (US), Qualcomm (US), Samsung Electronics (South Korea), SK Hynix (South Korea), Taiwan Semiconductors (Taiwan), Texas Instruments (US), Toshiba Corporation (Japan), Maxim Integrated Products, Inc. (US), Micron Technology (US), NVIDIA Corporation (US), and NXP Semiconductors N.V. (Netherlands) are prominent market leaders [14]. In the coming decade, we will almost certainly see many of the world's tech behemoths added to this list, as companies like Apple, Google, and Amazon enter the market [15].

The global semiconductor device market is anticipated to reach USD 911.13 billion near future, exhibiting a compound annual growth rate (CAGR) of 8.19% [16]. Here is a pinnacle-level view of the modern country of the semiconductor employer. With reduced chip scarcity peaking in 2021, the global chip shortage is ultimately beginning to decrease. This is because of the improved potential of semiconductor manufacturers, easing COVID-19 guidelines, and decreased demand for sure electronic additives [17]. Demand for semiconductors remains robust. Despite efforts to mitigate the chip scarcity, the demand for semiconductors remains robust, fueled by factors such as the the the ongoing advancement of the Internet of Things (IoT), the proliferation of electric vehicles (EVs), and the widespread adoption of 5G technology. The semiconductor industry is anticipated to continue developing. This growth is expected to be propelled by the increasing demand for semiconductors across a wide variety of applications. However, the semiconductor industry is confronted with a myriad of challenging situations [18]. Also, the semiconductor industry faces some worrying situations, including the continued chip scarcity, the growing cost of producing semiconductors, the growing complexity of semiconductor designs, and the dearth of expert employees in the semiconductor industry.

28.2.2 CURRENT TECHNOLOGICAL IMPROVEMENTS

The semiconductor industry is constantly evolving, with new technology being developed continuously. Some of the most promising technological upgrades inside the modern-day semiconductor enterprise include the following.

28.2.2.1 3-D Chip Stacking

This technology enables the stacking of multiple layers of transistors on the top of each other, potentially resulting in enhanced performance and energy efficiency [19].

28.2.2.2 Gate-all-round Transistors

These transistors are more green than conventional transistors, and they can be used to create smaller and quicker chips, shifting toward superior manufacturing strategies together with 7 nm and 5 nm nodes [20].

28.2.2.3 New Substances

New materials, along with graphene and gallium arsenide, are being explored for semiconductors. These materials offer potential benefits over silicon, improving velocity and performance.

28.2.2.4 Artificial Intelligence

AI is being used to enhance the layout and manufacturing of semiconductors. AI can assist in optimizing the layout of transistors and becoming aware of defects in chips. Additionally, there has been elevated emphasis on specialized chips like graphics processing units (GPUs) for AI and system getting-to-know programs.

28.2.3 Current Geopolitics in the Semiconductor Industry

The semiconductor enterprise is increasingly becoming a geopolitical battleground. The United States, China, and other countries are competing to secure the right of entry to brand-new semiconductor technologies, as those chips are vital for various industries, including defense, telecommunications, and production. The United States has taken several steps to try to steady its very own semiconductor delivery chain. In 2021, the U.S. Congress passed the CHIPS Act, which affords $52 billion in investment for semiconductor studies and development [21]. The U.S. authorities have also imposed sanctions on Chinese semiconductor businesses, including SMIC, to restrict China's admission to cutting-edge semiconductor technologies [22]. The geopolitical opposition over semiconductors is likely to hold in the future. The final results of this competition can have a massive impact on the global financial system and the safety of the U.S. and its allies. The key geopolitical trends that are shaping the semiconductor industry are as follows.

28.2.3.1 The Rise of China

China is the arena's largest semiconductor marketplace, and it is also one of the main producers of semiconductors. The Chinese authorities intend to make China a global chief in semiconductor manufacturing by 2025. China has been constructing new semiconductor foundries and is investing in studies and improvement, and it is looking to become an international leader in semiconductor production.

28.2.3.2 The U.S.-China Competition

The U.S.-China opposition is likewise playing out within the semiconductor enterprise. The U.S. government has imposed sanctions on Chinese semiconductor companies and is attempting to encourage major U.S. corporations to manufacture semiconductors inside the United States and Taiwan.

28.2.3.3 The Worldwide Chip Scarcity

The worldwide chip shortage has moreover highlighted the significance of semiconductors. The scarcity has disrupted delivery chains and caused expenses to upward thrust. This has caused calls for governments to ensure a more consistent and resilient semiconductor delivery chain. The geopolitical panorama of the semiconductor enterprise is complicated and ever-converting. However, the above tendencies will likely shape the employer in the coming years.

28.2.4 CURRENT INDUSTRY CONSOLIDATION

The semiconductor enterprise has seen enormous industry consolidation in recent years. This is because of the growing value of manufacturing semiconductors, the need to maintain economies of scale to be profitable, and the growing significance of semiconductors in various industries. In 2011, Intel purchased Altera for $16.7 billion [23]. In 2015, Avago Technologies bought Broadcom for $37 billion [24]. 2016, Qualcomm bought NXP Semiconductors for $44 billion [25]. In 2020, AMD acquired Xilinx for $35 billion [26]. These are some examples of industry consolidation in the semiconductor enterprise.

28.2.5 SUPPLY AND DEMAND DYNAMICS

The Semiconductor Industry Association (SIA) is an organization representing the semiconductor industry. The SIA publishes many reports and research on the semiconductor industry, including delivery reports and calls for dynamics within the modern scenario. The World Semiconductor Trade Statistics (WSTS) is an international company that tracks the semiconductor industry. The WSTS publishes a monthly report on the worldwide semiconductor market, including statistics on delivery and demand within the current scenario. The International Technology Roadmap for Semiconductors (ITRS) is a roadmap that outlines the destiny of the semiconductor enterprise [27]. The ITRS consists of statistics on the expected call for semiconductors in the future, in addition to the delivery and demand dynamics inside the modern situation. The Semiconductor Engineering website is a complete resource for information, evaluation, and statements on the semiconductor industry. The website includes a segment on delivery and a call for dynamics, which presents updated facts on the present-day scenario. The chip scarcity is easing; however, calls for semiconductors stay robust. The industry is predicted to keep growing within the coming years; however, it must overcome some challenges to meet this call.

28.3 RECENT ADVANCEMENTS IN THE SEMICONDUCTOR INDUSTRY

The semiconductor industry has witnessed remarkable advancements in recent years, transforming various aspects of technology and shaping our modern world. Semiconductors are the foundation of electronic devices, enabling the development of innovative products across industries such as telecommunications, computing, automotive, healthcare, and renewable energy.

28.3.1 MINIATURIZATION

The continual miniaturization of electronic components is one of the most important themes in the semiconductor industry. Moore's law, stating that the number of transistors on integrated circuits would double roughly every two years, has been true for many years. However, as transistor sizes approach atomic limits, researchers have faced significant challenges in further scaling down the size of semiconductor devices. The industry has investigated alternate strategies, including three-dimensional (3D) integration and new materials to tackle these difficulties. Multiple layers of transistors can be stacked thanks to 3D integration, boosting component density without further reducing their size. Additionally, the creation of novel materials holds promise for future miniaturization because they have outstanding electrical and thermal properties at nanoscale scales, such as graphene and carbon nanotubes [28].

28.3.2 ADVANCED MATERIALS

Creating and incorporating new materials has largely driven advancements in the semiconductor industry. Due to its superior electrical characteristics, silicon has been the most widely used material in the production of semiconductors. Alternative materials, however, have arisen to get around these restrictions as conventional silicon-based technologies approach their limits. One such substance is gallium nitride (GaN), which outperforms silicon in power efficiency, switching speed, and temperature tolerance. EVs, renewable energy systems, and power electronics increasingly employ GaN-based power devices. Indium gallium arsenide (InGaAs), which permits high-speed and low-noise operations in optical communications, is another substance that has attracted interest [29]. InGaAs-based photodetectors and lasers have revolutionized data transmission rates in fiber-optic networks, enabling quicker and more dependable data transport over vast distances.

28.3.3 ARTIFICIAL INTELLIGENCE INTEGRATION

New opportunities in numerous industries have emerged due to the combination of semiconductor technologies and AI. Due to the high computational demands of AI algorithms, semiconductor producers have created specialized AI processors, also called AI chips or accelerators. AI chips are made to efficiently carry out tasks linked to AI, like inference and training of neural networks. These chips have cutting-edge architectures that are tailored for AI workloads, including GPUs and field-programmable gate arrays (FPGAs) [30]. Computer vision, natural language processing, autonomous cars, and robotics have all seen substantial advancements due to incorporating AI chips into products and systems.

28.3.4 QUANTUM COMPUTING

The concepts of quantum mechanics are used in quantum computing, a paradigm shift that allows for computations that are impossible with conventional computers. Recent developments in the semiconductor sector have pushed quantum computing closer to being used in everyday life. Quantum computers use quantum bits, or qubits, with exponentially more processing capacity because they may exist simultaneously in several states [31]. In order to create scalable and trustworthy quantum systems, semiconductor-based techniques such as superconducting qubits and trapped ion qubits have shown promise. The development of quantum computers, which have the potential to revolutionize industries, including materials science, optimization, drug discovery, and cryptography, is currently being intensively pursued by leading technology businesses and research organizations.

28.3.5 PREPARATION OF NANO-HETEROJUNCTION PHOTOCATALYSTS USING MULTIFUNCTIONAL SEMICONDUCTORS

Burning conventional energy sources like coal, methane, and petrol emits several greenhouse gases, including CO, CO_2, and N_2O, a major danger to human health and the environment. As a result, there is a dire need for effective techniques for decreasing pollution and establishing clean, ecologically friendly energy sources.

Utilizing semiconductor photocatalysis to harness abundant solar energy presents a promising strategy for tackling the energy crisis and escalating environmental concerns. This approach entails producing hydrogen and detoxifying organic contaminants through solar energy. Recent years have witnessed extensive research endeavors to enhance the efficiency of semiconductor photocatalytic materials. A schematic of the photocatalytic process of a semiconductor is represented in Figure 28.2. To ensure an effective photocatalytic process, employing semiconductors characterized by a wide range of optical absorption, efficient separation of charge carriers, and robust chemical stability in wet environments is imperative. Originally, considerable attention was directed toward broad bandgap semiconductors like TiO_2 as typical photocatalytic materials [33]. However, the limited optical absorption of such materials prompted the realization of the need for narrow band gap semiconductors. These materials can significantly expand the range of optical absorption effectiveness of solar energy. In parallel with the ongoing advancement of visible-light-responsive semiconductor materials, the enhancement of their chemical stability and the reduction

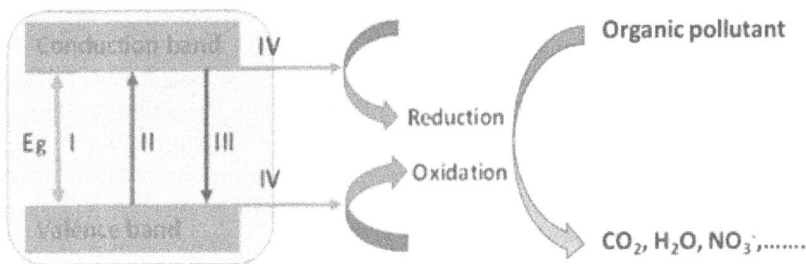

FIGURE 28.2 Photocatalytic process of a semiconductor. (Adapted with permission [32]. Copyright [2020] American Chemical Society.)

of photogenerated charge carrier recombination is of paramount importance. In recent studies, various functional semiconductors, including ferroelectric, ferromagnetic, and graphene-based photosensitizers, have emerged to address these challenges. For instance, ferroelectric materials offer the advantage of generating an internal electric field that aids the separation of photogenerated charge carriers efficiently.

Similarly, investigations into ferromagnetic semiconductors have unveiled heightened photocatalytic activity when subjected to external magnetic fields. Despite these promising developments, inherent limitations such as constrained solar spectrum absorption, limited carrier diffusion length, and heightened recombination rates have revealed the impracticality of achieving high efficiency and stability using a solitary semiconductor photocatalyst. To surmount these challenges, researchers have turned to the design of multicomponent semiconductor heterostructures. By combining two or more semiconductors with distinct functionalities, these heterostructures have demonstrated a remarkable capacity to enhance the efficiency and stability of photocatalytic materials.

Fascinating reviews have recently emerged in the realm of photocatalyst development, focusing on the utilization of organic semiconductor/semiconductor heterostructures, facet-dependent and interfacial plane-related photocatalytic heterostructures, heterostructure semiconductor nanowires, and graphitic carbon nitride-based metal sulfide heterojunctions [34–36]. Integrating ferroelectric and ferromagnetic materials into nano-heterostructures presents a promising avenue to enhance the efficiency of transferring charge through electrical and magnetic tunneling.

While significant strides have been taken in the realm of semiconductor heterostructure production, there remains a need for more refined research endeavors aimed at creating heterostructure photocatalysts that are simultaneously efficient and stable. Band edge positions of different semiconductor heterostructures are shown in Figure 28.3. Establishing optimal interfacial contacts

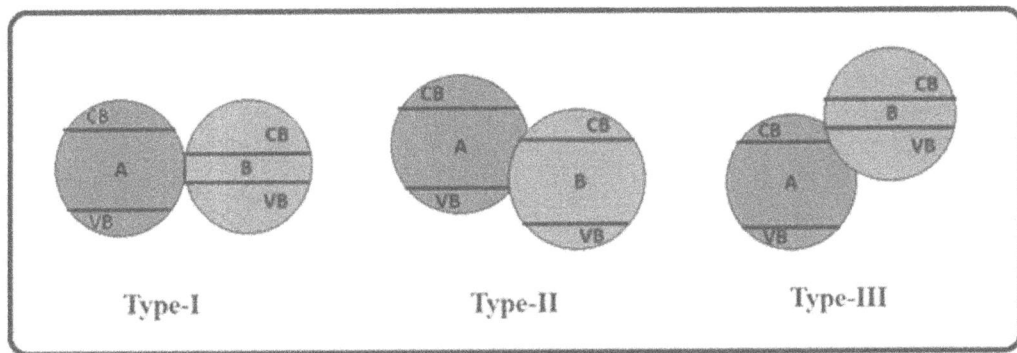

FIGURE 28.3 Band edge positions of different types of semiconductor heterostructures. (Adapted with permission [32]. Copyright [2020] American Chemical Society.)

among diverse constituent layers is pivotal in multicomponent heterostructure systems. Therefore, future progress in crafting heterojunction photocatalysts should emphasize interfacial engineering, a critical means to achieve favorable band alignment and proficient charge transport capabilities. Furthermore, exploring novel photocatalytic materials with suitable band alignment for visible light activation holds potential as an effective avenue for advancing this field. Integrating multifunctional semiconductors, like ferroelectric and ferromagnetic materials, into heterostructure photocatalysts offers a promising strategy to attain heightened efficiency. Nevertheless, it is imperative to undertake systematic investigations employing advanced spectroscopic techniques to gain a comprehensive understanding of photogenerated charge carrier migration pathways. Moreover, a comprehensive grasp of charge migration kinetics at the interface necessitates theoretical calculations and modeling approaches [37].

28.4 EFFICIENCY ENHANCEMENT TECHNIQUES OF SEMICONDUCTORS

Now let us explore various techniques used to enhance the efficiency of semiconductors and their impact on electronics.

28.4.1 MATERIALS ENGINEERING

Materials engineering is one of the primary methods to boost semiconductor efficiency. It is feasible to improve the qualities of the materials used in semiconductor production by carefully choosing and modifying them. For instance, scientists have created brand-new semiconductor materials with high carrier mobility, like silicon carbide (SiC) and gallium nitride (GaN). These materials enable semiconductors to operate at greater frequencies and power levels while reducing losses because of their improved thermal conductivity and higher breakdown voltages.

28.4.2 BAND GAP ENGINEERING

The energy band structure of semiconductors can be changed using band gap engineering to improve performance. Efficiency can be increased by modifying the band gap or the energy difference between the valence and conduction bands. In particular, superlattices and quantum wells can be used to build structures with different bandgaps. The absorption and emission properties of optoelectronic devices, such as light-emitting diodes (LEDs) and solar cells, have been improved using this technology, leading to increased energy conversion efficiency.

28.4.3 DOPING TECHNIQUES

To change the electrical characteristics of semiconductor materials, doping is adding impurities. The conductivity of semiconductors can be changed by deliberately adding dopant atoms. For instance, doping with impurities like boron or phosphorus can form p-type or n-type areas in silicon-based semiconductors. Due to its ability to precisely manage charge carriers and increase overall efficiency, this method is essential for creating transistors and other electronic components.

28.4.4 SURFACE PASSIVATION

Surface recombination in semiconductors, which can result in large efficiency losses, is reduced using the surface passivation approach. The surface flaws of the semiconductor can be efficiently passivated by coating them with a thin layer of a passivation substance, like silicon nitride or silicon dioxide. This enhances the device's overall efficiency by preventing charge carrier recombination at the surface. Surface passivation techniques are commonly used to improve the performance of photovoltaics and integrated circuits.

28.4.5 ADVANCED FABRICATION PROCESSES

Developments have greatly aided the enhancement of efficiency in semiconductor production methods. Material deposition, etching, and patterning can be precisely controlled using methods including atomic layer deposition (ALD), plasma-enhanced chemical vapor deposition (PECVD), and reactive ion etching (RIE) [38–40]. These procedures make it possible to produce nanoscale structures with extreme accuracy, which lowers parasitic resistance and capacitance and enhances overall device performance.

28.4.6 POWER MANAGEMENT

Power management strategy optimization is a part of improving semiconductor efficiency. Power management integrated circuits (PMICs) control voltage levels and distribute power to semiconductor devices effectively. The power consumption of semiconductors can be reduced using methods like clock gating, power gating, and dynamic voltage scaling. As a result, portable electronic devices' batteries last longer and are more energy efficient.

28.4.7 ADVANCED PACKAGING

Their packaging greatly influences the efficiency of semiconductors as a whole. To improve heat dissipation and reduce power losses, advanced packaging techniques are used, which enable semiconductors to function at higher power levels. Better electrical and thermal connections between the semiconductor chip and the package are made possible by flip-chip bonding, 3D integration, and wafer-level packaging, reducing parasitic resistance and capacitance. Higher operational efficiency and improved power supply result from this.

28.5 APPLICATIONS OF SEMICONDUCTORS

The wide range of applications of semiconductors across various industries and technologies has been very effective in the current situation. The common applications of semiconductors are briefly discussed here.

28.5.1 MANUFACTURING OF ELECTRONICS

Semiconductors form the foundation of modern electronics. They are used to manufacture transistors, diodes, and integrated circuits (ICs). Transistors are fundamental building blocks of electronic devices, serving as amplifiers and switches. Diodes are crucial for controlling the flow of current in electronic circuits. Integrated circuits, which are miniature electronic circuits integrated onto a single chip, enable the creation of complex electronic systems, such as microprocessors and memory chips. Figure 28.4 shows the several applications of semiconductors in the manufacturing of electronic devices.

28.5.2 COMPUTING AND INFORMATION TECHNOLOGY

Semiconductors play a pivotal role in the computing and information technology sectors. They are used to design and produce computer processors, memory chips, and storage devices. The advancements in semiconductor technology, particularly the miniaturization of transistors (as per Moore's law), have led to the development of increasingly powerful and efficient computer systems.

28.5.3 COMMUNICATION SYSTEMS

Semiconductors are extensively used in communication systems, including wired and wireless networks. They are vital for developing devices such as routers, switches, modems, and network

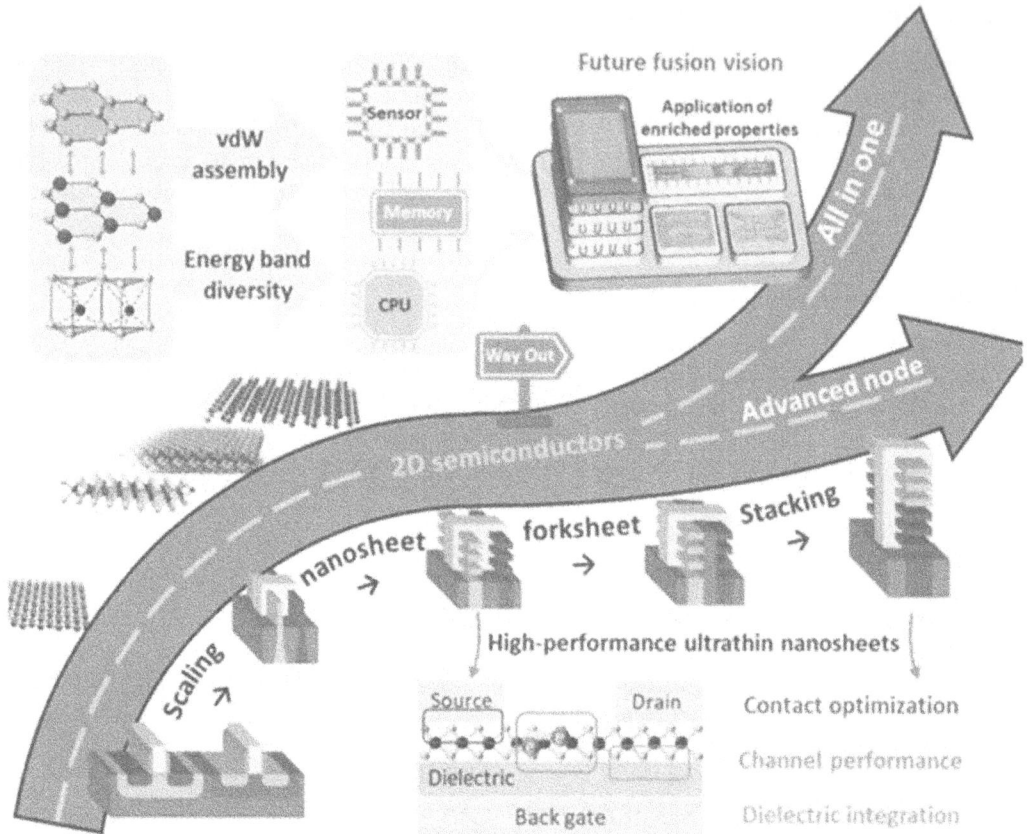

FIGURE 28.4 A schematic diagram of the potential road of specific electrical applications for 2D semiconductors. (Adapted with permission [41]. Open Access licensed under a Creative Commons Attribution 4.0 International License.)

interface cards. Semiconductors enable signal processing, amplification, and modulation/demodulation, allowing efficient data transmission and communication between devices.

28.5.4 OPTOELECTRONICS AND PHOTONICS

Semiconductors find extensive application in optoelectronic and photonic devices. LEDs are semiconductors that emit light when electricity passes through them. LEDs are used in various applications, including lighting, display panels, signage, and indicators. Semiconductor lasers, such as diode lasers, are crucial in optical communication systems, laser printing, and laser cutting. Photodetectors, solar cells, and image sensors rely on semiconductors for efficient light detection and energy conversion.

28.5.5 RENEWABLE ENERGY GENERATION

Semiconductors are important in renewable energy systems, especially solar energy generation. Solar cells, also known as photovoltaic cells, convert light (mainly sunlight) energy directly into electric energy using semiconductors. Silicon-based solar cells are the most prevalent, but other semiconductor materials, like gallium arsenide, are also used for specialized applications. Semiconductors are also employed in wind turbines and energy storage systems, contributing to the development of sustainable energy infrastructure.

28.5.6 AUTOMOTIVE INDUSTRY

Semiconductors have become increasingly important in the automotive industry. They are utilized in various applications, including engine control units (ECUs), navigation systems, advanced driver-assistance systems (ADAS), airbag systems, anti-lock braking systems (ABS), and entertainment systems [42]. The growth of EVs has further heightened the demand for semiconductors in power electronics and battery management systems.

28.5.7 MEDICAL DEVICES AND HEALTHCARE

Semiconductors are crucial in medical devices, where precision, miniaturization, and reliability are essential. They are used in devices like X-ray machines, MRI scanners, ultrasound systems, pacemakers, insulin pumps, and blood glucose monitors. Semiconductors enable precise sensing, imaging, signal processing, and control in medical equipment, contributing to better diagnostics, treatment, and patient care.

28.5.8 AEROSPACE AND DEFENSE

The aerospace and defense sectors heavily rely on semiconductors for various applications. Semiconductors are utilized in satellite communication, radar, guidance, avionics, and military systems.

These are just a few examples of the diverse applications of semiconductors. As technology advances, semiconductors will likely find new and innovative applications in various industries.

28.6 BARRIERS AND CHALLENGES OF THE SEMICONDUCTOR INDUSTRY

The semiconductor industry is a critical and complex sector foundational to modern technology. While it has driven incredible advancements, it also faces numerous barriers and challenges. Some of the key barriers and challenges faced by the semiconductor industry are:

a. *Technological Complexity and Miniaturization*: The semiconductor industry constantly drives to increase the number of transistors on a single chip, following Moore's law. As components shrink to nanoscale sizes, it becomes increasingly difficult to maintain performance, power efficiency, and reliability, leading to challenges in manufacturing and design.

b. *Cost of Research and Development*: Developing new semiconductor technologies requires substantial investments in research, development, and innovation. The cost of designing cutting-edge semiconductor processes and equipment can run into billions of dollars, making it a barrier for smaller players to compete.

c. *Capital-Intensive Manufacturing*: Establishing and maintaining semiconductor manufacturing facilities (fabs) is extremely capital-intensive. The costs associated with building, equipping, and maintaining these facilities can be overwhelming, limiting the number of companies that can enter the market.

d. *Global Supply Chain Vulnerabilities*: The semiconductor industry relies on a global supply chain for raw materials, equipment, and manufacturing. Any disruption in the supply chain, such as geopolitical tensions, natural disasters, or pandemics (as witnessed with COVID-19), can lead to shortages and production delays.

e. *Geopolitical Challenges*: The semiconductor industry operates across international boundaries, making it susceptible to geopolitical tensions and trade restrictions. Export controls and trade conflicts can disrupt the flow of materials, technologies, and products, affecting the entire industry.

f. *Intellectual Property Protection*: Protecting intellectual property (IP) is a significant concern in the semiconductor industry. Reverse engineering and counterfeiting can lead to the unauthorized replication of advanced technologies, undermining the competitive advantage of companies that invest heavily in R&D.

g. *Talent Shortages*: The semiconductor industry demands a highly skilled and specialized workforce. A shortage of qualified chip design, process engineering, and advanced manufacturing professionals can impede innovation and growth.

h. *Environmental Concerns*: The semiconductor manufacturing process involves using hazardous chemicals and producing large amounts of waste, raising environmental concerns. Stricter regulations and pressure to adopt more sustainable practices can increase operational costs and complexity.

i. *Complex Packaging and Integration*: Integrating diverse components into a single chip package requires advanced packaging technologies. Packaging challenges increase as chip designs become more intricate, affecting performance, heat dissipation, and reliability.

j. *Quality Control and Yield Enhancement*: Ensuring high yields (the proportion of functional chips from a batch) is essential to maintain profitability. The smaller feature sizes and complex manufacturing processes make it challenging to control defects, leading to lower yields and increased costs.

k. *Cyclical Nature of the Industry*: The semiconductor industry is known for its cyclical nature, characterized by periods of high demand and growth followed by downturns. This cyclicality can lead to overcapacity during boom times and financial strain during downturns.

l. *Security Vulnerabilities*: With the increasing connectivity of devices, cybersecurity is a significant concern. Malicious actors could exploit vulnerabilities in semiconductor designs to gain unauthorized access or control over systems, leading to data breaches and other security risks.

The semiconductor industry faces many barriers and challenges due to its technological complexity, high costs, supply chain vulnerabilities, geopolitical factors, environmental concerns, and more. However, the industry's ability to continuously innovate and adapt has historically allowed it to overcome these challenges and drive technological progress.

28.7   PROSPECTS OF THE SEMICONDUCTOR INDUSTRY

28.7.1   Continued Technological Advancements

The semiconductor industry has a history of pushing the boundaries of technology. While the traditional trends of shrinking transistor sizes and increasing chip density are becoming more challenging due to physical limitations, innovations in materials, 3D integration, and new architectures are expected to drive continued improvements in performance, power efficiency, and functionality.

28.7.2   Post-Moore's Law Era

Moore's law, which predicted the doubling of transistor density every couple of years, shows signs of slowing down due to fundamental physical limitations. The industry is entering an era where the focus has shifted from merely increasing transistor count to optimizing performance through various architectural and design innovations. This might involve leveraging heterogeneous architectures, combining different processing elements on a single chip, and optimizing for specific workloads.

28.7.3   Specialized Chip Designs

As applications like AI, machine learning, data analytics, and cryptocurrency mining gain prominence, there is a growing need for specialized hardware solutions to accelerate these workloads.

Custom-designed application-specific integrated circuits (ASICs) and field-programmable gate arrays (FPGAs) are expected to play a crucial role in meeting the demand for efficient and high-performance computing [43, 44].

28.7.4 More than Moore (MTM) Approach

The "More than Moore" (MTM) [45] approach involves developing technologies and functionalities beyond traditional scaling. This can include sensor innovations, power management, micro-electro-mechanical systems (MEMS), and 3D integration techniques [46]. This approach allows for integrating various functions onto a single chip, enabling more diverse and sophisticated applications.

28.7.5 Quantum Computing and Neuromorphic Computing

Quantum and neuromorphic computing represent radically different computing paradigms. Quantum computers can solve complex problems that normal computers cannot solve. At the same time, neuromorphic computing draws inspiration from the human brain's architecture and can excel at tasks like pattern recognition and optimization.

28.7.6 Advanced Packaging and Interconnects

Innovations in packaging and interconnect technologies are crucial for improving chip performance, power efficiency, and thermal management. Advanced packaging techniques like system-in-package (SiP), chip-on-wafer-on-substrate (CoWoS), and 3D stacking allow for greater integration of multiple chips and functionalities within a single package [47].

28.7.7 Sustainability and Environmental Concerns

Environmental considerations are gaining prominence in the semiconductor industry. There's a growing emphasis on developing more energy-efficient manufacturing processes, reducing the environmental impact of electronic waste, and utilizing eco-friendly materials [48].

28.7.8 Supply Chain Diversification

The COVID-19 pandemic highlighted vulnerabilities in global supply chains. Semiconductor companies are likely to invest in diversifying their supply chains and exploring more localized manufacturing to mitigate future disruptions.

28.7.9 Geopolitical Factors and Trade Dynamics

Geopolitical tensions can impact global trade and technology flows. Companies might need to navigate trade restrictions, export controls, and intellectual property concerns, which could influence their business strategies and global operations.

28.7.10 Skilled Workforce and Talent Acquisition

The semiconductor industry relies heavily on a skilled workforce for research, design, manufacturing, and innovation. Companies must invest in education and training programs to ensure a steady supply of skilled professionals.

28.7.11 Regulatory and Ethical Considerations

Advancements in technology, such as AI and autonomous systems, raise ethical and regulatory questions. The semiconductor industry must collaborate with policymakers to establish appropriate regulations and standards to ensure these technologically responsible development and deployment.

28.8 CONCLUSIVE REMARKS

The future of the semiconductor industry holds significant promise and challenges. As technology advances, semiconductors will remain the bedrock of modern electronics, powering AI, IoT, 5G, and beyond innovations. However, the industry must navigate complexities such as miniaturization limits, energy efficiency concerns, and geopolitical factors affecting supply chains. Continued investment in research, collaboration, and adaptation to new materials and manufacturing techniques will be crucial for sustaining growth and driving the next wave of technological breakthroughs. In conclusion, the future of the semiconductor industry appears to be characterized by a shift from traditional scaling to more specialized and diverse technological advancements. Innovations will likely focus on improving performance, efficiency, and functionality through various approaches while considering environmental, regulatory, and geopolitical factors.

ACKNOWLEDGMENTS

Author (S. Roy) would like to acknowledge "Scheme for Transformational and Advanced Research in Sciences (STARS)" (MoE-STARS/STARS-2/2023-0175) by the Ministry of Education, Govt. of India for promoting translational India-centric research in sciences implemented and managed by Indian Institute of Science (IISc), Bangalore, for their support.

REFERENCES

1. Irwin, D., & Klenow, P. (1994). Learning-by-doing spillovers in the semiconductor industry. Journal of Political Economy, 102(6), 1200–1227.
2. Rose, K., Eldridge, S., & Chapin, L. (2015). The internet of things: An overview. The Internet Society (ISOC), 80, 1–50.
3. Lundstrom, M. (2003). Moore's law forever? Science, 299(5604), 210–211.
4. Dahlben, L., Eckelman, M., Hakimian, A., Somu, S., & Isaacs, J. (2013). Environmental life cycle assessment of a carbon nanotube-enabled semiconductor device. Environmental Science & Technology, 47(15), 8471–8478.
5. Uryu, T., Yoshinaga, J., & Yanagisawa, Y. (2003). Environmental fate of gallium arsenide semiconductor disposal: A case study of mobile phones. Journal of Industrial Ecology, 7(2), 103–112.
6. Chanchani, R. (2009). 3D Integration Technologies–An Overview. Materials for Advanced Packaging, 1–50.
7. Mowery, D. (2009). Plus ca change: Industrial R&D in the "third industrial revolution". Industrial and Corporate Change, 18(1), 1–50.
8. Yeager, K. (2008). Striving for power perfection. IEEE Power and Energy Magazine, 6(6), 28–35.
9. Van Zant, P. (2014). Microchip Fabrication. McGraw-Hill Education.
10. Cusumano, M., & Yoffie, D. (2015). Extrapolating from Moore's law. Communications of the ACM, 59(1), 33–35.
11. Jordan, A. (2008). Frontiers of research and future directions in information and communication technology. Technology in Society, 30(3-4), 388–396.
12. De Alwis, C., Kalla, A., Pham, Q.V., Kumar, P., Dev, K., Hwang, W.J., & Liyanage, M. (2021). Survey on 6G frontiers: Trends, applications, requirements, technologies and future research. IEEE Open Journal of the Communications Society, 2, 836–886.
13. Moore, S. (2022, October 19). The Current State of the Global Semiconductor Market. AZoM. Retrieved on August 09, 2023 from https://www.azom.com/article.aspx?ArticleID=22111.

14. Garrou, P., Koyanagi, M., & Ramm, P. (2014). Handbook of 3D Integration, Volume 3: 3D Process Technology. John Wiley & Sons.
15. Cusumano, M., Yoffie, D., & Gawer, A. (2020). The future of platforms. MIT Sloan Management Review, 61, 26–34.
16. Mordor Intelligence LLP. (2022). Global Semiconductor Device Market - Growth, Trends, COVID-19 Impact, and Forecasts (2022–2027). Report ID: 6321515
17. Aboagye, A., Burkacky, O., Mahindroo, A., & Wiseman, B. (2022). When the chips are down: How the semiconductor industry is dealing with a worldwide shortage.
18. Infiniti Research Limited. (2022). Global Semiconductor Market in Military and Aerospace Industry 2023-2027. Report ID:5483776.
19. Sakuma, K., Andry, P., Dang, B., Maria, J., Tsang, C., Patel, C., Wright, S., Webb, B., Sprogis, E., & Kang, S., & others (2007). 3D chip stacking technology with low-volume lead-free interconnections. In 2007 Proceedings 57th Electronic Components and Technology Conference (pp. 627–632).
20. Thomas, S. (2020). Nature Electronics. Gate-all-around transistors stack up.
21. Capri, A., & Clark, R. (2022). Australia's Semiconductor National Moonshot. Australian Strategic Policy Institute.
22. Bown, C. (2020). How the United States marched the semiconductor industry into its trade war with China. East Asian Economic Review, 24(4), 349–388.
23. Debter, L. (2015). Intel Buying Chipmaker Altera For $16.7 Billion.
24. Gara, A. (2015). Avago To Buy Broadcom For $37 Billion As Semiconductor Dealmaking Heats Up.
25. Subba, N. (2018). China just approved the $44 billion merger between Qualcomm and NXP Semiconductor.
26. Bloomberg, I.K. (2020). AMD to buy Xilinx in $35 billion all-stock deal.
27. Arden, W. (2002). The international technology roadmap for semiconductors—Perspectives and challenges for the next 15 years. Current Opinion in Solid State and Materials Science, 6(5), 371–377.
28. Saito, S. (1997). Carbon nanotubes for next-generation electronics devices. Science, 278(5335), 77–78.
29. Chan, P.Y., Gogna, M., Suarez, E., Karmakar, S., Al-Amoody, F., Miller, B., & Jain, F. (2011). Nonvolatile memory effect in indium gallium arsenide-based Metal–Oxide–Semiconductor devices using II–VI tunnel insulators. Journal of Electronic Materials, 40, 1685–1688.
30. Brown, S., Francis, R., Rose, J., & Vranesic, Z. (1992). Field-Programmable Gate Arrays. (Vol. 180) Springer Science & Business Media.
31. Brassard, G., Chuang, I., Lloyd, S., & Monroe, C. (1998). Quantum computing. Proceedings of the National Academy of Sciences, 95(19), 11032–11033.
32. Singh, S., Faraz, M., & Khare, N. (2020). Recent advances in Semiconductor–Graphene and Semiconductor–Ferroelectric/Ferromagnetic nanoheterostructures for efficient hydrogen generation and environmental remediation. ACS Omega, 5(21), 11874–11882.
33. Chiarello, G., Dozzi, M., & Selli, E. (2017). TiO2-based materials for photocatalytic hydrogen production. Journal of Energy Chemistry, 26(2), 250–258.
34. Huang, M., & Madasu, M. (2019). Facet-dependent and interfacial plane-related photocatalytic behaviors of semiconductor nanocrystals and heterostructures. Nano today, 28, 100768.
35. Kavitha, R., Nithya, P., & Kumar, S. (2020). Noble metal deposited graphitic carbon nitride based heterojunction photocatalysts. Applied Surface Science, 508, 145142.
36. Maitra, S., Halder, S., Maitra, T., & Roy, S. (2021). Superior light absorbing CdS/vanadium sulphide nanowalls@TiO2 nanorod ternary heterojunction photoanodes for solar water splitting. New Journal of Chemistry, 45, 7353–7367.
37. Bai, S., Jiang, J., Zhang, Q., & Xiong, Y. (2015). Steering charge kinetics in photocatalysis: Intersection of materials syntheses, characterization techniques and theoretical simulations. Chemical Society Reviews, 44(10), 2893–2939.
38. George, S. (2010). Atomic layer deposition: An overview. Chemical Reviews, 110(1), 111–131.
39. Meyyappan, M. (2009). A review of plasma enhanced chemical vapour deposition of carbon nanotubes. Journal of Physics D: Applied Physics, 42(21), 213001.
40. Jansen, H., Gardeniers, H., Boer, M., Elwenspoek, M., & Fluitman, J. (1996). A survey on the reactive ion etching of silicon in microtechnology. Journal of Micromechanics and Microengineering, 6(1), 14.
41. Huang, X., Liu, C., & Zhou, P. (2022). 2D semiconductors for specific electronic applications: From device to system. Npj 2D Materials Applications, 6, 51.
42. Ferre, A., & Fontanilles, J. (2005). Devices and microsystems in the automotive industry. In Conference on Electron Devices, 2005 Spanish (pp. 19–22).

43. Einspruch, N. (2012). Application Specific Integrated Circuit (ASIC) Technology. (Vol. 23) Academic Press.

44. Trimberger, S. (2012). Field-Programmable Gate Array Technology. Springer Science & Business Media.

45. Ramm, P., Klumpp, A., Weber, J., & Taklo, M. (2010). 3D system-on-chip technologies for more than Moore systems. Microsystem Technologies, 16, 1051–1055.

46. Gatzen, H. (2001). Dicing challenges in microelectronics and micro electro-mechanical systems (MEMS). Microsystem Technologies, 7, 151–154.

47. Lin, L., Yeh, T.C., Wu, J.L., Lu, G., Tsai, T.F., Chen, L., & Xu, A.T. (2013). Reliability characterization of chip-on-wafer-on-substrate (CoWoS) 3D IC integration technology. In 2013 IEEE 63rd Electronic Components and Technology Conference (pp. 366–371).

48. Ding, Y., Maitra, S., Wang, C., Halder, S., Zheng, R., Barakat, T., Roy, S., Chen, L.H., & Su, B.L. (2022). Vacancy defect engineering in semiconductors for solar light-driven environmental remediation and sustainable energy production. Interdisciplinary Materials, 1(2), 213–255.

Index

For Product Safety Concerns and Information please contact our EU
representative GPSR@taylorandfrancis.com
Taylor & Francis Verlag GmbH, Kaufingerstraße 24, 80331 München, Germany